Stochastic Mechanics
Random Media
Signal Processing and Image Synthesis
Mathematical Economics and Finance
Stochastic Optimization
Stochastic Control

Applications of Mathematics

Stochastic Modelling and Applied Probability

47

Springer Science + Business Media,LLC

Applications of Mathematics

1 Fleming/Rishel, **Deterministic and Stochastic Optimal Control** (1975)
2 Marchuk, **Methods of Numerical Mathematics,** Second Ed. (1982)
3 Balakrishnan, **Applied Functional Analysis,** Second Ed. (1981)
4 Borovkov, **Stochastic Processes in Queueing Theory** (1976)
5 Liptser/Shiryayev, **Statistics of Random Processes I: General Theory,** Second Ed. (1977)
6 Liptser/Shiryayev, **Statistics of Random Processes II: Applications,** Second Ed. (1978)
7 Vorob'ev, **Game Theory: Lectures for Economists and Systems Scientists** (1977)
8 Shiryayev, **Optimal Stopping Rules** (1978)
9 Ibragimov/Rozanov, **Gaussian Random Processes** (1978)
10 Wonham, **Linear Multivariable Control: A Geometric Approach,** Third Ed. (1985)
11 Hida, **Brownian Motion** (1980)
12 Hestenes, **Conjugate Direction Methods in Optimization** (1980)
13 Kallianpur, **Stochastic Filtering Theory** (1980)
14 Krylov, **Controlled Diffusion Processes** (1980)
15 Prabhu, **Stochastic Storage Processes: Queues, Insurance Risk, Dams, and Data Communication,** Second Ed. (1998)
16 Ibragimov/Has'minskii, **Statistical Estimation: Asymptotic Theory** (1981)
17 Cesari, **Optimization: Theory and Applications** (1982)
18 Elliott, **Stochastic Calculus and Applications** (1982)
19 Marchuk/Shaidourov, **Difference Methods and Their Extrapolations** (1983)
20 Hijab, **Stabilization of Control Systems** (1986)
21 Protter, **Stochastic Integration and Differential Equations** (1990)
22 Benveniste/Métivier/Priouret, **Adaptive Algorithms and Stochastic Approximations** (1990)
23 Kloeden/Platen, **Numerical Solution of Stochastic Differential Equations** (1992)
24 Kushner/Dupuis, **Numerical Methods for Stochastic Control Problems in Continuous Time,** Second Ed. (2001)
25 Fleming/Soner, **Controlled Markov Processes and Viscosity Solutions** (1993)
26 Baccelli/Brémaud, **Elements of Queueing Theory** (1994)
27 Winkler, **Image Analysis, Random Fields, and Dynamic Monte Carlo Methods: An Introduction to Mathematical Aspects** (1994)
28 Kalpazidou, **Cycle Representations of Markov Processes** (1995)
29 Elliott/Aggoun/Moore, **Hidden Markov Models: Estimation and Control** (1995)
30 Hernández-Lerma/Lasserre, **Discrete-Time Markov Control Processes: Basic Optimality Criteria** (1996)
31 Devroye/Györfi/Lugosi, **A Probabilistic Theory of Pattern Recognition** (1996)
32 Maitra/Sudderth, **Discrete Gambling and Stochastic Games** (1996)

(continued after index)

Harold J. Kushner

Heavy Traffic Analysis of Controlled Queueing and Communication Networks

With 50 Illustrations

Springer

Harold J. Kushner
Division of Applied Mathematics
Brown University
Providence, RI 02912, USA
hjk@dam.brown.edu

Managing Editors

I. Karatzas
Departments of Mathematics and Statistics
Columbia University
New York, NY 10027, USA

M. Yor
CNRS, Laboratoire de Probabilités
Université Pierre et Marie Curie
4, Place Jussieu, Tour 56
F-75252 Paris Cedex 05, France

Mathematics Subject Classification (2000): 60K25, 90Bxx, 93C70, 93E02, 93E20

Library of Congress Cataloging-in-Publication Data
Kushner, Harold J. (Harold Joseph), 1933–
Heavy traffic analysis of controlled queueing and communication networks / Harold J. Kushner.
p. cm. — (Applications of mathematics ; 47)
Includes bibliographical references and index.

1. Queueing theory. I. Title. II. Series.
QA274.8 .K87 2001
519.8'2—dc21 2001020202

Printed on acid-free paper.

Production managed by Michael Koy; manufacturing supervised by Erica Bresler.
Photocomposed copy prepared from the author's TEX files.

9 8 7 6 5 4 3 2 1

SPIN 10797544

ISBN 978-1-4612-6541-2 ISBN 978-1-4613-0005-2 (eBook)
DOI 10.1007/978-1-4613-0005-2

Springer-Verlag New York Berlin Heidelberg
A member of BertelsmannSpringer Science+Business Media GmbH

To My Parents,
Harriet and Hyman Kushner

Preface

The aim of this book is the development of the heavy traffic approach to the modeling and analysis of queueing networks, both controlled and uncontrolled, and many applications to computer, communications, and manufacturing systems. The methods exploit the multiscale structure of the physical problem to get approximating models that have the form of reflected diffusion processes, either controlled or uncontrolled. These approximating models have the basic structure of the original problem, but are significantly simpler. Much of inessential detail is eliminated (or "averaged out"). They greatly simplify analysis, design, and optimization and yield good approximations to problems that would otherwise be intractable, under broad conditions.

Queueing-type processes are ubiquitous occurrences in operations research, and in communications and computer systems. Indeed, it is hard to avoid them in modern technology. The subject is now about 100 years old. and there is an enormous literature. Impressive techniques, many based on Markov chain and ergodic theory, have been developed to handle a great variety of models. A sampling of the numerous books includes [6, 8, 18, 27, 33, 46, 81, 86, 132, 133, 220, 243].

But the models of interest are growing fast in the face of the demands of new applications, particularly in communications and computer systems. One is concerned with rather general interarrival and service intervals, complicated network topologies and routing schemes, control or optimal control mechanisms, effective numerical methods, stability questions, state dependence, service interruptions, non-Markov models, and other complications such a those due to finite or shared buffers, feedback, blocking, or the balancing of the demands or the scheduling of several different job classes. Correlations in the service times for the same job as it proceeds from processor to processor and other forms of correlation are also current issues, as are randomly varying parameters, perhaps due to varying environmental factors. The demands of modern high-speed and high-performance communications pose very serious challenges, since they must operate with very small error rates. Although there is considerable effort devoted to these problems, a great many of the most significant remain intractable. When one wishes to apply optimal control methods, the exact physical model is often much too complicated for an exact solution to be obtained.

Although one would like to solve the problems without any type of approximation, asymptotic methods are often the best solution (or even the only hope). We will be concerned with the approximation of controlled or uncontrolled queueing networks by reflected diffusion or jump–diffusion processes. Diffusion approximations have long been an effective tool for simplifying or approximating physical or biological problems [35, 72, 148, 219]. The reflected diffusion approximation simplifies the queueing network problem in many ways. It is Markovian and lower (often much lower) dimensional, numerical methods are available, and the structure is exhibited more clearly. Even formal analyses of complex problems that are guided by the insights and approximations of heavy traffic analysis can lead to impressive results, as seen in Chapter 1.

The idea of diffusion approximations to processes arising in electronics, chemistry, mathematical genetics, and in many areas of physics is very old. It is essentially an exploitation of the multiscale structure of the system. When the quantities of interest are the result of very many small and not strongly dependent or only "locally" dependent effects, then central limit or law of large number ideas are valid, and convenient approximations by relatively simple diffusion-type processes are possible. For example, consider the evolution of the concentration of a reagent in a chemical reaction, where the evolution of the mean concentration is often virtually indistinguishable from the true random evolution, by an application of the law of large numbers.

For another example, consider the computation of the noise in an electrical circuit due to the random motions of electrons. Due to the large number of electrons and the small electronic charge, unless one is concerned with quantum effects one can aggregate the effects of the huge number of electrons and approximate the effect by the response of a circuit to white noise. This is an example of the central limit theorem. In this example, one is not concerned with the detail of the individual motions, but in the response of standard measuring devices attached to the circuit. The multiscale structure contrasts the local fast movement with the slower-moving *cumulative effects* in real time. The current in the actual physical circuit is a random process, which depends on the number of free electrons and their charge. In the physical model, the electrons are viewed as moving in independent "impulses" or "shots," with each shot being of short duration and having a random amount of energy, which depends on the charge. Loosely speaking [219], one proves that as the electronic charge goes to zero while the number of free electrons grows so that the total average spectral density remains constant, the actual physical current converges to that determined by a white noise driving voltage, with the same spectral density. Thus, the physical system is embedded in a family of systems that is parameterized by the charge per electron or the number of electrons. This might seem like a strange procedure, but it is very successful since the charge is so small that the actual performance is nearly the asymptotic performance, analogous to what one would get for the central limit theorem for a large enough number of summands. The white noise that appears in the limit equation depends only on the mean power, and not on the detailed structure of the individual random impulses. The form of the limit is robust, in that similar results can be obtained even if local interactions and structure (e.g., collisions, uneven metallic structure, surface effects) are accounted for. The approximations would not be of great interest if they were not robust to such variations. The exploitation of multiscale effects is of wide use in control theory; see, for example, [13, 142, 157, 159].

Analogous results can be obtained for a great variety of queueing networks, both controlled and uncontrolled. By heavy traffic, we mean that the average fraction of time at which the server or processor is free is small, or that the communications channel will have little spare capacity. Largely due to this "small idle time" assumption, suitably scaled queueing processes can be well approximated by a reflected diffusion process. The communications networks are large in the sense that there are many users and the channel capacity is large. This size parameterizes the system. Other types of queueing systems are large in the sense that the small idle time implies that the queue size is large. Then, with suitable scaling, the reflected diffusion approximation can be obtained, again under broad conditions.

The classical central limit theorem can provide useful and accurate estimates even if the number of random variables involved is not very large, depending on their distributions. The reflected diffusion approximation to a network can also provide accurate estimates even for "moderate" traffic. This would depend on the distributions that are involved. But simulations over a wide variety of practical problems and operating conditions show that it occurs very often [50, 94, 97, 126, 172, 211]. The central limit theorem gives an approximation that depends only on the means and variances (and the correlation function, if the random variables are correlated). Finer detail is eliminated, and this is a main reason for its power. The situation is similar with the diffusion or reflected diffusion approximation. It is only the means, variances and the internal mean routing structure (and the correlation function, if the interarrival times, service times, or routings are correlated) that appear in the limit. Because of this, the results are robust to variations, an important advantage in applications. The limit equations are usually much simpler to analyze than the original physical system, and numerical methods are available for them if the dimension is not large. The methods of [167, 159] can be used to compute optimal controls and values, or used simply for the numerical analysis under fixed controls, even for general nonlinear problems. The QNET methods of [50, 55, 97] can be used to get the first and second moments of the stationary distributions for non-state-dependent and uncontrolled systems, even in high dimensions. The basic structure of the physical system is preserved in the limit equations, and the parametric and other dependencies can be more readily seen.

The state space for the classical diffusion approximations in physics, say that for the noise effects due to electronic motion in a circuit, is some Euclidean space, since the currents can have any sign and the maximum or minimum values are not a priori bounded. On the other hand, the problems of interest in this book are usually defined on some constrained state space that is a subset of a Euclidean space. The simplest constraint occurs when the queues are nonnegative (i.e., no backlogging allowed). Then the state space is the positive orthant defined by the requirement that all components of the state vector be nonnegative. Another form of constraint occurs when there are finite buffers, so that the queues are bounded above. Other forms of the constraints occur if several queues share a finite buffer or if the jobs or messages circulate in the network without ever leaving. In other cases, it is linear combinations of the basic queue lengths that give the most natural or simplest approximations. Then the state takes the form of a convex polyhedron. Some components might be constrained and others not.

These constraints are carried over to the diffusion approximation, and the behavior of the diffusion approximation on the boundary is determined by the behavior of the physical problem on its boundary. Consider the simplest form of a single queue that is constrained to be nonnegative. In heavy traffic the mean rates of arrival and service are close, say parameterized by $1/\sqrt{n}$. Scale space so that $k/\sqrt{n}$ denotes that there are k in the queue. When the queue is not empty, the behavior is like a random walk with drift, and a standard diffusion approximation can be used as $n \to \infty$ (perhaps with an appropriate rescaling of time as well). The behavior when the queue is empty is different, since then there can be only arrivals, and this difference must be accounted for by the approximating or limit process. This physical "boundary behavior" leads to a "reflection" on the boundary for the diffusion limit.

The term heavy traffic comes from Kingman [128, 129, 131, 130], who analyzed various asymptotic approximations but did not develop a diffusion-type model. See also Kendall [127], who developed a diffusion approximation for a one-dimensional problem. The pioneering work on the diffusion approximations was done by Borovkov [20, 21, 22, 23, 24, 25, 26], Iglehart [106, 107, 108], Iglehart and Taylor [109], and Iglehart and Whitt [110, 111, 112], with the latter references containing many of the seeds of subsequent development. Surveys of early work are in [180, 250]. The references [110, 111] as well as [88] are also concerned with simple tandem networks. The reference [90] introduced the reflection map to model and simplify the limit process for the tandem network, and [98] did this for a more general network. The modern approach to networks was initiated by Reiman [213]. Following this, there was an explosion of interest in networks, and many further references appear throughout the book. Not all networks, even under heavy traffic, can be approximated by a reflected diffusion process [51, 53], but the counterexamples are not what one would ordinarily encounter.

Numerical data such as in [48, 49, 50, 55, 94, 97, 172, 186, 211] show that the heavy traffic methods can often be used to get very good estimates as well as information on parametric dependencies and control structure under conditions of moderate traffic. A useful way of exploiting the heavy traffic approach is illustrated in [94], where intuitive heavy traffic analysis is used to guess a good strategy for a particular network, which is then proved to be nearly optimal. More generally, the intuition provided by heavy traffic analysis is helpful throughout queueing theory.

Chapter 1 introduces the subject through a number of examples. The examples are selected to represent some classes of problems of current interest in queueing, communications and manufacturing, to introduce some of the methods, and illustrate the type of limit equations that are to be obtained. The simplifications provided by the approach can be seen. In particular, it is seen how heavy traffic analysis can provide powerful heuristic guides for obtaining good control policies, even without a rigorous analysis.

Chapters 2, 3, and 4 concern the mathematical background. They summarize the necessary material from the theory of stochastic processes, and where appropriate provide the details. Chapter 2 deals with martingale-type processes, stochastic integrals, and the theory of weak convergence. The primary approximation methods that are used in the book are applications of weak convergence theory, which is to process approximation what the central limit theorem is to the approximation of random variables. Martingales appear at virtually all stages of the analysis, as a component of the physical or the limit model. Chapter 3 introduces the theory of stochastic differential equations and contains a detailed survey of reflected stochastic differential equations (and the Skorohod problem), the fundamental limit models, either not controlled or controlled. It also outlines some of the methods from control theory that will be needed. Chapter 4 is concerned with the mathematics of the ergodic cost (average cost per unit time over an infinite time interval) problem, and introduces the methods from ergodic and control theory, as well as a maximum principle, that are necessary for the treatment. It is perhaps the most technical chapter, and its reading can be deferred until the results are needed. Indeed, the essential ideas of the subsequent chapters should be intuitively understandable after a quick survey of the mathematical background chapters.

The rest of the book gives a comprehensive development of the fundamental models and problems. The heavy traffic analysis of the uncontrolled problem is in Chapters 5 to 8. For expository simplicity, we start in Chapter 5 with the one-dimensional problem, where there is a single server or processor. Nevertheless, many of the techniques for the one-dimensional problem provide the basis for the more complex network analysis. A variety of network models, which exhibit the fundamental methods of analysis, are dealt with in Chapters 6 and 7. Chapter 8 concerns extensions of the methods and results of Chapters 5–7 to problems where the data (say, concerning interarrival or service times) depend on the state of the system. The rest of the book is concerned with the control problem. Chapter 9 deals with what might be called the classical control problem, as opposed to the singular control problem of Chapter 10. In these chapters we work with canonical models that describe a large variety of applications and illustrate the methods via concrete applications to problems arising in modern high-speed communications. Chapter 11 concerns the polling problem, where there might be either zero or nonzero switching times and cost or server breakdowns. The workload of a server can be defined as the total amount of work that the server has to do if there are no more inputs. Using the workload as the state often simplifies the modeling and analysis, and yields a limit process of (perhaps greatly) reduced dimension. The concept is dealt with in Chapters 5–7, but it is crucial to the results of Chapters 11 and 12.

The scheduling problem for multiclass problems is treated in Chapter 12. Here there are several classes of arrivals and several processors. Each arriving class might be servable by one or more than one processor, with the work depending on the assignment. There is a great deal of current activity concerning the problem of optimal assignment (e.g., [10, 93, 94, 126, 178, 255]), and the potential applications are enormous. The proper definition of workload was obvious in the problems treated in the previous chapters. While some multiclass problems are treated in other parts of the book, the appropriate definition of workload for the scheduling problems of Chapter 12 is not a priori obvious, and is a subtle and important recent achievement. Again, the great simplification provided by the heavy traffic approach will be apparent. Since the subject of Chapter 12 is still under active development, we concentrate on some of the fundamental concepts and formulate the assumptions to minimize details where possible.

In all of the examples, unless explicitly noted otherwise, it is supposed that within each queue the service is first come first served (FCFS) and that the processors will work if there is work to do. Owing to the length of the book and to the additional background that would need to be introduced, stability is not covered [7, 28, 39, 54, 52, 119, 60, 144, 150, 222]. Current results in stability caution us concerning the properties of multiclass feedback systems with fixed (hence uncontrolled) priorities.

Earlier forms of the book were used as a text for a graduate course in the Applied Mathematics Department at Brown University, and the comments of the students were very helpful. Parts or all of later versions were read by Eitan Altman, Robert Buche, Amarjit Budhiraja, and Michael Harrison, all of whom made many useful and well appreciated suggestions. The author gratefully acknowledges the long-term support by the Army Research Office and the National Science Foundation.

Numbering and cross referencing. Cross reference numbering *within* a chapter does not include the chapter number. For example, within Chapter 5, equation 4 of Section 3 of Chapter 5 is called equation (3.4), and Subsection 6 of Section 3 of Chapter 5 is called Subsection 3.6, with the analogous use for Theorem, Figure, and Assumption. Cross references *between* chapters do include the chapter number. For example, in Chapter 5, a reference to equation 4 of Section 3 of Chapter 2 is called equation (2.3.4), and Subsection 6 of Section 3 of Chapter 2 is called Subsection 2.3.6, with the analogous use for Theorem, Figure, and Assumption.

A glossary of some of the more frequently used symbols appears at the end of the book.

Providence, Rhode Island, USA — Harold J. Kushner

Contents

Contents

1
Introduction: Models and Applications

Chapter Outline

The purpose of this introductory chapter is the illustration of some of the wide variety of difficult and important problems that can be treated by means of heavy traffic analysis. Many good examples can be drawn from communications and computer systems and networks, manufacturing systems, and other queueing-type networks in many areas. Only a few examples can be given. Any selection depends on the interests and tastes of the author, and one can select these examples in many ways. But the examples selected are canonical in that each is representative of a large class of current interest. These examples should provide some flavor for the type of problems that are of interest.

The first few examples are more of the nature of general models, and help to set up some of the terminology as well as to illustrate in a relatively simple and informal manner the type of analysis that is to be used, as well as the resemblance of some of the ideas to those of the central limit theorem. The analysis tends to follow lines that will become increasingly familiar. The first step is the determination of the model and assumptions. Then one writes the equation for the state (say the queue size or scaled queue size) in terms of the input and service process and whatever controls might be involved. The next step is to manipulate the equation into a form that "looks like" a reflected diffusion process, whether controlled or not. One separates out the "conditional mean" terms from the "noise," and ends up with a form of the equation written in terms of clearly defined processes which are close to the noise and drift terms of the desired approximating or limit model. In order to get a useful representation, one needs to quantify the notion of heavy traffic. This is done by the so-called heavy traffic condition, which quantifies the notion that the server has little idle time, or the channel little spare capacity. It is essentially a condition on the asymptotic magnitude of the differences between the mean input rate and service rate at the various processors or servers. As with the central limit theorem or diffusion-type approximations in general, one works with a sequence of processes (parameterized by n) and makes statements about the limits as $n \to \infty$. These are the general theoretical results. But, as with the central limit theorem, one often uses it for a specific system and not a sequence, and the method for doing this is discussed.

Section 1 takes a simple one-dimensional queue and sets up the dynamical equation so that it looks like a reflected diffusion, and discusses some of the general factorizations and issues that appear in getting such representations. Two approaches are introduced to model the behavior when the queue is empty. The first uses the fictitious departure model. In the second, which is actually more convenient in general, the fictitious departures are replaced by the idle time of the server, and this becomes the reflection process in the limit. One type of model is based on the scaled number of queued jobs. An alternative is concerned with the equation for the scaled queued work, a form that has advantages when some processor has several arrival streams with different work requirements or there are several classes or jobs. Asymptotically, there is a nice linear relation between the results for the two approaches. Section 2 introduces networks of queues, and revisits the modeling and representation issues raised for the one-dimensional case. The reflection process is more complicated owing to the interactions among the queues, but can be simply deduced from the average routing probabilities. A network is called open if all jobs eventually leave. Closed networks, where jobs circulate but do not leave, are also important, and some such examples are discussed in Subsection 2.2.

So far, the examples have been of a general nature. Sections 3 to 5 concern specific applications. The data show that the results and methods of heavy traffic modeling can be useful even when the system is not in heavy traffic. Section 3 introduces the problem of polling, where there are several queues competing for the attention of a single server. The basic physical problem has been the subject of a huge literature. It is seen how the heavy traffic approach can greatly simplify analysis. One must decide when to stop serving one queue and switch to another, and there might be a switching cost or time. The approach of [211] to this control problem is discussed and it is seen how the insights of heavy traffic modeling can lead to substantial simplifications and excellent control policies and cost savings, even if a completely rigorous theory is not available, and even outside of the heavy traffic regime. Section 4 introduces the so-called statistical multiplexer, a device that is used to combine inputs from many independent sources and send them to a communications channel in some order. The sources are bursty in that they create data in a random way that is subject to sudden large changes in the total rate. Because of this burstiness, control over the inputs is introduced. Again, the heavy traffic modeling simplifies the problem. The control has a substantial effect on the performance, as can be seen from the continued development of this problem in Chapters 8 and 9. Section 5 is concerned with another class of examples from communications, that of an integrated services data network, where several types of users compete for a single channel. Some users are guaranteed a fixed amount of bandwidth, and others share what is left. The control problem is set up, and the limit is a well-defined singularly controlled reflected stochastic differential equation.

Many other examples could have been selected. A few possibilities will be listed here, and others are scattered throughout the text. A network of fork-join queues is dealt with in [203] and single fork-join queues are treated via matched asymptotic expansions in [134, 232]. There have been numerous attempts to model the TCP (transfer control protocol), which is used to manage file transfers into networks. A heavy traffic approach is in [226]. See [3] and Chapter 11 for the control of polling when there is a possibility of processor breakdown. In the problem of controlling the rerouting/rejection of arrivals to large trunk line networks, the control in the limit model is over the reflection directions [163, 176]. The papers [87, 101] were motivated by scheduling problems in manufacturing. They start with the limit model, which is obtained formally but with the help of insights gained from heavy traffic analysis. The optimal control problem is solved and useful controls obtained.

The paper [74] considers a system with a channel of a given bandwidth and a number of bursty and uncontrollable sources. There are also a number of "fluid" sources that can be controlled. These sources have an unlimited amount of data to send. They can be controlled in the sense that the rate at which they send information to the system can be regulated. There is a finite buffer and delays in the transmission of state information to the controllers. The point is to use heavy traffic analysis to develop an adaptive procedure for control that improves the throughput. The model is an approximation to a type of TCP for controlling flow for systems that are shared by many users. The example is typical of many that start by assuming a particular reflected diffusion form for the overall system. Storage processes with general release rules are discussed in [250, 256]. This leads to a limit process that has a singularity at the origin. Early works on routing for communications systems include [79, 80]. The problem of assigning jobs from many different classes to a bank of nonidentical processors is developed in [10, 93, 96, 94, 100, 255] and elsewhere. See also Chapter 12. These multiclass problems are usually nearly impossible to solve for the actual physical problems. The heavy traffic analysis for such problems provides a powerful illustration of the potential of the approach.

1.1 A Single Queue: Heavy Traffic Modeling

The purpose of this section is to give a relatively simple example of how one can obtain a reflected diffusion approximation to a queue when the traffic intensity is close to unity. The approximation is in the sense of the central limit theorem. The central limit theorem allows us to approximate the distribution of the sum $X_n = \sum_{l=1}^{n} \xi_l/\sqrt{n}$ by a normally distributed random variable when the ξ_l have zero mean, unit variance, and are mutually independent and identically distributed. The theorem says that the distribution of X_n converges to a normal distribution with zero mean and unit variance. Alternatively, we can say that X_n converges in distribution to a normally distributed random variable, with zero mean and unit variance. In applications, when we have just one sum X_n and not a sequence, we say that X_n is "approximately" normally distributed if n is large. How large n must be depends on the distribution of ξ_l. But the approximation is often used in applications even for n as small as 5 or 10. The development is motivational only and no proofs are given. Proofs of more general results appear starting in Chapter 5.

1.1.1 Simple One Dimensional Models.

Example 1.1. A Simple Single Queue. In order to illustrate the idea of the diffusion approximation with minimal notation and in a way that allows us to relate it to the central limit theorem, let us start by considering a queueing problem with a single server, where the arrivals and service completions can take place at the integer times $0, 1, 2, \ldots$ only. The interarrival and service times are assumed to be geometrically distributed, the discrete time analogue of the classical $M/M/1$ queue. This is one of the simplest queueing problems, for which virtually all needed information can be provided by classical methods. In this model, time is divided into intervals $[k-1, k)$, $k = 1, \ldots$, and at most one arrival and one departure can occur on each interval. For specificity, suppose that an arrival (if any) in the kth interval occurs at time $(k-1)+$, just after its start, and the departure (if any) occurs at time $k-$, just before the end of the interval. Let $Q(k)$ denote the number of customers in the system at real time k.

Let I_k^a denote the indicator function of the event that there is an arrival at time $(k-1)+$. Let I_k^d denote the indicator function of the event that there is a departure at time $k-$. Suppose that there are real numbers $\bar{\lambda}^a$ and $\bar{\lambda}^d$ in $(0,1)$ such that

$$P\left\{I_k^a = 1 \middle| \text{past data}\right\} = \bar{\lambda}^a, \tag{1.1}$$

$$P\left\{I_k^d = 1 \middle| \text{past data}, Q(k-1) + I_k^a > 0\right\} = \bar{\lambda}^d. \tag{1.2}$$

Equations (1.1) and (1.2) imply that the service and interarrival intervals are geometrically distributed with mean values $1/\bar{\lambda}^d$ and $1/\bar{\lambda}^a$, respectively, and that the intervals are mutually independent.

In order for there to be a departure at time $k-$, there must be a customer in the queue at time $(k-1)+$, hence the conditioning on $Q(k-1)+I_k^a > 0$ in (1.2). If $Q(k-1) + I_k^a = 0$, so that there is no actual customer that can complete service at $k-$, let us *redefine* I_k^d to be a $\{0,1\}$-valued random variable, independent of the others, and satisfying $P\{I_k^d = 1|\text{past data}\} = \bar{\lambda}^d$. Thus, $\{I_k^a, k < \infty\}$ and $\{I_k^d, k < \infty\}$ are mutually independent, and the members of each set are mutually independent and identically distributed, and independent of the initial state of the queue. The state equation for the queue can be written as

$$Q(k) = Q(0) + \sum_{l=1}^{k} I_l^a - \sum_{l=1}^{k} I_l^d I_{\{Q(l-1)+I_l^a>0\}}. \tag{1.3}$$

The redefined I_k^d create *fictitious departures* from the queue, and these are accounted for by using the indicator function $I_{\{Q(l-1)+I_l^a>0\}}$.

The next step is to approximate $Q(\cdot)$ by a diffusion-type process. Since the jumps in $Q(\cdot)$ are ± 1 and occur only at times $1, 2, \ldots$, $Q(\cdot)$ cannot itself be approximated by a diffusion process. The first step toward this goal is to rewrite the above expression in a more suggestive form by centering the random variables about their means as

$$\begin{aligned} Q(k) = Q(0) &+ \sum_{l=1}^{k} \left[I_l^a - \bar{\lambda}^a\right] - \sum_{l=1}^{k} \left[I_l^d - \bar{\lambda}^d\right] + k\left[\bar{\lambda}^a - \bar{\lambda}^d\right] \\ &+ \sum_{l=1}^{k} I_l^d I_{\{Q(l-1)+I_l^a=0\}}. \end{aligned} \tag{1.4}$$

Define the traffic intensity $\rho = \bar{\lambda}^a/\bar{\lambda}^d$. If $\rho < 1$, then the queue is a stable process and $\lim_k EQ(k) = 1/(1-\rho)$ [105, 132]. There are three types of terms in (1.4). The "noise" terms, which are due to the randomness in the arrival and departure processes, are represented by the first two sums. The "drift" term, which is due to the mean difference between the input and output "rates" (assuming that the queue is never empty), is represented by $k[\bar{\lambda}^a - \bar{\lambda}^d]$. The last term corrects for the fact that there can be no departure if there was no customer to service. The factorization in (1.4) is convenient, since the second and third terms are just sums of centered mutually independent and identically distributed random variables, and the corrections for the fact that the queue behaves differently when it is empty are all put into the right-hand term. The right-hand term has the important property that it is nondecreasing and can increase only when the queue is empty.

The central limit theorem tells us that if we divide the first two sums on the right of (1.4) by $\sqrt{k}$ and k is large, then they can be approximated by mutually independent normally distributed random variables with variances $(1-\bar{\lambda}^\alpha)\bar{\lambda}^\alpha$, $\alpha = a, d$, respectively. This consideration and our desire to get an approximation to the queue for *all* time and not just for one specific time suggest that we work with the rescaled process $x^n(t)$ defined by

$$\begin{aligned} x^n(t) = \frac{Q(nt)}{\sqrt{n}} &= x^n(0) + \frac{1}{\sqrt{n}}\sum_{l=1}^{nt}\left[I_l^a - \bar{\lambda}^a\right] - \frac{1}{\sqrt{n}}\sum_{l=1}^{nt}\left[I_l^d - \bar{\lambda}^d\right] \\ &+ \sqrt{n}t\left[\bar{\lambda}^a - \bar{\lambda}^d\right] + \frac{1}{\sqrt{n}}\sum_{l=1}^{nt} I_l^d I_{\{Q(l-1)+I_l^a=0\}}. \end{aligned} \tag{1.5}$$

In (1.5) we say that t represents *scaled time.* When nt is used as an index of summation, we always mean to use the *integer part.* Define the processes

$$\bar{w}^{\alpha,n}(t) = \frac{1}{\sqrt{n}}\sum_{l=1}^{nt}\left[I_l^\alpha - \bar{\lambda}^\alpha\right], \quad \alpha = a, d.$$

The classical multidimensional central limit theorem [32] implies that for any $t_1 < t_2 < \cdots$, the sets

$$\left\{\bar{w}^{\alpha,n}(t_{i+1}) - \bar{w}^{\alpha,n}(t_i),\ i < \infty\right\}, \quad \alpha = a, d,$$

converge to mutually independent sets of mutually independent and normally distributed random variables with mean zero and variances $(1 - \bar{\lambda}^\alpha)\bar{\lambda}^\alpha[t_{i+1} - t_i]$, $\quad i = 1, \cdots$. In fact, as discussed in Chapter 2, the convergence is even stronger. As $n \to \infty$, the pair of *processes* $\bar{w}^{\alpha,n}(\cdot), \alpha = a, d$, converges in distribution to a pair of mutually independent Wiener processes $\bar{w}^\alpha(\cdot)$, $\alpha = a, d$, with variances $(1 - \bar{\lambda}^\alpha)\bar{\lambda}^\alpha$, $\alpha = a, d$, respectively, in the sense that for a large class of real-valued functions $f(\cdot)$ that are defined on the *path space* of the $\bar{w}^{\alpha,n}(\cdot)$, the sequence $f(\bar{w}^{a,n}(\cdot), \bar{w}^{d,n}(\cdot))$ converges in distribution to $f(\bar{w}^a(\cdot), \bar{w}^d(\cdot))$. This is called weak convergence (Chapter 2). The last term on the right of (1.5) is due to the fact that there are not always customers in the queue. That term can increase only at times t where $x^n(t)$ is either zero or $1/\sqrt{n}$. Thus, it serves to keep the other terms from driving $x^n(\cdot)$ negative. It will be called the *reflection term*

The remaining problem concerns the drift $\sqrt{n}t[\bar{\lambda}^a - \bar{\lambda}^d]$. So far, the parameter n was used to scale time and space. It had no interpretation in terms of the parameters of the original problem. But it has an additional and crucial role. Our main interest is in approximating the queue process by a reflected diffusion. If the process defined by (1.5) is to have such a limit, then $\sqrt{n}[\bar{\lambda}^a - \bar{\lambda}^d]$ must converge to some constant as $n \to \infty$. This implies that as n grows the difference between the arrival and service rates goes to zero. This is what is meant by *heavy traffic.* For large n, the unscaled queue $Q(nt)$ will be large, on the order of $1/[\bar{\lambda}^d - \bar{\lambda}^a]$ in the mean if $\bar{\lambda}^d > \bar{\lambda}^a$. In particular, in order to get a limit theorem we require that there be a number b such that

$$\lim_n \sqrt{n}[\bar{\lambda}^a - \bar{\lambda}^d] = b. \tag{1.6}$$

The heavy traffic condition (1.6) implies that the fraction of time during which the processor is idle is small for large n. In fact, for this problem, the fraction of time at which the server is idle is $1 - \rho$ [132]. Thus, to analyze the original queue, we embed it in a sequence of queueing problems indexed by n, where $\rho \to 1$ as $n \to \infty$.

Finally, suppose that the initial condition $x^n(0)$ is approximately $x(0)$ for large n. It can be shown (Chapter 5) that the last term on the right of (1.5) also converges to a limit $z(\cdot)$ as $n \to \infty$. Then (1.5) is approximated by the reflected diffusion process

$$x(t) = x(0) + bt + \bar{w}^a(t) - \bar{w}^d(t) + z(t). \tag{1.7}$$

The process $z(\cdot)$ is the *reflection term,* which is nondecreasing and continuous, satisfies $z(0) = 0$, and can increase only when $x(t) = 0$. It keeps $x(\cdot)$ from becoming negative. The $\bar{w}^a(\cdot)$ and $\bar{w}^d(\cdot)$ represent the (asymptotic) effects of the randomness in the arrival and service processes, respectively. The parameter b represents the (scaled and asymptotic) difference between the input and service rates.

Remark on application to a particular system. In any particular application where one has a fixed queue and not a sequence, the scale parameter n is artificial. Then one chooses the value of n as large as possible so that the value of $\sqrt{n}[\bar{\lambda}^a - \bar{\lambda}^d]$ is still of "moderate" size and is *defined* as b. If this cannot be done, then the heavy traffic approximation might not be useful. The form (1.7) is used to get approximations to quantities of interest for the physical queue. For example, via either numerical or analytical methods, one can approximate the mean or the variance of the queue length or the fraction of time that the queue has more than some given number of customers. Other examples will be given in the following sections.

As noted earlier, the situation is analogous to what is done to get diffusion approximations in physics, mathematical genetics, etc. For example, circuit noise is caused by the sudden and discrete movements of electrons. If one supposes that the number of electrons goes to infinity and their charge goes to zero appropriately, then the complicated formulas for the exact physical current simplify to a linear diffusion process, which is the usual and very successful way of representing the effects of such noise in circuits (unless quantum phenomena are of interest) [56, 219]. Consider the problem of the evolution from generation to generation of genetic traits due to random selection. Then n is used to represent the population size, and time is scaled as in (1.5). As $n \to \infty$, the complicated combinatorial formulas simplify to yield a diffusion process. The relative simplicity of the diffusion approximation has enabled a much deeper understanding of the process [72, 148]. It is also the key to the power of the central limit theorem. What is important is that the results provide useful information about the particular systems of concern.

Recapitulation. Let us recapitulate the above procedure. We have taken a physical system and embedded it in a family of systems that is parameterized by n. The parameter n is used to scale both time and space, and also characterizes the mean difference in the arrival and service rates. As $n \to \infty$, the difference between these rates decreases. This leads to a buildup of the queue. But the scaling again comes to our rescue as a normalization parameter. The scaling automatically gives us the Wiener process limit for the "noise" terms. As noted, in an application one chooses the value of the scaling parameter n as large as possible so that $\sqrt{n}[\bar{\lambda}^a - \bar{\lambda}^d]$, which is then defined as b, is of moderate size. With the value of b given, (1.7) is used as an approximating model.

For the particular example dealt with above, the limit form (1.7) might not be too interesting, since the actual physical queue is simple to analyze. It becomes more interesting when more realistic assumptions are used on the arrival and service processes or controls (say, over the service rate or input) are included. Before further discussion of the general approach, let us redo the example from a more general point of view.

Example 1.2. An alternative development and a general model. We will now redo the problem of Example 1.1 in a more general setting, which is close to the approach to be used later in the book, and where the arrivals and service completions can occur at any time. Using the scaling that was motivated in Example 1.1, let $S^{a,n}(t)$ (respectively, $S^{d,n}(t)$) denote $1/n$ times the number of arrivals (respectively, departures) by real time nt. Define $A^n(t) = \sqrt{n}S^{a,n}(t)$ and $D^n(t) = \sqrt{n}S^{d,n}(t)$. With $x^n(t) = Q(nt)/\sqrt{n}$ as in (1.5), we can write

$$x^n(t) = x^n(0) + A^n(t) - D^n(t). \tag{1.8}$$

The approximation method proceeds by expanding the terms in (1.8) into a "random" and "mean" part, which will make it look more like a reflected diffusion of the form (1.7). The procedure is analogous to what was done in (1.4) and (1.5), but differs in that the arrivals and departures can occur at any time and the distributions of the intervals are more general. As in Example 1.1, we embed the system of concern into a sequence indexed by n. Let $\{\Delta_l^{a,n}, l < \infty\}$ and $\{\Delta_l^{d,n}, l < \infty\}$ denote the interarrival and service intervals, respectively. For some centering constants $\bar{\Delta}^{a,n} \equiv 1/\bar{\lambda}^{a,n}$,

$\alpha = a, d$, define the processes

$$w^{\alpha,n}(t) = \frac{1}{\sqrt{n}} \sum_{l=1}^{nt} \left[1 - \frac{\Delta_l^{\alpha,n}}{\bar{\Delta}^{\alpha,n}} \right]. \tag{1.9}$$

If the $\Delta_l^{\alpha,n}$ are identically distributed, whether correlated or not, then $\bar{\Delta}^{\alpha,n} = E\Delta_l^{\alpha,n}$. If the $\Delta_l^{\alpha,n}$ were deterministic, but periodic, then $\bar{\Delta}^{\alpha,n}$ would be the pathwise mean value (and $w^{\alpha,n}(\cdot)$ would converge to the "zero" process). With appropriate centering constants, as $n \to \infty$ and under broad conditions, the sequence of processes $w^{\alpha,n}(\cdot)$, $\alpha = a, d$, converges weakly (see Section 2.8) to a Wiener process, which will be called $w^{\alpha}(\cdot)$, $\alpha = a, d$.

If the $\Delta_l^{\alpha,n}$, $l < \infty$, are mutually independent and identically distributed for each n and α, then the variance of the summand is $[\text{var}\Delta_l^{\alpha,n}]/[\bar{\Delta}^{\alpha,n}]^2$ and is referred to as the *coefficient of variation*. If this converges to a constant σ_α^2 as $n \to \infty$, then the variance of $w^\alpha(\cdot)$ is σ_α^2.

Now let us take the convergence of $w^{\alpha,n}(\cdot)$, $\alpha = a, d$, to a Wiener process as an assumption in this discussion, and also suppose that

$$\bar{\Delta}^{\alpha,n} \to \bar{\Delta}^{\alpha} \equiv 1/\bar{\lambda}^{\alpha}, \quad \text{as } n \to \infty, \ \alpha = a, d. \tag{1.10}$$

It can be shown (Theorem 5.1.1) that the convergence assumption on the $w^{\alpha,n}(\cdot)$ and (1.14) imply that the $S^{\alpha,n}(\cdot)$ converge weakly to the processes with values $\bar{\lambda}^{\alpha} t$. Heuristically, the $\bar{\lambda}^{\alpha}, \alpha = a, d$, are the arrival and service rates in the limit, and we will refer to them as such (and to their inverses $\bar{\Delta}^{\alpha}$ as the limit mean intervals) even if the intervals are not identically distributed. The parameters $\rho^n = \bar{\Delta}^{d,n}/\bar{\Delta}^{a,n}$ and $\rho = \bar{\Delta}^{d}/\bar{\Delta}^{a}$ are called the *traffic intensities* for the physical and limit systems, respectively.

The development proceeds by rewriting and expanding (1.8) so that the random, or "noise, " effects are separated from the mean, or "drift," effects. Expansions of the type

$$\begin{aligned} \sqrt{n} S^{\alpha,n}(t) &= \frac{1}{\sqrt{n}} \sum_{l=1}^{nS^{\alpha,n}(t)} 1 \\ &= \frac{1}{\sqrt{n}} \sum_{l=1}^{nS^{\alpha,n}(t)} \left[1 - \frac{\Delta_l^{\alpha,n}}{\bar{\Delta}^{\alpha,n}} \right] + \frac{1}{\sqrt{n}\bar{\Delta}^{\alpha,n}} \sum_{l=1}^{nS^{\alpha,n}(t)} \Delta_l^{\alpha,n} \end{aligned} \tag{1.11}$$

will be used frequently. See Chapter 5 for full details. Now use (1.11) to expand (1.8) in the form

$$\begin{aligned} x^n(t) = x^n(0) &+ w^{a,n}(S^{a,n}(t)) - w^{d,n}(S^{d,n}(t)) \\ &+ \frac{1}{\sqrt{n}\bar{\Delta}^{a,n}} \sum_{l=1}^{nS^{a,n}(t)} \Delta_l^{a,n} - \frac{1}{\sqrt{n}\bar{\Delta}^{d,n}} \sum_{l=1}^{nS^{d,n}(t)} \Delta_l^{d,n}. \end{aligned} \tag{1.12}$$

The *sum* in the next to last term of (1.12) is

$$nt - [\text{difference between } nt \text{ and the last arrival before or at } nt].$$

Thus, modulo a small error, the next to last term of (1.12) is $\sqrt{n}t/\bar{\Delta}^{a,n}$. The last *sum* in (1.12) is

$$nt - \Big[\text{ idle time by real time } nt \\ + \text{time that job served at } nt \text{ has been in service}\Big].$$

Finally, we can write (modulo a small error)

$$x^n(t) = x^n(0) + w^{a,n}(S^{a,n}(t)) - w^{d,n}(S^{d,n}(t)) + \sqrt{n}\left[\frac{1}{\bar{\Delta}^{a,n}} - \frac{1}{\bar{\Delta}^{d,n}}\right] t \\ + \frac{1}{\sqrt{n}\bar{\Delta}^{d,n}} [\text{idle time by real time } nt]. \tag{1.13}$$

In order to complete the development, we require that the difference between the arrival and service rates go to zero as $n \to \infty$. In particular, suppose that there is a real number b such that (the heavy traffic condition)

$$\sqrt{n}\left[\frac{1}{\bar{\Delta}^{a,n}} - \frac{1}{\bar{\Delta}^{d,n}}\right] = \frac{\sqrt{n}}{\bar{\Delta}^{d,n}}[\rho^n - 1] \equiv b^n \to b \tag{1.14}$$

is satisfied.

Choosing the scale parameter *n*. As in Example 1.1, in an application one usually works with a fixed queue, there is no a priori scale parameter n, and the centering constants are just $\bar{\Delta}^\alpha$, and not the n-dependent quantities $\bar{\Delta}^{\alpha,n}$, $\alpha = a, d$. Then one chooses n as large as possible so that the term

$$\sqrt{n}\left[\frac{1}{\bar{\Delta}^a} - \frac{1}{\bar{\Delta}^d}\right]$$

is of moderate size, denotes it by b, and uses the resulting process (1.15) to get approximations to quantities of interest for the original queue.

Note that in this development we did not introduce "fictitious services" to account for an empty queue, as done in Example 1.1. Idle time was used instead. The last term on the right of (1.13) is a "reflection term." Formally speaking, it equals $1/\sqrt{n}$ times the average number of customers that could be served during the idle time. The term can increase only when $x^n(t) = 0$, and it prevents the other terms in (1.13) from driving $x^n(t)$ negative. It turns out that this reflection term also converges weakly as $n \to \infty$, and to a continuous limit reflection term. Thus, the limit can be represented as

$$x(t) = x(0) + bt + w^a(\bar{\lambda}^a t) - w^d(\bar{\lambda}^d t) + z(t), \tag{1.15}$$

which is (1.7), with the identification of $\bar{w}^\alpha(\cdot)$ with $w^\alpha(\bar{\lambda}^a\cdot)$, $\alpha = a, d$.

Discussion: heavy traffic modeling. Although the development in Example 1.2 was formal, it should be clear that the convergence of the suitably scaled queue process to a reflected diffusion occurs under very broad conditions. Because of the special "no memory" assumptions (1.1) and (1.2) that were used in Example 1.1, the process $Q(\cdot)$ of that example is a real-valued Markov process with a stationary transition probability, and is simple to analyze. However, the analysis of the queue for general service and interarrival times can be quite difficult. The formulas for even simple quantities such as the stationary mean value (if it exists) can be complicated and difficult to interpret. This is certainly true if the intervals are correlated. But it is also true if they are only mutually independent, but not exponentially (or geometrically) distributed. The situation is much worse for networks.

The limit equation (1.15), on the other hand, has a simple structure. It is particularly convenient, since it depends only on the drift parameter b and the variances of the Wiener processes. The other details of the data in the physical problem disappear in the limit, analogously to what happens in the central limit theorem for (normalized) sums of independent and identically distributed random variables, where the limit is expressed only in terms of the first two moments, and the finer detail disappears. This facilitates understanding of the main qualitative features. If the intervals are mutually independent and the $\Delta_l^{\alpha,n}$, $l < \infty$, are identically distributed for each n and $\alpha = a, d$, then the Wiener processes in (1.15) will depend on only the first and second moments. More generally, for stationary (and suitably mixing) processes, they will depend on the correlation function and mean in a simple way, and not on any more detail about the process. Finite buffers of order $O(\sqrt{n})$ could be added with little additional complication (see Chapter 5). The order $O(\sqrt{n})$ arises from the fact that buffers of a higher order will not appear in the limit, and buffers of smaller order will always be full in the limit.

Thus, the approximation of the original physical queue process by (1.7) or (1.15) can be quite useful, owing to their relative simplicity. Apart from the "central limit theorem" type scaling, all that we have paid is the heavy traffic assumption (1.14). In any application one has a single system and not an indexed sequence. In practice, simulation results show that the limit results provide quite good approximations for quantities of interest (for example, stationary means and variances) even for so-called moderate traffic intensities. One is given particular values of $\bar{\Delta}^{\alpha,n}$, $\alpha = a, d$, not a value of n. One chooses a value of n such that $\sqrt{n}[\rho^n - 1]$ is of moderate size. Then use the approximate relationship $Q(nt) \approx \sqrt{n}x(t)$ to approximate the properties of the actual physical process $Q(\cdot)$. The value is even greater for networks of queues, which are discussed in the next section. See, for example, [48, 49, 94, 97, 188, 211, 246].

Another advantage of heavy traffic approximation is the availability of good numerical methods. The QNET algorithms of [50, 55, 97] provide good estimates of the first two moments of the stationary distributions even for high-dimensional problems. If the drift or covariance depends on the state or if there are controls, then the Markov chain approximation methods of [167] can be used, provided that the dimension is not too high. These numerical methods can provide a great deal of useful information for the heavy traffic limits of queues that would be impossible to analyze otherwise. Typical queueing behavior such as balking, reneging, multiple processors, batch arrivals/services, periodic or random mean rates, non-Markovian data, breakdowns, shared buffers, blocking, state dependencies, several input streams, and priorities can all be treated under broad conditions. Suppose that we have several arrival processes with different work requirements, and each arrival is to be assigned to one of a bank of processors. The scheduling problem is difficult, but heavy traffic approximations yield solutions of great value [10, 94, 255]. See also [96, 100, 126, 239] and Chapter 12.

Suppose that the server breaks down or becomes unavailable from time to time in a random manner. The basic effect is to increase the queue size. The adjustment to the limit equation (1.15) depends on the "process of unavailability." If the periods of unavailability are brief and frequent, then one adds a new Wiener process and drift term to (1.15). These account for the scaled effects in the limit of the randomness and mean of the periods. If the periods of unavailability are infrequent, but long enough to increase the scaled queue noticeably, then the modification of (1.15) involves the addition of a random jump, analogous to a Poisson process, still leaving us with a nice limit equation; see Section 7.3. Such complications are substantially more difficult to treat for the actual physical system.

The heavy traffic limits can tell us the proper dimensioning of the system for good performance. For example, how fast must the servers work and what should the excess capacity be over what is needed to handle the mean loads. Control and optimal control problems are particularly difficult to treat, and the heavy traffic approximations yield well-defined limit systems that are controlled reflected diffusions. These can be used to obtain good controls for the physical problem under heavy traffic. See Sections 3–5 and Chapters 9–12.

Fast arrivals and services. In Examples 1.1 and 1.2, both magnitude and time were rescaled. In many problems arising in communications or computer systems the basic arrival and service rates are large, and it is only the magnitude that needs to be rescaled. (See the examples in [2, 172], and in Sections 4 and 5 in this chapter.) Consider a high-speed communications system where the average arrival rate of data is n, a large number. In general, the service speed will be at least what is needed to handle the average load. A development analogous to that in Example 1.2 can be carried out, but only the magnitude needs to be scaled and not the time. Then the heavy traffic analysis embeds the system in a sequence that is parameterized by the speed (or, equivalently, the channel capacity). The heavy traffic condition is equivalent to the mean arrival and service rates differing by $O(\sqrt{n})$, that is often enough for good performance. See Chapter 9.

Fast systems and the scale parameter n. In systems such as those described in the last paragraph, the scaling parameter n often does have a physical meaning, since such systems tend to "grow" in time, and it might be desirable to analyze a sequence of systems of increasing speed or capacity and that are appropriately "dimensioned."

1.1.2 The Workload Form

Example 1.3. The analysis in Example 1.2 was in terms of the numbers queued. Sometimes it is most convenient to work with the work queued. For the problem of Example 1.2, define the workload $WL^n(t)$ to be $1/\sqrt{n}$ times the real time that is required to complete the processing of all of the customers that are in the system at real time nt. This is $1/\sqrt{n}$ times the sum of the service times of all customers waiting to be served plus the residual service time of any customer currently in service. Let $W^{a,n}(t)$ denote $1/\sqrt{n}$ times the work that has *arrived* by real time nt. There is a remarkable asymptotic relation (1.22) between the workload and the scaled queue size. The convenience of such relations for analysis is another of the advantages of heavy traffic analysis. The result (1.22) is an "arbitrary time" form of Little's law [132] and holds under quite general conditions, as seen in Section 5.3.

The processor does a unit amount of work in a unit of real time. Thus, the scaled work that has been done by real time nt is

$$\frac{1}{\sqrt{n}}\left[nt - \text{ idle time by } nt\right].$$

Hence, the equation for the workload process is

$$WL^n(t) = WL^n(0) + W^{a,n}(t) - \frac{1}{\sqrt{n}}\left[nt - \text{ idle time by } nt\right]. \tag{1.16}$$

As in the last example, the procedure starts by using representations such as (1.11) to expand the terms in (1.16) into noise and drift terms, so that it looks more like a reflected diffusion. By the definition of $W^{a,n}(t)$, it can be written as

$$W^{a,n}(t) = \frac{1}{\sqrt{n}} \sum_{l=1}^{nS^{a,n}(t)} \Delta_l^{d,n}. \tag{1.17}$$

Analogously to (1.11), rewrite (1.17) as (an expansion that is to be used frequently)

$$-\bar{\Delta}^{d,n}\frac{1}{\sqrt{n}}\sum_{l=1}^{nS^{a,n}(t)}\left[1-\frac{\Delta_l^{d,n}}{\bar{\Delta}^{d,n}}\right]+\sqrt{n}\bar{\Delta}^{d,n}S^{a,n}(t). \qquad (1.18)$$

By (1.11), the second term in (1.18) can be written as

$$\frac{\bar{\Delta}^{d,n}}{\sqrt{n}\bar{\Delta}^{a,n}}\sum_{l=1}^{nS^{a,n}(t)}\Delta_l^{a,n}+\bar{\Delta}^{d,n}w^{a,n}(S^{a,n}(t)). \qquad (1.19)$$

The first term in (1.19) is (modulo a small error) $[\bar{\Delta}^{d,n}/\bar{\Delta}^{a,n}]\sqrt{n}t$. Thus, analogously to what was done in Example 1.2, (1.16) can be manipulated into the form

$$\begin{aligned} WL^n(t) &= WL^n(0)+\bar{\Delta}^{d,n}\left[w^{a,n}(S^{a,n}(t))-w^{d,n}(S^{a,n}(t))\right] \\ &\quad +\sqrt{n}\left[\frac{\bar{\Delta}^{d,n}}{\bar{\Delta}^{a,n}}-1\right]t+\frac{1}{\sqrt{n}}\left[\text{idle time by real time } nt\right]. \end{aligned} \qquad (1.20)$$

Under the assumptions of Example 1.2, and recalling that $\bar{\lambda}^a = \bar{\lambda}^d = 1/\bar{\Delta}^a = 1/\bar{\Delta}^d$, the limit (1.15) is replaced by

$$WL(t) = WL(0)+\bar{\Delta}^d bt+\bar{\Delta}^d\left[w^a(\bar{\lambda}^a t)-w^d(\bar{\lambda}^a t)\right]+z(t), \qquad (1.21)$$

where $z(\cdot)$ is the reflection term due to the idle time. Thus, in the heavy traffic limit, a comparison of (1.15) and (1.21) yields that

$$WL(t) = \bar{\Delta}^d x(t), \qquad (1.22)$$

which is the scaled queue size times the "mean service time." A similar relationship holds for networks. See Chapters 5 and 6 for a rigorous development. The linear asymptotic relationship (1.22), that the workload is the number of customers times the *mean work* per customer, holds only in the heavy traffic limit, but it provides a considerable simplification for many problems.

1.2 Networks

Modern communications, computer, and manufacturing systems frequently involve some sort of queueing network, often where there are several interacting queues and processors, and it is in such networks that the advantages of heavy traffic analysis are particularly apparent. The approximating processes are reflected diffusions, but obtaining the reflection directions and processes is more subtle. By way of introduction, we start by describing an extension of the ideas of Example 1.2 to a tandem queue. Then a more general form as well as some special examples will be given. The development will be purely heuristic, since the aim is to motivate a method of approximation by reflected diffusion processes and illustrate the type of results that are to be obtained as well as the simplifying role of heavy traffic analysis.

1.2.1 Simple Networks

Example 2.1: A tandem queue. Let us start with the problem of two queues in tandem. The main purpose of this example is a formal illustration of the reflection process and the interaction of the queues, so the parameters are selected to simplify the notation and computations and have no other significance. Queue 1 has inputs from the outside only, and the outputs of queue 1 either leave the system or else go to queue 2, whose outputs leave the system. For use later on the notation will be defined for a more general network.

As in Example 2.1, we will work with a sequence of queues that are indexed by n. As $n \to \infty$, the traffic intensities at the two processors go to unity. Later, we will see how to choose the parameters when only a particular queue is of interest. Let the interarrival intervals for exogenous arrivals to queue i be $\Delta_{i,l}^{a,n}$, and let $\Delta_{i,l}^{d,n}$ denote the service intervals for queue i. Let $S_i^{a,n}(t)$ (respectively, $S_i^{d,n}(t)$) denote $1/n$ times the number of exogenous arrivals to (respectively, completed services at) queue i by real time nt. Define the scaled number of inputs $A_i^n(t) = \sqrt{n}S_i^{a,n}(t)$ and outputs $D_i^n(t) = \sqrt{n}S_i^{d,n}(t)$. Let $D_{ij}^n(t)$ denote $1/\sqrt{n}$ times the number of outputs of queue i that have gone to queue j by real time nt. For some centering constants $\bar{\Delta}_i^{\alpha,n}$, $\alpha = a, d$, define the processes

$$w_i^{\alpha,n}(t) = \frac{1}{\sqrt{n}} \sum_{l=1}^{nt} \left[1 - \frac{\Delta_{i,l}^{\alpha,n}}{\bar{\Delta}_i^{\alpha,n}} \right]. \tag{2.1}$$

As in Example 1.2, the $\bar{\Delta}_i^{\alpha,n}$ are ergodic means, or averages, of the actual intervals $\Delta_{i,l}^{\alpha,n}$.

Following the scaling that was suggested in Examples 1.1 and 1.2, let $x_i^n(t)$ denote $1/\sqrt{n}$ times the number in queue i at real time nt. Then, for our tandem queue,

$$\begin{aligned} x_1^n(t) &= x_1^n(0) + A_1^n(t) - D_1^n(t), \\ x_2^n(t) &= x_2^n(0) + D_{12}^n(t) - D_2^n(t). \end{aligned} \tag{2.2}$$

The next step is to separate the "mean" and "noise" components in each term in (2.2) so that (2.2) resembles a reflected diffusion, analogously to what was done in the previous section. By using the expansions of Example 1.2, we can write

$$\begin{aligned} A_i^n(t) &= \frac{1}{\sqrt{n}} \sum_{l=1}^{nS_i^{a,n}(t)} 1 \\ &= \frac{1}{\sqrt{n}} \sum_{l=1}^{nS_i^{a,n}(t)} \left[1 - \frac{\Delta_{i,l}^{a,n}}{\bar{\Delta}_i^{a,n}}\right] + \frac{1}{\sqrt{n}\bar{\Delta}_i^{a,n}} \sum_{l=1}^{nS_i^{a,n}(t)} \Delta_{i,l}^{a,n}, \end{aligned} \tag{2.3}$$

and, analogously,

$$D_i^n(t) = \frac{1}{\sqrt{n}} \sum_{l=1}^{nS_i^{d,n}(t)} \left[1 - \frac{\Delta_{i,l}^{d,n}}{\bar{\Delta}_i^{d,n}}\right] + \frac{1}{\sqrt{n}\bar{\Delta}_i^{d,n}} \sum_{l=1}^{nS_i^{d,n}(t)} \Delta_{i,l}^{d,n}. \tag{2.4}$$

As in Example 1.2, the last term on the right of (2.3) is $\sqrt{n}t/\bar{\Delta}_i^{a,n}$, modulo a negligible error, which is $1/\sqrt{n}$ times the time between nt and the last arrival before or at nt. Let $y_i^n(t)$ denote $1/[\bar{\Delta}_i^{d,n}]$ times the cumulative idle time at queue i by real time nt. Then, as in Example 1.2 and modulo a negligible error, the last term on the right of (2.4) is $\sqrt{n}t/\bar{\Delta}_i^{d,n} - y_i^n(t)$. Let $I_{12,l}^{r,n}$ denote the indicator function of the event that the lth output of queue 1 goes to queue 2 and suppose that there are $q_{12}^n \to q_{12}$ such that (Markov routing)

$$P\left\{I_{12,l}^{r,n} = 1 \middle| \text{past data}\right\} = q_{12}^n. \tag{2.5}$$

The centering constants play the same role here that they did in Example 1.2, where they were the long-term averages of the intervals.

A heavy traffic assumption. Define

$$\begin{aligned} b_1^n &= \sqrt{n}\left[\frac{1}{\bar{\Delta}_1^{a,n}} - \frac{1}{\bar{\Delta}_1^{d,n}}\right], \\ b_2^n &= \sqrt{n}\left[\frac{q_{12}^n}{\bar{\Delta}_1^{d,n}} - \frac{1}{\bar{\Delta}_2^{d,n}}\right]. \end{aligned}$$

In order to get the reflected diffusion approximating process, it will be assumed that the b_i^n converge to some real numbers $b_i, i = 1, 2$. Note that this implies that the difference between the mean input and output rates for each processor is close to zero for large n.

Selecting the drift parameter for a given queue. Suppose for the moment that we have only some particular queue, where $\bar{\Delta}_i^{\alpha,n} = \bar{\Delta}_i^{\alpha}$ and $q_{12}^n = q_{12}$ are given. Then, following the argument in Example 2.1, suppose that there is large n such that the b_i^n are of "moderate" size, call them

b_i, and use the limit (2.8) to approximate the quantities of interest for the original physical queue.

Now, returning to the limit and approximation procedure, we need to expand the term $D_{12}^n(t)$ in (2.2). Define

$$w_{12}^{r,n}(t) = \frac{1}{\sqrt{n}} \sum_{l=1}^{nt} \left[I_{12,l}^{r,n} - q_{12}^n \right] \tag{2.6}$$

and write

$$D_{12}^n(t) = \frac{1}{\sqrt{n}} \sum_{l=1}^{nS_1^{d,n}(t)} \left[I_{12,l}^{r,n} - q_{12}^n \right] + \frac{1}{\sqrt{n}} \sum_{l=1}^{nS_1^{d,n}(t)} q_{12}^n. \tag{2.7}$$

The right-hand term is $q_{12}^n D_1^n(t)$. Adding up the approximations (2.3)–(2.7) yields, modulo negligible errors,

$$x_1^n(t) = x_1^n(0) + b_1^n t + w_1^{a,n}(S_1^{a,n}(t)) - w_1^{d,n}(S_1^{d,n}(t)) + y_1^n(t), \tag{2.8a}$$

$$\begin{aligned} x_2^n(t) = x_2^n(0) + b_2^n t + q_{12}^n w_1^{d,n}(S_1^{d,n}(t)) \\ - w_2^{d,n}(S_2^{d,n}(t)) + w_{12}^{r,n}(S_1^{d,n}(t)) + y_2^n(t) - q_{12}^n y_1^n(t). \end{aligned} \tag{2.8b}$$

Discussion of (2.8). As noted in Example 1.2, under broad conditions the $w_i^{\alpha,n}(\cdot)$ will converge weakly to Wiener processes $w_i^\alpha(\cdot)$, $\alpha = a, d$. Let us suppose that this is true. Then it can be shown that the $S_i^{\alpha,n}(\cdot)$ converge to the linear functions with values $\bar{\lambda}_i^\alpha t$. Under (2.5), $w_{12}^{r,n}(\cdot)$ converges weakly to a Wiener process with variance $q_{12}(1 - q_{12})$. The third and fourth terms on the right-hand side of (2.8a) are the "noise" effects of the randomness in the arrival and service intervals, respectively. The randomness in the input to queue 2 can be split into two parts, that due to the randomness of the output process of queue 1 and that due to the randomness of the selection of these outputs as inputs to queue 2. These are the $q_{12}^n w_1^{d,n}(S_1^{d,n}(t))$ and the $w_{12}^{r,n}(S_1^{d,n}(t))$ terms, respectively, in (2.8b). This representation is both convenient and revealing, since it separates distinct events and allows us to assess their effects. The $y_i^n(\cdot)$ can increase only when $x_i^n(t) = 0$. The reflection term $y_1^n(\cdot)$ in (2.8a) is just what we had in Example 1.2. The reflection term $y_2^n(t) - q_{12}^n y_1^n(t)$ in (2.8b) accounts for the effects of the idle time of the first processor on the second queue, and will be discussed further below. The limit equations have the form

$$\begin{aligned} x_1(t) &= x_1(0) + b_1 t + w_1^a(\bar{\lambda}_1^a t) - w_1^d(\bar{\lambda}_1^a t) + y_1(t), \\ x_2(t) &= x_2(0) + b_2 t + q_{12} w_1^d(\bar{\lambda}_1^a t) - w_2^{d,n}(\bar{\lambda}_2^d t) + w_{12}^{r,n}(\bar{\lambda}_1^a t) + y_2(t) - q_{12} y_1(t). \end{aligned}$$

The process $y_i(\cdot)$ is continuous and nondecreasing, $y_i(0) = 0$, and the process can increase only at t where $x_i(t) = 0$. For future use, let us rewrite

(2.9) as

$$x(t) = x(0) + bt + w(t) + [I - Q']\, y(t), \quad Q = \begin{bmatrix} 0 & q_{12} \\ 0 & 0 \end{bmatrix}. \tag{2.9}$$

Note that $Q = \{q_{ij}; i, j\}$ is the (limit) internal routing matrix of the tandem network; i.e., q_{ij} is the (limit) probability (given all past data) that a departure from the ith queue goes to the jth queue. The $y_i(\cdot)$ can increase only at values of t at which $x_i(t) = 0$.

See Figure 2.1, where the directions of reflection d_1 and d_2 are indicated. Note that the reflection on the $x_1 = 0$ axis is at an angle pointing down. This follows from the way that the y_i terms appear in (2.9): The directions are determined by the structure of the routing. Loosely speaking, the idle time of processor 1 reduces its scaled output by $1/\sqrt{n}$ times the number that would be served during that idle time, and the result is $y_1^n(t)$. The number of customers that would be served during that idle time is (loosely speaking) the number of fictitious outputs of Example 1.1. Hence, the average reduction (over what would be the case if queue 1 was never empty) that the idleness of processor 1 causes in the input to queue 2 is just $q_{12}^n y_1^n(t)$. From the fictitious output point of view, the fictitious outputs from queue 1 must be canceled as outputs of queue 1. But if such a fictitious output is routed to queue 2, then it must also be canceled as an input to queue 2. The proportion of the fictitious outputs of queue 1 that go to queue 2 is q_{12}^n. Thus the reflection on the $x_1 = 0$ axis keeps x_1 from going negative and also increases x_2.

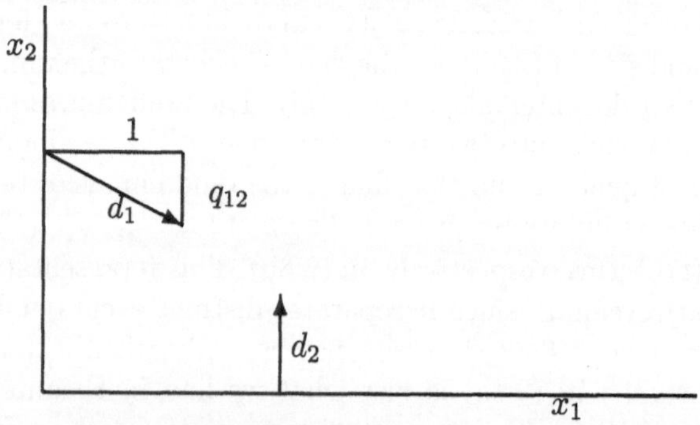

Figure 2.1. Reflection directions: Tandem queue.

Example 2.2. A form of processor sharing. Now consider the problem where there are two parallel queues, each with its own input process and processor, where the inputs and services and initial condition for each queue satisfy the conditions of Example 1.2, and where the times for queue 1 are independent of those for queue 2.

Suppose that if one queue is empty but the other has a customer that is not being served, then a customer will transfer to the empty queue and be served immediately. If one is interested only in the total number of customers in the system, then the problem can be reduced to a one-dimensional model in the limit. For the two dimensional queue, as long as the state is not near the origin the limit process has the form

$$x(t) = x(0) + bt + w(t) + [I - Q']\,y(t) = 0, \qquad Q = \begin{bmatrix} 0 & 1 \\ 1 & 0 \end{bmatrix}, \tag{2.10}$$

where $w(\cdot)$ is a two dimensional Wiener process with independent components, each being representable in the form used in (1.15). The reflection directions are illustrated in Figure 2.2. The directions point at each other, owing to the routing from one to the other when one queue is empty and the other not. Thus, at the origin, the $y_i(\cdot)$ are not well defined, since they can cancel one another. Such physical systems are difficult to analyze. In any case, (2.10) needs to be replaced by $x(t) = x(0) + bt + w(t) + z(t)$, where $z(\cdot)$ is the reflection process. It is continuous and has the form in (2.10) when $x(t)$ is not "at" the origin. The direction of change of $z(\cdot)$ when $x(t) = 0$ can be any convex combination of the directions on the two sides. The process $z(\cdot)$ will not be of bounded variation owing to its behavior when $x(t) = 0$. Nevertheless, the limit process is still much simpler than the original physical problem. See [66, 65] for a detailed analysis of such problems and associated results in large deviations.

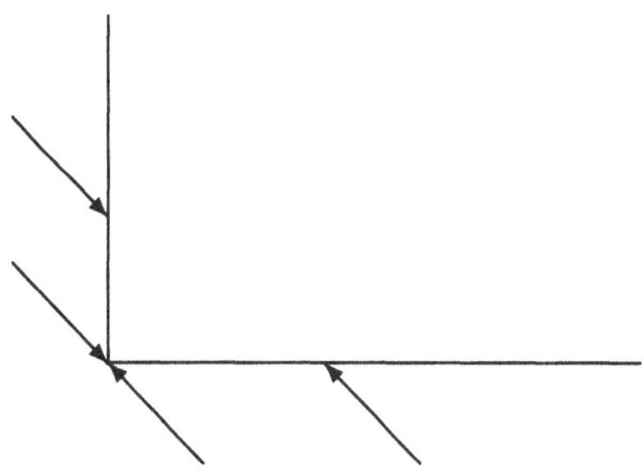

Figure 2.2. Reflection directions, processor sharing.

Example 2.3. A canonical network model. The model used in Example 2.1 is a special case of the general network form of [213]. In the system of concern, there are K processors, denoted by P_i, $i \leq K$. Each can have an exogenous input stream. The output of any P_i can either leave the system or be routed to another processor, and the conditional probability (given the "past") that an output from P_i goes to P_j is q^n_{ij}. Suppose that

$q_{ij}^n \to q_{ij}$. Thus, the queue at each P_i can contain both exogenous arrivals (coming directly from the outside) and arrivals from other processors. As in the previous examples, the system of concern is embedded in a sequence parameterized by n, and the system scale and parameters depend on n. The full setup is given in Chapter 6, and we will only make the following comments here. Define the routing matrix $Q = \{q_{ij}; i, j\}$. If customers can leave the system from P_i, then $\sum_j q_{ij}^n < 1$. In this case, the state i is transient, as are all states from which it can be reached with positive probability. In order to ensure that all of the customers eventually leave the system, we require that the spectral radius of Q be less than unity. Since Q is the transition matrix of a finite-state Markov chain, the spectral radius condition guarantees that all states of the chain are transient.

In order to get a limit approximation, an assumption analogous to the heavy traffic assumption (1.14) must be made. The condition asserts that the mean input rate to each P_i (over all sources) equals its mean service rate, plus a difference that goes to zero as $n \to \infty$. Now, following the procedure of that section, which is analogous to that which led to (2.8), we would get the expression, for $i \le K$,

$$\begin{aligned} x_i^n(t) = x_i^n(0) + b_i t + w_i^{a,n}(S_i^{a,n}(t)) + \sum_j q_{ji}^n w_j^{d,n}(S_j^{d,n}(t)) - w_i^{d,n}(S_i^{d,n}(t)) \\ + \sum_j w_{ji}^{r,n}(S_j^{d,n}(t)) + y_i^n(t) - \sum_j q_{ji}^n y_j^n(t). \end{aligned} \tag{2.11}$$

The process $w_i^{a,n}(\cdot)$ (respectively, $w_i^{d,n}(\cdot)$) represents the "randomness" in the exogenous arrival sequence to P_i (respectively, in the departure sequence from P_i). The $q_{ji}^n w_j^{d,n}(S_j^{d,n}(\cdot))$ represent the effect of the randomness in the service times at P_j on the input to P_i. The $w_{ji}^{r,n}(S_j^{d,n}(t))$ represent the effect on P_i of the randomness in the routing from P_j. The term $q_{ji}^n y_j^n(\cdot)$ represents the effect on P_i of the idle time at P_j, and b_i quantifies the difference between the mean total input and service rates at P_i.

Under broad conditions, $w_i^{a,n}(\cdot), w_i^{d,n}(\cdot)$ and $w_{ij}^{r,n}(\cdot)$ converge weakly to Wiener processes and $S_i^{\alpha,n}(t) \to \bar{\lambda}_i^\alpha t$, $\alpha = a, d$, where $\bar{\lambda}_i^a$ (respectively, $\bar{\lambda}_i^d$) is the asymptotic mean exogenous arrival (respectively, service) rate at P_i. Then, if the initial condition $x^n(0)$ converges to some random variable $x(0)$ in probability, $x^n(\cdot)$ will converge weakly to a limit process $x(\cdot)$ that satisfies the vector-valued reflected diffusion

$$x(t) = x(0) + bt + w(t) + [I - Q']\, y(t). \tag{2.12}$$

The process $w(\cdot) = (w_i(\cdot), i \le K)$ is a vector-valued Wiener process. The right-hand term $[I - Q']y(t)$ is the reflection process. It is a consequence of the idle time. Indeed, without this reflection term, the process (2.12) would represent the limit of the (scaled) initial condition plus the (scaled) number of inputs minus the number of outputs that would have been created if the processor worked continuously (i.e., no idle time). The process $y_i(\cdot)$ is nonincreasing, satisfies $y_i(0) = 0$, and can increase only at times t where $x_i(t) = 0$.

The form of the reflection term can be heuristically justified by an extension of the argument that was used for Example 2.1. The reflection term in Example 2.1 was obtained via the introduction of the idle time at P_i in the representation of the number of customers served there and the number sent to other processors. Note that (in the limit) a proportion q_{ij} of the output of P_i is sent to P_j. Suppose that, formally speaking, N jobs could have been served during this idle time at P_i. Then the mean number *not sent* to P_j due to the idle time at P_i on that time interval is $q_{ij}N$. The effect on the queue at P_j of a unit idle time at P_i is essentially $-q_{ij}$ for $j \neq i$ and is $1 - q_{ii}$ for $j = i$. This explains the form $[I - Q']y(t)$.

Comment on computation. A queueing network with K processing stations and infinite buffers, Markov routing, and where each customer eventually leaves, is called a *Jackson network* if station i has s_i identical processors, service is FCFS, the exogenous arrival processes are mutually independent and Poisson, the service times are mutually independent and exponentially distributed, independent of the input processes with their parameters depending only on the processor, and the service capacity at each processor is greater than its total arrival rate. Then the stationary distribution exists and can be written as the product of distributions $\Pi_{i=1}^{K} P_i(Q_i)$, where $P_i(\cdot)$ is the distribution for an $M/M/s_i$ queue whose parameter is the effective traffic intensity at station i [18, 115, 116, 125]. The product form distribution was shown to exist for an analogous class of closed networks in [85] and was extended to a broader class of networks and distributions in [9, 18]. In such cases, one does not need to approximate if the only interest is in the values of stationary moments. This nice product form is lost when there are controls, finite buffers, processor sharing, state-dependent rates, or general service or interarrival times.

Under certain conditions, the limit system (2.12) will have a stationary distribution with a density $p(x)$ of the *product form* [102] $p(x) = \Pi_{i=1}^{K} \gamma_i e^{-\gamma_i x_i}$, where the γ_i depend only on the value of b, the covariance of the Wiener process, and the matrix Q. Thus, the product form for (2.12) might hold even when it does not for the original physical system.

More generally, one must resort to a numerical method. The QNET method [50, 55, 97] is designed to solve for the low-order moments of the stationary distribution of (2.12). These papers contain many examples where the solution via QNET is compared to an estimation via simulation, under a range of traffic intensities. It is seen that the estimates obtained from the heavy traffic approximations can be very good even when operating far from the heavy traffic regimes. One simple example, taken from [97], is in Table 2.1. The system is a two-stage tandem queue. The input to the first processor is Poisson with rate $\bar{\lambda}_1^a$, and the service time there is deterministic, and of unit length. The output of the first processor goes to the second, where the service is exponential with rate unity. The output of the second processor leaves the system. The traffic intensity is controlled by varying the rate of the input Poisson process. The numbers give the mean stationary waiting time at the second station. The limit or approximating model is (2.9) with $q_{12} = 1$. The parameters b_i are obtained as discussed above (2.6), and the variances of the Wiener processes $w_i^\alpha(\bar{\lambda}_i^\alpha \cdot), \alpha = a, d$, $i = 1, 2$, are just $\bar{\lambda}_1^a E[1 - \Delta_{i,l}^\alpha/\Delta_i^\alpha]^2$. This is zero for $\alpha = d$, $i = 1$, since the service at P_1 is deterministic.

ρ	Simulation	QNET estimate
.5	.72	.75
.6	1.12	1.13
.7	1.74	1.75
.8	3.01	3.00
.9	6.76	6.75

Table 2.1. Average waiting time at the second processor.

1.2.2 Closed and Open/Closed Networks

The following examples illustrate some of the many forms of communication and computer networks for which heavy traffic analysis has considerable benefits.

Example 2.3. Refer to Figure 2.3, a model of an interactive computer system. There are n independent users ($n = 2$ in the figure) who send jobs to the CPU. Each user can have at most one active job at a time. The CPU services them in some way (sharing its time among all of the active jobs in some way, or processing the jobs sequentially). When a job is completed, an acknowledgment is sent to the user, who then waits for a (random) interval, a latency period, before sending another job to the CPU. During the latency period, we say that the job resides at the user. The system is said to be closed, since (counting the number of jobs residing at the users) the number of jobs in it is fixed at n. From a formal point of view, the jobs can be said to circulate at random in the network, and never leave it, whereas in an open network such as those in Examples 2.1 and 2.2, any job will eventually leave. In an open/closed network, some jobs circulate without leaving, but other jobs enter exogenously and eventually leave.

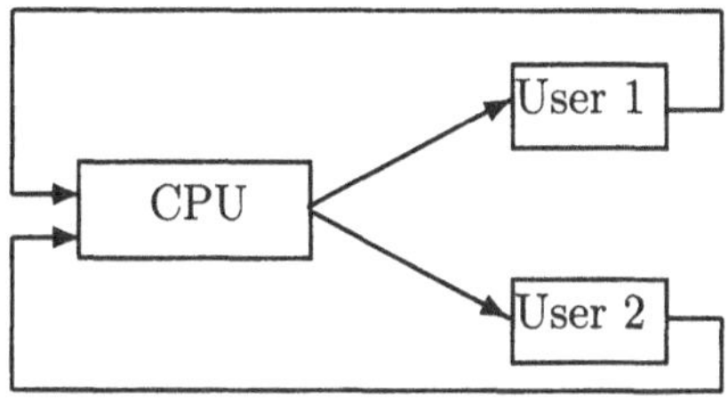

Figure 2.3. A closed interactive computing network.

Such systems are hard to analyze for general service and latency time distributions. If the number of users is large, as is often the case, and if the average fraction of time that the CPU is free is small, then heavy traffic analysis can greatly simplify the analysis See, e.g., [40, 50, 101, 103, 139, 141]. It turns out that that the analysis for the open and closed networks is very similar. See Chapter 6.

Example 2.4. Closed and open/closed networks occur in manufacturing systems. Consider the example of Figure 2.4. Material for a job enters P_1, then the resulting product goes to either P_2 or P_3 for further processing. From P_3 it goes to P_4 and then leaves the system. On completing service at P_2, it might either return to P_1 and repeat the process or else go on to P_4, from which it leaves the system. Owing to limited buffer space, only n jobs are allowed in the system at a time. When a job leaves P_4, a new one immediately enters P_1, and the arrow from P_4 to P_1 indicates this event. Thus, from a formal point of view, it can be said that n jobs circulate in the system.

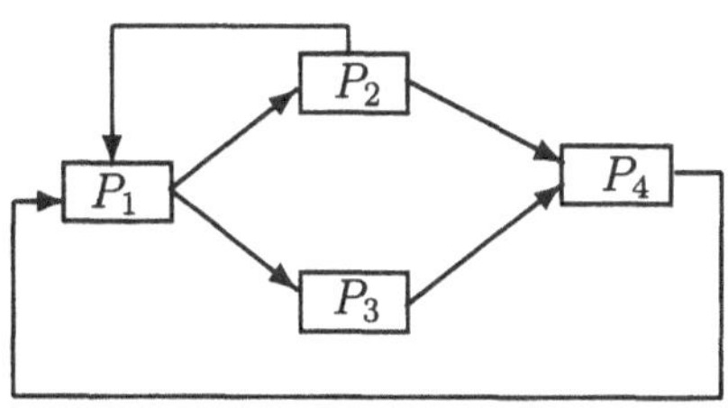

Figure 2.4. A closed manufacturing system.

1.3 Server Scheduling and Assignment Problems

Introductory remarks. The problem of assigning incoming jobs to one of a bank of processors is one of the oldest control problems for queues. The jobs might be assigned on arrival or just at the time that they are to be processed. They might be divided into distinct classes, each with its own requirements. The processors might not be fully interchangeable, in that the distribution of the required time might depend on the processor to which assignment is made as well as on the job class. The cost of waiting before leaving the system might also depend on the class. Suppose that there is a single processor and several classes of jobs, which are queued until served. Let the holding costs be linear (and class dependent) and with no cost for the processor to switch the class of jobs being worked on. Then (under some broad additional conditions) the classical $c\mu$-rule [243] says that one should serve the queue with the largest value of $c_i \bar{\lambda}_i^d$ and that is nonempty. The $c\mu$-rule seeks the most rapid immediate decrease of cost, where the decrease is measured by the cost per unit waiting time per job times the service rate. This was extended to nonlinear costs, in the heavy traffic limit, in [239].

The references [10, 94, 96, 255] and Chapters 11 and 12 show how to use heavy traffic modeling to simplify the assignment problem when there are many job classes and processors and they exhibit what is called *state space collapse,* in that in the heavy traffic limit, the problem can be represented in terms of a process of lower (often much lower) dimension. See also [30, 100, 208, 254].

Consider the case of a single server. The problem of assigning that server to the queues is known as *polling,* and is the subject of a large literature [27, 230]. The processor is said to *poll* the queues. Generally, the analysis is restricted to fixed control policies, with no optimization. The simplifying nature of the heavy traffic approximations allows the possibility of actually applying optimal control methods. Example 3.2 below is taken from [211] and shows that very useful results can be obtained by using the intuition and general guiding principles from heavy traffic analysis even for a formal analysis of a control problem. Other forms of the assignment problem are in [166]. See also [182] for an analysis of the multiserver problem, which does not involve heavy traffic assumptions. Related works in the heavy traffic context include [40, 187, 188, 210]. The reference [126] contains much insight for scheduling in networks. Processor sharing is another form of scheduling.

1.3.1 Multiple Input Classes and Controlled Polling

Example 3.1. Multiple input classes and a single processor: an averaging principle. Consider the problem where there is a single server but two distinct arrival processes, each queued separately. The single processor must divide its attention between the two queues. There is no switching time in going from one queue to the other, and the processor does not idle if there is work to be done.

The control problem (also called the scheduling problem) for the processor will be discussed in Example 3.2 (and in Chapter 11). An issue in the scheduling of the processor is whether or not to allow the processor to switch in the middle of a job, and if it can do so, what time is needed for the completion of the work in the stopped job. In the interest of simplicity at this time, we suppose that if the processor does switch in the middle of a job, then the completion of that job later requires only its residual service time (the so-called "preempt-resume" strategy). In fact, it is a consequence of the heavy traffic analysis that the results will be the same if the current job is completed before switching. In the limit we cannot distinguish between such policies. This is an example of a detail that can complicate a direct analysis of the physical problem but that does not appear in the limit.

Polling occurs in computer and communication systems where there are many buffers, each receiving data from some individual source according to some stochastic rule, and the processor connects the buffers to the channel in some order. The switching time might be small. In manufacturing systems, the classes might correspond to different processing requirements, and a nonzero switching time might be required to reset the machinery for work on the new class. But unless otherwise noted, we assume that the switching time is zero.

Let us work with a sequence of systems in heavy traffic. Using the terminology of Example 2.2, let $\Delta_{i,l}^{a,n}$, (respectively, $\Delta_{i,l}^{d,n}$) denote the lth interarrival interval (respectively, service time) for class i. Let the centering parameters be $\bar{\Delta}_i^{\alpha,n}$, $i = 1, 2, \alpha = a, d$, and suppose that they converge to $\bar{\Delta}_i^{\alpha} \equiv 1/\bar{\lambda}_i^{\alpha}$. Set $S_i^{a,n}(t)$ (respectively, $S_i^{d,n}(t)$) to be $1/n$ times the number of arrivals (respectively, service completions) from class i by real time nt. With $W_i^{a,n}(t)$ denoting $1/\sqrt{n}$ times the work from class i that has arrived by real time nt, the equation for the evolution of the total workload $WL^n(\cdot)$ is

$$WL^n(t) = WL^n(0) + \sum_i W_i^{a,n}(t) - \frac{1}{\sqrt{n}}\left[nt - \text{idle time by } nt\right]. \tag{3.1}$$

Define the *traffic intensity* $\rho_i^n = \bar{\Delta}_i^{d,n}/\bar{\Delta}_i^{a,n}$ for class i, and set $\rho_i = \bar{\lambda}_i^a/\bar{\lambda}_i^d$. Suppose that the $w_i^{\alpha,n}(\cdot)$, $\alpha = a, d$, $i = 1, 2$, converge weakly to Wiener processes $w_i^{\alpha}(\cdot)$ with variances $\sigma_{\alpha,i}^2$.

Applying the expansions that were used in Example 1.3, but to both classes, we can write (modulo an asymptotically negligible error)

$$\begin{aligned} WL^n(t) = WL^n(0) &+ \sum_i \bar{\Delta}_i^{d,n}\left[w_i^{a,n}(S_i^{a,n}(t)) - w_i^{d,n}(S_i^{a,n}(t))\right] \\ &+ \sqrt{n}\left[\sum_i \rho_i^n - 1\right]t + \frac{1}{\sqrt{n}}\left[\text{idle time by real time } nt\right]. \end{aligned} \tag{3.2}$$

Note the simplicity of (3.2). The form is the same and is one-dimensional, no matter what the number of sources. It is simply the sum of the effects of the randomness associated with each source, the difference between the (scaled) mean input minus mean output rates, assuming that the processor works all of the time, and a correction for the time that the processor does not work. The heavy traffic condition is now the existence of a real number b such that

$$\lim_n \sqrt{n}\left[\sum_i \rho_i^n - 1\right] = b. \tag{3.3}$$

Thus, for large n the scaled "mean rate" at which work from all classes arrives is nearly unity, the scaled maximum rate at which the server can work. As for the examples in Section 1, (3.3) and the assumed convergence of the $w_i^{\alpha,n}(\cdot)$ to Wiener processes imply that (Theorem 5.1.1) $S_i^{\alpha,n}(\cdot)$ converges to the process defined by $\bar\lambda_i^a t$. The limit process is

$$WL(t) = WL(0) + \sum_i \bar\Delta_i^d \left[w_i^a(\bar\lambda_i^a t) - w_i^d(\bar\lambda_i^a t)\right] + bt + z(t), \tag{3.4}$$

where $z(\cdot)$ is the reflection term at the origin. It is continuous and nondecreasing, $z(0) = 0$, and it can increase only when $WL(t) = 0$.

If we are concerned with just one queue and not with a sequence, then suppose there is a large n such that $\sqrt{n}[\sum_i \rho_i^n - 1]$ is of moderate size, denote it by b, and use the limit system to approximate the quantities of interest. This is the procedure used to get the policies for the data in Table 3.1.

Let $x_i^n(t)$ (respectively, $WL_i^n(t)$) denote $1/\sqrt{n}$ times the number of jobs (respectively, the workload) in queue i (including any job in service) at real time nt. Under the assumption that the switchover time is zero and that the processor works if there is work to do, the value of $WL^n(t)$ does not depend on how the processor divides its time between the two queues. From the perspective of the total workload, the processor can allocate its time in any way at all, without changing (3.1) or the limit equation (3.4). Thus, for large n the total workload changes smoothly in (scaled) time, illustrating an advantage of the workload formulation. On the other hand, in scaled time the individual workloads $WL_i^n(\cdot)$ change very fast for large n.

Let us examine the paths of the individual workloads $WL_i^n(\cdot)$ more closely. Recall that the time scale that defines the $WL_i^n(\cdot)$ is real time compressed by a factor of n. Also, the workloads are the real times required to complete all the currently queued jobs, divided by $\sqrt{n}$. Thus, in *scaled time and work* the processor works on the queued individual workloads at rate $\sqrt{n}$, and the "mean arrival rate" of work from class i is $\rho_i^n\sqrt{n}$, which we will approximate by $\rho_i\sqrt{n}$. First consider the following particular strategy. Suppose that the processor works on queue 1 until the next time that it is empty, then switches immediately to queue 2, works on that until the next time that it is empty, etc. Suppose that at scaled time t the processor has just switched to queue 1, whose scaled content is $w = WL_1^n(t) = WL^n(t)$. On "average" (indeed, in the sense of probability), for large n queue 1 will be served for a (scaled) length of time δ such that $w + \sqrt{n}\rho_1\delta - \sqrt{n}\delta \approx 0$. Such results can be made rigorous, as in Theorem 5.6.4 and Chapter 11. Hence, it takes an average amount of (scaled) time $w/[(1-\rho_1)\sqrt{n}]$ to empty the queue. The weak convergence of (3.1) to (3.4) tells us that during this small time interval the (scaled) total workload has barely changed. During this time, the workload in queue 2 has increased at a rate of $\rho_2\sqrt{n} = (1-\rho_1)\sqrt{n}$.[1] Thus, for large n, the "mean" workload at queue 2 at scaled time $t+\delta$ is approximately w, as expected.

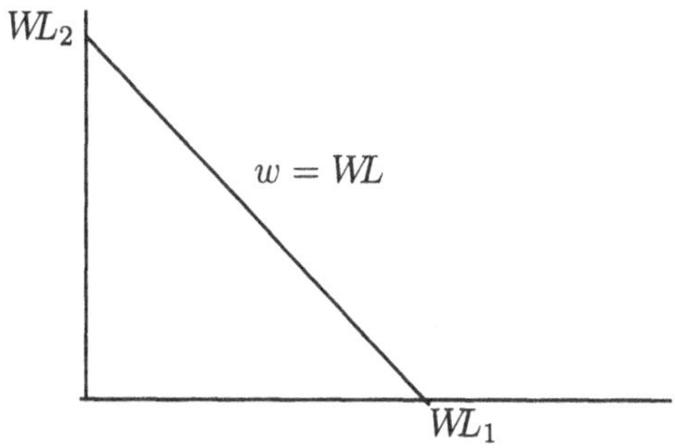

Figure 3.1. The $(WL_1(\cdot), WL_2(\cdot))$ path at constant total workload w.

Let us recapitulate the above discussion with the aid of Figure 3.1 and try to follow the individual workloads. As noted above, as time increases, the total workload $WL^n(\cdot)$ does change, but very slowly in comparison with the individual workloads $WL_i^n(\cdot)$, which move at an approximate rate of order $O(\sqrt{n})$. This smoothness of the path $WL^n(\cdot)$ and the time scale separation can be fruitfully exploited to yield good scheduling policies.

Let the current total workload value be w, and suppose that n is large enough so that $WL^n(\cdot)$ does not change much over small time intervals. For large n, the path of $(WL_1^n(\cdot), WL_2^n(\cdot))$ moves at an essentially constant rate, in the following sense. Suppose that the processor switched to queue 1 at scaled time τ. Then $WL_2^n(\tau + t/\sqrt{n}) - WL_2^n(\tau) \sim \rho_2 t$ and $WL_1^n(\tau + t/\sqrt{n}) - WL_1^n(\tau) \sim -(1-\rho_1)t = -\rho_2 t$, as long as there is work to do in queue 2, and analogously for queue 1. Thus, over many cycles, each one being of "approximate" duration in scaled time

$$\frac{w}{\sqrt{n}\rho_1} + \frac{w}{\sqrt{n}\rho_2} = \frac{w}{\sqrt{n}\rho_1\rho_2}, \tag{3.5}$$

and during which the value of $WL^n(\cdot)$ changes very little, the location of $WL_i^n(t)$ is approximately *uniformly distributed* on the line of constant total

[1] $\rho_1 + \rho_2 = 1$, since by (3.3), $\sum_i \rho_i^n \approx 1 + b/\sqrt{n}$.

workload w in Figure 3.1. This logic was developed in [42, 43] to show that for any bounded and continuous function $f(\cdot)$ and any $T > 0$,

$$\int_0^T f(WL_i^n(s))ds \to \int_0^T \left[\int_0^1 f(uWL(s))du \right] ds \tag{3.6}$$

in the sense of distribution, as $n \to \infty$.

We will use these comments in Example 3.2, which concerns the problem of optimizing the scheduling of the processor. The result (3.6) was derived under the assumption that each class was served until its queue was empty, and then the processor switched immediately to the other class. But the general logic of the derivation of (3.6) suggests that it holds for a much wider class of scheduling policies, and this is proved in Chapter 11. In the next example it is assumed that (3.6) holds for all of the policies that are being considered there.

The above discussion shows how the problem can be simplified in heavy traffic. The development can be made rigorous (Chapter 11 and [42]). This rigorization and the associated simplification of the problem illustrates an important advantage of heavy traffic modeling. The rigorization provides an intuition for the physical problem, even going deeper into the behavior than implied by the basic convergence of (3.1) to (3.4). This intuition can be exploited to get good scheduling policies, even when there is no rigorous convergence result. The following example emphasizes the importance of the previous discussion. It shows how to use the intuition gained from the theoretical analysis to obtain effective scheduling policies for the processor for problems that could not otherwise be solved, even for particular queues that are not necessarily in heavy traffic. See the data at the end of the section.

Example 3.2. Controlled Polling. A problem of optimal scheduling for the setup of Example 3.1 was thoroughly treated in [211]. We will discuss part of the motivation and development in [211], which, although heuristic, is compelling. It is a good illustration of the use of the insights of heavy traffic analysis on a hard problem. Let $S^{s,n}(t)$ denote $1/\sqrt{n}$ times the number of completed polling cycles by real time nt. We continue to assume the conditions of Example 3.2, unless otherwise mentioned.

The cost function is, for $c_i > 0$, $k > 0$,

$$\limsup_T \frac{1}{T} E \left[\int_0^T \sum_i c_i x_i^n(t)dt + kS^{s,n}(t) \right], \tag{3.7}$$

which we wish to minimize. It is assumed that the arrival and service processes are mutually independent and that switching between queues is instantaneous. The complicating factor is that there is a cost for switching. The time required to switch from serving one class to serving another is not always negligible, and it is very hard to deal with such nonzero switching times. (A heuristic analysis of one such problem is also in [211]. See also Chapter 11.) The penalty $kS^{s,n}(\cdot)$ associated with the number of switching cycles acts as a surrogate for the nonzero switching time, in that it discourages "excessive" switching between classes. Owing to the penalty $kS^{s,n}(\cdot)$, if the queue being served becomes empty and the other queue has little work, then the processor might choose not to switch immediately but to work on whatever jobs arrive or to idle if there are no new jobs, allowing the other queue to grow. Indeed, even without allowing for switching times, the problem is too difficult for analytic solution at this time. Nevertheless, the intuition gained from Example 3.1 suggests a very reasonable heuristic approach to getting a good scheduling policy. Simulations [211] have verified that the policy that is obtained is nearly optimal with respect to a large class of reasonable alternatives.

The cost rate in (3.7) is in terms of the number queued. Since the work required by any job might not be known until that job is completed, the queued workloads might not be known. But asymptotically, an analysis of the type that leads to (1.22) yields the asymptotic relationships (see Section 5.3)

$$WL_i^n(t) \approx \bar{\Delta}_i^d x_i^n(t), \tag{3.8}$$

so that the cost function and the control can be written (asymptotically) in terms of the total workload. The control will be restricted to be of the (non time dependent) feedback form. It is convenient to write the control in terms of the workloads; since the total workload changes slowly, it is easier to exploit the time scale differences in developing the heuristics. If the switching time is zero and there is no switching cost, and the arrival streams are Poisson, then the so-called $c\mu$-rule is optimal [243]. Little is known concerning optimal strategies when there is a switching cost (see the references in [211]).

The reference [211] exploits the difference in the rates of change of $WL^n(\cdot)$ and $WL_i^n(\cdot)$. This allows a two-step "hierarchical" development, where part of the optimization is done by assuming that the total workload $WL^n(\cdot)$ is essentially fixed over many switching cycles. A reasonable form for the optimal control is first motivated, and then the optimization is done (approximately) within this class of policies. At each time there are three possibilities for the control action: continue serving the current queue if there are jobs to be served, switch to the other queue and start service, or idle (while serving whatever jobs of the current class that arrive) without switching. For specificity, the preempt–resume strategy is assumed.

Let $c_1\bar{\lambda}_1^d > c_2\bar{\lambda}_2^d$. Suppose that the processor is currently set up to serve queue 1. Since the holding cost is reduced faster by serving queue 1 than by serving queue 2, it seems reasonable to keep serving queue 1 at least until it is empty, and we impose this restriction on the allowed policies. While this restriction is reasonable, the argument is not rigorous, and the control does not use all the information that is available, since it is restricted to be a function of the current individual workloads only. (For example, it does not use the information on the length of time that the current job has been in service.) This strategy would make sense in a "memoryless" system where the $\Delta_{i,l}^{\alpha,n}$, $\alpha = a, d$, $i = 1, 2$, are mutually independent and exponentially distributed with mean values depending only on α and i. However, intuitively, the restriction should have negligible effect for large n because the original and the "memoryless" system have the same limit (3.4).

Now, continue to suppose that the server is currently set up to serve queue 1, but queue 1 is empty. Then the processor has two choices: idle (but serve whatever class 1 customers might arrive) or switch to queue 2. Note that under the heavy traffic hypothesis, queue 1 will remain empty or nearly empty until after the processor switches. It is reasonable to suppose that there is a threshold $\bar{w}_2$ such that the processor would switch only when $WL_2^n(t) \geq \bar{w}_2$. (Such a restriction will not necessarily be fully optimal for each n under the general assumptions that are used, which are only those required for $WL^n(\cdot)$ to converge to $WL(\cdot)$.) Then, at the moment of switching to queue 2, $WL_2^n(t) = w_2 \geq \bar{w}_2$. The server will, of course, start serving queue 2 immediately. The server now needs to decide when to return to queue 1.

According to the discussion in Example 3.1, while queue 2 is being served, $WL_1^n(\cdot)$ increases at a mean rate close to $\rho_1\sqrt{n}$ (for large n). Recall that in the time elapsed before returning to queue 1, the *total workload* changes only negligibly. It is argued in [211] that a reasonable policy is defined by a function $u(\cdot)$ such that the processor switches back to queue 1 when $WL_2^n(\cdot)$ is reduced by $u(w_2)$. Following this, queue 1 will be served until it is empty and $WL_2^n(t) \geq \bar{w}_2$, etc. But suppose that $u(w_2) = w_2$, so that queue 2 is reduced to zero before switching. If the value of $WL_1^n(t)$ is very small when the queue for class 2 is reduced to zero, then it might be inefficient to switch immediately. Thus, in this case, the processor has to decide whether to switch or idle (while serving whatever class 2 customers might arrive). It is reasonable to suppose that there is a threshold $\bar{w}_1$ such that switching to queue 1 will take place only when $WL_1^n(t) \geq \bar{w}_1$. Thus, the policy form suggested in [211] involves three quantities: $u(\cdot), \bar{w}_1, \bar{w}_2$. Having fixed the form of the policy, one needs to determine the unknowns.

Determining the controls and thresholds. If $\bar{w}_i = 0$, $i = 1, 2$, then the cost will be infinite, since as the total workload goes to zero, the rate of switching (in scaled time) would go to infinity. Thus, some $\bar{w}_i$ must be positive.

Suppose that queue 1 is being served and that $\bar{w}_2 > w$, the total workload level (hence the value of the workload in queue 2) at the time that queue 1 becomes empty. Then, the server serves whatever class 1 customers arrive (and idles if queue 1 is empty), until the next time that $WL_2^n(\cdot)$ reaches

$\bar{w}_2$. But this happens virtually instantaneously in scaled time, the delay being approximately $\rho_2[\bar{w}_2 - w]/\sqrt{n}$ when n is large. If class 2 is being served, then the processor will not switch until $WL_1^n(t) \geq \bar{w}_1$. Thus, we can (heuristically and asymptotically) view the total workload as being constrained by the thresholds so that it does not fall below $\max\{\bar{w}_1, \bar{w}_2\} = \bar{w}$. This argument implies that the limit workload process satisfies (3.4), but with the reflection being at level $\bar{w}$ and not at zero. Thus, the pair $(\bar{w}_1, \bar{w}_2)$ is replaced by a single value $\bar{w}$. It is important to recall that the control $u(\cdot)$ does not enter into (3.4), since it does not affect the idle time.

For some number $w \geq \bar{w}$, suppose that $WL^n(t) \approx w$ during several cycles. Then, during that time, the value of $WL_2^n(\cdot)$ is approximately w at the time that the server switches to queue 2. Using the averaging result (3.6) on these cycles, we can assume (for large n) that the workload path $WL_1^n(\cdot)$ is uniformly distributed on its mean trajectory $[0, u(w)]$ and $WL_2^n(\cdot)$ is uniformly distributed on its mean trajectory $[w - u(w), w]$. See Figure 3.2.

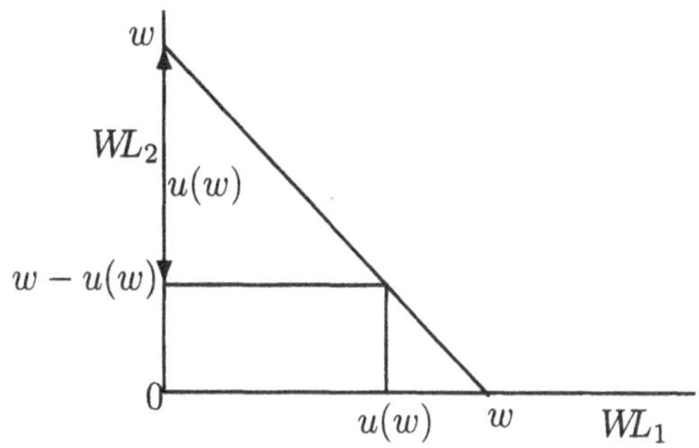

Figure 3.2. The control policy, workload $= w$.

To compute $u(\cdot)$ and $\bar{w}$, we need to estimate the cost (3.7) in terms of these quantities. The argument of the last paragraph and the approximation (3.6) imply that we can replace the holding cost rate $\sum_i c_i x_i^n(t)$ by

$$c_2 \bar{\lambda}_2^d w + \frac{u(w)}{2}\left[c_1 \bar{\lambda}_1^d - c_2 \bar{\lambda}_2^d\right]. \tag{3.9}$$

The asymptotic holding cost over a polling cycle is (3.9) times the duration of the cycle. Since the asymptotic mean (in scaled time) rate at which $WL_1^n(\cdot)$ (respectively, $WL_2^n(\cdot)$) decreases when it is being served is $(1 - \rho_1)\sqrt{n}$ (respectively, $(1 - \rho_2)\sqrt{n}$), for large n the mean (scaled) cycle time is approximately

$$\frac{u(w)}{\sqrt{n}}\left[\frac{1}{1-\rho_1} + \frac{1}{1-\rho_2}\right] = \frac{u(w)}{\sqrt{n}}\frac{1}{\rho_1 \rho_2}, \tag{3.10}$$

where we used the fact that $\rho_1 + \rho_2 = 1$. The switching cost per polling cycle in (3.7) is $k/\sqrt{n}$, and this switching cost per cycle divided by the (scaled) cycle time is $\rho_1 \rho_2 k / u(w)$.

We now determine the control $u(w)$ by minimizing the total average cost over a cycle divided by the cycle time, namely,

$$\min_{u(w)\le w} \left[c_2\bar{\lambda}_2^d w + \frac{u(w)}{2}\left[c_1\bar{\lambda}_1^d - c_2\bar{\lambda}_2^d\right] + \frac{\rho_1\rho_2 k}{u(w)}\right]. \tag{3.11}$$

This yields the minimizer

$$u(w) = \min\{w, \hat{w}\},$$

where

$$\hat{w} = \sqrt{\frac{2\rho_1\rho_2 k}{[c_1\bar{\lambda}_1^d - c_2\bar{\lambda}_2^d].}} \tag{3.12}$$

Thus, the control has been determined by minimizing the average cost over a polling cycle, working on the fast scale.

The problem is now reduced to choosing $\bar{w} \le \hat{w}$. This is to be determined via an analysis on the slow scale, that of $WL^n(\cdot)$. Recall that the limit equation for the total workload is (3.4), but with reflection at $\bar{w}$, and that it does not depend explicitly on the control. The stationary density of (3.4) has an exponential form whose parameters are a function of $\bar{w}, \hat{w}, b$ and the variance of the Wiener process. Assume that this holds (approximately, for large n) for $WL^n(\cdot)$. The cost (3.7) has been reduced to the expectation of the average (under the assumed approximate stationary distribution) cost rate per cycle. This cost rate/cycle (times $\sqrt{n}$) is

$$\begin{aligned} &c_2\bar{\lambda}_2^d WL^n(t) + \frac{WL^n(t)\left[c_1\bar{\lambda}_1^d - c_2\bar{\lambda}_2^d\right]}{2} + \frac{\rho_1\rho_2 k}{WL^n(t)}, \quad \text{for } WL^n(t) \le \hat{w},\\ &c_2\bar{\lambda}_2^d WL^n(t) + \sqrt{2k\rho_1\rho_2[c_1\bar{\lambda}_1^d - c_2\bar{\lambda}_2^d]}, \qquad \text{for } WL^n(t) \ge \hat{w}. \end{aligned} \tag{3.13}$$

The final details [211] involve the minimization of the expectation of the function defined by (3.13), using the stationary distribution as a function of $\bar{w}$, and are unimportant for our purposes in this introduction. The use of the limit *stationary* distribution is also heuristic, since there is no a priori guarantee that this can yield a good approximation to the cost (3.7) for large n. But this step is justified by the functional occupation measure results in Section 4.7, which can be used to show that

$$\lim_{n,T} \frac{1}{T} E\left[\int_0^T \sum_i c_i x_i^n(t)dt + kS^{s,n}(t)\right]$$

converges to the stationary cost (uniformly in $\bar{w}$) for the limit system when n, T go to their limits in any way at all (if $WL^n(0)$ is bounded).

Discussion and data. The data in [211] bear out the value of the approach. This article contains an extensive numerical study, comparing the optimal cost with the cost for the suggested policy and with a standard alternative, which is to poll each source until its queue is empty, and then switch, provided that the other queue is not empty. The data include comparisons for total traffic intensities in the range [.5, .9], and for a large set of cost coefficients. Under quite broad conditions and even for moderate traffic, the suggested control performs nearly optimally and is superior to the standard alternative, sometimes being substantially better.

The reference [211] also deals with the much harder case where the switching time is a constant, not zero, but there is no cost of switching. In this case, the polling policy affects the time devoted to switching. Hence it affects the idle time, and the limit process (3.4) contains a drift term that depends nonlinearly on the control functions. Nevertheless, a similar heuristic analysis, motivated by the intuition obtained from heavy traffic analysis, yields a policy (and supporting data) that is nearly optimal under a broad range of operating conditions. The results use a heuristic extension of (3.6) to estimate the rate of switching. See Chapter 11. In all of the cases discussed above, the limit system is one-dimensional, and this dimensionality is not affected by the cost function that is used. Thus, the approach is feasible even with nonlinear holding costs. Then one can either extend the above analysis or use a numerical method directly (such as [167]) to get the optimal policy for the limit model.

The following data are taken from Table II of [211]. Let $\bar{\lambda}_i^d = 1, \bar{\lambda}_i^a = \rho/2$ and $c_2 = 1$. The interarrival and service intervals are exponentially distributed and mutually independent. Suboptimality is defined as 100×[actual cost minus optimal cost]/optimal cost.

c_1	k	ρ	opt. cost	Subopt. of proposal	Subopt. of alt.
1.5	2	.5	1.382	.6%	2.8%
1.5	2	.7	2.880	.2%	7.4%
1.5	2	.9	9.691	.7%	17.2%
1.5	20	.5	2.886	3.1%	3.4%
1.5	20	.7	4.611	.6%	1.6%
1.5	20	.9	11.319	.4%	9.3%
5	2	.5	2.557	.0%	20%
5	2	.7	4.771	.0%	50.4%
5	2	.9	12.557	.3%	115.4%
5	20	.5	4.368	4.6%	7.8%
5	20	.7	7.200	.4%	21.8%
5	20	.9	15.408	1.0%	82.1%

Table 3.1. The switching cost problem.

The references [246, 247] give other interesting examples of the use of insights from heavy traffic analysis for obtaining good control policies for difficult constrained scheduling problems in manufacturing, even outside of the heavy traffic regime. A crucial contribution of these works is the use of the "state space collapse" phenomena of heavy traffic analysis to isolate the important variables (such as "workload imbalance") on which good controls should be based, and then to validate via both analysis and simulation.

1.4 Communication and Computer Networks: The Multiplexer

Modern communications and computer networks provide some of the most challenging problems in queueing. The systems tend to be complicated and their operating requirements stringent. Two canonical classes of problems are described in this and in the next section. The models can be elaborated in many ways, and more detail is given in Chapters 8–10. At this time, even the basic forms of the controlled examples that will be given are too complicated to be analyzed effectively by means other than heavy traffic analysis. In these examples the parameter n denotes the size or speed of the system, which increase as n increases. The systems are in heavy traffic in that the [system capacity minus average need]/[average need] is small. Thus, we are concerned with large systems. In applications, one wishes to know the optimal performance and desired excess capacity as the size and speed grow, and heavy traffic analysis is a useful tool for dealing with such questions.

Consider a system where there are n independent sources (i.e., users), with each source creating data (which is grouped in "cells" of small size) in some random and bursty way. The data from all of the sources are to be sent over a single channel, which has a flow-smoothing buffer. A multiplexer is a device that collects the information from the individual users by polling them in some organized (and very rapid) way, and then passes it into the buffer, from which service is first come first served, and the cells are sent to the channel at the rate at which it can handle them. Since the process of cell generation by each source is random, the total instantaneous rate of cell generation will sometimes exceed the channel capacity, unless the channel has a capacity that is substantially larger than the average requirement. Unless excessive buffer sizes are acceptable, the cell loss problem due to buffer overflow remains, and there is always the problem of excessive delays. Because of the variety of applications situations, there are many possible models for such systems, different buffering and control philosophies, and different models for the data creation process. One particular model will be dealt with, which illustrates the general principles of a powerful approach to the issues of analysis and control.

The basic references are [162, 165, 168, 169, 172]. Approaches based on matched asymptotic expansions as in [136] are also useful for certain forms of the problem. When they can be used, they get higher order approximations, although they are less robust and not (to date) appropriate for the optimal control problem.

Formulating the problem of design as an optimal control problem forces us to assess the values of particular tradeoffs and competing criteria of performance. In this sense, the control approach is important for the *exploration* of the various possibilities for the performance. The control of such systems is usually realized by controlling what flows into the buffers, and this can be done in many ways. One common approach to control uses the "leaky bucket" or token-type controller [34]. There, cells are deleted (or simply "slowed down") at the individual sources according to a rule determined by the token arrival rates and token buffer size. To date, the rule has not been state-dependent. An alternative approach identifies and marks lower-priority cells at the source, and deletes them either at the source or at multiplexer according to some state-dependent rule if a buffer overflow problem seems to be developing. It is this latter view that will be taken, although an advantage of the overall "aggregation" approach is that many types of control mechanisms fall into the framework. For example, the final limit equations are identical to what one would get if the available channel capacity were controlled, by the purchase of additional (marginal) capacity when needed.

The more complex systems serve a variety of needs simultaneously, where the sources can be divided into statistically distinct classes; for example, data, voice, and video, and these classes might be subdivided even further. Figures 4.1 and 4.2 illustrate systems with one and two source classes, respectively. Figure 4.3 illustrates a small feedforward network. We will concentrate on the case where all n users are statistically identical and independent (i.e., there is only one user class, as in Figure 4.1), although the methods are applicable to any number of classes and even allow random movement among them [169, 171, 172].

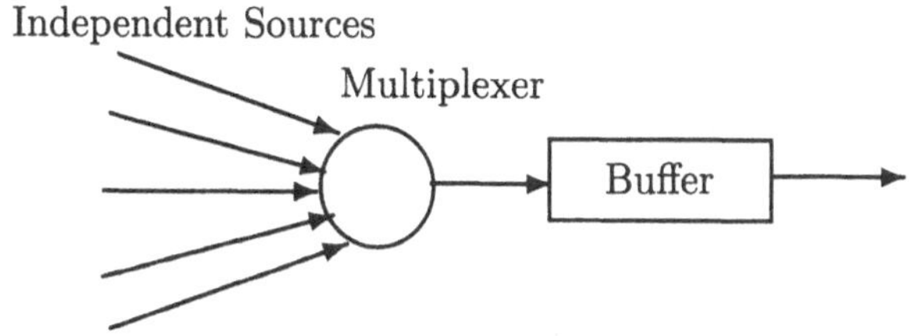

Figure 4.1. A multiplexing system: One user class.

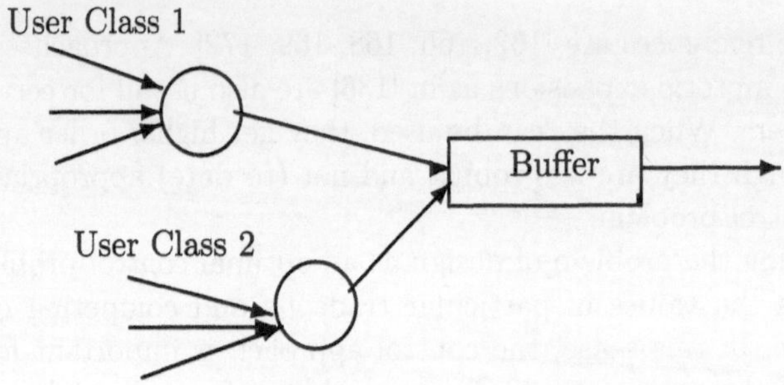

Figure 4.2. Two user classes.

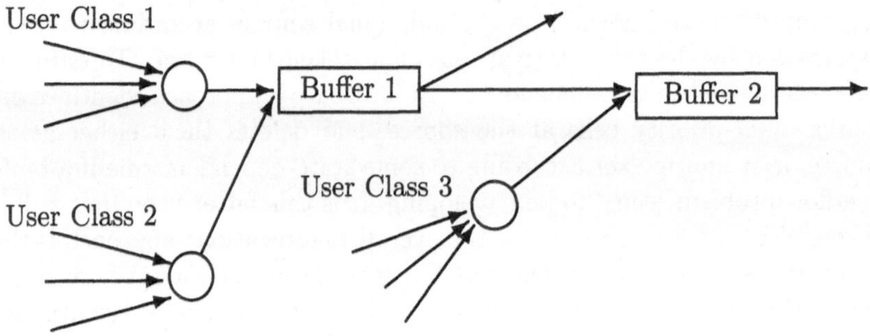

Figure 4.3. A small network.

A general model for the data creation process. We need to specify how the sources create data and how the transmitter (i.e., channel or server) deals with it. A common and versatile model will be used. When there is a single user class, a useful model for the data creation process at a source supposes that the source can be in one of K distinct states, and the "rate" at which it generates cells depends on its state. (If there are several user classes, then each will have its own model.) Let $S^{d,n}(t)$ denote $1/n$ times the total number of cells actually transmitted by time t and define $D^n(t) = \sqrt{n}S^{d,n}(t)$. The transmitter speed is assumed to be roughly proportional to the mean number of users. Essentially, it equals the mean rate at which data are created plus some surplus to handle "most" bursts, and will be made more precise below. Let $x^n(t)$ denote $1/\sqrt{n}$ times the number of cells in the buffer at real time t, and let the buffer size be $\sqrt{n}B$, for some $B < \infty$. This scaling is suggested by the central limit theorem. It is the usual scaling in heavy traffic analysis, and the numerical data in Chapter 9 and [169] show that such a buffer size is often adequate for good performance. A frequent advantage of heavy traffic analysis is that the scalings indicate reasonable system sizing. Let $U^n(t)$ denote $1/\sqrt{n}$ times the number of cells denied entry by real time t due to the buffer being full at the time of their arrival.

Suppose that the actual number of users in state $k \le K$ at time t can be written as $\bar{v}_k^n(t) = nc_k + \sqrt{n}v_k^n(t)$, where the $v_k^n(\cdot)$ are zero-mean random processes that model the deviation about the mean, which is nc_k. Let the total (unscaled) number of cells created by all sources by time t be represented as

$$\sqrt{n}A^n(t) = \int_0^t \sum_k \gamma_k \bar{v}_k^n(s)ds + \sqrt{n}\sum_k w_k^{a,n}(t), \tag{4.1}$$

where γ_k is the average rate of cell creation per user in state k and the random processes $w_k^{a,n}(\cdot)$ measure deviations from the mean. Let $F^n(t)$ denote $1/\sqrt{n}$ times the number of cells from all sources that have been deleted by the controller by t. Suppose that the restrictions on the control mechanism are such that $F^n(t) \approx \int_0^t u^n(s)ds$, for some bounded function $u^n(\cdot)$. More will be said about this in Chapters 8 and 9, and it will be seen to be a reasonable restriction for large n. The control process $u^n(\cdot)$ is assumed to be admissible in that it is measurable and its value at time t depends only on the data that are available up to time t, before the value $u^n(t)$ is chosen. Then the system equation is

$$\begin{aligned} x^n(t) &= x^n(0) + \sqrt{n}\sum_k \gamma_k c_k t \\ &+ \int_0^t \sum_k \gamma_k v_k^n(s)ds + \sum_k w_k^{a,n}(t) - F^n(t) - D^n(t) - U^n(t). \end{aligned} \tag{4.2}$$

The development in Chapters 8 and 9 is for the case where the state process for each source is a first-order Markov process that takes two values. For example, each source is alternatively *fully on* and *completely off*. The corresponding cell creation process is called *Markov modulated* [5].

Depending on the application, each cell might take the same time to be transmitted, or the time might be random. Since the transmitter speed is roughly proportional to the number of users, it is convenient to scale the required times and write the actual time required for the lth cell to be transmitted as $\Delta_l^{d,n}/n$. Most often, the cells have the same size and $\Delta_l^{d,n}$ does not depend on l. Then we write it as $\bar{\Delta}^{d,n}$, where $n\bar{\Delta}^{d,n}$ is the actual transmitter speed. Then (modulo a negligible roundoff error) $D^n(t) = \sqrt{n}t/\bar{\Delta}^{d,n} - y^n(t)$, where $y^n(t)$ is defined as $\sqrt{n}/\bar{\Delta}^{d,n}$ times the idle time of the channel by time t. If there is some randomness in the processing of each cell, then we write the scaled number transmitted as

$$D^n(t) = \frac{\sqrt{n}t}{\bar{\Delta}^{d,n}} + w^{d,n}(S^{d,n}(t)) - y^n(t), \tag{4.3}$$

where $w^{d,n}(t)$ represents the scaled randomness in the service times of the first nt cells.

Under the heavy traffic condition, the channel capacity is close to the mean arrival rate of cells, and the randomness terms are of a smaller order than the mean terms. This is quantified by supposing that there is a constant b such that

$$\sqrt{n}\left[\sum_k \gamma_k c_k - \frac{1}{\bar{\Delta}^{d,n}}\right] = b^n \to b. \tag{4.4}$$

In Chapter 8, the form (4.4) is made more precise for the particular setup of concern.

Of course, in any given application there is no parameter n, just some particular physical system. But the parameterization is very useful as an effective approximation device. In an application, for large but fixed n, the value of b would be $b = b^n$, and the solution of the limit problem would then be used to approximate the properties of the actual physical system.

Typically, one tries to model such that the centered and scaled "deviation" processes $v^n(\cdot) = (v_k^n(\cdot), k \le K)$ converge weakly (as $n \to \infty$) to some limit diffusion-type process $v(\cdot) = (v_k(\cdot), k \le K)$ and the $(w_k^{a,n}(\cdot), k \le K)$ and $w^{d,n}(\cdot)$ converge weakly to some mutually independent Wiener processes $(w_k^a(\cdot), k \le K)$ and $w^d(\cdot)$, respectively. Such approximations hold under a great variety of realistic assumptions on the physical problem. Let $\bar{\lambda}^d = \lim_n 1/\bar{\Delta}^{d,n}$, the scaled mean service rate in the limit. Then the limit scaled model takes the form of the controlled reflected diffusion process, for $x(t) \in [0, B]$,

$$x(t) = x(0) + bt + \sum_k \gamma_k v_k(t) + \sum_k w_k^a(t) - w^{d,n}(\bar{\lambda}^d t) - F(t) + y(t) - U(t), \tag{4.5}$$

where $y(\cdot)$ is the reflection term at the origin, $U(\cdot)$ the reflection term at B, and $F(t) = \int_0^t u(s)ds$, the limit control term. By the heavy traffic condition (4.4), $\bar{\lambda}^d = \sum_k \gamma_k c_k$. The quantity $U(\cdot)$, which is the limit of the scaled number of unadmitted cells, is of great interest. The limit scaled idle time term $y(\cdot)$ is what keeps the solution of (4.5) from becoming negative, since it is a consequence of the transmitter not working when the buffer is empty. Otherwise, it is always assumed that the transmitter works if there is work to be done. The basic state variables in (4.2) or (4.5) are the scaled buffer content and the scaled numbers of active sources in the various states of each of the user classes, or their limits. The "limit" variables can be interpreted as "aggregated" states. They contain the data of main interest. Good results can often be obtained even when operating far from the heavy traffic region, and when the probabilities of buffer overflow are small [169, 172]. For other work on scaling and numerics for queueing-type problems see [47, 70, 97, 167, 170, 175, 197, 214, 215, 245].

One important control problem concerns the tradeoffs between the losses that are due to the deliberate discarding of cells by the controller (called control losses) and the losses due to buffer overflow (called buffer losses). The buffer overflow losses are more important, particularly when the controller can select lower priority cells for deletion, and the control approach allows us to understand the possible tradeoffs, as well as to penalize delay. A relatively small rate of cell deletion at times when the use is heavy can reduce the overflow losses and delays substantially.

There is no theoretical impediment to letting the number of source classes or the number of possible states for each class be arbitrary, but the so-called "curse of dimensionality" applies to the numerics. The multiple source class case is interesting, particularly when each class can be controlled separately, since the control for each class affects the environment for the other class, and the results are not always intuitive, depending on the relative arrival and service rates and the cost coefficients. See [169] for a discussion and data.

1.5 Controlled Admission in a Multiservice System: Formulation

The so-called ISDN (integrated systems data network) type systems support several types of services, each with their own requirements. For simplicity in the exposition, we consider only two types of users. The first type of user, called *guaranteed performance* (GP), requires a guarantee on delay, loss or throughput, and each user that is admitted is guaranteed some fixed bandwidth for the duration of its session. The other type of user, called *best effort* (BE), is less sensitive to delays and throughput and to the variations caused by fluctuating available bandwidth, and they share the remaining bandwidth (that not used by the GPs). The internet is an example of BE service. Despite the lack of guarantees for the quality of BE service, the random variations in the number of GP sessions can cause unacceptably poor behavior for the BE customers, particularly if there are far more than the average number of GPs in the system. One can manage this by controlling the admission of the GP requests, with some cost associated with the unadmitted requests. It turns out that small controls on the GP admissions can have substantial effects on the quality of service for the BE users.

In this introduction it is supposed that all the GP users share the available bandwidth equally, that the channel is time shared among all users, and that the BE requests are always admitted, unless some very large limit is reached. The control problem for the physical model is quite hard, and the heavy traffic approach offers much simplification. In the heavy traffic regime, the system has little spare bandwidth over what is needed to handle effectively the average load. The heavy traffic formulation allows a systematic exploration of the best tradeoffs between gains in quality of service for BE versus the cost of not admitting some GP. Numerical data show that reductions (say, 30% or more) in BE delays can be obtained with only very small levels of rejection of GP customers, when appropriate and at times determined by the system state. The percentage rejected will depend on the system parameters, arrival rates and current utilization, but (under the heavy traffic assumption) it goes to zero as the system grows, for any fixed percentage improvement in BE delay.

Two of the many forms of the problem will be discussed in more detail in Section 10.3. In the basic model, the BE users can take advantage of *all* of the available bandwidth. From the perspective of modeling, this is similar to supposing that each BE arrival constitutes the entire work for a "session," which arrives essentially instantaneously, is buffered, and then fed to the channel at any allowable rate up to the channel capacity. Thus, there is no limit to the rate at which work can be done on the set of BE customers except for the available capacity. Sometimes, however, there might be a maximum rate at which data can be fed to the channel, irrespective of the available bandwidth. In this case, if there are not many BE customers in the system, then not all of the available bandwidth will be used. This changes the dynamics and the scaling, but the analysis is very similar. One can handle multiple classes of BE customers, the members of which might be allocated bandwidth in class-dependent ways. These are among the many extensions that can be treated with an essentially similar analysis, as seen in [2].

The basic model. We use the subscripts b and g, respectively, to denote data associated with the BE and GP users. In heavy traffic, the mean service capacity is slightly greater than the mean demand. The system is parameterized by the scale or size parameter n, which is the order of the mean number of GP users and the arrival rates for the BE and GP users. The system capacity, excess capacity (over what is required for the average demands), and demand grow as $n \to \infty$, with the *relative* excess capacity going to zero, as quantified by (5.3). A consequence of the heavy traffic analysis is that such relationships can represent good design. The bandwidth is normalized so that each GP user gets one unit of bandwidth.

Define $S_\alpha^{a,n}(t)$, $\alpha = b, g$, to be $1/n$ times the number of class α customers (e.g., sessions) that have arrived by time t. Define $\Delta_{\alpha,l}^{a,n}$ such that $\Delta_{\alpha,l}^{a,n}/n$, $l = 1, \ldots,$ are the interarrival times for the members of class α. Suppose that for each α, the $\Delta_{\alpha,l}^{a,n}$ are mutually independent and identically distributed with mean $\bar{\Delta}_\alpha^a = 1/\bar{\lambda}_\alpha^a$ and that $E\left[1 - \Delta_{\alpha,l}^{a,n}/\bar{\Delta}_\alpha^a\right]^2 = \sigma_{a,\alpha}^2$. Define

$$w_\alpha^{a,n}(t) = \frac{1}{\sqrt{n}} \sum_{l=1}^{nt} \left[1 - \frac{\Delta_{\alpha,l}^{a,n}}{\bar{\Delta}_\alpha^a}\right]. \tag{5.1}$$

Then $w^{a,\alpha}(\cdot)$ converges weakly to a Wiener process with variance $\sigma^2_{a,\alpha}$. If the arrival process for class α sessions is Poisson, then $\sigma^2_\alpha = 1$. While we confine our interest to the independence case, one could let the intervals be correlated. It is always assumed that the arrival processes are mutually independent and independent of the initial condition and the service times for the GP users.

The service times for the sessions of class GP are assumed to be mutually independent, exponentially distributed with rate $\bar\lambda^d_g$, and independent of the initial condition. The restriction to the exponential distribution is unfortunate, especially in view of the weaker conditions placed on the interarrival times. But in such "self service" systems, it is hard to model the departure process in a simple way otherwise.

The GP arrivals can be controlled in that any requested admission can be denied by the controller. Any arrival that is denied admission disappears from the system. As usual in heavy traffic scaling, the basic system variables are scaled by $1/\sqrt{n}$. Thus, let $F^n_g(t)$ denote $1/\sqrt{n}$ times the number of GP customers not admitted by time t. It is supposed that there is a $B_b < \infty$ such that the maximum number of BE users in the system at any time is $B_b\sqrt{n}$. This upper bound is inconsequential for large enough B_b. Let $U^n_b(t)$ denote $1/\sqrt{n}$ times the number of BE that were rejected by time t because on their arrival there were $\sqrt{n}B_b$ BE users already in the system.

The service time requirements for the BE depend on the history of the available bandwidth during their stay in the system and are defined as follows. Suppose that $B(t)$ is the total bandwidth that is unused by the GP at time t, and there are $N(t) > 0$ BE customers in the system. Then there is $\bar\lambda^d_b > 0$ such that the probability (conditioned on the data up to t) that any particular BE user will depart on the interval $[t, t+\delta)$ is $\bar\lambda^d_b B(t)\delta/N(t) + o(\delta)$. The (conditional) probability that there will be only one departure in that interval is $\bar\lambda^d_b B(t)\delta + o(\delta)$, and the (conditional) probability of more than one departure is $o(\delta)$.

Suppose for the moment that the channel capacity is unlimited, and all GP users are admitted. Then the mean stationary number is $n\bar\lambda^a_g/\bar\lambda^d_g$ [132]. Define $x^n_b(t)$ to be $1/\sqrt{n}$ times the number of BE users in the system at time t, and set

$$x^n_g(t) = \frac{1}{\sqrt{n}}\left[\text{number of } GP \text{ users in system at } t - n\frac{\bar\lambda^a_g}{\bar\lambda^d_g}\right]. \tag{5.2}$$

Thus $x^n_g(t)$ is the scaled number of GP users, centered about the mean that would hold if there were no rejections and an infinite channel capacity. For the heavy traffic condition, suppose that there is a constant b such that the channel capacity is defined by

$$C_n = n\left[\frac{\bar\lambda^a_g}{\bar\lambda^d_g} + \frac{\bar\lambda^a_b}{\bar\lambda^d_b}\right] + b\sqrt{n}. \tag{5.3}$$

For appropriate $b > 0$, this will be sufficient for good behavior for large n. The term $n\bar{\lambda}_g^a/\bar{\lambda}_g^d$ accounts for the mean (unconstrained and uncontrolled) GP usage. If the bandwidth component $n\bar{\lambda}_b^a/\bar{\lambda}_b^d$ were replaced by a quantity that is smaller by more than $O(\sqrt{n})$, then the number (times $1/\sqrt{n}$) of the BE users in the system would go to infinity. In any particular application, with a fixed n, define b by (5.3).

Define $A_\alpha^n(t)$ (respectively, $D_\alpha^n(t)$) to be $1/\sqrt{n}$ times the number of arrivals (respectively, departures) of class α by time t. The control process $F_g^n(\cdot)$ is assumed to be admissible in that it is measurable and its value at time t depends only on the data that are available at time t, before the control actions at that time are taken. The system equations are

$$x_b^n(t) = x_b^n(0) + A_b^n(t) - D_b^n(t) - U_b^n(t), \tag{5.4a}$$

$$x_g^n(t) = x_g^n(0) + A_g^n(t) - D_g^n(t) - F_g^n(t) - U_g^n(t), \tag{5.4b}$$

where $U_g^n(t)$ is $1/\sqrt{n}$ times the number of class GP customers that could not be admitted by time t due the *entire* channel being occupied by GP. The $U_g^n(\cdot)$ will disappear in the limit. See Chapter 10 or [2] for full details.

The system (5.4) can be put into the following form (see Chapter 10), which "looks like" a controlled reflected stochastic differential equation (RSDE):

$$\begin{aligned} x_b^n(t) = x_b^n(0) + w_b^{a,n}(S_b^{a,n}(t)) - \bar{\lambda}_b^d \int_0^t g_0(x^n(s))ds \\ - w_b^{d,n}(t) + y_b^n(t) - U_b^n(t) \end{aligned} \tag{5.5a}$$

$$\begin{aligned} x_g^n(t) = x_g^n(0) + w_g^{a,n}(S_g^{a,n}(t)) - \bar{\lambda}_g^d \int_0^t x_g^n(s)ds \\ - w_g^{d,n}(t) - F_g^n(t) - U_g^n(t), \end{aligned} \tag{5.5b}$$

where $g_0(x) = b - x_g$ and $x_b^n(t) \in [0, B_b]$. The $w_\alpha^{d,n}(\cdot), \alpha = b, g$, are martingales that represent the randomness in the service processes, and they converge weakly to Wiener processes. The term $y_b^n(\cdot)$ is the reflection process. It can increase only at t where $x_b^n(t) = 0$. Note the way that $x_g^n(\cdot)$ affects the $x_b^n(\cdot)$ process. The weak sense limit processes have the form

$$x_b(t) = x_b(0) - \bar{\lambda}_b^d \int_0^t g_0(x(s))ds + w_b^a(\bar{\lambda}_b^a t) - w_b^d(t) + y_b(t) - U_b(t), \tag{5.6a}$$

$$x_g(t) = x_g(0) - \bar{\lambda}_g^d \int_0^t x_g(s)ds + w_g^a(\bar{\lambda}_g^a t) - w_g^d(t) - F_g(t), \tag{5.6b}$$

and $x_b(t)$ is constrained to the set $[0, B_b]$ by the reflection process $y_b(\cdot)$. The various w−processes are Wiener processes, whose properties will be given in Theorem 10.3.1. The process $F_g(\cdot)$ in (5.6b) is known as a *singular* control [170, 179, 231]. It is called singular, since it cannot in general be represented as an integral of a bounded function. Such controls appear in heavy traffic problems when the control consists in deleting inputs with no constraint on the rate [189].

The heavy traffic limit (5.6) is substantially simpler than the original physical system, and can be used to compute nearly optimal controls for the physical system, under heavy traffic. Sometimes, the rate at which the channel can accept data from any source is limited by factors other than the available channel capacity. Such a variation is discussed in Section 10.3 also. Other variations are in [2]. The analysis is similar for the various cases, and the results differ mainly in the form of the drift term $g_0(\cdot)$. The GP and BE can be divided into independent subclasses each with its own arrival process, service rate parameter, and costs, and only some of the GP might be controllable. The dimension of the heavy traffic limit does not depend on the number of BE subclasses, a remarkable simplification. Such asymptotic reductions in the dimension of the state space are called "state space collapse" [30, 100, 208, 254]. Since the dimension of the state is a major issue in both analysis and numerical approximations, such a reduction is of considerable value.

represented by an analogue of a transfer function. Such controls appear in [illegible] when [illegible] control [illegible] in [illegible] with an [illegible] (1.6.1) [illegible].

The theory [illegible] is [illegible] than the original physical system and [illegible] introduced for the physical system [illegible] the control [illegible] from [illegible] that [illegible]. Such a [illegible] in Section [illegible]. The [illegible] for the [illegible] The [illegible] the [illegible] of the [illegible] does not [illegible] a [illegible] [illegible] the state [illegible] space are [illegible] state is a [illegible] and [illegible] [illegible].

2

Martingales and Weak Convergence

Chapter Outline

The basic questions dealt with in this book concern approximations of relatively complex processes by simpler and more tractable processes. These simpler processes will usually be diffusions, reflected diffusions, or reflected jump-diffusions. They might be controlled or not controlled. They are of interest to facilitate numerical or mathematical analysis and practical design, or to expose the more important structural properties and parametric dependencies. Some of the basic processes and methods will be introduced in this and in the next chapter. The martingale and Wiener processes are among the most fundamental in stochastic analysis, and the basic ideas are reviewed in Section 1. In the course of the analysis in the succeeding chapters it will be necessary to characterize the various processes that are obtained as limits of the approximation procedures. Of particular concern are methods for verifying that such a process is a Wiener process, and some standard and useful methods are described in Subsection 1.2 and in Section 8. Section 2 gives the basic definitions of the Itô stochastic integral with respect to a Wiener process, in preparation for the discussion of stochastic differential equations in the next chapter. Poisson and other jump processes will occur, for example, in the state-dependent models in Chapter 8, when the servers are subject to interruptions, and in "impulsive" models of arrival processes. The necessary background is dealt with in Section 3, as is the martingale decomposition of a Poisson-type jump process. In Section 4 we state some important results concerning the decomposition of the square of a martingale, and these will be employed in the proofs that certain weak-sense limits are Wiener processes or stochastic integrals.

The theory of weak convergence of measures is a fundamental and widely used tool for approximation and limit theorems for sequences of stochastic processes. It plays a role similar to that of the central limit theorem for sequences of real or vector-valued random variables. The basic ideas that will be useful in the sequel are discussed in Sections 5 and 6. The basic techniques discussed will be further developed in connection with the applications in the succeeding chapters. The concepts and criteria have a rather abstract flavor as they are initially introduced, but when applied in the succeeding chapters, they take on concrete and readily usable forms. See, for example, the techniques, examples, and references in [71, 157, 160, 167]. The application of the theory requires that certain compactness or tightness results be proved. This can be difficult for the so-called singular control and related problems. Section 6 describes a useful time transformation method for alleviating this difficulty.

Measure-valued random variables and processes appear in dealing with the so-called relaxed control problem and in the study of ergodic problems (in the form of sample occupation measures). The use of such random variables and processes provides a unifying approach to many important problems, and actually simplifies their development considerably. Some of the appropriate concepts are introduced in Section 7. In Section 8 we return to the issue of effective criteria for verifying that a process is indeed a Wiener process. This is simplest when the "driving" random variables are mutually independent. Otherwise, some sort of mixing condition or stochastic averaging method is required. The perturbed test function method is a versatile tool for such purposes, and a useful result derived with it in [157, 158] is stated at the end of Section 8. See [157, 158, 160] for further results and many applications. Due to considerations of simplicity of exposition, we do not deal with the so-called state-dependent and other complicated driving noise processes. But all of the processes considered in the above references can be accommodated under appropriate conditions.

On first reading or if the reader is not concerned with the general control or state-dependent problem, familiarity with Sections 1, 2, and 5 only is needed.

2.1 Martingales and the Wiener Process

2.1.1 Martingales

Definitions. For a topological space S, let $\mathcal{B}(S)$ denote the σ-algebra of Borel subsets of S. Let $(\Omega, \mathcal{F}, P)$ denote a probability space. It will be assumed throughout the book that $\mathcal{F}$ is *complete;* i.e., it contains all subsets of P-null sets. Let $\overline{\mathcal{F} \times \mathcal{B}([0,\infty))}$ denote the completion of the product σ-algebra with respect to the product measure (Lebesgue measure used on $\mathcal{B}([0,\infty))$). A function $\phi(\cdot)$ on $\Omega \times [0,\infty)$ and with values $\phi(\omega,t)$ in some metric space S is said to be a *measurable process* if it is a measurable mapping from $(\Omega \times [0,\infty), \overline{\mathcal{F} \times \mathcal{B}([0,\infty))}$ to $(S, \mathcal{B}(S))$. All processes are assumed to be measurable and separable. (For the concept of separability, see [59]. Any process has a separable version, and it is only separable versions that are used in applications.)

A family of σ-algebras $\{\mathcal{F}_t, t \geq 0\}$ is called a *filtration* on this probability space if $\mathcal{F}_s \subset \mathcal{F}_t \subset \mathcal{F}$ for all $0 \leq s \leq t$. Unless otherwise mentioned, throughout the book we will always assume that the $\mathcal{F}_t$ are complete in that $\mathcal{F}_t$ contains all the subsets of null sets in $\mathcal{F}$. Often $\mathcal{F}_t$ will be defined to be the minimal $\sigma-$algebra that measures $\{V(s), s \leq t\}$, for some given random process $V(\cdot)$. Then we suppose that $\mathcal{F}_t$ is complete with respect to the measure of $V(\cdot)$. If A is a collection of random variables defined on the probability space $(\Omega, \mathcal{F}, P)$, then we use $\mathcal{F}(A)$ to denote the σ-algebra generated by A. Let $E_{\mathcal{F}_t}$ and $P_{\mathcal{F}_t}$ denote the expectation and probability, respectively, conditioned on the σ-algebra $\mathcal{F}_t$. For a complete and separable metric space S, let $C(S;0,T)$ denote the space of S-valued continuous functions on $[0,T]$, and let $D(S;0,T)$ denote the set of S-valued functions on $[0,T]$ that are continuous from the right and have limits from the left. Let $C(S;0,\infty)$ and $D(S;0,\infty)$ denote the analogous path spaces for the time interval $[0,\infty)$. These spaces may be metrized so they are complete separable metric spaces ([15, 71] and Section 5). Most commonly S will be $I\!R^k$, k-dimensional Euclidean space, or a compact subset of it. But other spaces will appear as well.

Let $M(\cdot)$ be a stochastic process defined on $(\Omega, \mathcal{F}, P)$ with filtration $\{\mathcal{F}_t, t \geq 0\}$. If $M(t)$ is $\mathcal{F}_t$-measurable for each t, then $M(\cdot)$ is said to be $\mathcal{F}_t$-adapted. Let $M(\cdot)$ be $\mathcal{F}_t$-adapted and take values in the path space $D(I\!R^k;0,\infty)$. Then $M(\cdot)$ is said to be an *$\mathcal{F}_t$-martingale* if $E|M(t)| < \infty$ for all $t \geq 0$ and

$$E_{\mathcal{F}_t} M(t+s) = M(t) \text{ w.p.1 for all } s,t \geq 0. \tag{1.1a}$$

If the filtration is unimportant or obvious, then we will simply say that $M(\cdot)$ is a martingale. If $M(\cdot)$ is an $\mathcal{F}_t$-martingale, then it is also an $\mathcal{F}(M(s), s \leq t)$-martingale. We say that $\{\mathcal{F}(M(s), s \leq t), t \geq 0\}$ is the filtration generated by $M(\cdot)$. If the $\mathcal{F}_t$-adapted vector-valued process $M(\cdot)$ satisfies the inequality

$$E_{\mathcal{F}_t} M(t+s) \geq M(t) \text{ w.p.1 for all } s,t \geq 0, \tag{1.1b}$$

then it is called a *submartingale*.

Martingales are a fundamental tool in stochastic analysis, partly due to the bounds and inequalities associated with them. Processes can often be decomposed into a sum of a process of bounded variation and a martingale. This decomposition can be used to facilitate the analysis, since the bounded-variation term is often relatively easy to handle, and there are many useful techniques for the analysis of martingales. The following inequalities will be useful. Let $M(\cdot)$ be a real or vector-valued $\mathcal{F}_t$-martingale with paths in $D(\mathbb{R}^k; 0, \infty)$ for some $k \geq 1$. Then [32, 59, 122, 201] for any $c > 0$ and $0 \leq t \leq T$,

$$P_{\mathcal{F}_t}\left\{ \sup_{t \leq s \leq T} |M(s)| \geq c \right\} \leq E_{\mathcal{F}_t}|M(T)|^2/c^2 \text{ w.p.1,} \tag{1.2}$$

$$E_{\mathcal{F}_t} \sup_{t \leq s \leq T} |M(s)|^2 \leq 4E_{\mathcal{F}_t}|M(T)|^2 \quad \text{w.p.1.} \tag{1.3}$$

Usually, (1.2) and (1.3) are proved without the conditional expectation. But the proof is essentially the same in general.

Definition: Stopping time and local martingale. Let $\{\mathcal{F}_t, t \geq 0\}$ be a filtration. A random variable τ with values in $[0, \infty]$ is called an $\mathcal{F}_t$*-stopping time* if $\{\tau \leq t\} \in \mathcal{F}_t$ for all $t < \infty$. If $M(\cdot)$ is an $\mathcal{F}_t$-martingale and τ is an $\mathcal{F}_t$-stopping time, then the "stopped" process defined by $M(t \wedge \tau)$ is also an $\mathcal{F}_t$-martingale [32, 201]. Let $\mathcal{F}_\tau$ denote the "stopped" σ-algebra that is composed of the sets $A \in \mathcal{F}$ such that $A \cap \{\tau \leq t\} \in \mathcal{F}_t$ for all t.

If there exists a nondecreasing sequence $\{\tau_n\}$ of $\mathcal{F}_t$-stopping times such that $\tau_n \to \infty$ with probability one and each of the stopped processes $M(t \wedge \tau_n)$ is an $\mathcal{F}_t$-martingale, then $M(\cdot)$ is called an $\mathcal{F}_t$*-local martingale.*

Verifying that a process is a martingale. We now give a method that will be useful in the sequel for verifying that a process is a martingale. It is simply a rewording of the definition of a martingale in terms of conditional expectations. This "verification" method is used in this book mainly to show that a process that is obtained as a weak-sense limit is a Wiener process. See the next subsection and Section 8 for more detail.

Let Y be a vector-valued random variable with $E|Y| < \infty$, and let $V(\cdot)$ be a process with paths in $D(S; 0, \infty)$, where S is a complete and separable metric space. Suppose that for some given $t > 0$, each integer p and each set of real numbers $0 \leq s_i \leq t$, $i = 1, \ldots, p$, and each bounded and continuous real-valued function $h(\cdot)$,

$$Eh(V(s_i), i \leq p)Y = 0.$$

This fact and the arbitrariness of the function $h(\cdot)$ implies that for each measurable set H,

$$EI_{\{(V(s_i), i \leq p) \in H\}}Y = 0.$$

Thus $E[Y|V(s_i), i \leq p] = 0$ with probability one. Continuing, the arbitrariness of p and of the set $\{s_i, i \leq p\}$ and of $h(\cdot)$ now imply that

$$E[Y|V(s), s \leq t] = 0$$

with probability one [32].

Next, let $U(\cdot)$ be a random process with $E|U(t)| < \infty$ for each t, with values in $D(S;0,\infty)$, and such that for all p, $h(\cdot)$, $s_i \leq t, i \leq p$, as given above and a given real $\tau > 0$,

$$Eh(U(s_i), i \leq p)\,[U(t+\tau) - U(t)] = 0. \tag{1.4}$$

Then $E[U(t+\tau) - U(t)|U(s), s \leq t] = 0$. If this holds for all t and $\tau > 0$, then by the definition (1.1a) of a martingale, $U(\cdot)$ is a martingale with respect to the filtration generated by $U(\cdot)$. It is often more convenient to work with the following more general setup, whose proof follows from the preceding argument.

Theorem 1.1. *Let $U(\cdot)$ be a random process with paths in $D(I\!R^k;0,\infty)$ and with $E|U(t)| < \infty$ for each t. Let $V(\cdot)$ be a process with paths in $D(S;0,\infty)$, where S is a complete and separable metric space. Let $U(t)$ be measurable on the σ-algebra $\mathcal{F}_t^V$ determined by $\{V(s), s \leq t\}$. Suppose that for each real $t \geq 0$ and $\tau \geq 0$, each integer p, and each set of real numbers $s_i \leq t$, $i = 1, \ldots, p$, and each bounded and continuous real-valued function $h(\cdot)$,*

$$Eh(V(s_i), i \leq p)\,[U(t+\tau) - U(t)] = 0. \tag{1.5}$$

Then $U(\cdot)$ is an $\mathcal{F}_t^V$-martingale.

The following theorem will be useful. The proof uses the martingale properties to show that $E|M(t) - M(0)|^2 = 0$.

Theorem 1.2. *For some filtration $\{\mathcal{F}_t, t \geq 0\}$, let $M(\cdot)$ be an $\mathcal{F}_t$-martingale with continuous paths and suppose that $M^2(\cdot)$ is also a martingale. Then $M(t) = M(0)$ for all $t \geq 0$ with probability one.*

2.1.2 The Wiener Process

Definition. Let $(\Omega, \mathcal{F}, P)$ be a probability space and let $\{\mathcal{F}_t, t \geq 0\}$ be a filtration defined on it. Let Σ be a nonnegative definite symmetric matrix. A vector-valued process $w(\cdot)$ is called an *$\mathcal{F}_t$- Wiener process* with covariance matrix Σ if it satisfies the following conditions: $w(0) = 0$ with probability 1, $w(t)$ is $\mathcal{F}_t$-adapted, $\{w(s) - w(t) : s \geq t\}$ is independent of $\mathcal{F}_t$ for all $t \geq 0$, and the increments $w(s) - w(t)$ are normally distributed with mean 0 and covariance $(s-t)\Sigma$ for all $s \geq t \geq 0$.

There is a version of $w(\cdot)$ that is continuous with probability 1, and this is the one that is always used. If $\Sigma = I$, the identity matrix, then $w(\cdot)$ is called a *standard Wiener process*. Full details of the theory of the Wiener process and the proofs of the assertions of this subsection are in [32, 122, 228]. If $\mathcal{F}_t$ is simply $\mathcal{F}(w(s), 0 \leq s \leq t)$ or if it is understood, then the $\mathcal{F}_t$ prefix is usually suppressed, and we refer to $w(\cdot)$ simply as a Wiener process.

A process $w(\cdot)$ is a Wiener process with a time-varying covariance matrix if it satisfies the Gaussian and independence properties, but $w(s)-w(t), s > t$, has covariance $\int_t^s \Sigma(u)du$ for some bounded and measurable nonnegative definite symmetric matrix-valued and nonrandom function $\Sigma(\cdot)$.

Any $\mathcal{F}_t$-Wiener process is an $\mathcal{F}_t$-martingale. If $\Sigma > 0$ (positive definite), then the fact that $w(\cdot)$ has continuous sample paths and is also a martingale implies that the sample paths are of unbounded variation over any nonzero time interval (with probability one) [122].

Verifying that a process is a Wiener process. The following definition of a Wiener process is equivalent to the one given above. Let the $I\!R^r$-valued process $w(\cdot)$ have continuous paths and satisfy $w(0) = 0$ with probability one, and $E|w(t)|^2 < \infty$ for each $t < \infty$. Let $\{\mathcal{F}_t, t \geq 0\}$ be a filtration such that $w(\cdot)$ is $\mathcal{F}_t$-adapted and let $E_{\mathcal{F}_t}[w(t+s) - w(t)] = 0$ with probability one for each t and each $s \geq 0$. (Hence $w(\cdot)$ is an $\mathcal{F}_t$-martingale.) Let there be a nonnegative definite symmetric matrix Σ such that for each t and each $s \geq 0$,

$$E_{\mathcal{F}_t}\left[w(t+s) - w(t)\right]\left[w(t+s) - w(t)\right]' = \Sigma s, \text{ with probability 1.}$$

Then $w(\cdot)$ is an $\mathcal{F}_t$-Wiener process with covariance matrix Σ [122], [184, Volume 1, Theorem 4.1].

The criterion of Theorem 1.1 for verifying that a process is a martingale can be used to verify that it is an $\mathcal{F}_t$-Wiener process for appropriate $\mathcal{F}_t$, as in the following theorem.

Theorem 1.3. *Suppose that the vector-valued process $w(\cdot)$ has continuous paths with $E|w(t)|^2 < \infty$ for each t. Let $V(\cdot)$ be a random process with paths in $D(S; 0, \infty)$, where S is a complete and separable metric space, and let $\mathcal{F}_t^V$ be the smallest σ-algebra that measures $\{V(s), w(s), s \leq t\}$. Let $h(\cdot), p, t, \tau > 0, s_i \leq t$, be arbitrary but satisfy the conditions put on these quantities in Theorem 1.1. Suppose that*

$$Eh(V(s_i), w(s_i), i \leq p)\left[w(t+\tau) - w(t)\right] = 0 \tag{1.6}$$

and that there is a nonnegative definite symmetric matrix-valued and non-random function $\Sigma(\cdot)$ such that

$$\begin{aligned} &Eh(V(s_i), w(s_i), i \leq p) \\ &\quad \times \left[(w(t+\tau) - w(t))\,(w(t+\tau) - w(t))' - \int_t^{t+\tau} \Sigma(s)ds\right] = 0. \end{aligned} \tag{1.7}$$

Then $w(\cdot)$ is an $\mathcal{F}_t^V$-Wiener process, with time-varying covariance matrix $\Sigma(\cdot)$.

Proving that (1.7) holds can be hard in the case of greatest interest to us, namely when $w(\cdot)$ is the limit of a sequence of processes $\{w^n(\cdot)\}$, since it usually requires showing that $\{|w^n(t)|^2\}$ is uniformly integrable for each t. This is often just a technical issue, and the difficulty can be avoided by using the following equivalent characterization.

For a matrix $\Sigma = \{a_{ij}; i,j\}$ and smooth function $f(\cdot)$ with Hessian matrix $f_{ww}(w) = \{f_{w_i w_j}(w); i,j\}$, note that

$$\text{trace}\,[f_{ww}(w)\Sigma] = \sum_{i,j} a_{ij} f_{w_i w_j}(w). \tag{1.8}$$

Theorem 1.4. *Let $f(\cdot)$ be an arbitrary continuous real-valued function on $I\!R^k$ that has compact support and whose mixed partial derivatives up to second order are continuous and bounded. Let $V(\cdot)$ be a random process with paths in $D(S;0,\infty)$, where S is a complete and separable metric space. Let the $I\!R^k$-valued process $w(\cdot)$ have continuous paths with probability one and let $\Sigma(\cdot) = \{a_{ij}(\cdot); i,j\}$ be a bounded nonnegative definite symmetric matrix-valued and nonrandom function. Suppose that for each real $t \geq 0$ and $\tau \geq 0$, each integer p, and each set of real numbers $s_i \leq t, i = 1, \ldots, p$, and each bounded and continuous real-valued function $h(\cdot)$,*

$$\begin{aligned} &Eh\left(V(s_i), w(s_i), i \leq p\right) \\ &\quad\times\left[f(w(t+\tau)) - f(w(t)) - \frac{1}{2}\int_t^{t+\tau} \text{trace}\,[f_{ww}(w(s))\Sigma(s)]\, ds\right] = 0. \end{aligned} \tag{1.9}$$

Then $w(\cdot)$ is an $\mathcal{F}_t^V$-Wiener process with time-varying covariance matrix $\Sigma(\cdot)$, where $\mathcal{F}_t^V$ is the smallest σ-algebra that measures $\{V(s), w(s), s \leq t\}$. Now, suppose that (1.9) holds as asserted, except that $\Sigma(\cdot)$ is random and $\mathcal{F}_t^V$-adapted and its components have paths in $D(I\!R;0,\infty)$. Then both $w(\cdot)$ and the process defined by

$$w(t)w(t)' - \int_0^t \Sigma(s) ds$$

are $\mathcal{F}_t^V$-local martingales.

2.2 Stochastic Integrals

2.2.1 *Definition and Properties*

Let the probability space be $(\Omega, \mathcal{F}, P)$ and suppose that $\mathcal{F} = \cup_{t>0}\mathcal{F}_t$, where $\{\mathcal{F}_t, t \geq 0\}$ is a filtration. In this section we will briefly review the definition and some the properties of the stochastic integral with respect to a standard (k-dimensional) $\mathcal{F}_t$-Wiener process $w(\cdot)$. Full details of the theory can be found in many places, e.g., [59, 82, 83, 114, 122, 184, 204, 228]. Stochastic differential equations such as

$$x(t) = x(0) + \int_0^t b(x(s))ds + \int_0^t \sigma(x(s))dw(s) \tag{2.1}$$

(and their "reflected" counterparts) will be reviewed in the next chapter. Only a brief description of the main ideas will be given. The only quantity in (2.1) that needs explanation is the stochastic integral term $\int_0^t \sigma(x(s))dw(s)$. Since the paths of $w(\cdot)$ are of unbounded variation, the standard Lebesgue–Stieltjes definition of the integral cannot be used. Instead, a direct approximation procedure due to Itô that exploits the martingale property of the Wiener process is used to get the standard definition and properties of the integral.

We next state the conditions on the integrand $\phi(\cdot)$ that are needed for the stochastic integral

$$\psi(t) = \int_0^t \phi(u)dw(u) \tag{2.2}$$

to be well-defined.

If $w(\cdot)$ is an $\mathcal{F}_t$-Wiener process and $f(\cdot)$ is $\mathcal{F}_t$-adapted, then $f(\cdot)$ is said to be *nonanticipative* with respect to $w(\cdot)$, since $\{f(u), u \le t\}$ and $\{w(s) - w(t), s \ge t\}$ are independent for each t.

Let $0 < T < \infty$. Let $\mathcal{H}_i(T)$, $i = 1, 2$, denote the set of $\mathcal{F}_t$-adapted, real-valued processes $\phi(\cdot)$ that satisfy $\int_0^T |\phi(u)|^i du < \infty$, with probability one, and let $\mathcal{H}_i^*(T)$ denote the subset such that $E\int_0^T |\phi(u)|^i du < \infty$. Let $\mathcal{H}_b(T)$ denote the subset of $\mathcal{H}_1(T)$ of bounded functions. If the restriction of a function $\phi(\cdot)$ to $[0, T]$ is in $\mathcal{H}_i(T)$ for all $T < \infty$, then we write $\phi(\cdot) \in \mathcal{H}_i$, and define $\mathcal{H}_i^*$ and $\mathcal{H}_b$ similarly. If all of the components of some vector-valued process $\Phi(\cdot)$ are in $\mathcal{H}_i$, we simply say that $\Phi(\cdot) \in \mathcal{H}_i$, and similarly for the other spaces.

We say that a random process $\phi(\cdot)$ is a simple function if there exists a sequence of $\mathcal{F}_t$-stopping times $\{t_n, n = 0, 1, \ldots\}$ such that $0 = t_0 < t_1 < \cdots < t_n \to \infty$ with probability one, and such that $\phi(t) = \phi(t_n)$ for $t \in [t_n, t_{n+1})$. The set of all simple functions in $\mathcal{H}_b$ will be denoted by $\mathcal{H}_{sb}$.

We are now in a position to give the Itô definition of the integral (2.2). The integral is defined for an arbitrary function in $\mathcal{H}_2$ via an approximation argument. In general, the stochastic integral defined below will be unique only in the sense that any two versions will have sample paths that agree with probability one. We always consider any two processes that agree with probability one to be identical. Until further notice, let $w(t)$ be real-valued.

Definition and properties of the stochastic integral (2.2). First, let $\phi(\cdot) \in \mathcal{H}_{sb}$. Then there are stopping times $t_n \to \infty$ with probability one such that $\phi(\cdot)$ is constant on each $[t_n, t_{n+1})$. Because of this, (2.2) is defined simply as a sum, analogously to what is done for the Lebesgue or Stieltjes integral, as follows. For $t \in [t_n, t_{n+1})$, define

$$\int_0^t \phi(u)dw(u) = \sum_{i=0}^{n-1} \phi(t_i)\left[w(t_{i+1}) - w(t_i)\right] + \phi(t_n)\left[w(t) - w(t_n)\right]. \quad (2.3)$$

Properties. Let $\phi_i(\cdot) \in \mathcal{H}_{sb}$, $i = 1, 2$. Then the integral defined by (2.3) has the following properties for any $0 < s < t < \infty$:

$$E_{\mathcal{F}_s} \int_0^t \phi(u)dw(u) = \int_0^s \phi(u)dw(u), \quad (2.4)$$

$$E_{\mathcal{F}_s}\left[\int_s^t \phi_1(u)dw(u) \int_s^t \phi_2(u)dw(u)\right] = \int_s^t E_{\mathcal{F}_s}\left[\phi_1(u)\phi_2(u)\right] du, \quad (2.5)$$

$$\int_0^t \phi_1(u)dw(u) + \int_0^t \phi_2(u)dw(u) = \int_0^t \left[\phi_1(u) + \phi_2(u)\right] dw(u), \quad (2.6)$$

$\int_0^t \phi(u)dw(u)$ is continuous with probability one, and is an $\mathcal{F}_t$-martingale. (2.7)

The general definition. The integral defined by (2.3) is extended to $\phi(\cdot) \in \mathcal{H}_2$ by an approximation procedure. First, let $\phi(\cdot) \in \mathcal{H}_2^*$ and fix T. It can be shown [122, Proposition 3.2.6] that for each $\phi(\cdot) \in \mathcal{H}_2^*(T)$ there exist $\phi_n \in \mathcal{H}_{sb}(T)$ such that

$$\int_0^T E|\phi_n(u) - \phi(u)|^2 du \le 2^{-n}. \quad (2.8)$$

By (1.2) and (2.5)–(2.8), for $\epsilon > 0$ and $m > n$,

$$\begin{aligned} P\left\{\sup_{t\le T}\left|\int_0^t \phi_n(u)dw(u) - \int_0^t \phi_m(u)dw(u)\right| \ge \epsilon\right\} \\ \le \frac{1}{\epsilon^2} E\int_0^T \left|\phi_n(u) - \phi_m(u)\right|^2 du \le \frac{2^{-n+2}}{\epsilon^2}. \end{aligned} \quad (2.9)$$

We now define $\int_0^t \phi(u)dw(u)$ to be the limit of the sequence of processes $\int_0^t \phi_n(u)dw(u)$. The limit exists (with probability one), and is independent (with probability one) of the approximating sequence. Full details are in [122, Section 3.2] and in the other references cited above. Furthermore, (2.4)–(2.7) continue to hold.

Now, let $\phi(\cdot) \in \mathcal{H}_2$, but not in $\mathcal{H}_2^*$. By working with $\int_0^{t\wedge\tau_n} \phi(u)dw(u)$ for stopping times $\tau_n = \min\left\{t : \int_0^t |\phi(u)|^2 du = n\right\}$, the definition of the stochastic integral can be extended uniquely to $\mathcal{H}_2$. This is true since

$$\int_0^{t\wedge\tau_n} \phi(u)dw(u) = \int_0^{t\wedge\tau_{n+1}} \phi(u)dw(u)$$

with probability one for $t \le \tau_n$. Then "martingale" in (2.7) is replaced by "local martingale."

The vector case. Suppose that $w(\cdot)$ is an $I\!R^k$-valued standard $\mathcal{F}_t$-Wiener process, and let $\phi(\cdot) = \{\phi_{ij}(\cdot); i, j\}$ be a matrix with k columns, with each $\phi_{ij}(\cdot) \in \mathcal{H}_2$. Then (2.2) is defined in terms of its components

$$\psi_i(t) = \int_0^t \sum_j \phi_{ij}(u)dw_j(u).$$

A representation theorem for continuous martingales. A stochastic integral is a martingale or local martingale. Conversely, a large class of continuous martingales can be represented as stochastic integrals, as in the following theorem.

Theorem 2.1. [122, Theorem 3.4.2]. *Let $M(\cdot)$ be a continuous $I\!R^k$-valued $\mathcal{F}_t$-martingale, for some filtration $\{\mathcal{F}_t, t \ge 0\}$. Suppose that, for an $\mathcal{F}_t$-adapted nonnegative definite symmetric matrix-valued process $\Sigma(\cdot)$,*

$$M(t)M'(t) - \int_0^t \Sigma(s)ds$$

is also an $\mathcal{F}_t$-martingale. Then, with the possible augmentation of the probability space by the addition of a vector-valued standard Wiener process that is independent of all processes defined on the space, $M(\cdot)$ can be written as a stochastic integral

$$M(t) = M(0) + \int_0^t \Sigma^{1/2}(s)dw(s), \tag{2.10}$$

for some $\mathcal{F}_t$-standard Wiener process $w(\cdot)$. Here $\Sigma^{1/2}(s)$ is an adapted square root of $\Sigma(s)$. Suppose that $\Sigma(\cdot)$ is block diagonal, with the blocks being of size n_i, $i \le q$. Then the components $M_i(\cdot)$, $i \le k$, can be divided into blocks of n_i components, $i \le q$, and the Wiener processes that are used in the representation of the different blocks are mutually independent.

2.2.2 Itô's Lemma

Let $C^i(I\!R^k)$ denote the set of real-valued functions on $I\!R^k$ that are continuous together with their mixed partial derivatives up to and including order

i. Let $C_0^i(I\!R^k)$ (respectively, $C_b^i(I\!R^k)$) denote the subset of such functions with compact support (respectively, that are bounded).

Real-valued processes. Let $f(\cdot) \in C^1(I\!R)$, and let $\bar{b}(\cdot)$ be a real-valued function that is integrable on each interval $[0, T]$. Define $x(t) = x(0) + \int_0^t \bar{b}(s)ds$. By the rules of calculus,

$$f(x(t)) - f(x(0)) = \int_0^t f_x(x(s))\bar{b}(s)ds.$$

Let $b(\cdot) \in \mathcal{H}_1$ and $\phi(\cdot) \in \mathcal{H}_2$. The analogous formula for functions of processes $x(\cdot)$ defined by

$$x(t) = x(0) + \int_0^t b(s)ds + \int_0^t \phi(s)dw(s) \tag{2.11}$$

plays an equally important role in stochastic analysis, although it is a little more complicated. All processes are real-valued until further notice. *Itô's lemma* [59, 82, 83, 114, 122, 184, 204, 228] states that for any $f(\cdot) \in C^2(I\!R)$,

$$f(x(t)) - f(x(0)) = \int_0^t f_x(x(s))dx(s) + \frac{1}{2}\int_0^t f_{xx}(x(s))\phi^2(s)ds, \tag{2.12}$$

where

$$\int_0^t f_x(x(s))dx(s) = \int_0^t f_x(x(s))b(s)ds + \int_0^t f_x(x(s))\phi(s)dw(s).$$

We will sometimes write this relationship symbolically in differential form as

$$df(x(t)) = \left[f_x(x(t))b(t) + \frac{1}{2}f_{xx}(x(t))\phi^2(t)\right]dt + f_x(x(t))\phi(t)dw(t).$$

The vector form of Itô's lemma. All vectors are column vectors and prime denotes transpose. Let $w(\cdot)$ be a k-dimensional vector-valued $\mathcal{F}_t$-standard Wiener process. Let $\Phi(\cdot) = \{\phi_{ij}(\cdot); i = 1, \ldots, r, j = 1, \ldots, k\}$, where $\phi_{ij}(\cdot) \in \mathcal{H}_2$. Let $b(\cdot) = \{b_i(\cdot), i = 1, \ldots, r\}$, where $b_i(\cdot) \in \mathcal{H}_1$. Define $x(\cdot)$ by

$$x(t) = x(0) + \int_0^t b(s)ds + \int_0^t \Phi(s)dw(s). \tag{2.13}$$

This will be written alternatively in the symbolic differential form

$$dx(t) = b(t)dt + \Phi(t)dw(t).$$

Then the *vector version* of Itô's lemma is as follows. For any $f(\cdot) \in C^2(I\!R^r)$, let $f_x(\cdot) = \{f_{x_i}(\cdot), i = 1, \ldots, r\}$ and $f_{xx}(\cdot) = \{f_{x_i x_j}(\cdot); i, j = 1, \ldots, r\}$

denote the gradient and Hessian matrix of $f(\cdot)$, respectively. Define $a(s) = \Phi(s)\Phi'(s) = \{a_{ij}(s); i, j\}$. Then

$$f((x(t)) - f(x(0)) = \int_0^t f_x'(x(s))dx(s) + \frac{1}{2}\int_0^t \sum_{i,j=1}^r a_{ij}(s) f_{x_i x_j}(x(s))ds, \tag{2.14}$$

where

$$\int_0^t f_x'(x(s))dx(s) = \int_0^t f_x'(x(s))b(s)ds + \int_0^t f_x'(x(s))\Phi(s)dw(s).$$

The sum in (2.14) can be written as trace$[f_{xx}(x(s))a(s)]$.

The form of Itô's lemma for processes such as (2.13) but with jumps added will be given in Section 4.

2.3 Poisson Measures

2.3.1 The Poisson Process and Poisson Random Measures

In this section we are given a probability space $(\Omega, \mathcal{F}, P)$ on which a filtration $\{\mathcal{F}_t, t \geq 0\}$ is defined.

Finite jump rate. Define the $I\!R^k$-valued $\mathcal{F}_t$-adapted piecewise constant and right continuous jump process $p(\cdot)$ with independent increments in the following way. For each $t, s \geq 0$, $p(t)$ is finite with probability one, $p(t+s) - p(t)$ is independent of $\mathcal{F}_t$, and the distribution of $p(t+s) - p(t)$ does not depend on t. Furthermore, there is a Borel set $\Gamma \subset I\!R^k$, such that the jumps of $p(\cdot)$ are in Γ. If the closure of Γ does not contain the origin, then the jumps have a minimum positive size. If the closure contains the origin, then there is the possibility of arbitrarily small jumps.

Let $N(\sigma, B)$ denote the *number of jumps* on the time interval σ whose value is in the Borel set B. If $\sigma = [0, t]$, then we write simply $N(\sigma, B) = N(t, B)$. Suppose first that the origin is not in the closure of Γ. Then there are [82] a $\lambda < \infty$ (called the *jump rate*, or simply the rate) and a probability measure $\Pi(\cdot)$ (the probability distribution of the jumps) on the Borel subsets of Γ such that $E_{\mathcal{F}_t}[N(t+s, B) - N(t, B)] = \lambda\Pi(B)s$. Since $N(t, B)$ counts jumps with magnitudes in the Borel set $B \subset \Gamma$ on the time interval $[0, t]$, it is sometimes called a counting measure. If the closure of Γ contains the origin, then the general theory allows the jump rate to go to infinity as the jump size goes to zero. However, this possibility will not be considered in this book. We will always assume that the process can be described in terms of a finite jump rate λ and a jump distribution $\Pi(\cdot)$.

The process $p(\cdot)$ can be described alternatively in terms of the interjump times and the jump values. The interjump times are mutually independent and exponentially distributed with mean $1/\lambda$, and the set of jumps are independent of the jump times and are mutually independent, each having distribution $\Pi(\cdot)$. If Γ contains only the single value unity, then $p(\cdot)$ is called a Poisson process.

The Poisson random measure. Finite jump rate. When processes such as $p(\cdot)$ above are to be used as "driving" processes for differential equations (see Chapter 3), it is more convenient to work with the derived random measure $N(\cdot)$, and we now formalize the definition of a Poisson random measure from that point of view. We start with $N(\cdot)$ as the primary object, although it is clearly equivalent to $p(\cdot)$.

Define $R^+ = [0, \infty)$. An $\mathcal{F}_t$-*Poisson random measure* with *intensity measure* $h(dt\, d\gamma) = \lambda dt \times \Pi(d\gamma)$ is defined to be a measurable mapping $N(\cdot)$ from $(\Omega, \mathcal{F}, P)$ into the space of nonnegative integer-valued measures on $(I\!R^+ \times \Gamma, \mathcal{B}(I\!R^+ \times \Gamma))$ with the following properties:

1. For every $t \geq 0$ and every Borel subset A of $[0,t] \times \Gamma$, $N(A)$ is $\mathcal{F}_t$-measurable.
2. For every $t \geq 0$ and every collection of Borel subsets A_i of $[t, \infty) \times \Gamma$, $\{N(A_i)\}$ is independent of $\mathcal{F}_{t-}$.
3. $EN(A) = h(A)$ for every Borel subset A of $I\!R^+ \times \Gamma$.
4. For each $s > 0$ and Borel $B \subset \Gamma$, the distribution of $N([t, t+s], B)$ does not depend on t.

These properties are satisfied by the $N(\cdot)$ obtained from $p(\cdot)$ above. Conversely, a Poisson random measure $N(\cdot)$ yields a jump process with independent increments via

$$p(t) = \int_0^t \int_\Gamma \gamma N(ds\, d\gamma).$$

If $B \subset \Gamma$ is a Borel set, then the process defined by

$$p(t, B) = \int_0^t \int_B \gamma N(ds\, d\gamma)$$

includes only jumps with values in B.

Remark. Processes with independent and stationary increments. Let $Y(\cdot)$ be an $\mathcal{F}_t$-adapted process with paths in $D(I\!R^k; 0, \infty)$ such that for each t, $Y(t + \cdot) - Y(t)$ is independent of $\mathcal{F}_t$, and its distribution does not depend on t. Then [82] $Y(\cdot)$ can be represented as the sum of a constant, an $I\!R^k$-valued $\mathcal{F}_t$-Wiener process with a constant covariance matrix, and an $I\!R^k$-valued $\mathcal{F}_t$-adapted jump process $p(\cdot)$ with independent increments, where the distribution of $p(t + \cdot) - p(t)$ does not depend on t, and the Wiener and jump processes are mutually independent.

2.3.2 Martingale Decomposition of a Jump Process

Definition. Predictable process. Let $\{\mathcal{F}_t, t \geq 0\}$ be a filtration. The minimal σ-algebra over the Borel subsets of $[0, \infty) \times \Omega$ that contains sets of the form $(s,t] \times B, 0 \leq s \leq t, B \in \mathcal{F}_s$, and $[0,t] \times B, B \in \mathcal{F}_0$, is called the *predictable* σ-algebra and is denoted by $\mathcal{P}(\mathcal{F}_\cdot)$. It is the minimal σ-algebra that measures the $\mathcal{F}_t$-adapted left continuous processes. A $\mathcal{P}(\mathcal{F}_\cdot)$-measurable process is called *predictable* [33].

Definition. Stochastic intensity and compensator [33]. Let $p(\cdot)$ be a real-valued $\mathcal{F}_t$-adapted jump process with bounded jumps and satisfying $p(0) = 0$ and $Ep(t) < \infty$ for each t. Let $\lambda(\cdot)$ be a bounded and nonnegative process that is $\mathcal{F}_t$-adapted, with paths in $D(I\!R; 0, \infty)$, and satisfying $E\int_0^t \lambda(s)ds < \infty$ for each $t < \infty$.[1] If

$$p(t) - \int_0^t \lambda(s)ds \tag{3.1}$$

is an $\mathcal{F}_t$-martingale, then $\lambda(\cdot)$ is said to be the *intensity* (or $\mathcal{F}_t$-intensity, if the filtration is not clear) of $p(\cdot)$. The integral $\int_0^t \lambda(s)ds$ will be referred to as the *compensator* or $\mathcal{F}_t$-compensator. Thus, to verify that $\lambda(\cdot)$ is the intensity, we need only verify that

$$E_{\mathcal{F}_t} p(t+\delta) - p(t) = E_{\mathcal{F}_t} \int_t^{t+\delta} \lambda(s)ds, \tag{3.2}$$

for each $t, \delta > 0$.

Suppose that $p(\cdot)$ has only (positive) unit jumps. Then $\lambda(t)$ has the interpretation of the "jump rate at t" in that

$$E_{\mathcal{F}_t} \{\text{number of jumps in } [t, t+\delta)\} = E_{\mathcal{F}_t} \int_t^{t+\delta} \lambda(s)ds, \tag{3.3}$$

and

$$P_{\mathcal{F}_t} \{\text{one jump in } [t, t+\delta)\} = \lambda(t)\delta + o(\delta), \tag{3.4}$$

where $o(\delta)/\delta \to 0$ with probability one and in the mean, uniformly in t on any bounded interval.

The compensator for the jump process is often obtained via a local computation, as illustrated in the following theorem. In the theorem, we "subdivide" the jump process $p(\cdot)$ according to the size of the jumps. The theorem can be proved by a direct verification, and the details are omitted. The process $p(t, B)$ is obtained from $p(\cdot)$ by including only jumps in the set B. The

[1] The integrability and boundedness properties are stronger than necessary for the general theory, but they are sufficient for our purposes.

(possibly random) measure $\bar{\lambda}(t, d\gamma)$ in the next theorem is a generalization of the $\lambda\Pi(d\gamma)$ of Subsection 3.1, and is called the *jump rate measure.*

Theorem 3.1. *For some filtration $\{\mathcal{F}_t, t \geq 0\}$, let $p(\cdot)$ be an $I\!R^k$-valued jump process with right continuous paths, $Ep(t) < \infty$ for each t and where $p(\cdot)$ is $\mathcal{F}_t$-adapted and the jumps take values in a Borel set Γ. Suppose that there is a measure-valued process $\bar{\lambda}(\cdot)$ such that for each t, $\bar{\lambda}(t, \cdot)$ is a measure on the Borel subsets $\mathcal{B}(\Gamma)$ of Γ. Write $\bar{\lambda}(t, B)$ for its values. Let $\sup_t \bar{\lambda}(t, \Gamma)$ be bounded. Suppose that for each $B \in \mathcal{B}(\Gamma)$, $\bar{\lambda}(\cdot, B)$ is $\mathcal{F}_t$-adapted and has paths in $D(I\!R^k; 0, \infty)$. Then*

$$\lim_{\delta \downarrow 0} \left|\bar{\lambda}(t + \delta, B) - \bar{\lambda}(t, B)\right| = 0 \tag{3.5}$$

in the mean and uniformly on any bounded time interval $[0, T]$. Suppose that

$$\frac{P_{\mathcal{F}_t}\{p(t + \delta) - p(t) \in B\}}{\delta} \to \bar{\lambda}(t, B), \tag{3.6}$$

in the mean and uniformly for $t \leq T$. Alternatively, replace (3.6) by the following: for each $T > 0$ and Borel set B,

$$\lim_{\delta \to 0} \sum_{i\delta \leq T} \left|E_{\mathcal{F}_{i\delta}} p(i\delta + \delta, B) - p(i\delta, B) - \delta\bar{\lambda}(i\delta, B)\right| = 0, \tag{3.7}$$

where the limits are in the sense of probability. Let $f(\cdot)$ be a bounded and continuous real-valued function. Then

$$p(t) - \int_0^t \int_\Gamma \gamma\bar{\lambda}(s, d\gamma)ds, \quad f(p(t)) - \int_0^t A_p f(p(s))ds \tag{3.8}$$

are martingales, where

$$A_p f(p(s)) = \int_\Gamma \left[f(p(s) + \gamma) - f(p(s))\right] \bar{\lambda}(s, d\gamma). \tag{3.9}$$

The right side of (3.9) is the integral of the "conditional mean rate of change" of $f(p(\cdot))$. The following weaker form of the operation is also useful [157, Subsection 3.2.2], [147].

Theorem 3.2. *Assume the conditions of Theorem 3.1 but replace (3.6)–(3.7) by the following. For each T and t,*

$$\sup_{\delta > 0, t \leq T} E\left|\frac{E_{\mathcal{F}_t} f(p(t + \delta)) - f(p(t))}{\delta}\right| < \infty, \tag{3.10}$$

$$\lim_{\delta \to 0} E\left|\frac{E_{\mathcal{F}_t} f(p(t + \delta)) - f(p(t))}{\delta} - A_p f(p(t))\right| = 0, \tag{3.11}$$

$$\lim_{\delta \to 0} E\left|A_p f(p(t + \delta)) - A_p f(p(t))\right| = 0. \tag{3.12}$$

Then the processes in (3.8) are martingales.

2.4 The Doob–Meyer Process

2.4.1 Introduction and Itô's Formula

Let $\{\mathcal{F}_t, t \geq 0\}$ be a filtration and $H(\cdot)$ an $\mathcal{F}_t$-local submartingale on $[0, \infty)$ with paths in $D(I\!R; 0, \infty)$. Suppose that for each sequence of nondecreasing and uniformly bounded stopping times τ^n, $EH(\tau^n) \to EH(\tau)$, where $\tau = \lim \tau^n$. Then[2] [69, 122, 146, 194] there is a unique continuous nondecreasing $\mathcal{F}_t$-adapted (hence $\mathcal{F}_t$-predictable) and locally integrable process $A(\cdot)$ such that $A(0) = 0$ and $H(\cdot) - A(\cdot)$ is a local $\mathcal{F}_t$-martingale. $A(\cdot)$ is called the *Doob–Meyer* process for the submartingale $H(\cdot)$. The forms for several special cases will be described below.

Now, let $M_i(\cdot)$, $i = 1, \ldots, q$, be real-valued $\mathcal{F}_t$-martingales with paths in $D(I\!R; 0, \infty)$. In our applications, $M(\cdot)$ will take the special form $M(\cdot) = M^c(\cdot) + M^d(\cdot)$, where $M^c(\cdot)$ is continuous, $M^d(\cdot)$ is a compensated jump process of the types in (3.8), and $E|M(t)|^2 < \infty$ for each t. The following remarks will be confined to this case. The $M_i^2(\cdot)$ are then $\mathcal{F}_t$-submartingales. We can suppose, via a localization argument[3] if necessary, that $EM_i^2(\tau^n) \to EM^2(\tau)$, where τ^n, τ are as above.

Let $M(\cdot)$ denote the vector $(M_i(\cdot), i = 1, \ldots, q)$. The following facts are proved in [146]. There are unique continuous and nondecreasing $\mathcal{F}_t$-adapted processes $\langle M_i \rangle(\cdot)$ such that $M_i^2(\cdot) - \langle M_i \rangle(\cdot)$ is an $\mathcal{F}_t$-martingale. Sometimes, $\langle M_i \rangle(\cdot)$ will be written as $\langle M_i, M_i \rangle(\cdot)$. Then $\langle M_i \rangle(\cdot)$ is the Doob–Meyer process for the submartingale $M_i^2(\cdot)$. For any i, j, define

$$\langle M_i, M_j \rangle(\cdot) = \frac{1}{4} \left[\langle M_i + M_j, M_i + M_j \rangle(\cdot) - \langle M_i - M_j, M_i - M_j \rangle(\cdot) \right].$$

Then $M_i(\cdot)M_j(\cdot) - \langle M_i, M_j \rangle(\cdot)$ is an $\mathcal{F}_t$-martingale. If $\langle M_i, M_j \rangle(\cdot) = 0$, we say that $M_i(\cdot)$ and $M_j(\cdot)$ are *orthogonal*. This definition of orthogonality will also be used for vector-valued martingales.

Define the matrix–valued process $\langle M \rangle(\cdot) = \{\langle M_i, M_j \rangle(\cdot); i, j \leq q\}$, which is called the Doob–Meyer process for the $\mathcal{F}_t$-martingale $M(\cdot)$. If $M(\cdot)$ is continuous, then the Doob–Meyer process is also known as the *quadratic variation*. We will sometimes have need to identify the quadratic variation process. The following criterion [146] is useful for this and also gives a representation that will be needed.

Theorem 4.1. *Let $M_i(\cdot)$ and $\mathcal{F}_t$ be as defined above. Let $\Sigma(\cdot) = \{a_{ij}(\cdot); i, j\}$ be an $\mathcal{F}_t$-adapted nonnegative definite, symmetric $(q \times q)$ matrix-valued*

---

[2] If the expectations do not exist, then use a localization argument. We are presenting only a special case that will be sufficient or our needs.

[3] With "localization," one works with $M(\cdot \wedge \tau_k)$ for appropriate stopping times τ_k, gets the desired representation for the stopped process for each k, and then gets the desired result by letting $\tau_k \to \infty$.

integrable process such that, for each real-valued $f(\cdot)$ with compact support and that is continuous together with its partial derivatives up to second order,

$$f(M(t)) - \frac{1}{2}\int_0^t \sum_{i,j} f_{m_i m_j}(M(s))a_{ij}(s)ds \tag{4.1}$$

is an $\mathcal{F}_t$-martingale. Let the paths of the $a_{ij}(\cdot)$ be in $D(I\!R; 0, \infty)$. Then

$$\langle M\rangle(t) = \int_0^t \Sigma(s)ds. \tag{4.2}$$

Theorem 2.1 *holds if $M(\cdot)$ is continuous.*

Stochastic integrals with respect to a continuous martingale. The development of the stochastic integral with respect to a Wiener process in Section 2 can be extended to general local martingales [122]. In particular, suppose that $M(\cdot)$ is a continuous $\mathcal{F}_t$-martingale with Doob–Meyer process of the form $\langle M\rangle(t) = \int_0^t \Sigma(s)ds$ for an integrable and $\mathcal{F}_t$-adapted process $\Sigma(\cdot)$. Let $\phi(\cdot)$ be $\mathcal{F}_t$-adapted and satisfy $\int_0^t |\phi(s)|^2 d\langle M\rangle(s) < \infty$ for each t w.p.1. Then the development of Section 2 can be carried over with essentially no change to uniquely (w.p.1) define the stochastic integral $\int_0^t \phi(s)dM(s)$. Its properties are analogous to those proved in Section 2 where $M(\cdot) = w(\cdot)$.

Itô's lemma. The following extension of (2.14) will be sufficient for our purposes. For a filtration $\{\mathcal{F}_t, t \geq 0\}$, define

$$X(t) = X(0) + M^c(t) + p(t) + A(t), \tag{4.3}$$

where $X(0)$ is $\mathcal{F}_0$-measurable, all of the processes are $\mathcal{F}_t$-adapted, $M^c(\cdot)$ is a continuous local $\mathcal{F}_t$-martingale, the paths of $A(\cdot)$ are continuous and of bounded variation w.p.1 on each bounded time interval, and $p(\cdot)$ is a jump process of the form of the previous subsection where $\sup_t \bar{\lambda}(t,\Gamma) < \infty$ w.p.1. Then, for any real–valued function $f(\cdot)$ that is continuous together with its partial derivatives up to second order [114],

$$\begin{aligned} f(X(t)) = f(X(0)) &+ \int_0^t f_x'(X(s))dM^c(s) + \int_0^t f_x'(X(s))dA(s) \\ &+\frac{1}{2}\int_0^t \sum_{i,j} f_{x_i x_j}(X(s))d\langle M_i^c, M_j^c\rangle(s) \\ &+ \int_0^t \int_\Gamma [f(X(s)+\gamma) - f(X(s))]\,\bar{\lambda}(s,d\gamma)ds + \bar{M}(t), \end{aligned} \tag{4.4}$$

where $\bar{M}(\cdot)$ is the local $\mathcal{F}_t$-martingale

$$\bar{M}(t) = \sum_{0<s\le t} [f(X(s)) - f(X(s-))] - \int_0^t \int_\Gamma [f(X(s)+\gamma) - f(X(s))]\,\bar{\lambda}(s,d\gamma)ds. \tag{4.5}$$

2.4.2 The Doob–Meyer Process for a Jump Process

Let $p(\cdot)$ be an $\mathcal{F}_t$-adapted vector-valued jump process with uniformly bounded jumps and with jump rate measure $\bar{\lambda}(\cdot)$, where $\sup_t \bar{\lambda}(t,\Gamma)$ is bounded and $\bar{\lambda}(\cdot,B)$ has paths in $D(I\!\!R;0,\infty)$ for each Borel set $B \in \Gamma$. The boundedness of $\bar{\lambda}(\cdot)$ implies that $E|p(t)|^2 < \infty$ for each $t < \infty$. Define $A(t) = \int_0^t \gamma\bar{\lambda}(s,d\gamma)ds$ and $M(t) = p(t) - A(t) = \{M_i(t)\}$, where $M_i(\cdot)$ are the real-valued components.

Itô's lemma (4.3) implies that for any smooth function $f(\cdot)$,

$$f(M(t)) = -\int_0^t \int_\Gamma f'_m(M(s))\gamma\bar{\lambda}(s,d\gamma)ds + \int_0^t \int_\Gamma [f(M(s)+\gamma) - f(M(s))]\,\bar{\lambda}(s,d\gamma)ds \tag{4.6}$$

plus a local $\mathcal{F}_t$-martingale. Now, let $f(M) = M_iM_j$. Then, we see that

$$M(t)M'(t) = \int_0^t \int_\Gamma \gamma\gamma'\bar{\lambda}(s,d\gamma)ds + \hat{M}(t), \tag{4.7}$$

where $\hat{M}(\cdot)$ is an $\mathcal{F}_t$-martingale. Thus, the Doob–Meyer process for $M(\cdot)$ is

$$\langle M\rangle(t) = \int_0^t \int_\Gamma \gamma\gamma'\bar{\lambda}(s,d\gamma)ds. \tag{4.8}$$

The next two theorems summarize this discussion for two special cases. Both forms will be needed in Chapter 8.

Theorem 4.2. *Consider the vector case, where $p_i(\cdot)$, $i \le q$, are $\mathcal{F}_t$-adapted jump processes of the type dealt with at the beginning of this subsection, no two of which have any jumps in common (with probability one), and with jump rate measures $\bar{\lambda}_i(\cdot)$. Define the vector-valued martingale $M(\cdot) = (M_i(\cdot),\ i \le q)$, where $M_i(t) = p_i(t) - \int_0^t \lambda_i(s)ds$, $\lambda_i(s) = \int_\Gamma \gamma\bar{\lambda}_i(s,d\gamma)$. Then,*

$$M(t)M'(t) - \int_0^t \Sigma(s)ds \tag{4.9}$$

is an $\mathcal{F}_t$-martingale, where $\Sigma(s)$ is diagonal with entries $\int_\Gamma \gamma^2\bar{\lambda}_i(s,d\gamma), i \le q$.

The following form of the vector case will be needed in Chapter 8.

Theorem 4.3. *Let $p(\cdot)$ denote a real-valued jump process of the type that was used at the beginning of this subsection, but with only (positive) unit jumps and bounded jump rate $\lambda(\cdot)$. Let $I_i(t), i \leq q$, be indicator functions of disjoint events, with each $I_i(\cdot)$ being $\mathcal{F}_t$-adapted. Let $\sum_{i\leq q} I_i(t) = 1$ if there is a jump at t, and equal to zero otherwise. On the event where $p(t) - p(t-) = 1$, define*

$$q_i(t) = E\left[I_i(t)\middle|\mathcal{F}_{t-}, p(t) - p(t-) = 1\right],$$

and set $q_i(t) = 0$ if $p(t) - p(t-) = 0$. Define the $\mathcal{F}_t$-martingales

$$P_i(t) = \int_0^t [I_i(s) - q_i(s)]\, dp(s).$$

Define $P(\cdot) = (P_i(\cdot), i \leq q)$. Then the Doob–Meyer process for $P(\cdot)$ is $\langle P\rangle(t) = \int_0^t \Sigma(s)ds$, where $\Sigma(s) = \{a_{ij}(s); i, j\}$ and $a_{ij}(s) = -\lambda(s)q_i(s)q_j(s)$ for $i \neq j$, and $a_{ii}(s) = \lambda(s)(1 - q_i(s))q_i(s)$.

2.5 Weak Convergence

2.5.1 Motivation and Examples

Let $\{A_n\}$ be a sequence of $I\!R^k$-valued random variables on a probability space $(\Omega, \mathcal{F}, P)$, with $(a_{n,i}, i = 1, \ldots, k)$ being the real-valued components of A_n. Let P_n denote the measure on the Borel sets of $I\!R^k$ that is determined by A_n, and let $x = (x_1, \ldots, x_k)$ denote the canonical point of $I\!R^k$. If there is an $I\!R^k$-valued random variable A with real-valued components $(a_1, \ldots, a_k)$ such that

$$P\{a_{n,1} < \alpha_1, \ldots, a_{n,k} < \alpha_k\} \to P\{a_1 < \alpha_1, \ldots, a_k < \alpha_k\} \tag{5.1}$$

for each $\alpha = (\alpha_1, \ldots, \alpha_k) \in I\!R^k$ at which the right side of (5.1) is continuous, then it is said that A_n *converges to A in distribution*. Let P_A denote the measure on the Borel sets of $I\!R^k$ that is determined by A. An equivalent definition to (5.1) is that [32]

$$Ef(A_n) = \int f(x)dP_n(x) \to Ef(A) = \int f(x)dP_A(x) \tag{5.2}$$

for each bounded and continuous real-valued function $f(\cdot)$ on $I\!R^k$. We say that the sequence $\{P_n\}$ is *tight* or *bounded in probability* if

$$\lim_{K\to\infty} \sup_n P\{|A_n| \geq K\} = 0. \tag{5.3a}$$

An equivalent definition of boundedness in probability is this: Let $|A_n| < \infty$ with probability one for each n, and for each small $\mu > 0$ let there be finite M_μ and K_μ such that

$$P\{|A_n| \geq K_\mu\} \leq \mu, \quad \text{for } n \geq M_\mu. \tag{5.3b}$$

Given a sequence of random variables $\{A_n\}$ with values in some Euclidean space, tightness is a necessary and sufficient condition that any subsequence have a further subsequence that converges in distribution [32, 71]. Convergence in distribution is also called *weak convergence*.

The general theory of weak convergence is an extension of the idea of convergence in distribution to sequences of random variables that take values in more abstract spaces than $\mathbb{R}^k$, in particular to random processes. This theory provides powerful tools for the approximation of random processes and for obtaining useful limit theorems for sequences of random processes. The heavy traffic limit theorems, where we wish to characterize the limits as the traffic intensity tends to unity, are impressive applications of these results.

As an illustration of the idea, consider the following form of one of the classical illustrations of weak convergence. Let $\{\xi_n\}$ be a sequence of real-valued random variables that are mutually independent and identically distributed, with mean zero and unit variance. Then, by the classical central limit theorem $\sum_{i=1}^n \xi_i/\sqrt{n}$ converges in distribution to a normally distributed random variable with zero mean and unit variance. Now, for each t, define

$$w^n(t) = \frac{1}{\sqrt{n}} \sum_{i=1}^{nt} \xi_i, \tag{5.4}$$

where nt always denotes the integer part, when used as a limit of summation. Then the central limit theorem tells us that for each t, $w^n(t)$ converges in distribution to a normally distributed random variable with mean zero and variance t. For an integer k, let $0 = t_0 < t_1 < \cdots < t_{k+1}$ be real numbers, and let $w(\cdot)$ be a real-valued Wiener process with unit variance parameter. Then, by the multivariate central limit theorem [32], the set $\{w^n(t_{i+1}) - w^n(t_i), i \leq k\}$ converges in distribution to $\{w(t_{i+1}) - w(t_i), i \leq k\}$. But we can also consider $w^n(\cdot)$ to be a random process with piecewise constant paths (constant on the intervals $[i/n, (i+1)/n)$), and it is natural to ask whether, when considered as a process, $w^n(\cdot)$ converges to $w(\cdot)$ in a stronger sense. For example, will the distribution of the first passage time for $w^n(\cdot)$ defined by $\min\{t : w^n(t) \geq 1\}$ converge in distribution to the first passage time for $w(\cdot)$ defined by $\min\{t : w(t) \geq 1\}$? Will the maximum $\max\{w^n(t) : t \leq 1\}$ converge in distribution to $\max\{w(t) : t \leq 1\}$ and similarly for other useful functionals? In general, we would like to know the class of real-valued functionals $f(\cdot)$ for which $f(w^n(\cdot))$ converges in distribution to $f(w(\cdot))$. Donsker's theorem states that this convergence occurs for a large class of functionals [15, 71], indeed for measurable $f(\cdot)$ that are continuous almost everywhere with respect to the measure of $w(\cdot)$. This will be spelled out in more detail in the rest of this section and in Section 8.

For another example, consider the following more general form of (5.4). For ξ_n as given above, a real-valued $U(0)$, and $\Delta > 0$, define real-valued random variables U_n^Δ by $U_0^\Delta = U(0)$ and for $n \geq 0$,

$$U_{n+1}^\Delta = U_n^\Delta + \Delta g(U_n^\Delta) + \sqrt{\Delta}\sigma(U_n^\Delta)\xi_n,$$

where $g(\cdot)$ and $\sigma(\cdot)$ are bounded and continuous functions. Define the interpolated process $U^\Delta(\cdot)$ by $U^\Delta(t) = U_n^\Delta$ on $[n\Delta, n\Delta + \Delta)$. Then in what sense will $U^\Delta(\cdot)$ converge to the process $U(\cdot)$ defined by the stochastic differential equation (see Chapter 3 for a discussion of such equations)

$$dU = g(U)dt + \sigma(U)dw$$

as $\Delta \to 0$? We expect that $U(\cdot)$ is the "natural" limit of $U^\Delta(\cdot)$.

More challenging questions arise when the random variables ξ_n are correlated. The central limit theorem and the law of large numbers are very useful for the approximation of random variables that are the "sums" of many "small effects" whose mutual dependence is "local," by the simpler normally distributed random variable or by some constant, respectively. The theory of weak convergence is concerned with analogous questions when the random variables are replaced by random processes as in the above examples. The two main steps in getting the limit theorems are analogous to what is done for proving the central limit theorem: First show that there are appropriately convergent subsequences and then identify the limits. For vector-valued random variables, the necessary and sufficient condition (5.3a) for the first step says that, neglecting an n-dependent set of small probability (small, uniformly in n), the values of the random variables A_n are confined to some compact set. There will be an analogous condition, ensuring appropriate compactness with a "high probability," when random processes replace random variables.

2.5.2 Basic Theorems of Weak Convergence

Definitions. Let S denote a metric space with metric $\rho(\cdot)$ and let $C(S)$ denote the set of real-valued continuous functions defined on S. Let $C_b(S)$ and $C_0(S)$ denote the subsets of $C(S)$ of functions that are bounded and have compact support, respectively. Let $X_n, n < \infty$, and X be S-valued random variables that might possibly be defined on different probability spaces. Let P_n and P denote the measures induced by X_n and X, respectively. The sequence $\{X_n, n < \infty\}$ is said to *converge in distribution* to X

if $Ef(X_n) \to Ef(X)$ for all $f \in C_b(S)$. Convergence in distribution is a property of the P_n and P, since for any $f \in C_b(S)$,

$$Ef(X_n) \to Ef(X) \Leftrightarrow \int_S f(s)P_n(ds) \to \int_S f(s)P(ds).$$

We will refer to this form of convergence of probability measures as *weak convergence* and use the notation $P_n \Rightarrow P$. However, we will often abuse terminology and say that the sequence of random variables X_n that are associated as above with the P_n converges weakly to X, and denote this by $X_n \Rightarrow X$ as well. The X will be said to be the *weak-sense limit.* Let $g(\cdot)$ be any continuous function from S into any metric space. By the definition of weak convergence, $X_n \Rightarrow X$ implies $g(X_n) \Rightarrow g(X)$. Comprehensive references for the general theory of weak convergence are [15, 71], and proofs of the statements in this section can be found there. Applications to problems in control and stochastic systems theory can be found in [157, 160, 167, 177].

Let $\mathcal{P}(S)$ denote the space of probability measures on $(S, \mathcal{B}(S))$, and let P_1 and P_2 be in $\mathcal{P}(S)$. For $A \in \mathcal{B}(S)$, define $A^\varepsilon = \{s' : \rho(s', s) < \varepsilon \text{ for some } s \in A\}$. The *Prohorov metric* $\pi(\cdot)$ on $\mathcal{P}(S)$ is defined by

$$\pi(P_1, P_2) = \inf\{\varepsilon > 0 : P_1(A) \leq P_2(A^\varepsilon) + \varepsilon \text{ for all closed } A \in \mathcal{B}(S)\}. \tag{5.5}$$

Let $P_\lambda \in \mathcal{P}(S)$ for $\lambda \in \Lambda$, an arbitrary index set. The set $\{P_\lambda, \lambda \in \Lambda\}$ is called *tight* if for each $\varepsilon > 0$ there is a compact set $K_\varepsilon \subset S$ such that

$$\inf_{\lambda \in \Lambda} P_\lambda(K_\varepsilon) \geq 1 - \varepsilon. \tag{5.6}$$

If P_λ is the measure defined by some random variable X_λ, then we will also say that $\{X_\lambda, \lambda \in \Lambda\}$ is tight. If all the random variables are defined on the same probability space, then (5.6) can be written as

$$\inf_{\lambda \in \Lambda} P\{X_\lambda \in K_\varepsilon\} \geq 1 - \varepsilon. \tag{5.7}$$

Theorem 5.1. [71, page 101.] *If S is complete and separable, then $\mathcal{P}(S)$ is complete and separable.*

Theorem 5.2. [71, Theorem 3.2.2, Prohorov's theorem.] *If S is complete and separable, then a set $\{P_\lambda, \lambda \in \Lambda\} \subset \mathcal{P}(S)$ has compact closure in the topology induced by the Prohorov metric if and only if $\{P_\lambda, \lambda \in \Lambda\}$ is tight.*

Suppose that S is complete and separable and that some given sequence of probability measures has compact closure with respect to the Prohorov metric. Theorem 5.2 then implies the existence of a convergent subsequence

[61, Theorem 13, page 21]. Via the criterion of tightness, Prohorov's theorem provides an effective method for verifying the compact closure property. This is due to the fact that tightness is also a property of the random variables associated with the measures P_n. These random variables (or processes) often have explicit representations that allow a convenient verification of the tightness property.

An extension to product spaces.

Corollary 5.1. *Let S_1 and S_2 be complete and separable metric spaces, and define $S = S_1 \times S_2$ with the usual product space topology. For $\{P_\lambda, \lambda \in \Lambda\} \subset \mathcal{P}(S)$, let $P_{\lambda,i}$ be the marginal distribution of P_λ on S_i. Then $\{P_\lambda, \lambda \in \Lambda\}$ is tight if and only if $\{P_{\lambda,i}, \lambda \in \Lambda\}$, $i = 1, 2$, are tight.*

The next theorem contains some statements that are equivalent to weak convergence, and which will be useful. Let ∂B be the boundary of the set $B \in \mathcal{B}(S)$.

Theorem 5.3. [71, Theorem 3.3.1] *Let S be a metric space and let P_n, $n < \infty$, and P be elements of $\mathcal{P}(S)$. Then statements (i)-(iv) below are equivalent and are implied by (v). If S is separable, then (i)-(v) are equivalent:*

(*i*) $P_n \Rightarrow P$,
(*ii*) $\limsup_n P_n(F) \le P(F)$ *for closed sets* F,
(*iii*) $\liminf_n P_n(O) \ge P(O)$ *for open sets* O,
(*iv*) $\lim_n P_n(B) = P(B)$ *if* $P(\partial B) = 0$,
(*v*) $\pi(P_n, P) \to 0$.

Thus, for separable S, convergence in the Prohorov metric is equivalent to weak convergence. Part (iv) of Theorem 5.3 implies the following important extension of weak convergence.

Theorem 5.4. [15, Theorem 5.1.] *Let S be a metric space, and let $P_n, n < \infty$, and P be probability measures on $\mathcal{P}(S)$ satisfying $P_n \Rightarrow P$. Let $f(\cdot)$ be a real-valued measurable function on S and define D_f to be the measurable set of points at which $f(\cdot)$ is not continuous. Let X_n and X be random variables that induce the measures P_n and P on S, respectively. Then $f(X_n) \Rightarrow f(X)$ whenever $P\{X \in D_f\} = 0$.*

The Skorohod representation. Suppose that $X_n \Rightarrow X$. Whether or not X_n and X are defined on the same probability space is unimportant, since weak convergence is a statement on the measures of the random variables. But for analytic purposes, such as characterizing the weak-sense limit X, it is often useful to have all processes defined on the same space and weak convergence replaced by probability one convergence. That this can be done without changing the distributions of the X_n or X is quite important. The result is known as the *Skorohod representation* [71], and it is a very useful tool for proving weak convergence.

Theorem 5.5. [71, Theorem 3.1.8.] *Let S be a separable metric space, and assume that the probability measures $P_n \in \mathcal{P}(S)$ converge weakly to $P \in \mathcal{P}(S)$ as $n \to \infty$. Then there exists a probability space $(\tilde{\Omega}, \tilde{\mathcal{F}}, \tilde{P})$ on which there are defined random variables $\tilde{X}_n, n < \infty$, and $\tilde{X}$ such that for all Borel sets B and all $n < \infty$,*

$$\tilde{P}\left\{\tilde{X}_n \in B\right\} = P_n(B), \qquad \tilde{P}\left\{\tilde{X} \in B\right\} = P(B), \tag{5.8}$$

and such that

$$\tilde{X}_n \to \tilde{X} \text{ with probability one.} \tag{5.9}$$

2.5.3 The Function Space $D(I\!R^k; 0, \infty)$

In the applications in this book, the "physical" and limit processes of interest have their paths in $D(S; 0, \infty)$ for some complete and separable metric space S. Most commonly, $S = I\!R^k$ for some k. The weak-sense limit processes will frequently have continuous paths. For example, the processes defined by (5.4) are discontinuous and constant on the intervals $[i/n, (i+1)/n)$, although the weak-sense limit process is a Wiener process that is continuous. In such cases, it is possible to work with the piecewise linear interpolations of the piecewise constant $w^n(\cdot)$, which are continuous processes. The weak-sense limits would not change in this example. Using the piecewise linear interpolation can be notationally awkward. But more importantly, it is usually much easier to prove tightness in the space $D(I\!R^k; 0, \infty)$ than in the space of continuous functions, and this can be an important consideration even for (5.4) if the ξ_n are correlated. In fact, if a sequence of processes $\{x^n(\cdot)\}$ converges weakly to a process $x(\cdot)$ in $D(I\!R^k; 0, \infty)$, and if the supremum of the discontinuities of the $x^n(\cdot)$ on any finite time interval go to zero in probability as $n \to \infty$, then the paths of $x(\cdot)$ *must be continuous with probability one.* Many of the applications will have weak-sense limit processes with discontinuities; for example, they might be caused by control actions or by machine breakdowns. We now define the metric to be used on $D(I\!R^k; 0, \infty)$.

The Skorohod metric on $D(I\!R^k; 0, T)$ [71, Chapter 3.5], [15, Chapter 3]. Let Λ_T denote the space of continuous and strictly increasing functions from the interval $[0, T]$ onto the interval $[0, T]$. The functions in this set will be "allowable time scale distortions." Actually, there are two equivalent (in the sense of inducing the same topology on $D(I\!R^k; 0, T)$) metrics that are known as the Skorohod metric.

The first metric $d_T(\cdot)$ is defined, for $\lambda(\cdot) \in \Lambda_T$, by

$$d_T(f(\cdot), g(\cdot)) =$$
$$\inf\left\{\epsilon : \sup_{0\le s\le T} |s - \lambda(s)| \le \epsilon, \ \sup_{0\le s\le T} |f(s) - g(\lambda(s))| \le \epsilon \text{ for some } \lambda(\cdot)\right\}.$$

This metric has the following properties. If $f_n(\cdot) \to f(\cdot)$ in $d_T(\cdot)$ where $f(\cdot)$ is continuous, then the convergence must be uniform on $[0, T]$. If there are $\eta_n \to 0$ such that the discontinuities of $f_n(\cdot)$ are less than η_n in magnitude and if $f_n(\cdot) \to f(\cdot)$ in $d_T(\cdot)$, then the convergence is uniform on $[0, T]$ and $f(\cdot)$ must be continuous. Because of the "time scale distortion" that is involved in the definition of the metric $d_T(\cdot)$, we can have (loosely speaking) convergence of a sequence of discontinuous functions where there are only a finite number of discontinuities, where both the locations and the values of the discontinuities converge, and a type of "equicontinuity" condition holds between the discontinuities. For example, define $f(\cdot)$ by $f(t) = 1$ for $t < 1$ and $f(t) = 0$ for $t \ge 1$. Define the function $f_n(\cdot)$ by $f_n(t) = f(t + 1/n)$. Then $f_n(\cdot)$ converges to $f(\cdot)$ in the Skorohod topology, but not in the sup norm. Such properties of the Skorohod metric can be found in [15, 71].

Under the metric $d_T(\cdot)$, the space $D(\mathbb{R}^k; 0, T)$ is separable but not complete [15, Chapter 3]. Owing to Theorem 5.1, it is important to have the completeness property. There is an equivalent metric $d'_T(\cdot)$ under which the space is both complete and separable, and it is this latter one that will be used, as is usual. The metric $d'_T(\cdot)$ weights the "derivative" of the time scale change $\lambda(t)$ as well as its deviation from t. For $\lambda(\cdot) \in \Lambda_T$ define

$$|\lambda| = \sup_{s<t<T} \left|\log\left\{\frac{\lambda(t) - \lambda(s)}{t - s}\right\}\right|.$$

The metric $d'_T(\cdot)$ is defined by, for $\lambda(\cdot) \in \Lambda_T$,

$$d'_T(f(\cdot), g(\cdot)) = \inf\{\epsilon : |\lambda| \le \epsilon, \ \sup_{0\le s\le T} |f(s) - g(\lambda(s))| \le \epsilon, \text{ for some } \lambda(\cdot)\}. \tag{5.10}$$

On the space $D(\mathbb{R}^k; 0, \infty)$, the metric is defined by

$$d'(f(\cdot), g(\cdot)) = \int_0^\infty e^{-t} \min\left[1, d'_t(f(\cdot), g(\cdot))\right] dt. \tag{5.11}$$

2.5.4 The Function Spaces $D(S; 0, T)$ and $D(S; 0, \infty)$

Let S denote a metric space with metric $\rho(\cdot)$. Henceforth, $D(S; 0, T)$ is the space of S-valued functions on the interval $[0, T]$ that are right continuous and have left-hand limits, and equipped with the Skorohod metric defined by the $d'_T(\cdot)$ above, but with $\rho(f(s), g(\lambda(s)))$ used in place of $|f(s) - g(\lambda(s))|$, where both $f(\cdot)$ and $g(\cdot)$ are now points in $D(S; 0, T)$.

Define the space $D(S;0,\infty)$ analogously. If S is complete and separable, then so are $D(S;0,T)$ and $D(S;0,\infty)$ [71].

The following criterion for tightness will be used. Recall that for a filtration $\{\mathcal{F}_t, t \geq 0\}$, the random time τ is an $\mathcal{F}_t$-stopping time if $\{\tau \leq t\} \in \mathcal{F}_t$ for all $t \in [0,\infty)$.

Theorem 5.6. [148, Theorem 2.7b.] *Let $x^n(\cdot)$ be processes with paths in $D(S;0,\infty)$, where S is a complete and separable metric space with metric $\rho(\cdot)$. For each $\delta > 0$ and rational $t < \infty$, let there be a compact set $S_{\delta,t} \subset S$ such that*

$$\sup_n P(x^n(t) \notin S_{\delta,t}) \leq \delta. \tag{5.12}$$

Let $\mathcal{F}_t^n$ be the σ-algebra determined by $\{x^n(s),\ s \leq t\}$, and $\mathcal{T}_n(T)$ the set of $\mathcal{F}_t^n$-stopping times that are no bigger than T. Suppose that

$$\lim_{\delta\to 0} \limsup_n \sup_{\tau \in \mathcal{T}_n(T)} E \min\{1, \rho(x^n(\tau+\delta), x^n(\tau))\} = 0 \tag{5.13}$$

for each $T < \infty$. Then $\{x^n(\cdot),\ n < \infty\}$ is tight in $D(S;0,\infty)$.

If $S = I\!R^k$, then $|x^n(\tau+\delta) - x^n(\tau)|$ replaces $\rho(x^n(\tau+\delta), x^n(\tau))$ in (5.13).

Example of Theorem 5.4: The countable set of points of discontinuity. Let S be a complete and separable metric space. Let $\{x^n(\cdot)\}$ be a sequence of processes with paths in $D(S;0,\infty)$, and that converges weakly to a process $x(\cdot)$. Suppose that $x(\cdot)$ is either continuous or just continuous with probability one at each fixed t. For given $t > 0$, the real-valued function $F_t(\cdot)$ on $D(S;0,\infty)$ defined by $F_t(\psi(\cdot)) = \psi(t)$ is not continuous at all $\psi(\cdot)$ in the Skorohod topology [71], but it is continuous at each $\psi(\cdot) \in D(S;0,\infty)$ that is continuous at t. Thus, we can apply Theorem 5.4 to get that for each t, $x^n(t) \Rightarrow x(t)$. Let $x(\cdot)$ have its paths in $D(S;0,\infty)$ but not necessarily be continuous. The set $\mathcal{T}_d$ of points $t \geq 0$ at which the probability of a discontinuity is positive is countable. A consequence of the continuity of $F_t(\psi(\cdot))$ at all t at which $\psi(\cdot)$ is continuous is that for $t_i \notin \mathcal{T}_d$, $i \leq k$,

$$(x^n(t_i), i \leq k) \Rightarrow (x(t_i), i \leq k).$$

2.5.5 Truncated Processes

In trying to prove the weak convergence of a sequence $x^n(\cdot)$, the first step is to prove tightness. The main difficulties in this step often stem from the possible unboundedness in the sense that it might be hard to prove directly that for each $t > 0$,

$$\lim_{k\to\infty} \sup_n P\left\{\sup_{s\leq t} |x^n(s)| \geq k\right\} = 0.$$

Truncation methods are a useful approach to circumventing this problem [157, 177]. Typically, for each $0 < K < \infty$ one defines a process $\bar{x}^{K,n}(\cdot)$ such that $x^n(t) = \bar{x}^{K,n}(t)$ until at least the first time that $|x^n(t)| \geq K$, and where the weak convergence of $\bar{x}^{K,n}(\cdot)$ is easier to establish for each K. Suppose that $\{\bar{x}^{K,n}(\cdot), n < \infty\}$ is tight for each $K < \infty$. Let $\bar{x}^K(\cdot)$ denote the weak-sense limit of a weakly convergent subsequence of $\bar{x}^{K,n}(\cdot)$. Suppose that for each K, the weak-sense limit process does not depend on the chosen subsequence at least up to the first time that the limit process equals or exceeds K and that for each $t < \infty$,

$$\lim_{k\to\infty} \sup_K P\left\{\sup_{s\leq t} |\bar{x}^K(s)| \geq k\right\} = 0. \tag{5.14}$$

Then, if we stop the processes $\bar{x}^{K+1}(\cdot)$ and $\bar{x}^K(\cdot)$ at the first time that they exceed K in absolute value, they are equal in distribution. These facts imply that there is a subsequence n_i such that as $i \to \infty$, $x^{n_i}(\cdot)$ converges weakly to a process $x(\cdot)$. Furthermore, if we stop the processes $\bar{x}^K(\cdot)$ and $x(\cdot)$ at the first time that they exceed K in absolute value, they are equal in distribution. Indeed, owing to the assumption that the process $\bar{x}^K(\cdot)$ (stopped on first exceeding K in magnitude) does not actually depend on the selected subsequence, $x^{n_i}(\cdot)$ can be replaced by the original sequence $x^n(\cdot)$.

The truncation method is quite flexible, and can take many forms. To illustrate some possibilities, consider the following example. Suppose that the (real-valued) system takes the form $x^n(t) = x_i^n$ on $t \in [i/n, (i+1)/n)$, where

$$x_{i+1}^n = x_i^n + \frac{1}{n} b(x_i^n) + \frac{1}{\sqrt{n}} \xi_i^n + \frac{1}{n} \epsilon_i^n. \tag{5.15}$$

Here, $b(\cdot)$ is unbounded and continuous, ξ_i^n represents a noise sequence, and ϵ_i^n is an "error term" that goes to zero as $n \to \infty$ as long as the x_i^n remain bounded. Suppose that the process defined by $\sum_{i=1}^{nt} \xi_i^n/\sqrt{n}$ converges weakly to a Wiener process and that $x^n(0) \Rightarrow x(0)$. The proof of tightness and the characterization of the weak-sense limit processes are usually much simpler if $b(\cdot)$ is bounded. Define the real-valued smooth *truncation function* $q_K(\cdot)$ such that $0 \leq q_K(x) \leq 1$, $q_K(x) = 1$ for $|x| \leq K$, and $q_K(x) = 0$ for $|x| \geq K + 1$. Now define the truncated process $\bar{x}^{K,n}(\cdot)$, where $\bar{x}^{K,n}(0) = x^n(0)$, $\bar{x}^{K,n}(t) = \bar{x}_i^{K,n}$ on $[i/n, (i+1)/n)$, and

$$\bar{x}_{i+1}^{K,n} = \bar{x}_i^{K,n} + \frac{1}{n} b(\bar{x}_i^{K,n}) q_K\left(\bar{x}_i^{K,n}\right) + \frac{1}{\sqrt{n}} \xi_i^n + \frac{1}{n} \epsilon_i^n q_K\left(\bar{x}_i^{K,n}\right). \tag{5.16}$$

The sequence $\{\bar{x}^{K,n}(\cdot)\}$ is tight for each K. One then needs to characterize the weak-sense limits of the weakly convergent subsequences and try to prove the assertions in the previous paragraph. The weak-sense limit process of any weakly convergent subsequence can be represented as

$$\bar{x}^K(t) = x(0) + \int_0^t b(\bar{x}^K(s))q_K(\bar{x}^K(s))ds + w(t), \tag{5.17}$$

where $w(\cdot)$ is a Wiener process, and the problem concerns the possible explosion of the solution to (5.17) when the truncation function is dropped. That is, define $x(\cdot)$ by the solution to the stochastic differential equation (such equations are discussed in Chapter 3)

$$x(t) = x(0) + \int_0^t b(x(s))ds + w(t). \tag{5.18}$$

If for the given initial condition $x(0)$ and each t,

$$\lim_{k\to\infty} \sup_{\text{all solns. to (5.18)}} P\left\{\sup_{s\le t}|x(s)| \ge k\right\} = 0, \tag{5.19}$$

then the weak convergence implies that the $\limsup_n$ of the probability that $\sup_{s\le t}|x^n(s)|$ would exceed some large number k can be made small (uniformly in n) by making k large enough. This, together with the weak convergence of $\bar{x}^{K,n}(\cdot)$ for each K, implies that $x^n(\cdot)$ is tight and that there is a subsequence converging weakly to a solution of (5.18).

The obvious analogue of this method can be used if the process $x^n(\cdot)$ changes at random times (rather than at times i/n) or if it evolves "continuously" rather than discretely in time. If the noise $\{\xi_j^n, j > i\}$ depends on the values of the states x_j^n, $j \le i$, in that the distribution of $\{\xi_j^n, j > i\}$ given $\{\xi_j^n, j \le i\}$ is not the same as that given $\{\xi_j^n, x_j^n, j \le i\}$, then the state dependence of the ξ_i^n might have to be truncated also. There is an extensive discussion of truncation methods in [157, 160].

2.6 The Time Transformation Method

The processes that appear in our heavy traffic analyses cannot always be shown to be tight in the Skorohod topology of Subsection 5.4, as the following example shows. Let $F_0^n(\cdot)$ be piecewise constant, where $F_0^n(t) = 0$ for $t < 1$, then jumps up an amount $1/\sqrt{n}$ at times $1 + i/n$, $i = 0, \ldots,$ until it first reaches or exceeds the value unity, after which it is constant. Thus, the time interval on which $F_0^n(\cdot)$ increases has approximate length $1/\sqrt{n}$. The sequence $F_0^n(\cdot)$ converges to the step function with a unit jump at $t = 1$ in an obvious manner. But it converges via smaller and smaller (size $1/\sqrt{n}$) steps over smaller and smaller total time intervals (length $1/\sqrt{n}$), rather than in one large jump. Such a sequence is not tight in the Skorohod topology. This problem is common in heavy traffic analysis for singular control problems or where there are processor breakdowns or shutdowns.

There is an alternative Skorohod topology (the so-called M_1 topology [223]) that can frequently be used. But for our general purposes, it is convenient to use an essentially equivalent "time transformation" method, which is simpler with the control problems. The method will be illustrated via a model problem.

We will use the following example, where all processes have paths in $D(I\!R;0,\infty)$. Although it is a special case, the illustration gives all of the essential details for the general problem. Let $w^n(\cdot)$ converge weakly to a continuous process $w(\cdot)$, and let $b(\cdot)$ be bounded and continuous. Suppose that $x^n(\cdot)$ is defined by

$$x^n(t) = x(0) + \int_0^t b(x^n(s))ds + w^n(t) + F^n(t) + \epsilon^n(t), \tag{6.1}$$

where $\epsilon^n(\cdot)$ converges weakly to the "zero" process. Let $F^n(\cdot)$ be nonnegative, right continuous, and nondecreasing with $F^n(0) = 0$. To simplify the development, we will suppose that there is $K < \infty$ such that

$$F^n(\infty) \leq K, \text{ all } n, \text{ w.p.1.} \tag{6.2}$$

The sequence $\{F^n(\cdot)\}$ might or might not be tight in the Skorohod topology. For example, it could be the $F_0^n(\cdot)$ discussed above. For $\beta > 0$, $c > 0$, and $x^n(0) = x$, define

$$W_\beta^n(x, F^n) = E\left[\int_0^\infty e^{-\beta t} k(x^n(t))dt + c\int_0^\infty e^{-\beta t} dF^n(t)\right], \tag{6.3}$$

where $k(\cdot)$ is bounded and continuous.

We wish to show that there is a subsequence n_i and processes $x(\cdot), w(\cdot), F(\cdot)$ with paths in $D(I\!R;0,\infty)$ such that

$$W_\beta^{n_i}(x, F^{n_i}) \to W_\beta(x, F), \tag{6.4}$$

where $F(\cdot)$ is nondecreasing, right continuous, and nonnegative;

$$W_\beta(x, F) = E\left[\int_0^\infty e^{-\beta t} k(x(t))dt + c\int_0^\infty e^{-\beta t} dF(t)\right], \tag{6.5}$$

and $x(\cdot)$ satisfies

$$x(t) = x + \int_0^t b(x(s))ds + w(t) + F(t). \tag{6.6}$$

In the applications, where $w(\cdot)$ might be a Wiener process, we will also need to show that $x(\cdot), F(\cdot)$ are nonanticipative. In the language of control theory, $F(\cdot)$ might be a *singular* control, in that it is not necessarily representable as an integral.

A time transformation method for proving convergence. In [167, Chapter 11.1.2] and [170, Section 4] a very useful time rescaling idea was introduced that greatly simplified the treatment of weak convergence issues for "singular" control problems such as ours, and it will be applied here (see also [173]). Basically, one "stretches out" time such that the "poorly behaved" processes are tight, gets the appropriate weak-sense limit, shows that (6.4) and (6.6) hold for the rescaled processes and its limit, and then inverts the (limit of the) time rescaling to get (6.4) and (6.6) for the original process and the correct limit process. We now define the appropriate rescaling for this example.

Define

$$T^n(t) = t + F^n(t), \tag{6.7}$$

and its inverse

$$\hat{T}^n(t) = \inf\{s : T^n(s) > t\}. \tag{6.8}$$

Define "hat" processes by the time transformation $\hat{T}^n(\cdot)$ as, for example, in $\hat{F}^n(t) = F^n(\hat{T}^n(t))$, $\hat{x}^n(t) = x^n(\hat{T}^n(t))$, etc. Note that the transformation *stretches out* time and that, owing to the forms (6.7), (6.8), the piecewise linear interpolations of $\hat{F}^n(\cdot)$ are Lipschitz continuous with Lipschitz constant no greater than unity. Thus, $\{\hat{F}^n(\cdot)\}$ is tight in $D(I\!R; 0, \infty)$. Since $\hat{T}^n(t)$ grows at a rate no faster than unity, $\{\hat{w}^n(\cdot)\}$ continues to be tight and $\hat{\epsilon}^n(\cdot)$ still converges weakly to the "zero" process. The time transformed processes satisfy

$$\begin{aligned}\hat{x}^n(t) &= x + \int_0^{\hat{T}^n(t)} b(x^n(s))ds + w^n(\hat{T}^n(t)) + F^n(\hat{T}^n(t)) + \epsilon^n(\hat{T}^n(t)) \\ &= x + \int_0^t b(\hat{x}^n(s))d\hat{T}^n(s) + \hat{w}^n(t) + \hat{F}^n(t) + \hat{\epsilon}^n(t).\end{aligned} \tag{6.9}$$

An example. Figures 6.1 to 6.5 illustrate the time transformation for the case where $F^n(\cdot) = F_0^n(\cdot)$, where $F_0^n(\cdot)$ was defined above. Figure 6.1 graphs $T^n(\cdot)$, and Figure 6.2 the inverse function $\hat{T}^n(\cdot)$. The function $\hat{F}^n(\cdot) = F^n(\hat{T}^n(\cdot))$ is graphed in Figure 6.3, and the limits $\hat{T}(\cdot)$ and $\hat{F}(\cdot)$ are in Figures 6.4 and 6.5, respectively.

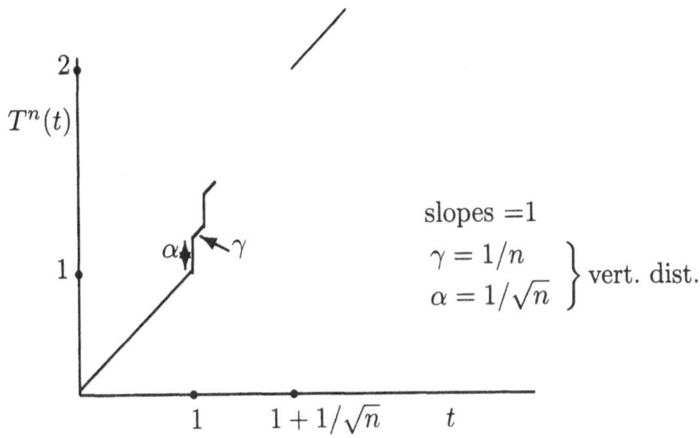

Figure 6.1. The function $T^n(\cdot)$ for $F_0^n(\cdot)$.

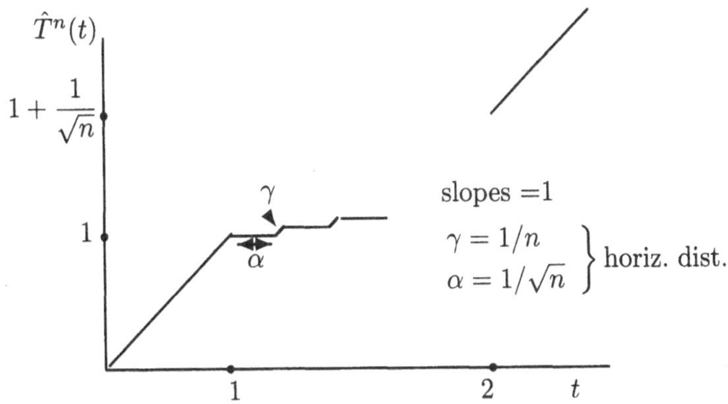

Figure 6.2. The inverse (time transformation) function $\hat{T}^n(\cdot)$.

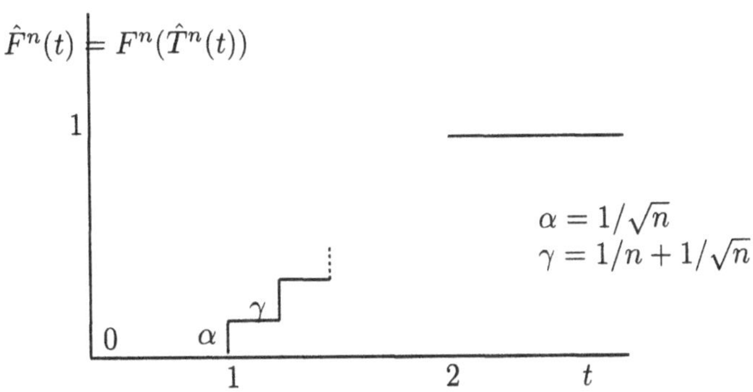

Figure 6.3. The function $\hat{F}^n(\cdot)$.

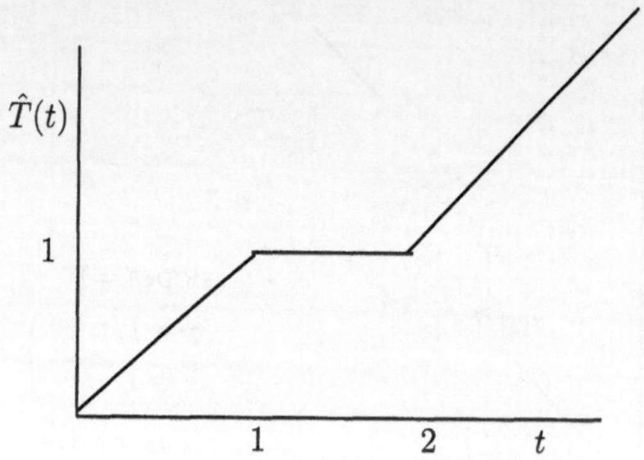

Figure 6.4. The limit $\hat{T}(\cdot)$.

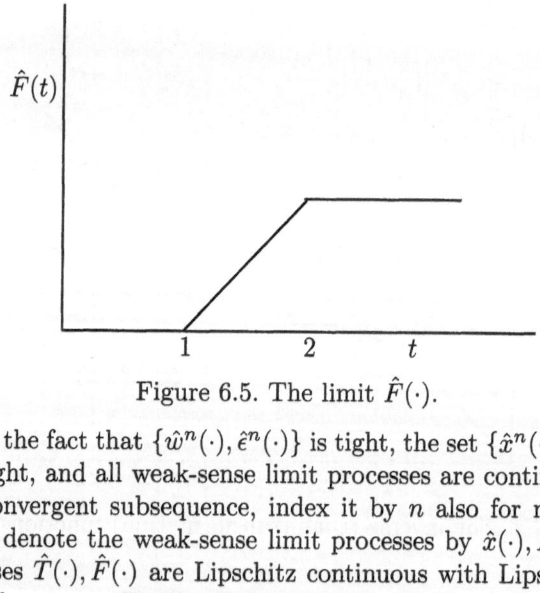

Figure 6.5. The limit $\hat{F}(\cdot)$.

Owing to the fact that $\{\hat{w}^n(\cdot), \hat{\epsilon}^n(\cdot)\}$ is tight, the set $\{\hat{x}^n(\cdot), \hat{F}^n(\cdot), \hat{T}^n(\cdot), w^n(\cdot)\}$ is tight, and all weak-sense limit processes are continuous. Extract a weakly convergent subsequence, index it by n also for notational simplicity, and denote the weak-sense limit processes by $\hat{x}(\cdot), \hat{F}(\cdot), \hat{T}(\cdot), w(\cdot)$. The processes $\hat{T}(\cdot), \hat{F}(\cdot)$ are Lipschitz continuous with Lipschitz constant no greater than unity. By the weak convergence and the continuity of $\hat{x}(\cdot)$ and $w(\cdot)$, the weak-sense limit processes satisfy

$$\hat{x}(t) = x + \int_0^t b(\hat{x}(s))d\hat{T}(s) + w(\hat{T}(t)) + \hat{F}(t). \tag{6.10}$$

By the definition of $\hat{T}^n(\cdot)$, we can write

$$W_\beta^n(x, F^n) = E\int_0^\infty e^{-\beta\hat{T}^n(t)} \left[k(\hat{x}^n(t))d\hat{T}^n(t) + cd\hat{F}^n(t)\right]. \tag{6.11}$$

By the weak convergence and the continuity of the limit process,

$$\begin{aligned} \lim_n E\int_0^\infty & e^{-\beta\hat{T}^n(t)} \left[k(\hat{x}^n(t))d\hat{T}^n(t) + cd\hat{F}^n(t)\right] \\ &= E\int_0^\infty e^{-\beta\hat{T}(t)} \left[k(\hat{x}(t))d\hat{T}(t) + cd\hat{F}(t)\right]. \end{aligned} \tag{6.12}$$

Define the inverse $T(t) = \inf\{s : \hat{T}(s) > t\}$. (See [167, page 312] or [170, Theorem 5.3] for a similar transformation and application.) We need to show that the inverse is defined for all $t \geq 0$. This will be true if $\hat{T}(t) \to \infty$ with probability one as $t \to \infty$, and this is implied by (6.2). (A sample path of the process $\hat{T}^n(t)$ can remain bounded as $t \to \infty$ only if (for that sample path) $T^n(t)$ goes to infinity at some finite t.) Thus $T(t)$ is defined and finite for each $t < \infty$.

Define the rescaled processes by the inverse scaling $x(t) = \hat{x}(T(t))$, $F(t) = \hat{F}(T(t))$, etc. By (6.10), these rescaled processes satisfy (6.6). Similarly, the right side of (6.12) equals

$$E \int_0^\infty e^{-\beta t} \left[k(x(t))dt + cdF(t)\right],$$

which yields (6.4).

Remark: Extensions. This method will be applied in Chapter 10 and elsewhere when the tightness of the control or reflection terms cannot be otherwise ascertained, or when they might not be tight. The example implies that a particular functional (the integral in (6.3)) converges weakly to that functional of a "natural" limit process. The functional depended on the sections of the processes that were tight in the Skorohod topology, but asymptotically negligibly on the behavior of the processes near the discontinuity point. Any other functional with such properties could have been used as well.

2.7 Measure-Valued Processes

Let S denote a complete and separable metric space with metric $\rho(\cdot)$. Then (Theorem 5.1) $\mathcal{P}(S)$ is a complete and separable metric space, where we recall that $\mathcal{P}(S)$ is the space of probability measures on $(S, \mathcal{B}(S))$ with the Prohorov metric. Let Q_n and Q be $\mathcal{P}(S)$-valued *random variables*. Their probability distributions are, of course, elements of $\mathcal{P}(\mathcal{P}(S))$. Suppose that $\{Q_n\}$ is tight (in the sense of a sequence of random variables) and let $Q_n \Rightarrow Q$. Then Q is a measure-valued random variable. Suppose that it is defined on a probability space $(\Omega', \mathcal{F}', P')$ with Ω' having generic variable ω'. Write the sample values either as $Q^{\omega'}$ or as $Q^{\omega'}(\cdot)$. Suppose now that the Skorohod representation of Theorem 5.5 is used, so we can assume that $\{Q_n(\cdot),\ n < \infty,\ Q(\cdot)\}$ are defined on the same sample space. For notational simplicity, we also denote this space by $(\Omega', \mathcal{F}', P')$. By the assumed weak convergence and the Skorohod representation, $Q_n^{\omega'}(\cdot) \to Q^{\omega'}(\cdot)$ for almost all ω', where the convergence is in the topology of $\mathcal{P}(S)$. That is,

$$\pi(Q_n^{\omega'}, Q^{\omega'}) \to 0 \tag{7.1}$$

for almost all ω', where $\pi(\cdot)$ denotes the Prohorov metric. For each ω', the value $Q_n^{\omega'}(\cdot)$ is a measure on $(S, \mathcal{B}(S))$. Let ω denote the canonical point of S. Then, for each $\omega' \in \Omega'$, the probability measure $Q_n^{\omega'}(\cdot)$ induces an S-valued random variable, which we denote by $X_n^{\omega'}$ and whose sample values are denoted by $X_n^{\omega'}(\omega)$. Note that the ω' and n index the *measure* and, hence, the *random variable*. With the measure $Q_n^{\omega'}(\cdot)$ given, ω gives the *sample value* of the *random variable associated with that measure.*

Now, by convergence of the probability measures in (7.1) for almost all ω', we have (for almost all ω')

$$X_n^{\omega'} \Rightarrow X^{\omega'}. \tag{7.2}$$

We can use the Skorohod representation for the set of *random variables* $(\{X_n^{\omega'}, n < \infty\}, X^{\omega'})$ for each *fixed* ω' not in some null set N_0. Thus, with an appropriate choice of the probability space, we can assert that for almost all ω',

$$\rho(X_n^{\omega'}, X^{\omega'}) \to 0, \text{ with } Q^{\omega'} \text{ probability one.} \tag{7.3}$$

The following theorem will be quite useful in proving tightness of $\{Q^n(\cdot)\}$.

Theorem 7.1. [160, Theorem 1.6.1.] *Let S be a complete and separable metric space, and Q_n a $\mathcal{P}(S)$-valued random variable. Then tightness of $\{Q_n\}$ is implied by the tightness of $\{EQ_n\}$.*

The following examples illustrate some ways that the theorem will be used. Let $\mathcal{M}(S)$ denote the set of (not necessarily probability) measures on the Borel subsets of the complete and separable metric space S.

Example 1. Let $\mathcal{U}$ be a compact set in some Euclidean space and define $S = \mathcal{U} \times [0, T]$, $T < \infty$, with the Euclidean metric used on S. Then, since S is compact, any set $\{Q_\lambda\}$ of $\mathcal{M}(S)$-valued random variables (with the Prohorov metric used) has a weakly convergent subsequence if

$$\lim_{k\to\infty} \sup_\lambda P\{Q_\lambda(S) \geq k\} = 0.$$

Example 2. Define $S = \mathcal{U} \times [0, \infty)$, where $\mathcal{U}$ is as in Example 1 but where we use the "compactified" metric on $\mathcal{M}(S)$ defined by $v_n \to v$ if and only if $\int_S f(s)v_n(ds) \to \int_S f(s)v(ds)$ for each $f(\cdot) \in C_0(S)$, the space of continuous real-valued functions with compact support. Then $\mathcal{M}(S)$ is a complete and separable metric space. Any set $\{Q_\lambda\}$ of $\mathcal{M}(S)$-valued random variables has a weakly convergent subsequence if for each $T < \infty$,

$$\lim_{k\to\infty} \sup_\lambda P\{Q_\lambda(\mathcal{U} \times [0, T]) \geq k\} = 0.$$

Example 3. Let $S = D(\mathbb{R}^k; 0, \infty)$. Let $\{Q_\lambda\}$ be a set of random variables with values in $\mathcal{P}(S)$. If for each $\delta > 0$ there is a compact set $S_\delta \in S$ with

$$\sup_\lambda EQ_\lambda(S - S_\delta) \leq \delta,$$

then Q_λ will have a weakly convergent subsequence in $\mathcal{P}(S)$. To see how tightness might be proved in this example, first note that $\overline{Q}_\lambda \equiv EQ_\lambda$ is a probability measure on $(S, \mathcal{B}(S))$. Let $\overline{x}_\lambda(\cdot)$ denote the process induced on S by $\overline{Q}_\lambda$. Then tightness of $\{Q_\lambda\}$ is implied by the tightness of $\{EQ_\lambda\}$, which, in turn, is equivalent to tightness of $\{\overline{x}_\lambda(\cdot)\}$.

2.8 Tightness and Convergence to Wiener Processes

Because of the importance of the Wiener process as a weak-sense limit of processes in heavy traffic analysis, a selection of results concerning weak convergence to the Wiener process will be useful. First, some general results concerning weak convergence to martingales will be stated. The basic criterion is in Theorem 8.2. Various special cases will be described in Subsection 8.2.

2.8.1 Criteria for Weak Convergence to Martingales

Theorem 8.1 is the weak convergence analogue of Theorems 1.1 and 1.3. A sequence $\{X_n\}$ is said to be *uniformly integrable* if

$$\lim_{K\to\infty} \sup_n E\,|X_n|\, I_{\{|X_n|\geq K\}} = 0.$$

Suppose that the nonnegative X_n converge weakly to X. Then $EX_n \to EX$ if and only if $\{X_n\}$ is uniformly integrable [15, Theorem 5.4].

Theorem 8.1. *Let $w^n(\cdot)$ and $V^n(\cdot)$ be random processes with paths in $D(\mathbb{R}^k; 0, \infty)$ and $D(S; 0, \infty)$, respectively, where S is a complete and separable metric space. Let $\mathcal{F}_t^n$ denote the minimal σ-algebra that measures $\{w^n(s), V^n(s), s \leq t\}$, and let E_t^n be the associated conditional expectation. Suppose that for each $t, \tau \geq 0$,*

$$E_t^n\left[w^n(t+\tau) - w^n(t)\right] \to 0$$

in the mean. Let $\{w^n(t); t \leq T, n\}$ be uniformly integrable for each $T > 0$, and suppose that $(w^n(\cdot), V^n(\cdot)) \Rightarrow (w(\cdot), V(\cdot))$, which engenders the filtration $\{\mathcal{F}_t, t \geq 0\}$. Then $w(\cdot)$ is an $\mathcal{F}_t$-martingale.

Proof. Although the result is well known, the proof will be given, since similar results will frequently be used. As noted in Section 5, since the paths are right continuous, there is a set $\mathcal{T}_d$, at most countable, such that $(w(\cdot), V(\cdot))$ is continuous with probability one at each $t \notin \mathcal{T}_d$. Let $h(\cdot), s_i, p, t, \tau$, be arbitrary, but satisfy the conditions in Theorem 1.1, with $s_i, t, t+\tau \notin \mathcal{T}_d$. Then,

$$\lim_n Eh(w^n(s_i), V^n(s_i), i \leq p)\,[w^n(t+\tau) - w^n(t)] = 0.$$

This together with the weak convergence and uniform integrability yield

$$Eh(w(s_i), V(s_i), i \leq p)\,[w(t+\tau) - w(t)] = 0.$$

Now, as in Theorem 1.1, this expression and the arbitrariness of $h(\cdot), p, s_i, t$, and τ imply that

$$E\left[w(t+\tau) - w(t) \big| w(s), V(s), s \leq t, s \notin \mathcal{T}_d\right] = 0$$

with probability one. The right continuity implies that the result holds for all $t, \tau, s \leq t$. ∎

In the next theorem, S_q denotes the space of q-dimensional square matrices that are nonnegative definite and symmetric.

Theorem 8.2. *Suppose that $(V^n(\cdot), M^n(\cdot), \Sigma^n(\cdot))$ converges weakly to $(V(\cdot), M(\cdot), \Sigma(\cdot))$, where $M^n(\cdot)$ has values in $D(I\!R^q; 0, \infty)$, $\Sigma^n(\cdot) = \{a^n_{ij}(\cdot); i, j = 1, \ldots, q\}$ has values in $D(S_q; 0, \infty)$, and $V^n(\cdot)$ has values in $D(S; 0, \infty)$ for some complete and separable metric space S. Suppose that $\{\Sigma^n(\cdot)\}$ is bounded on each bounded interval. Let $M(\cdot)$ and $\Sigma(\cdot)$ be continuous. Let $\mathcal{F}^n_t$ (respectively, $\mathcal{F}_t$) denote the minimal σ-algebra that measures $\{V^n(s), M^n(s), \Sigma^n(s), s \leq t\}$ (respectively, $\{V(s), M(s), \Sigma(s), s \leq t\}$). Let*

$$E^n_t \left[f(M^n(t+\tau)) - f(M^n(t)) - \frac{1}{2}\int_t^{t+\tau} \sum_{i,j} f_{m_i m_j}(M^n(s)) a^n_{ij}(s)ds \right] \to 0 \tag{8.1}$$

in the mean for each real-valued $f(\cdot)$ with compact support that is continuous together with its partial derivatives up to second order and each $t, \tau \geq 0$. Then $M(\cdot)$ is an $\mathcal{F}_t$-martingale with Doob–Meyer process $\int_0^t \Sigma(s)ds$. If $\Sigma(\cdot)$ is not random, then $M^n(\cdot)$ is an $\mathcal{F}_t$-Wiener process with $M(t)$ having covariance $\int_0^t \Sigma(s)ds$.

Now drop the weak convergence assumptions and the convergence of (8.1). But suppose that $M^n(\cdot)$ is an $\mathcal{F}^n_t$-martingale with Doob–Meyer process $\int_0^t \Sigma^n(s)ds$ for bounded $\{\Sigma^n(\cdot)\}$, and that there are real $\delta_n \to 0$ such that the maximum discontinuity of $M^n(\cdot)$ is less than δ_n. Then $\{M^n(\cdot)\}$ is tight with continuous weak-sense (martingale) limits and (8.1) holds for $\{M^n(\cdot)\}$.

Comment on the proof. Let $\mathcal{T}_d$ denote the (at most countable) set of time points at which $V(\cdot)$ is not continuous with probability one. The purpose of (8.1) is to verify that (see the notation of Theorem 1.4)

$$Eh(M^n(s_i), V^n(s_i), i \le p)$$
$$\times \left[f(M^n(t+\tau)) - f(M^n(t)) - \frac{1}{2}\int_t^{t+\tau} \sum_{i,j} f_{m_i m_j}(M^n(s)) a_{ij}(s) ds \right] \to 0.$$

Here, $s_i, t, t+\tau \notin \mathcal{T}_d$. This expression implies that

$$Eh(M(s_i), V(s_i), i \le p)$$
$$\times \left[f(M(t+\tau)) - f(M(t)) - \frac{1}{2}\int_t^{t+\tau} \sum_{i,j} f_{m_i m_j}(M(s)) a_{ij}(s) ds \right] = 0,$$

which, in turn, implies the first part of the theorem.

Now assume the situation in the last paragraph of the theorem. Then Theorem 5.6 implies the tightness of $\{M^n(\cdot)\}$ as well as the continuity of any weak-sense limit process, since for any $T > 0$ and stopping time $\tau^n \le T$,

$$E\left[M_i^n(\tau^n + \delta) - M_i^n(\tau^n)\right]^2 = E\left[\langle M^n\rangle(\tau^n + \delta) - \langle M^n\rangle(\tau^n)\right].$$

The limit (8.1) can be shown to hold by a second-order Taylor series expansion and the fact that $\int_0^t \Sigma^n(s) ds$ is the Doob–Meyer process for $M^n(\cdot)$. ■

Remark on using an approximating process. Suppose that there are processes $\tilde{M}^n(\cdot)$ such that for each $T > 0$,

$$\lim_n \sup_{s \le T} \left| M^n(s) - \tilde{M}^n(s) \right| = 0$$

in the sense of probability. Then it is sufficient to verify the theorems with $\tilde{M}^n(\cdot)$ replacing $M^n(\cdot)$.

Tightness of a sequence of martingales based on jump processes. The following theorem is a jump process version of part of Theorem 8.2.

Theorem 8.3. *Suppose that we are given a bounded sequence $\{v_n\}$ of real numbers and a sequence of real-valued jump processes $p^n(\cdot)$ with jumps of magnitude v_n, and uniformly bounded jump rates $\lambda^n(\cdot)$. The different processes might be defined on different probability spaces. Define the martingales $M^n(t) = p^n(t) - \int_0^t \lambda^n(s) ds$. Then $\{M^n(\cdot)\}$ is tight.*

More generally, for each n let $p^n(\cdot)$ be a vector-valued jump process as in Theorem 3.1 with jump rate measure $\overline{\lambda}^n(\cdot)$ on the range space Γ_n, both of *which are bounded for each n. Then the Doob–Meyer process of the martingale* $M^n(t) = p^n(t) - \int_0^t \int_{\Gamma_n} \gamma \bar{\lambda}^n(s, d\gamma) ds$ *is*

$$\langle M^n \rangle(t) = \int_0^t \int_{\Gamma_n} \gamma\gamma' \bar{\lambda}^n(s, d\gamma) ds.$$

If

$$\limsup_n \sup_t \int_{\Gamma_n} |\gamma|^2 \bar{\lambda}^n(t, d\gamma) < \infty,$$

then $\{M^n(\cdot)\}$ *is tight.*

Proof. The proof follows from the representation (4.8) and the tightness criterion of Theorem 5.6, since for any bounded stopping time τ^n

$$E\,|M^n(\tau^n + \delta) - M^n(\tau^n)|^2 \leq E \int_{\tau^n}^{\tau^n+\delta} \int_{\Gamma_n} |\gamma|^2\, \bar{\lambda}^n(s, d\gamma) ds.$$

■

2.8.2 Tightness and Wiener Process Limits

Theorem 8.4. [15, Theorem 10.1, Donsker's theorem.] *Let* $\{\xi_i\}$ *be mutually independent and identically distributed vector-valued random variables with mean zero and covariance matrix* $\Sigma > 0$. *Then the* $w^n(\cdot)$ *defined by (5.4) converges weakly to a Wiener process with covariance matrix* Σ.

Theorem 8.5. [15, Problem 1, page 77.] *Let* $\{\xi_i^n, i \leq k_n\}$ *be mutually independent and real-valued for each n, where* $k_n \to \infty$ *as* $n \to \infty$. *Let* $E\xi_i^n = 0$ *and* $E[\xi_i^n]^2 = \sigma_{n,i}^2$. *Define* $s_{n,i}^2 = \sum_{l=1}^{i} \sigma_{n,l}^2$ *and* $s_n^2 = s_{n,k_n}^2$. *Define*

$$w^n(t) = \frac{1}{s_n} \sum_{l=1}^{i} \xi_l^n, \quad \text{for } t \in \left[\frac{s_{n,i-1}^2}{s_n^2}, \frac{s_{n,i}^2}{s_n^2}\right). \tag{8.2}$$

Assume the Lindeberg condition

$$\lim_{n\to\infty} \frac{1}{s_n^2} \sum_{l=1}^{k_n} \int_{\{|\xi_l^n| \geq \epsilon s_n\}} |\xi_l^n|^2 dP = 0, \text{ for each } \epsilon > 0. \tag{8.3}$$

Then $w^n(\cdot)$ *converges weakly in* $D(I\!R; 0, 1)$ *to a standard real-valued Wiener process.*

Remark. Let $\lim_n s_n^2 = \infty$, with k_n/s_n^2 bounded in n. Then (8.3) is implied by the uniform integrability of $\{|\xi_i^n|^2; n, i\}$. If $\sigma_{n,i}^2 \to \sigma^2 < \infty$, uniformly in

i as $n \to \infty$, then (8.3) and the theorem implies that the process defined by

$$w^n(t) = \frac{1}{\sqrt{n}} \sum_{l=1}^{nt} \xi_l^n \tag{8.4}$$

converges weakly in $D(I\!R; 0, \infty)$ to a Wiener process with variance σ^2.

The following is one vector version of Theorem 8.5, which follows from Theorems 1.4 and the tightness criterion of Theorem 5.6.

Theorem 8.6. *Let the ξ_i^n in (8.4) take values in $I\!R^k$ and be mutually independent for each n, with mean zero and covariance matrix $\Sigma_{n,i}$. Suppose that there is a matrix Σ such that $\Sigma_{n,i} \to \Sigma$, as $n \to \infty$, uniformly in i. Let $\{|\xi_l^n|^2; l, n\}$ be uniformly integrable. Then the process defined by (8.4) converges weakly in $D(I\!R^k; 0, \infty)$ to a Wiener process with covariance matrix Σ.*

Martingale processes. The results for $w^n(\cdot)$ defined in terms of mutually independent sequences can be extended to martingale processes in various ways. We simply cite one result from [71]. Other results can be found in [118] or as special cases of the results in [157, 158], such as Theorem 8.9 below. The conditions (8.5) and (8.6) ensure that the discontinuities disappear in the limit.

Theorem 8.7. [71, Theorem 1.4, page 339.] *Let $w^n(\cdot)$ be a local martingale with respect to a filtration $\{\mathcal{F}_t^n, t \geq 0\}$, and with paths in $D(I\!R^k; 0, \infty)$. Let $a^n(\cdot)$ be a $k \times k$ matrix-valued process whose components are adapted to $\mathcal{F}_t^n$ and have paths in $D(I\!R; 0, \infty)$. Suppose that $a^n(t+s) - a^n(t)$ is symmetric and nonnegative definite for each $s, t \geq 0$ and that $w^n(0) = 0$, $a^n(0) = 0$. For each $t > 0$, suppose that*

$$\lim_n E \sup_{s \leq t} |a^n(s) - a^n(s-)| = 0, \tag{8.5}$$

$$\lim_n E \sup_{s \leq t} |w^n(s) - w^n(s-)| = 0. \tag{8.6}$$

Let

$$w^n(t)[w^n(t)]' - a^n(t)$$

be a local $\mathcal{F}_t^n$ martingale. Suppose that there is a measurable (nonrandom) symmetric nonnegative definite matrix-valued function $\Sigma(\cdot)$ such that

$$a^n(t) \to \int_0^t \Sigma(s) ds \text{ in probability, for each } t, \tag{8.7}$$

as $n \to \infty$. Then $w^n(\cdot)$ converges weakly to a Wiener process $w(\cdot)$ with the covariance of $w(t)$ being $\int_0^t \Sigma(s) ds$.

An important application of Theorem 8.2 is to processes defined by forms such as (8.4), as in the next theorem.

Theorem 8.8. *Let $w^n(\cdot)$ be defined by (8.4), where the ξ_i^n are $I\!R^k$-valued. Let $\mathcal{F}_t^n$ denote the minimal σ-algebra that measures $\{\xi_j^n, j/n \leq t; V^n(s), s \leq t\}$, where $V^n(\cdot)$ takes values in $D(S; 0, \infty)$ for some complete and separable metric space S. Let E_t^n denote the associated conditional expectation. Assume that there is a matrix $\Sigma = \{a_{ij}; i, j\}$ such that the ξ_i^n satisfy*

$$E_{i/n}^n \xi_{i+1}^n = 0, \tag{8.8}$$

$$\lim_{n,l,i\to\infty} E_{i/n}^n \xi_{i+l}^n [\xi_{i+l}^n]' \to \Sigma, \tag{8.9}$$

where the limit is in the mean. Suppose that

$$\{|\xi_i^n|^2; n, i\} \text{ is uniformly integrable.} \tag{8.10}$$

Then $w^n(\cdot)$ converges weakly to a Wiener process with covariance matrix Σ. Suppose that $(V^n(\cdot), w^n(\cdot))$ converges weakly to $(V(\cdot), w(\cdot))$. Then $w(\cdot)$ is an $\mathcal{F}_t$-Wiener process with covariance matrix Σ, where $\{\mathcal{F}_t, t \geq 0\}$ is the filtration engendered by $(V(\cdot), w(\cdot))$.

Proof. The theorem can be obtained from Theorem 8.7, but we will use Theorem 8.2. Conditions (8.8), (8.9) and the criterion of Theorem 5.6 imply that $\{w^n(\cdot)\}$ is tight. Given tightness, to prove the "asymptotic" continuity, we need only show that the maximum of the jumps on any finite interval goes to zero in probability as $n \to \infty$. This is implied by (8.10) and Chebyshev's inequality as follows:

$$P\left\{\sup_{i\leq nt} \frac{|\xi_i^n|}{\sqrt{n}} \geq \epsilon\right\} \leq \sum_{i=1}^{nt} P\left\{|\xi_i^n| \geq \epsilon\sqrt{n}\right\} \leq \sum_{i=1}^{nt} \frac{1}{n\epsilon^2} \int_{\{|\xi_i^n| \geq \epsilon\sqrt{n}\}} |\xi_i^n|^2 \, dP, \tag{8.11}$$

which goes to zero as $n \to \infty$ due to (8.10). Thus, any weak-sense limit process has continuous paths with probability one. Then Theorem 8.1 can be used to verify that any weak-sense limit is a continuous martingale.

We will next verify the Wiener property. Let $f(\cdot)$ be continuous together with its partial derivatives up to second order and with compact support. Write

$$f(w^n(t+\tau)) - f(w^n(t)) = \sum_{i=nt}^{n(t+\tau)-1} \left[f(w^n(i/n + 1/n)) - f(w^n(i/n))\right]. \tag{8.12}$$

Expanding and taking conditional expectations yields

$$
\begin{aligned}
E_t^n f(w^n(t+\tau)) - f(w^n(t)) &= \frac{1}{\sqrt{n}} E_t^n \sum_{i=nt}^{n(t+\tau)-1} f_w'(w^n(i/n))\xi_{i+1}^n \\
&+ \frac{1}{2n} E_t^n \sum_{i=nt}^{n(t+\tau)-1} [\xi_{i+1}^n]' f_{ww}(w^n(i/n))\xi_{i+1}^n + E_t^n e^n(t,\tau),
\end{aligned}
\tag{8.13}
$$

where the error term is bounded by

$$
\begin{aligned}
&|e^n(t,\tau)| \leq \\
&\frac{1}{2n} \sum_{i=nt}^{n(t+\tau)-1} \left|\xi_{i+1}^n\right|^2 \left| f_{ww}(w^n(i/n) + c_{ni}\xi_{i+1}^n/\sqrt{n}) - f_{ww}(w^n(i/n))\right|,
\end{aligned}
\tag{8.14}
$$

where $c_{ni} \in [0,1]$. Condition (8.8) implies that the first term on the right of (8.13) is zero. The error term goes to zero in the mean as $n \to \infty$ by the asymptotic continuity of the $w^n(\cdot)$, and the uniform integrability (8.10). The continuity and boundedness of $f_{ww}(\cdot)$, the asymptotic continuity of $w^n(\cdot)$, and the uniform integrability imply that the arguments i/n in the second term on the right of (8.13) can be replaced by $i/n - l_n/n$, where $l_n \to \infty$, $l_n/n \to 0$ as $n \to \infty$, without changing the limits. Doing this and using (8.9) yields that the second term on the right of (8.13) behaves asymptotically as

$$
\frac{1}{2n} \sum_{i=nt}^{n(t+\tau)-1} \sum_{i,j} f_{w_i w_j}(w^n(i/n)) a_{ij},
$$

which proves the theorem. ■.

Correlated noise. The random variables in sums such as (5.4) and (8.4) are not always mutually independent or martingale differences. Under appropriate "mixing" conditions, one can still get weak convergence to a Wiener process. Many results are in [15, Chapter 4], [71, Chapter 7.3]. They are stated basically for processes defined as in (5.4) where the ξ_i are stationary. The perturbed test function method of the type developed in [157, 158, 160, 177] is a powerful set of techniques that can handle quite complicated dependencies and nonstationarities. It is perhaps the most effective method for treating "state-dependent" and correlated sequences. The full method as used in [157, 158, 160, 177] is a development of ideas in [17, 147, 216]. See the references for proofs and treatments of state-dependent and other more general problems. One useful result from [157, 158] is given in the next theorem. It is a rewriting of the "direct averaging-perturbed test function" method of Theorem 5.9 of [157]. One example of its use is in Section 5.5.

Theorem 8.9. *Let E_i^n denote the expectation conditioned on $\{\xi_j^n, j \le i\}$. Define $w^n(\cdot)$ by (8.4). Suppose that for each $T < \infty$, the following sets are uniformly integrable:*

$$\left\{|\xi_i^n|^2 ; n, i\right\}, \tag{8.15}$$

$$\left\{\left|\sum_{j=i+1}^{nT} E_i^n \xi_j^n\right|^2 ; i < nT, n > 0\right\}. \tag{8.16}$$

Then $\{w^n(\cdot)\}$ is tight, and all weak-sense limit processes have continuous paths with probability one. Suppose that there are matrices Σ_0 and Σ_1 and real numbers $N_n \to \infty$ as $n \to \infty$ such that $N_n/n \to 0$ and

$$\frac{1}{N_n} \sum_{l=i+1}^{i+N_n} E_i^n \xi_l^n [\xi_l^n]' \to \Sigma_0 \tag{8.17}$$

in probability as $n, i \to \infty$. Also, let (8.18)–(8.19) hold as $j - a \to \infty$, $m - j \to \infty$, and $a, n, j \to \infty$:

$$\sum_{k=j+1}^{m} E \xi_k^n [\xi_j^n]' \to \Sigma_1, \tag{8.18}$$

$$E\left|\sum_{k=j+1}^{m} E_a^n \xi_k^n [\xi_j^n]' - \Sigma_1\right| \to 0. \tag{8.19}$$

Then $w^n(\cdot)$ converges weakly to a Wiener process with covariance matrix $\Sigma = \Sigma_0 + \Sigma_1 + \Sigma_1'$.

Remarks. The proof in [157] uses the combined perturbed test function–direct averaging methods of that reference. See also [158]. Conditions (8.15) and (8.16) are used to prove the tightness and the continuity of the weak-sense limit processes.

Conditions (8.15) and (8.16) arise from the use of the test function perturbation method in the reference for proving tightness, where the perturbation is $f_w(w^n(i/n))' \sum_{j=i+1}^{nT} E_i^n \xi_j^n / \sqrt{n}$. The condition is quite weak owing to the use of the conditional expectations E_i^n. For example, if the ξ_i^n are martingale differences, then the sum in (8.16) is zero. To motivate (8.17) and (8.18), replace the conditional expectations by the expectation and suppose that the ξ_i^n are second-order stationary with covariance function $R(\cdot)$. Then $\Sigma_0 = R(0)$, and the sum in (8.18) converges to $\sum_{j=1}^{\infty} R(i) = \Sigma_1$.

Alternative conditions. Theorem 5.8 of [157] provides a sufficient, but quite general, condition. One usually works with sufficient conditions for that sufficient condition, as done in Theorem 5.9 of that reference and in Theorem 8.9 above. Depending on how one arranges the terms in the sums that are involved in the proof, one can get useful alternative sufficient conditions. For example, the following consequence of [157, Theorem 5.8] or [158, Theorem 5] is an alternative to (8.18)–(8.19), where N_n is as above (8.17):

$$\frac{1}{N_n} \sum_{j=i+1}^{i+N_n} E_a^n \left[\sum_{k=j+1}^{j+m} \xi_k^n \right] \left[\xi_j^n\right]' \to \Sigma_1, \tag{8.20}$$

in the mean, as a, n, i, m, and $i - a$ all go to infinity.

3
Stochastic Differential Equations: Controlled and Uncontrolled

Chapter Outline

This chapter introduces the basic models for the controlled and uncontrolled (limit) processes that are to be used. These are all some form of stochastic differential equation (SDE). Section 1 deals with the classical diffusion process, without reflections. A brief summary of the Itô construction of a solution is given, and that method will be adapted to the problem with reflections in Sections 4 and 5. Controlled diffusions are introduced in Subsection 1.2. Part of the book, as well as most of the work on heavy traffic analysis published to date, concerns limits of uncontrolled processes, or where the control is fixed in some simple way. Then the material on controls in this chapter is not needed. It is more efficient to deal with the controlled and uncontrolled problems together. But if the reader is concerned only with the uncontrolled problem, all references to controls in the definitions, conditions, and theorems can be safely ignored.

The subject of stochastic control is vast, and controls can occur in many forms. We deal with a few canonical forms that cover the applications of interest in the book. Subsection 1.2 concerns what might be called "classical" controls in that they appear in the SDEs as integrals of some bounded or integrable function. The so-called impulsive and singular control forms are also of great importance, and some remarks are given in Section 3. The Girsanov transformation, discussed in Subsection 1.4, is a widely used method for varying the control so that optimality theorems can be proved.

It is also required for the ergodic results for the control problem in Chapter 4.

For most applications of the "classical" form that have appeared to date in heavy traffic analysis, the control terms appear in a linear fashion in the dynamics and cost function, and take values in a convex set. The linearity is largely traditional, since nonlinear problems could rarely be solved analytically. But nonlinear forms allow more versatility and will certainly be of great interest in the future, and there are good numerical methods for many such problems. In the nonlinear control case, or where the dynamics contain nonlinear terms, there are serious questions concerning the existence of optimal controls and their approximations by controls of simple forms. Some such results will be needed in the sequel to establish the convergence of the optimally controlled physical systems under heavy traffic to optimally controlled limit processes, and to justify using the limit system to get good controls for the physical system. Because of this, in Section 2 we discuss the standard extension of the classical control, called a "relaxed" control, which is a very convenient and effective way of dealing with these difficulties. Several examples illustrate the need for and intuition behind this extension. The reader might wish to skip this material on first reading.

Reflected diffusions are covered in Sections 4 and 5. Section 4 deals with the so-called reflected Brownian motion, a currently widely used model for the heavy traffic analysis of queueing systems. The name "reflected Brownian motion" is misleading, since the approach can handle more general reflected diffusions as well. Subsection 4.1 introduces reflecting Brownian motion via some simple queueing-type examples, and shows how the reflection terms originate. The examples are for motivation only. In Subsection 4.2, a needed Lipschitz condition is proved, and existence and uniqueness of the solutions to the reflected diffusion process are discussed. Section 5 introduces an alternative model of a reflecting diffusion, the so-called Skorohod problem. It is more general than that of Section 4 in that it includes those models, and there are processes of interest that can be modeled by the method of Section 5, but not by that of Section 4. Where the two can be used they yield the same process. Subsection 5.1 gives the basic definitions and discusses an important Lipschitz continuity property of the solutions and reflection term as functions of the "driving data." Subsection 5.2 discusses the reflected SDEs. Section 6 develops the "contradiction method" for proving tightness or "asymptotic continuity" of the sequence of reflection terms, that will be used frequently in the sequel.

Section 7 gives the extensions of many of the previous results to reflecting jump–diffusions. Section 8 provides some results on approximations of nearly optimal controls by controls of a special simple form. These are not practical controls. They are of technical interest only, and will play an important role in the proofs that the limit of a sequence of optimal processes converges to an optimal limit as the traffic intensity goes to unity.

3.1 Stochastic Differential Equations and Diffusion Processes

3.1.1 Definitions: Uncontrolled SDEs

Let the probability space be $(\Omega, \mathcal{F}, P)$, with a filtration $\{\mathcal{F}_t, t \geq 0\}$ defined on it. Let $x(0) \in I\!R^r$ be given. For simplicity in the notation, we suppose that $x(0)$ is not random, unless otherwise mentioned. But all the results hold for random and $\mathcal{F}_0$-measurable $x(0)$. Let $b(\cdot) = \{b_i(\cdot), 1 \leq r\}$ and $\sigma(\cdot) = \{\sigma_{ij}(\cdot); i \leq r, j \leq k\}$ be measurable functions on $I\!R^r$ whose components are real-valued. Let $w(\cdot)$ be a standard $I\!R^k$-valued $\mathcal{F}_t$-Wiener process. We will not usually specify the dimension of $w(\cdot)$. It is always assumed that it is compatible with the dimensions of $\sigma(\cdot)$.

We now discuss the solution to the uncontrolled stochastic differential equation

$$x(t) = x(0) + \int_0^t b(x(s))ds + \int_0^t \sigma(x(s))dw(s), \tag{1.1}$$

which will often be written in symbolic differential form as

$$dx(t) = b(x(t))dt + \sigma(x(t))dw(t).$$

We call $b(\cdot)$ the *drift or drift vector*, and $a(\cdot) = \sigma(\cdot)\sigma'(\cdot)$ the *diffusion matrix*. By a solution to (1.1) we mean a continuous $\mathcal{F}_t$-adapted process $x(\cdot)$ that satisfies (1.1) with probability one. The process defined by (1.1) is one of the classical ones in stochastic analysis, owing to the large set of applications that it models or approximates and to the large literature concerning its analytical and numerical properties. It is the canonical model of finite-dimensional Markov processes in continuous time and with continuous paths.

The fundamental methods used in the construction and analysis of solutions to (1.1) are due to K. Itô. The basic theory is covered in many books; for example, [59, 82, 83, 113, 122, 204]. There are various senses in which the solution to (1.1) is said to exist or to be unique.

Strong-sense existence [122]. Let $\{\mathcal{F}_t, t \geq 0\}$ denote the filtration defined by the standard Wiener process $w(\cdot)$. A solution $x(\cdot)$ to (1.1) is said to exist in the *strong sense* if it is adapted to $\mathcal{F}_t$ and satisfies (1.1) with probability one for any such probability space and Wiener process.

Strong-sense and pathwise uniqueness. Suppose that for any given probability space $(\Omega, \mathcal{F}, P)$, and a filtration $\{\mathcal{F}_t, t \geq 0\}$ with an $\mathcal{F}_t$-standard Wiener process $w(\cdot)$ defined on it, there are two solutions $x^i(\cdot)$, $i = 1, 2$, to (1.1) with initial conditions $x^i(0) = x(0)$. We say that *strong-sense uniqueness* holds if the $x^i(\cdot)$ are equal with probability one for all t. Suppose that there are two filtrations $\{\mathcal{F}_t^i, t \geq 0\}$, $i = 1, 2$, $w(\cdot)$ is a standard $\mathcal{F}_t^i$-

Wiener process for each i, and the initial conditions satisfy $x^1(0) = x^2(0)$. If the corresponding solutions to (1.1) are equal, we say that there is *pathwise uniqueness.*

Weak-sense existence. A solution to (1.1) is said to exist in the *weak-sense* if for any probability measure μ on $I\!R^r$ there exists a probability space $(\Omega, \mathcal{F}, P)$, with a filtration $\{\mathcal{F}_t, t \geq 0\}$, an $\mathcal{F}_t$-standard Wiener process $w(\cdot)$, and an $\mathcal{F}_t$-adapted process $x(\cdot)$ satisfying (1.1) for all $t \geq 0$ defined on it, and μ is the distribution of $x(0)$.

Weak-sense uniqueness. Suppose that we have two probability spaces $(\Omega^i, \mathcal{F}^i, P^i)$, filtrations $\{\mathcal{F}_t^i, t \geq 0\}$, on them, and processes $w^i(\cdot), x^i(\cdot)$, where $w^i(\cdot)$ is a standard $\mathcal{F}_t^i$-Wiener process, $x^1(0) = x^2(0)$, and $(x^i(\cdot), w^i(\cdot))$ solve (1.1). We say that there is weak-sense uniqueness if the distributions induced by the $x^i(\cdot)$ on $D(I\!R^r; 0, \infty)$ are the same.

The weak-sense definition will be the most important for our purposes. In this book the solutions to (1.1) and related equations will usually be obtained via the theory of weak convergence, as weak-sense limits. This theory will give the existence of a measure over the path space that induces the processes $x(\cdot), w(\cdot)$ satisfying (1.1). Since our main interest is in approximations of cost functions, optimal cost functions, and in the expectation of functions or probabilities of events of particular importance or in numerical methods for computing them, it is the distributions only that matter to us, and not the exact probability space.

Weak-sense uniqueness is often called *uniqueness in the sense of probability law.* Two relationships between these concepts that are useful are the following: pathwise uniqueness implies weak-sense uniqueness, and weak-sense existence together with pathwise uniqueness implies strong-sense existence [122, Chapter 5.3].

Itô's lemma for (1.1). Let $x(\cdot)$ be any solution to (1.1), and suppose that $b(x(\cdot)) \in \mathcal{H}_1$ and $\sigma(x(\cdot)) \in \mathcal{H}_2$. For any $f(\cdot) \in C^2(I\!R^r)$ define the *differential generator* $\mathcal{L}$ of $x(\cdot)$ by

$$\mathcal{L}f(x) = f_x'(x)b(x) + \frac{1}{2}\sum_{i,j} f_{x_i x_j}(x)a_{ij}(x), \quad a(x) = \sigma(x)\sigma'(x). \tag{1.2}$$

Then Itô's lemma (2.2.14) states that

$$f(x(t)) = f(x(0)) + \int_0^t \mathcal{L}f(x(s))ds + \int_0^t f_x'(x(s))\sigma(x(s))dw(s). \tag{1.3}$$

3.1.2 Controlled Diffusions

In this and in the next section we will use the following assumption.

A1.1. *$\mathcal{U}$ is a compact set in some Euclidean space.*

Let $u(\cdot)$ be a $\mathcal{U}$-valued random process. Recall that (Subsection 2.2.1) the control $u(\cdot)$ is said to be *nonanticipative* with respect to an $\mathcal{F}_t$-standard Wiener process $w(\cdot)$ if $u(\cdot)$ is $\mathcal{F}_t$-adapted. Then $u(\cdot)$ is said to be an *admissible control* with respect to $w(\cdot)$, or (alternatively) that the pair $(u(\cdot), w(\cdot))$ is admissible. In this and in the next section we consider controlled diffusions of the form

$$dx(t) = b(x(t), u(t))dt + \sigma(x(t))dw(t), \tag{1.4}$$

where $w(\cdot)$ is a standard Wiener process.

In many of the applications in this book $b(x,u)$ will take the special form $b(x,u) = b(x) + Du$, for some matrix D, and similarly for the controlled reflected diffusion model. This is the way that the control occurs in the bulk of applications in queueing and communications to date. In Section 3, we will discuss the singular and impulsive control forms where (1.4) is replaced by (3.2), and that are very important in applications as well. The following definitions are analogous to those for the uncontrolled case.

Strong-sense existence. Let $u(\cdot)$ be a $\mathcal{U}$-valued random process. Let $\{\mathcal{F}_t, t \geq 0\}$ denote the filtration defined by $\{u(\cdot), w(\cdot)\}$ and suppose that $w(\cdot)$ is a standard $\mathcal{F}_t$-Wiener process. (Hence $u(\cdot)$ is admissible with respect to $w(\cdot)$.) A solution $x(\cdot)$ to (1.4) is said to exist in the *strong sense* if it is adapted to $\mathcal{F}_t$ and satisfies (1.4) with probability one, no matter what the probability space.

Strong-sense uniqueness. Let $u(\cdot)$ be a $\mathcal{U}$-valued random process. Let $\{\mathcal{F}_t, t \geq 0\}$ denote the filtration generated by $\{u(\cdot), w(\cdot)\}$ and suppose that $w(\cdot)$ is an $\mathcal{F}_t$-Wiener process. Let $x^i(\cdot), i = 1, 2$, solve (1.4) for the given Wiener and control processes, where $x^1(0) = x^2(0)$. Then we say that there is uniqueness in the strong sense if $x^1(\cdot) = x^2(\cdot)$ with probability one, no matter what the probability space.

Weak-sense existence. Suppose that on some probability space, we have an initial condition $\tilde{x}(0)$ and random processes $\tilde{u}(\cdot), \tilde{w}(\cdot)$, where $\tilde{u}(t)$ is $\mathcal{U}$-valued. Let $\tilde{w}(\cdot)$ be a standard Wiener process with respect to the filtration generated by $\{\tilde{u}(\cdot), \tilde{w}(\cdot)\}$. We say that there is weak-sense existence of a solution to (1.4) if there is a probability space on which are defined $(x(\cdot), u(\cdot), w(\cdot))$ solving (1.4), where $\tilde{x}(0) = x(0)$, $(u(\cdot), w(\cdot))$ has the distribution of $(\tilde{u}(\cdot), \tilde{w}(\cdot))$, and $w(\cdot)$ is a standard Wiener process with respect to the filtration generated by $\{x(\cdot), u(\cdot), w(\cdot)\}$.

Weak-sense uniqueness. Suppose that we are given probability spaces $(\Omega^i, \mathcal{F}^i, P^i), i = 1, 2$, with filtrations $\{\mathcal{F}^i_t, t \geq 0\}$, standard $\mathcal{F}^i_t$-Wiener processes $w^i(\cdot)$, and processes $x^i(\cdot)$, where $(x^i(\cdot), u^i(\cdot), w^i(\cdot))$ solves (1.4) and $x^1(0) = x^2(0)$. We say that weak-sense uniqueness holds if for $i = 1, 2$, the equality of the distributions of $(u^i(\cdot), w^i(\cdot))$ implies the equality of the distributions of $(x^i(\cdot), u^i(\cdot), w^i(\cdot))$.

Remark. We do not always define the probability space, but when writing equations such as (1.4), it is assumed implicitly that there is a filtration $\{\mathcal{F}_t, t \geq 0\}$ such that $w(\cdot)$ is a standard real or vector-valued $\mathcal{F}_t$-Wiener process, and $(u(\cdot), x(\cdot))$ is $\mathcal{F}_t$-adapted, with $u(t)$ being $\mathcal{U}$-valued. Then we will simply say, with an abuse of terminology but obvious meaning, that weak-sense uniqueness implies that the distribution of the "driving processes" $(u(\cdot), w(\cdot))$ determines the distribution of $(x(\cdot), u(\cdot), w(\cdot))$.

3.1.3 Construction of a Strong Solution

The classical construction of a strong solution uses the following Lipschitz condition.

A1.2. *$b(\cdot)$ is continuous. There exists $C \in (0, \infty)$ such that*

$$|b(x,u) - b(y,u)| \leq C|x-y|, \quad ||\sigma(x) - \sigma(y)|| \leq C|x-y|$$

for all $x, y \in \mathbb{R}^r$, $u \in \mathcal{U}$, where $||\sigma||^2 = \sum_{i=1}^r \sum_{j=1}^k \sigma_{ij}^2$.

Theorem 1.1. [59, 82, 83, 113, 122, 204]. *Assume* (A1.1) *and* (A1.2), *let $\{\mathcal{F}_t, t \geq 0\}$ be the filtration, and let $(u(\cdot), w(\cdot))$ be admissible, where $u(\cdot)$ is $\mathcal{U}$-valued and $w(\cdot)$ is a standard $\mathcal{F}_t$-Wiener process. Then for every initial condition $x(0)$,* (1.4) *has a strong-sense solution that is unique in the strong sense.*

Remarks on the proof. Itô's classical method of constructing a strong solution to (1.4) with $x(0)$ given and under (A1.1) and (A1.2) uses a form of Picard iteration. One generates a sequence of processes $\{x^n(\cdot)\}$ by $x^0(t) = x(0)$ and, for $n \geq 0$,

$$x^{n+1}(t) = x(0) + \int_0^t b(x^n(s), u(s))ds + \int_0^t \sigma(x^n(s))dw(s). \tag{1.5}$$

Then estimates for the differences $\max_{s \leq t} |x^{n+1}(s) - x^n(s)|$ are obtained using the Lipschitz condition (A1.2), the Schwarz inequality, and the martingale estimates (2.1.2) and (2.1.3) (to deal with the martingales $\int_0^t [\sigma(x^{n+1}(s)) - \sigma(x^n(s))]dw(s)$). We will give a few of the well known details, since they will be useful in the discussion of reflected diffusions in Section 4. Full details are in any of the references.

Applying the Lipschitz condition (A1.2), Schwarz's inequality, and (2.1.3), and working recursively from $n-1$ to n, it can readily be shown that each

$x^n(\cdot)$ satisfies $\int_0^t E\left|x^n(s)\right|^2 ds < \infty$ for each t. Then we see (Section 2.2) that the stochastic integral in (1.5) is continuous with probability one, and thus $x^n(\cdot)$ is also. Furthermore, by the Lipschitz condition, Schwarz's inequality and (2.1.3), for each bounded time interval there is a constant K, not depending on n, such that

$$E\sup_{s\leq t}\left|\int_0^s \left[b(x^n(v),u(v)) - b(x^{n-1}(v),u(v))\right] dv\right|^2 \\ \leq Kt\int_0^t E\sup_{v\leq s}\left|x^n(v) - x^{n-1}(v)\right|^2 ds,$$

$$E\sup_{s\leq t}\left|\int_0^s \left[\sigma(x^n(v)) - \sigma(x^{n-1}(v))\right] dw(v)\right|^2 \\ \leq K\int_0^t E\sup_{v\leq s}\left|x^n(v) - x^{n-1}(v)\right|^2 ds.$$

Define $\nu_n(t) = E\sup_{s\leq t}|x^{n+1}(s) - x^n(s)|^2$. Then the above inequalities imply that $\nu_{n+1}(t) \leq 4K(t+1)\int_0^t \nu_n(s)ds$. Now iterating backward to $n=0$ and evaluating the resulting integral yields, for any $T<\infty$,

$$E\sup_{0\leq t\leq T}\left|x^{n+1}(t) - x^n(t)\right|^2 \leq K_0(T)\frac{(4K(T+1))^n}{n!},$$

where

$$K_0(T) = E\sup_{0\leq t\leq T}\left[\left|\int_0^t b(x(0),u(s))ds\right| + \left|\int_0^t \sigma(x(0))dw(s)\right|\right]^2$$

is finite. By Chebyshev's inequality, for each $T>0$,

$$P\left\{\sup_{0\leq t\leq T}|x^{n+1}(t) - x^n(t)| \geq 2^{-n}\right\} \leq K_0(T)\frac{(4K(T+1))^n 4^n}{n!}. \tag{1.6}$$

The summability of the right-hand side and the Borel–Cantelli lemma imply that with probability one there are at most a finite number of occurrences of the event

$$\sup_{0\leq t\leq T}|x^{n+1}(t) - x^n(t)| \geq 2^{-n}.$$

It follows that the sequence

$$x(t) = x^0(t) + \sum_{n=0}^{m}\left[x^{n+1}(t) - x^n(t)\right] \tag{1.7}$$

converges to a nonanticipative process $x(\cdot)$ that is a solution of (1.4) and that satisfies $E\sup_{s\leq t}|x^n(s) - x(s)|^2 \to 0$ and $E\sup_{s\leq t}|x(s)|^2 < \infty$, for each $t<\infty$. Uniqueness in the strong sense also holds under the Lipschitz condition. To show this, one lets $x^i(\cdot)$, $i=1,2$, be two solutions, and then uses the Lipschitz condition, Schwarz's inequality, and (2.1.3) to show that $x^1(t) = x^2(t)$ for all t with probability one. Full details are in the references.

3.1.4 The Girsanov Transformation

Let $(\Omega, \mathcal{F}, P)$ denote the probability space, with a filtration $\{\mathcal{F}_t, t \geq 0\}$ such that $\mathcal{F} = \cup_t \mathcal{F}_t$. Let $w(\cdot)$ be a k-dimensional standard $\mathcal{F}_t$-Wiener process. The so-called Girsanov transformation method for obtaining weak-sense existence and uniqueness is widely used for control problems when the Lipschitz continuity property (A1.2) fails. Given a solution (either in the weak or strong sense) to (1.1) or (1.4), the Girsanov transformation method defines another solution, with a drift term that we are able to choose, by simply transforming the probability measure P on the original probability space to a new measure, that we denote by $\tilde{P}$. The new drift term contains the desired control. The key is the construction of the Radon–Nikodym derivative of the new measure with respect to P. This method of constructing a solution is of great importance in control theory, particularly where we might wish to use controls of the feedback form $u(x)$, but that are not Lipschitz continuous or where $b(\cdot)$ is not Lipschitz continuous. See [122] and the references therein.

Let $v(\cdot)$ be a k-dimensional process in $\mathcal{H}_b$ (hence, it is adapted to $\mathcal{F}_t$). Define

$$R(t) = \exp\left(\int_0^t v'(s)dw(s) - \frac{1}{2}\int_0^t |v(s)|^2 ds\right). \tag{1.8}$$

By Itô's lemma we have

$$R(t) = 1 + \int_0^t R(s)v'(s)dw(s).$$

Since $v(\cdot)$ is bounded, $ER(t) < \infty$. In fact [113, 122, 204], for $s > 0, t \geq 0$,

$$ER(t) = 1, \quad E\left[R(t+s)\middle|v(u), w(u), u \leq t\right] = R(t), \text{ w.p.1.}$$

Thus the process $R(\cdot)$ is an $\mathcal{F}_t$-martingale.

Now fix $T \in (0, \infty)$, and define a probability measure $\tilde{P}_T$ on $(\Omega, \mathcal{F}_T)$ by the Radon–Nikodym derivative $d\tilde{P}_T/dP = R(T)$. More explicitly,

$$\tilde{P}_T(A) = E\left[I_A R(T)\right], \text{ for } A \in \mathcal{F}_T, \tag{1.9}$$

where I_A denotes the indicator function of the event A. The equality $ER(T) = 1$ guarantees that $\tilde{P}_T$ is indeed a probability measure. There is a unique extension of the $\tilde{P}_T$, $T < \infty$, to a measure $\tilde{P}$ on $(\Omega, \mathcal{F})$. The most important consequence is the following theorem.

Theorem 1.2. (The Girsanov transformation [122, 184, 204].) *Assume that $R(\cdot)$ defined by equation* (1.8) *is a martingale. Then*

$$\tilde{w}(t) = w(t) - \int_0^t v(s)ds \tag{1.10}$$

is a standard $\mathcal{F}_t$-Wiener process on the probability space $(\Omega, \mathcal{F}, \tilde{P})$.

Application of the Girsanov transformation: Construction of a solution of an SDE. In control theory this remarkable result is often used in the following way. Let $w_1(\cdot)$ and $w_2(\cdot)$ denote the first m and last $k-m$ components of $w(\cdot)$, respectively. Let $\sigma_1(\cdot)$ be an $m \times m$ matrix-valued process with the property that $\sigma_1^{-1}(x)$ is uniformly bounded in x. Suppose that the stochastic differential equation

$$\begin{aligned} dx_1(t) &= b_1(x(t))dt + \sigma_1(x(t))dw_1, \\ dx_2(t) &= b_2(x(t))dt + \sigma_2(x(t))dw_2, \end{aligned} \tag{1.11}$$

has a unique weak-sense solution with the given initial condition. Let $\tilde{b}_1(\cdot)$ be a bounded, measurable, and $I\!R^m$-valued function. Define

$$z_1(\cdot) = \sigma_1^{-1}(x(\cdot))\tilde{b}_1(x(\cdot)).$$

Define $R(\cdot)$ and $\tilde{P}_T, \tilde{P}$ as in (1.8) and (1.9) above, but using $(z_1(\cdot), w_1(\cdot))$ in lieu of $(v(\cdot), w(\cdot))$, and define

$$\tilde{w}_1(t) = w_1(t) - \int_0^t z_1(s)ds.$$

Then, under the new measure $\tilde{P}$, $x(\cdot)$ satisfies the equation

$$\begin{aligned} dx_1(t) &= b_1(x(t))dt + \tilde{b}_1(x(t))dt + \sigma_1(x(t))d\tilde{w}_1(t), \\ dx_2(t) &= b_2(x(t))dt + \sigma_2(x(t))dw_2(t), \end{aligned} \tag{1.12}$$

and $(\tilde{w}_1(\cdot), w_2(\cdot))$ is a standard $\mathcal{F}_t$-Wiener process. We thereby obtain weak-sense existence for such a class of equations.

The process $\tilde{b}_1(x(\cdot))$ can be replaced by the more general form $b_c(x(t), u(t))$, where $b_c(\cdot)$ is bounded and measurable, $u(t)$ is $\mathcal{U}$-valued, and $(u(\cdot), w(\cdot))$ is an admissible pair. The solution to (1.12) is weak-sense unique (under the new measure $\tilde{P}$) [113, 122, 152], since the original solution to (1.11) is weak-sense unique. More generally, the $v(\cdot)$ in (1.10) need not be bounded, provided that $R(\cdot)$ is a martingale, and this is guaranteed by the *Novikov condition* [122, page 198]

$$E \exp\left[\int_0^t |v(s)|^2 ds/2\right] < \infty, \quad \text{for each } t < \infty. \tag{1.13}$$

3.2 Relaxed Controls

For many processes that arise at this time as weak-sense limits under heavy traffic, there is either no explicit control, or else the control occurs in an affine form in the stochastic differential equation and in the cost function. For example, $b(x,u) = b(x) + Du$, and analogously for the cost rate. Then the material of this section is not needed. It is also not needed for the singular and impulsive control problems.

This book is concerned with limit theorems for systems under conditions of heavy traffic. If the original physical process is uncontrolled, then we wish to use the weak-sense limit process to get estimates of quantities of interest for the original physical systems that they approximate. When the original physical system is controlled, we have the additional need to use the controlled weak-sense limit processes to get good or nearly optimal controls for the physical systems. Part of the procedure is the demonstration that the sequence of optimal physical processes converges weakly to an optimal "limit" process, and that good controls for this limit process will also be good for the physical process, under heavy traffic. There might be a nonlinear penalization of the control effort. For example, the control might measure the (scaled) rate at which customers are not admitted to the system, with the cost increasing nonlinearly as the rate increases. Additionally, whenever the system data (say, arrival or service statistics) depend nonlinearly and nonadditively on the control, the limit system will have drift rates $b(x,u)$ that are not affine in u. The problem is complicated further if $\mathcal{U}$ is not convex. In these cases there are serious questions concerning the existence of an optimal control and methods for its approximation. These questions require an extension of the classical notion of control as simply a $\mathcal{U}$-valued adapted process. The extension is known as the "relaxed control." All of the needed results are well known in the control theory literature, and we will only discuss some of the key points.

We will discuss the control of the drift term only. There are extensions of the ideas to variance control, but that is somewhat more complicated [159, 164].

3.2.1 Existence of a Deterministic Optimal Control

This subsection surveys some standard ideas in deterministic control with an ODE model, and is for motivation and intuition only. It introduces some of the issues and ideas that are also fundamental to the stochastic problem. It concerns one of the fundamental theoretical questions in control, that of the existence of an optimal control. Of course, in applications one rarely needs an optimal control, even if available and conveniently implementable. Nevertheless, the theoretical, as opposed to the practical, questions of approximation and characterization of limits do involve the notion of optimality. Even if the optimal control itself is not of direct interest, the question of existence is important in applications, since it is necessary for the dynamic programming and variational equations to be meaningful. These equations often serve as the basis of analysis of qualitative properties and of numerical methods. In order to discuss optimality, one needs to fix the class of admissible controls. Even for very innocent-looking (although perhaps artificial) problems, there might not be an optimal control in the basic class of $\mathcal{U}$-valued admissible controls. Then the class must be enlarged. This enlargement must not alter the infimum of the costs. Also, any control in the enlarged class must be able to be approximated by a control of the original class. We emphasize that the questions are only for the theoretical development. The conditions used are selected for convenience in the discussion. In applications, it is hard to find problems where the optimal control is not of the classical form.

The ODE of concern for this motivational discussion is

$$\dot{x}(t) = b(x(t), u(t)), \quad x \in I\!R^r, \ u(\cdot) \text{ admissible}, \tag{2.1}$$

where $b(\cdot)$ is bounded and measurable. By admissible we mean that $u(\cdot)$ is measurable and $\mathcal{U}$-valued. A uniqueness condition on the solution will be added below. Let $\beta > 0$ be a discount factor. The cost function of interest is

$$W_\beta(x(0), u) = \int_0^\infty e^{-\beta s} k(x(s), u(s)) ds. \tag{2.2}$$

The function $k(\cdot)$ is continuous and has at most polynomial growth in x. Define $V_\beta(x) = \inf_{\text{admissible } u} W_\beta(x, u)$.

Remarks on optimal controls and limits of sequences of controls. The following remarks and examples are intended to illustrate the definitions and structures that need to be introduced to properly handle the question of existence of an optimal control. The following example shows that there need not exist an optimal control in the sense that there are admissible $\bar{u}(\cdot)$ such that $V_\beta(x) = W_\beta(x, \bar{u})$.

Example 1. Let $\mathcal{U} = [-1, 1]$, and $\dot{x}(t) = u(t)$, $x(0) = 0$, and $W_\beta(x, u) = \int_0^\infty e^{-\beta t} \left[x^2(t) + (u^2(t) - 1)^2\right] dt$. The infimum of the costs is obviously zero. But there is no optimal control in the usual sense. A minimizing sequence alternates increasingly rapidly between the values ± 1.

The next example will suggest what one must do to establish a useful general existence result.

Example 2. For integer n, let $\dot{x}^n = b(x^n, u^n)$, $x^n(0) = x(0)$, $\mathcal{U} = \{\alpha_1, \alpha_2\}$, $b(\cdot)$ is bounded and continuous and $u^n(\cdot) = \alpha_1$ on $[2i/n, 2i/n + 1/n)$, $i = 0, 1, \ldots$, and equals α_2 on the complementary intervals. Then $\{x^n(\cdot)\}$ is equicontinuous, and the limit $x(\cdot)$ of any convergent subsequence satisfies

$$\dot{x} = \left[b(x, \alpha_1) + b(x, \alpha_2)\right]/2. \tag{2.3}$$

Thus the limit problem has taken us out of the class of solutions of (2.1) corresponding to ordinary controls. But the limit can be arbitrarily well approximated by the solution to (2.1) with a simply piecewise constant ordinary control.

The problem of existence of an optimal control is one of "closure." One needs to enlarge the sense of solution and control so that each sequence of (solution, control) in this enlarged class has a convergent subsequence, and so that the limit solution satisfies the ODE in some appropriate extended sense and with the control being the limit of the control sequence. Loosely speaking, we wish to embed the problem in a larger class that is compact in an appropriate sense and in which the classical (solutions, controls) are dense. In extending the definition of solution, we need to be certain that we have not altered the infimum of the costs, and that any problem (solution, control) in the extension can be well approximated by a classical one. The limits of the paths constructed in the above examples should be included in this extended class of solutions. These considerations concerning "closure" lead to the concept of *relaxed* control [76, 244, 257].

The representation of an ordinary control as a measure. Let $u(\cdot)$ be an admissible ordinary control and recall that $\mathcal{B}(\mathcal{U})$ and $\mathcal{B}(\mathcal{U} \times [0, \infty))$ are the σ−algebras over the Borel sets in $\mathcal{U}$ and $\mathcal{U} \times [0, \infty)$, respectively. Define the Dirac measure $r_t(\cdot)$ on $\mathcal{B}(\mathcal{U})$ and the measure $r(\cdot)$ on $\mathcal{B}(\mathcal{U} \times [0, \infty))$ by

$$r_t(A) = I_A(u(t)), \quad r(A \times [0, t]) = \int_0^t r_s(A)ds, \quad A \in \mathcal{B}(\mathcal{U}),$$

where $I_A(u)$ is the indicator function of the set A. The $r_t(\cdot)$ is a Dirac measure, since it is concentrated on a single point, which is the control value at t. The function $r(A \times [0, t])$ is just the amount of time on $[0, t]$ that the control $u(\cdot)$ takes values in the set A. We can now write (2.1) in the equivalent form

$$\dot{x}(t) = \int_{\mathcal{U}} b(x(t), \alpha) r_t(d\alpha), \tag{2.4a}$$

or, alternatively, as (2.4b) with cost (2.5), where $r(d\alpha\, ds) = r_s(d\alpha)ds$:

$$x(t) = x(0) + \int_0^t \int_{\mathcal{U}} b(x(s), \alpha) r(d\alpha\, ds) = x(0) + \int_0^t \int_{\mathcal{U}} b(x(s), \alpha) r_s(d\alpha) ds \tag{2.4b}$$

$$W_\beta(x(0), r) = \int_0^\infty e^{-\beta s} \int_{\mathcal{U}} k(x(s), \alpha) r(d\alpha\, ds). \tag{2.5}$$

Relaxed deterministic controls. The constructions to follow have been standard in deterministic control for a long time (e.g., [257]). An *admissible relaxed control*, or simply a *relaxed control*, $r(\cdot)$ is a measure on $\mathcal{B}(\mathcal{U} \times [0, \infty))$ such that $r(\mathcal{U} \times [0, t]) = t$ for all $t < \infty$. Given a relaxed control $r(\cdot)$, there is a derivative $r_t(\cdot)$ such that $r(d\alpha\, dt) = r_t(d\alpha)dt$; it follows that (considered as a function of t) $r_t(A)$ is measurable for each $A \in \mathcal{B}(\mathcal{U})$, and $r_t(\cdot)$ is a (probability) measure on $\mathcal{B}(\mathcal{U})$ for each t. In fact, we can define the derivative (for almost all t) by

$$r_t(A) = \frac{\lim_{\delta \to 0} r(A \times [t-\delta, t])}{\delta}. \tag{2.6}$$

We make the unrestrictive assumption that (2.4) has a unique solution for each relaxed control. Let $\mathcal{R}(\mathcal{U} \times [0, \infty))$ denote the set of all relaxed controls.

Example 2 revisited. Return to the example that led to (2.3). Let $r_t(\cdot)$ be the measure that takes the value 0.5 at α_i, $i = 1, 2$. Then the limit can be written as

$$\dot{x}(t) = \int_{\mathcal{U}} b(x(t), \alpha) r_t(d\alpha).$$

With the use of a relaxed control, the set of possible velocities and cost rates $(b(x, \mathcal{U}), k(x, \mathcal{U}))$ is replaced by its convex hull. (See (2.3) for an example of this "convexification.") Indeed, if this set is convex for each x and satisfies an upper semicontinuity property in x, then one need not introduce relaxed controls, and an implicit function theorem can be used [155, 192] to establish the existence of an optimal control. But there are few savings in doing so.

3.2.2 The Topology on the Space of Relaxed Controls

The weak topology is used on $\mathcal{R}(\mathcal{U} \times [0, \infty))$. That is, a sequence of relaxed controls $r^n(\cdot)$ converges to a relaxed control $r(\cdot)$ if and only if for any bounded and continuous function $\phi(\cdot)$ with compact support

$$\int_0^\infty \int_{\mathcal{U}} \phi(s, \alpha) r^n(d\alpha\, ds) \to \int_0^\infty \int_{\mathcal{U}} \phi(s, \alpha) r(d\alpha\, ds). \tag{2.7}$$

Thus, any sequence $\{r^n(\cdot), x^n(\cdot)\}$ generated in Example 2 has a convergent subsequence, and any limit $(x(\cdot), r(\cdot))$ satisfies (2.4), where $r_t(\alpha_i) = .5, i = 1, 2$. Since $\mathcal{U}$ is compact, the weak topology can be metrized so that $\mathcal{R}(\mathcal{U} \times [0, \infty))$ is a complete and separable metric space, and we always suppose that this is the case. Due to the compactness of $\mathcal{U}$, the space is compact. In this example $b(\cdot)$ is bounded. Hence the set of all solutions is equicontinuous. Thus, the space of (solution, control) for (2.4) is compact. The situation is the same for the vector case.

The cost values. We need to be sure that the extended definitions of control and solution have not altered the infimum of the cost values: That is, we need to ensure that

$$\inf_{\text{admissible } u} W_\beta(x(0), u) = \inf_{\text{admissible } r} W_\beta(x(0), r). \tag{2.8}$$

The following classical result [257] says that any admissible relaxed control can be well approximated by a "nice" ordinary admissible control and that (2.8) holds. Let $x(t, r)$ denote the solution to (2.4) under the relaxed control $r(\cdot)$.

Approximation theorem: The "Chattering" Theorem [257]. *Let $b(\cdot)$ be continuous, with at most linear growth in x, uniformly in u, and let $k(\cdot)$ be continuous with at most polynomial growth in x, uniformly in u. Suppose that (2.4a) has a unique solution for each initial condition and relaxed control. Then, for any $\epsilon > 0$, $T > 0$, there is a $\delta > 0$, a finite set $\mathcal{U}_\epsilon \subset \mathcal{U}$, and a $\mathcal{U}_\epsilon$-valued ordinary control $u^\epsilon(\cdot)$ that is constant on the intervals $[i\delta, i\delta + \delta)$ such that*

$$\sup_{t \le T} |x(t, r) - x(t, u^\epsilon)| \le \epsilon, \quad |W_\beta(x(0), r) - W_\beta(x(0), u^\epsilon)| \le \epsilon. \tag{2.9}$$

Hence (2.8) holds.

By the theorem, the class of relaxed controls is a natural extension of the class of ordinary controls. It can be arbitrarily well approximated by a classical control. It guarantees that the set of (solution, control) is compact, and does not change the infimum of the costs.

3.2.3 Stochastic Relaxed Controls

We now extend the ideas in the last subsection to the stochastic case. The basic SDE is

$$dx = b(x, u)dt + \sigma(x)dw. \tag{2.10}$$

Stochastic relaxed controls play a role analogous to that of deterministic relaxed controls. A stochastic relaxed control is a measure-valued process that is a deterministic relaxed control for each element ω of the underlying probability space, and that satisfies the nonanticipativeness condition that is also required of ordinary stochastic controls. Let $(\Omega, P, \mathcal{F})$ denote the probability space, $\{\mathcal{F}_t, t \ge 0\}$ a filtration, $w(\cdot)$ an $\mathcal{F}_t$-Wiener process, and $x(0)$ the initial condition.

We say that $r(\cdot)$ is an *admissible relaxed control* for $w(\cdot)$, or that the pair $(r(\cdot), w(\cdot))$ is *admissible,* if the sample value of $r(\cdot)$ is a deterministic relaxed control for almost all ω, and if $r(A \times [0,t])$ is $\mathcal{F}_t$-adapted (it is continuous in t) for all $A \in \mathcal{B}(\mathcal{U})$. Then there exists an $\mathcal{F}_t$-adapted derivative $r_t(A)$ for all $A \in \mathcal{B}(\mathcal{U})$, and it can be defined by (2.6). We continue to use the weak topology on $\mathcal{R}(\mathcal{U} \times [0, \infty))$. Thus, since $\mathcal{R}(\mathcal{U} \times [0, \infty))$ can be considered to be a compact metric space, *any* set of stochastic relaxed controls is tight.

Let $r(\cdot)$ be an admissible relaxed control. We will also use the following simplified notation. For $t \geq 0$, let $r(\cdot, t)$ denote the random measure with values $r(A, t) = r(A \times [0, t])$ for $A \in \mathcal{B}(\mathcal{U})$.

Now, rewrite (2.10) in terms of the relaxed control:

$$dx(t) = \int_{\mathcal{U}} b(x(t), \alpha) r_t(d\alpha) dt + \sigma(x(t)) dw(t),$$

or, equivalently,

$$x(t) = x(0) + \int_0^t \int_{\mathcal{U}} b(x(s), \alpha) r(d\alpha\, ds) + \int_0^t \sigma(x(s)) dw(s). \tag{2.11}$$

Unique strong-sense solutions. Under the conditions (A1.1) and (A1.2), the procedure of Subsection 1.4 can be repeated virtually word for word to obtain a strong-sense unique strong-sense solution to (2.11) for each admissible pair $(r(\cdot), w(\cdot))$.

Remark on the Girsanov transformation. The Girsanov transformation can be used to construct weak-sense unique solutions with relaxed controls in exactly the same way that it was used for ordinary stochastic controls in Subsection 1.4.

A cost function. We will use subsets of the following conditions. Condition (A2.3) or (A2.4) is used to ensure that there is an optimal control, and (A2.3) ensures that it can be approximated as in Theorem 2.2. They indicate some of the possibilities, and weaker conditions will occasionally be used. There is no space for a full development of the control problem.

A2.1. *$b(\cdot)$ and $k(\cdot)$ are continuous with $|b(x,u)| + |k(x,u)| \leq C(1 + |x|)$ for some $C < \infty$. The matrix $\sigma(\cdot)$ is bounded and continuous.*

A2.2. *There is a unique weak-sense solution to* (2.11) *for each admissible pair $(r(\cdot), w(\cdot))$ and nonanticipative initial condition.*

A2.3. *Either $b(\cdot)$ or $k(\cdot)$ is bounded.*

A2.4. *$k(x,u) \geq 0$, and there is some admissible control $\tilde{r}(\cdot)$ under which $W_\beta(x(0), \tilde{r}) < \infty$, where $W_\beta(x(0), \tilde{r})$ is defined in* (2.13).

Two types of cost functionals will be treated in the book, the discounted and the ergodic. The restriction to these two is done to avoid overcomplicating the analysis and because these are the cost functionals of greatest current interest in heavy traffic analysis. The methods can be readily adapted to treat optimal stopping, finite time and the other common cost functions that are used in stochastic control. We let the "cost rate" $k(\cdot)$

below depend on the current state value $x(t)$ (as well as on the current control value), but under appropriate continuity conditions, it could be a more general (nonanticipative) path functional.

In this chapter we use the discounted cost function, with discount factor $\beta > 0$. The ergodic cost function is discussed in Chapter 4. Let E_x^u denote expectation given that the control is $u(\cdot)$ and initial condition $x(0) = x$. For ordinary controls, the discounted cost function is

$$W_\beta(x(0), u) = E_{x(0)}^u \int_0^\infty e^{-\beta s} k(x(s), u(s)) ds. \tag{2.12}$$

In terms of the relaxed control, it is

$$W_\beta(x(0), r) = E_{x(0)}^r \int_0^\infty \int_{\mathcal{U}} e^{-\beta s} k(x(s), \alpha) r_s(d\alpha) ds. \tag{2.13}$$

Remark on (A2.3). We wish to give the essential ideas without overcomplication. In many applications, both $b(\cdot)$ and $k(\cdot)$ might be unbounded. But then we need to ensure that the tails

$$E_{x(0)}^r \int_T^\infty \int_{\mathcal{U}} e^{-\beta s} k(x(s), \alpha) r_s(d\alpha) ds$$

go to zero as $T \to \infty$ uniformly for all members of the class of controls with which we need to work. Any condition that guarantees (A2.5) would suffice and could replace (A2.3) in the following theorems.

A2.5. *There are positive C_0 and p such that for any $\epsilon > 0$ and admissible control $\tilde{r}(\cdot)$, there is an admissible control $r(\cdot)$ such that $W_\beta(x(0), r) \leq W_\beta(x(0), \tilde{r}) + \epsilon$ and $E_{x(0)}^r \int_{\mathcal{U}} |k(x(t), \alpha)| r_t(d\alpha) \leq C_0[1 + t^p]$.*

The existence of an optimal relaxed control and approximations of relaxed controls. We need to know that there is an optimal relaxed control and that

$$V_\beta(x(0)) = \inf_r W_\beta(x(0), r) = \inf_u W_\beta(x(0), u), \tag{2.14}$$

where the infimums are always over the admissible controls. Thus, we require that the infimum in the class of admissible ordinary controls be the same as that over admissible relaxed controls, thus emphasizing once again that the use of relaxed controls does not change the infimum of the possible costs. A remark on the inf notation in (2.14) is necessary. If (A1.1) and (A1.2) hold, so that the solution to (2.11) can be defined for any fixed $w(\cdot)$ and all admissible relaxed controls, then the infimum is taken over $r(\cdot)$ or $u(\cdot)$ as the notation in (2.14) indicates, since we can fix the Wiener process. However, if the solutions to (2.11) are defined indirectly or in a weak sense, say by a Girsanov transformation, then the Wiener process might not be fixed. For example, as we saw in Subsection 1.4 where the Girsanov transformation method was used, the Wiener process itself depended on the chosen control. In such cases, it is the distribution of the pair $(u(\cdot), w(\cdot))$ or $(r(\cdot), w(\cdot))$ that determines the cost, and properly speaking the infimizations in (2.14) should actually read $\inf_{r,w}$ and $\inf_{u,w}$, where the $(r(\cdot), (w(\cdot))$ and $(u(\cdot), w(\cdot))$ are admissible pairs. But to simplify the notation we write it simply as in (2.14).

Theorem 2.1. Existence of an optimal control. [167, Chapter 10.] *Assume* (A1.1) *and* (A2.1)–(A2.3). *Let* $\{x^n(\cdot), r^n(\cdot), w^n(\cdot)\}$ *be a sequence of solutions to* (2.11), *with* $|x^n(0)|$ *bounded. Then* $\{x^n(\cdot), r^n(\cdot), w^n(\cdot)\}$ *is tight. Let* $(x(\cdot), r(\cdot), w(\cdot))$ *denote the weak-sense limit of a weakly convergent subsequence, indexed by* n_i. *Define* $\mathcal{F}_t = \mathcal{F}(x(s), r(s), w(s), s \leq t)$. *Then* $w(\cdot)$ *is a standard* $\mathcal{F}_t$*-Wiener process,* $r(\cdot)$ *is admissible with respect to* $w(\cdot)$, *and* $x(\cdot)$ *satisfies* (2.11). *Furthermore,* $W_\beta(x^{n_i}(0), r^{n_i}) \to W_\beta(x(0), r)$.

If we replace (A2.3) *by* (A2.4), *then tightness still holds and* $\liminf_i W_\beta(x^{n_i}(0), r^{n_i}) \geq W_\beta(x(0), r)$. *Thus, under either* (A2.3) *or* (A2.4), *if we let* n *index a minimizing sequence, then* $r(\cdot)$ *is an optimal relaxed control.* (A2.5) *can replace* (A2.3).

The following theorem asserts that any stochastic relaxed control can be arbitrarily well approximated by an ordinary stochastic control.

Theorem 2.2. Approximation of an optimal control. [76], [160, Theorem 3.5.2.]. *Assume* (A1.1) *and* (A2.1)–(A2.3), *with initial condition* $x(0)$. *Given* $\epsilon > 0$ *and an admissible pair* $(w(\cdot), r(\cdot))$, *there is a finite set* $\{\alpha_1^\epsilon, \ldots, \alpha_{p_\epsilon}^\epsilon\} = \mathcal{U}^\epsilon \subset \mathcal{U}$ *and a* $\delta > 0$ *with the following properties. There is a probability space on which are defined processes* $(x^\epsilon(\cdot), u^\epsilon(\cdot), w^\epsilon(\cdot))$, *with associated filtration* $\{\mathcal{F}_t^\epsilon, t \geq 0\}$, *where* $w^\epsilon(\cdot)$ *is a standard* $\mathcal{F}_t^\epsilon$*-Wiener process,* $x^\epsilon(0) = x(0)$, *and* $u^\epsilon(\cdot)$ *is an* $\mathcal{F}_t^\epsilon$*-adapted and* $\mathcal{U}^\epsilon$*-valued ordinary stochastic control that is constant on the intervals* $[i\delta, i\delta + \delta)$. *Furthermore, the processes satisfy* (2.11) *and* $W_\beta(x(0), u^\epsilon) \leq W_\beta(x(0), r) + \epsilon$. (A2.5) *can replace* (A2.3).

3.3 Impulsive and Singular Control Problems

Impulsive control problems. By an impulsive control we mean a control that has the effect of a Dirac delta function: It forces an instantaneous change in the system [14, 41, 99, 121, 193]. Such controls often occur naturally as limits of systems operating in heavy traffic environments, as the traffic intensity goes to unity. See, for example, [167, Example 5, Section 8.1.5] and [38, 123, 174]. They would occur if the control action or external interruptions shut down part of the system for some period of time or if there were a sudden large addition or reduction of resources or of inputs. They are also used as approximations to problems where a large "force" can be applied over a relatively short interval (and the integral of which over that interval is bounded) in such a way that there is a very rapid change in some state of the system.

Let the sequence of random variables $\{\nu_i, \tau_i\}$ denote the values and the times of the impulses, and define

$$F(t) = \sum_{i:\tau_i \leq t} \nu_i. \tag{3.1}$$

It is always assumed that $\tau_n \leq \tau_{n+1} \to \infty$ with probability one but some τ_n might be infinite with a positive probability. It can happen that a jump occurs "just after" another, so that $\tau_{n+1} = \tau_n$ with a positive probability for some n. But it is always assumed that there is a well defined order. An example where the physics of the problem lead naturally to jumps occurring "simultaneously" is in [174], and interesting questions arise in the treatment. Also, see the problem of control during processor interruptions in Chapter 10.

Following the setup at the beginning of Subsection 1.2, let $(\Omega, \mathcal{F}, P)$ be the probability space with a filtration $\{\mathcal{F}_t, t \geq 0\}$. Suppose that $w(\cdot)$ is an $\mathcal{F}_t$-Wiener process, the initial condition is $x(0)$, the τ_i are $\mathcal{F}_t$-stopping times and ν_i is $\mathcal{F}_{\tau_i}$-measurable. Note that $F(t)$ is $\mathcal{F}_t$-measurable and the jump process $F(\cdot)$ is *nonanticipative* with respect to $w(\cdot)$. We then say that $F(\cdot)$ is an admissible control with respect to $w(\cdot)$, or, alternatively, that the pair $(F(\cdot), w(\cdot))$ is admissible. Write $F_i(\cdot)$, $i \leq r$, for the real-valued components of $F(\cdot)$.

For a matrix D, the model for the impulsively controlled SDE is

$$dx(t) = b(x(t))dt + DdF(t) + \sigma(x(t))dw(t). \tag{3.2}$$

For an appropriate function $g(\cdot)$, an analogue of the discounted cost function (2.12) is

$$W_\beta(x(0), F) = E_{x(0)}^F \left[\int_0^\infty e^{-\beta t} k(x(t))dt + \sum_i e^{-\beta \tau_i} g(x(\tau_i -), \nu_i) \right]. \tag{3.3}$$

We will also use subsets of the following conditions.

A3.1. *$\sigma(\cdot)$ is bounded and continuous, $b(\cdot)$ is continuous, and $|b(x)| \leq C[1 + |x|]$ for some $C < \infty$. The real-valued functions $g(\cdot)$ and $k(\cdot)$ are continuous and $|k(x)| + |g(x, v)| \leq C[1 + |x|]$.*

A3.2. *$\mathcal{U}(x) \subset \mathcal{U}$ are compact sets in some Euclidean space, and $\mathcal{U}(\cdot)$ is continuous in x.*

A3.3. *ν_i is $\mathcal{U}(x(\tau_i-))$-valued.*

A3.4. *The system $dx = b(x)dt + \sigma(x)dw$ has a unique weak-sense solution for each initial condition.*

A3.5. $\inf_{x,\alpha\in\mathcal{U}(x)} g(x,\alpha) > 0$.

A3.6. *$k(\cdot)$ is bounded.*

A3.7. *$k(x) \geq 0$ and there is some admissible control $F(\cdot)$ for which $W_\beta(x(0), F) < \infty$, where $W_\beta(x(0), F)$ is defined in* (3.3).

If the uncontrolled system has a unique strong or weak-sense solution for each initial condition, then so does (3.2) (since $F(\cdot)$ is a pure jump process, and there will be only a finite number of jumps on any finite time interval, with probability one). The following can replace (A3.6). See Chapter 10, where such assumptions are shown to hold under more verifiable conditions.

A3.8. *There are positive C_0 and p such that for any $\epsilon > 0$ and admissible control $\tilde{F}(\cdot)$, there is a bounded admissible control $F(\cdot)$ and a $T_\epsilon < \infty$ such that $F(\cdot)$ is constant after T_ϵ, $W_\beta(x(0), F) \leq W_\beta(x(0), \tilde{F}) + \epsilon$, and $E^F_{x(0)}|k(x(t))| \leq C_0[1 + t^p]$.*

The singular control problem. Consider the SDE (3.2), but where the components $F_i(\cdot)$ of $F(\cdot)$ are nondecreasing right-continuous functions, with $F(0) = 0$, and $F(t)$ is $\mathcal{F}_t$-measurable. Such a process is called an *admissible singular control* with respect to $w(\cdot)$; equivalently, the pair $(F(\cdot), w(\cdot))$ is admissible. For the impulsive control problem, the $F_i(\cdot)$ were simply step functions. Here, the $F_i(\cdot)$ might have continuous (but not necessarily absolutely continuous) sections as well, thus the name "singular" [167, 170, 179, 189, 231]. Such controls arise naturally in heavy traffic modeling, for example in routing and admission control. The discounted cost function normally takes the form

$$W_\beta(x(0), F) = E^F_{x(0)}\left[\int_0^\infty e^{-\beta t}k(x(t))dt + \int_0^\infty e^{-\beta t}\bar{c}' dF(t)\right], \qquad (3.4)$$

where $\bar{c}$ is a vector with positive components.

Existence and approximations. The existence of an optimal control for the impulsive control problem can be proved under (A3.1)–(A3.5) and either (A3.6) or (A3.7) via an adaptation of the proof of Theorem 2.1 in [167, Chapter 10]. A similar program is carried out in [189, Section 5] for the singular control problem. The analogue of Theorem 2.2 can be proved under (A3.1)–(A3.6), where the jumps $F(t) - F(t-)$ are confined to a given finite set of values. Condition (A3.8) can replace (A3.6). Under the Lipschitz condition (A1.2), there is strong-sense unique solution for each admissible control.

The conditions (A3.6)–(A3.7) or the positivity of the components of $\bar{c}$ are often restrictive. But special features of particular problems often allow proofs of existence and approximations. The conditions are used to avoid problems of rapid growth of the cost rate, or wild variations of the path (as might occur, for example, if $\bar{c} = 0$ and D allowed cancellation of the effects of the control actions). But these possibilities can often be eliminated in particular problems. For example, the components of $x(t)$ might be constrained to be nonnegative with the control actions able only to decrease their values, thus restricting the control actions. Additional detail will be provided where necessary. See Chapter 10. Both ordinary (or relaxed) controls and singular/impulsive controls might occur together.

3.4 Reflected Diffusions

3.4.1 Introduction and Examples

In applications, reflected diffusion processes in $I\!R^r$ are a common form of the weak-sense limit. Frequently, as seen in Chapter 1, the state in such problems is the (suitably scaled) queue size, and nonnegative. Then the state space is the nonnegative orthant $G = (I\!R^+)^r = \{x : x_i \geq 0, i \leq r\}$. On taking the heavy traffic limits, the weak-sense limit is a process that is a diffusion interior to G, but (owing to the queue dynamics) if it tries to leave G, it is reflected back onto G in a direction determined by the routing within the network. Recall the examples of Chapter 1.

Some of the standard results concerning diffusion processes that are "instantaneously" reflected back into a closed domain G when they "try to leave" G will be discussed in this and in the next section. To motivate the definitions of a reflecting diffusion that will be given below, refer to the examples of Chapter 1. The examples concern processes whose components are constrained in some way by the physics of the problem being modeled. The processes were rescaled, and as the scaling goes to its limit, a law of large numbers or a central limit theorem comes into play and the associated weak-sense limit process is a reflected or constrained Wiener process or SDE.

Two approaches to modeling a reflected diffusion process will be discussed. The method discussed in Subsection 4.2 has been called *reflected Brownian motion* or RBM, [91, 98, 101, 102, 213]. The second method, discussed in the next section, is based on the so-called *Skorohod problem*. Originated by Skorohod in the case of one dimension [224], it has been extended to the multidimensional case in, e.g., [44, 63, 64, 98, 183, 218, 221, 233]. The approaches differ mainly in the point of view to the reflection term. Where both methods can be used they yield the same process. The Skorohod problem approach is more general in that it can handle a wider variety of state spaces and reflection directions, but the alternative is useful for many queueing problems, where the reflection terms can frequently be obtained directly from the routing structure of the network.

There is a third approach, called the *submartingale problem* [227]. This is analogous to the martingale method for diffusion processes [228]. This approach is the most appropriate when there are "sticky" boundary conditions, and with certain classes of reflecting Wiener process models [237]. Numerical methods for the submartingale problem can be found in [154, 155]. These works do not cover controls, but can be extended to the controlled problem by obvious adaptations of the applicable algorithmic and proof methods in [167]. Before proceeding with the formal development, let us recall some of the motivating models and heuristic ideas from the beginning of Chapter 1.

Example 4.1: A one-dimensional reflecting random walk. Let $\{\xi_i\}$ be a sequence of independent and identically distributed real-valued random variables with $E\xi_i = 0$ and $E\xi_i^2 = \sigma^2 < \infty$. For any sequence of real numbers $\{a_i\}$, define the continuous-time interpolation $a^n(\cdot)$ by setting $a^n(t) = a_i$ on the interval $[i/n, (i+1)/n)$. In analogy to (2.5.4), for some $x(0)$ (independent of $\{\xi_i\}$) define the process

$$\psi^n(t) = x(0) + \frac{1}{\sqrt{n}} \sum_{l=1}^{nt} \xi_l. \tag{4.1}$$

Note that there are two scalings used in the definition of $\psi^n(\cdot)$. Time is "squeezed" by a factor of n, and the amplitudes of the ξ_k are divided by $\sqrt{n}$. As in Subsection 2.5.1, such scalings are used to force the model into the framework of a central limit theorem.

Now consider the "reflected" (equivalently, "constrained") process $\{x_k^n, k < \infty\}$ that evolves as follows. As long as it remains nonnegative, it evolves as $\psi^n(k/n)$ does. But if x_k^n ever tries to become negative, then it is pushed immediately back to zero. More formally, with $x_0^n = x(0) > 0$, for $k \geq 1$ this reflected process can be written as

$$x_k^n = \max\left\{0, x_{k-1}^n + \frac{\xi_k}{\sqrt{n}}\right\} = \left[x_{k-1}^n + \frac{\xi_k}{\sqrt{n}}\right] + \delta y_k^n, \tag{4.2}$$

where

$$\delta y_k^n = \max\left\{0, -\left[x_{k-1}^n + \frac{\xi_k}{\sqrt{n}}\right]\right\}. \tag{4.3}$$

Define the "correction" or "reflection" term,

$$y_k^n = \sum_{l=1}^{k} \delta y_l^n. \tag{4.4}$$

The interpolated process can now be written as

$$x^n(t) = \psi^n(t) + y^n(t). \tag{4.5}$$

The process $y^n(\cdot)$ is called the *reflection term* and it satisfies

$$y^n(t) = \max\left\{0, -\min_{s\le t} \psi^n(s)\right\}, \tag{4.6}$$

which is the one-dimensional form of the *reflection map*.

Note that (4.5) is written as the sum of the unconstrained "driving function" $\psi^n(\cdot)$ plus the reflection term. Generalizations of this form to multidimensional processes will be a foundation stone of the queue models. It turns out that $x^n(\cdot)$ converges weakly to a reflected Wiener process $x(\cdot)$ with initial value $x(0)$, that we can write as

$$x(t) = x(0) + w(t) + y(t), \tag{4.7}$$

where

$$y(t) = \max\left\{0, -\min_{s\le t} [x(0) + w(s)]\right\} \tag{4.8}$$

and $w(\cdot)$ is a Wiener process with variance σ^2. See Chapter 5.

The above development parallels that done in Example 1.1 of Chapter 1. In the terminology of that example, define $\xi_k^n = [I_k^a - \bar{\lambda}^a] - [I_k^d - \bar{\lambda}^d] + [\bar{\lambda}^a - \bar{\lambda}^d]$. Then (1.1.4) can be written as

$$x^n\left(\frac{k}{n}\right) = x^n\left(\frac{k-1}{n}\right) + \frac{\xi_k^n}{\sqrt{n}} + \delta y_k^n, \tag{4.9}$$

where δy_k^n is defined analogously to (4.3).

Comments on the $y^n(\cdot)$ process in Example 2.1 of Chapter 1. Write (1.2.8) as

$$x^n(t) = x(0) + w^n(t) + [I - Q']y^n(t) + \text{small error}. \tag{4.10}$$

It is worth emphasizing the following points. The $y^n(\cdot)$ have three important properties. First, $y_i^n(\cdot)$ can change only at times t such that $x_i^n(t) = 0$; that is, on the boundary of the state space $G = \{x : x_i \geq 0, i \leq r\}$. Thus, the evolution of $x^n(\cdot)$ is the same as that of $w^n(\cdot)$ when $x^n(\cdot)$ is away from the boundary (where no processor will idle). The process $x^n(\cdot)$ can be viewed as a constrained $w^n(\cdot)$ with the constraint mechanism being the fact that a processor incurs idle time if its queue is empty. Second, the direction of change of the constraining term $[I-Q']y^n(\cdot)$ is determined by the constraint mechanism; the constraint is just what is required to correct for the idle time. The reflection direction on the horizontal axis $\{x : x_2 = 0\}$ is just the inward normal. This is because the correction for the idle time of processor 2 affects only queue 2. The reflection direction on the vertical axis $\{x : x_1 = 0\}$ is "down and inward" at an angle whose tangent is q_{12}. This is because a unit of idle time of processor 1 reduces the input to processor 2 by q_{12} "on average."

The third property is that the increment in $y^n(\cdot)$ at any given time is the minimal amount needed to keep $x(0) + w^n(\cdot)$ from pushing $x^n(\cdot)$ out of G.

It is common for models of scaled queueing systems to have analogous decompositions into an essentially "unconstrained" part for which it is often relatively easy to find a diffusion approximation, plus a term that keeps track of the effects of the constraint. The properties of $y^n(\cdot)$ in this example suggest the properties that should be expected of the reflection direction for any diffusion $x(\cdot)$ that is a weak-sense limit of such models $x^n(\cdot)$.

3.4.2 Reflected Brownian Motion and Related Models

The method for modeling a reflected diffusion process discussed in this subsection is based on the *reflected Brownian motion or RBM* model of [91, 98, 101, 102, 213]. The state space will be $G = \{x : x_i \geq 0, i \leq r\}$. Let $Q = \{q_{ij}; i, j\}$ be a matrix of real numbers. Consider the equation

$$x(t) = \psi(t) + [I - Q']\, y(t), \quad x(0) \in G, \tag{4.11}$$

where $\psi(\cdot) \in D(\mathbb{R}^r; 0, \infty)$. We seek solutions in $D(\mathbb{R}^r; 0, \infty)$ to (4.11) such that $x(t) \in G$ for all t, $y(0) = 0$, $y(\cdot)$ is nondecreasing, $y_i(\cdot)$ can increase only at t where $x_i(t) = 0$, and $y(t)$ depends only on the values $\{\psi(s), s \leq t\}$. Such a process $z(\cdot) = [I - Q']y(\cdot)$ is called the *reflection term,* and it serves to keep the process $x(\cdot)$ in G and to reflect it in the "correct" direction if it tries to leave. The following theorem is a minor modification of a result of Harrison and Reiman [98]. Theorem 5.1 contains a more general result.

Theorem 4.1. *Let the spectral radius of the matrix $\{|q_{ij}|; i, j\}$ be less than unity. Let $\psi(\cdot) \in D(\mathbb{R}^r; 0, \infty)$. Then, there are unique $x(\cdot)$ and $y(\cdot)$ in $D(\mathbb{R}^r; 0, \infty)$ such that (4.11) holds. Furthermore, $y(\cdot)$ and $x(\cdot)$ are Lipschitz continuous functions of $\psi(\cdot)$ in that there is $C < \infty$ depending only on Q and such that*

$$|x(t)| + |y(t)| \leq C \sup_{s \leq t} |\psi(s)|, \tag{4.12}$$

and for any $\psi^i(\cdot) \in D(\mathbb{R}^r; 0, \infty), i = 1, 2,$ and corresponding solutions $(x^i(\cdot), y^i(\cdot))$ we have

$$|x^1(t) - x^2(t)| + |y^1(t) - y^2(t)| \leq C \sup_{s \leq t} |\psi^1(s) - \psi^2(s)|. \tag{4.13}$$

Let $x(\cdot), y(\cdot)$ *solve* (4.11). *Fix* $\tau > 0$, *and define* $\delta x(t) = x(\tau + t) - x(\tau)$, $\delta y(t) = y(\tau + t) - y(\tau)$. *Then, for* $t \geq 0$,

$$\delta x(t) = [\psi(t + \tau) - \psi(\tau)] + [I - Q']\,\delta y(t),$$

and for some C *depending only on* Q,

$$\sup_{0 \leq s \leq t} |x(\tau + s) - x(\tau)| \leq C \sup_{0 \leq s \leq t} |\psi(t + \tau) - \psi(\tau)|. \tag{4.14}$$

Proof. The second paragraph of the theorem is a consequence of the first paragraph. Without loss of generality, the proof will be done for $q_{ij} \geq 0$. All functions $r(\cdot), y^n(\cdot)$, etc., introduced below are in $D(I\!R^r; 0, \infty)$. We follow the proof in [98], except for minor notational changes. Define $\alpha = \sup_j \sum_i q_{ij}$ and, until further notice, suppose that $\alpha < 1$. For $T > 0$ and a vector-valued function $v(\cdot)$, define

$$\|v(\cdot)\|_T = \sup_i \sup_{s \leq T} |v_i(s)|.$$

Define

$$\pi_j(\psi, v)(t) = \max\left\{0, \sup_{s \leq t}\left[\sum_i q_{ij} v_i(s) - \psi_j(s)\right]\right\}. \tag{4.15}$$

Define the vector $\pi(\psi, v)(t) = \{\pi_j(\psi, v)(t), j \leq r\}$. For any two vector-valued functions $v^i(\cdot)$, $i = 1, 2$, and any $T < \infty$, we have

$$\|\pi(\psi, v^2)(\cdot) - \pi(\psi, v^1)(\cdot)\|_T \leq \alpha \|v^2(\cdot) - v^1(\cdot)\|_T. \tag{4.16}$$

Assuming that (4.11) has a solution, it can be rewritten as

$$x(t) = [\psi(t) - Q'y(t)] + y(t). \tag{4.17}$$

The right-hand term $y(\cdot)$ in (4.17) can be defined (implicitly and component by component) in terms of the reflection map (4.8), with the components of $\psi(\cdot) - Q'y(\cdot)$ replacing $x(0) + w(\cdot)$. That is, we can write

$$x(t) = [\psi(t) - Q'y(t)] + \pi(\psi, y)(t). \tag{4.18}$$

The contraction (4.16) and the representation (4.18) provide the basis for the construction of the solution to (4.11). Use the iteration $y^0(t) = 0$ and define $y^n(\cdot)$, $n \geq 1$, recursively by

$$x^{n+1}(t) = [\psi(t) - Q'y^n(t)] + y^{n+1}(t), \tag{4.19}$$

where $y^{n+1}(t) = \pi(\psi, y^n)(t)$. By (4.16) and (4.19),

$$\|y^{n+1}(\cdot) - y^n(\cdot)\|_T \leq \alpha \|y^n(\cdot) - y^{n-1}(\cdot)\|_T. \tag{4.20}$$

By (4.20) and the fact that $\alpha < 1$, there is a function $y(\cdot) \in D(I\!R^r; 0, \infty)$ such that $\|y^n(\cdot) - y(\cdot)\|_T \to 0$ geometrically as $n \to \infty$. Hence, there exists $x(\cdot)$ such that $\|x^n(\cdot) - x(\cdot)\|_T \to 0$, also geometrically. It is straightforward to verify that $y(t) = \pi(\psi, y)(t)$, that $x(\cdot)$ and $y(\cdot)$ solve (4.11), and that $y_i(\cdot)$ can increase only at t where $x_i(t) = 0$.

We now derive the bound (4.12) and the global Lipschitz condition (4.13). For $\psi(\cdot)$ and $\bar{\psi}(\cdot)$ in $D(I\!R^r; 0, \infty)$, write the corresponding solutions as

$$x(t) = \psi(t) + [I - Q']\, y(t),$$

$$\bar{x}(t) = \bar{\psi}(t) + [I - Q']\, \bar{y}(t).$$

Then, using the above approximation method, we have

$$x^{n+1}(t) = [\psi(t) - Q'y^n(t)] + y^{n+1}(t),$$

$$\bar{x}^{n+1}(t) = \left[\bar{\psi}(t) - Q'\bar{y}^n(t)\right] + \bar{y}^{n+1}(t).$$

A computation like that which led to (4.20) yields

$$\|y^{n+1}(\cdot) - \bar{y}^{n+1}(\cdot)\|_T \leq \alpha \|y^n(\cdot) - \bar{y}^n(\cdot)\|_T + \|\psi(\cdot) - \bar{\psi}(\cdot)\|_T.$$

This, together with the geometric convergence of $y^n(\cdot)$ to $y(\cdot)$ and of $\bar{y}^n(\cdot)$ to $\bar{y}(\cdot)$ from the first part of the proof yields that

$$\|y(\cdot) - \bar{y}(\cdot)\|_T \leq \frac{1}{1-\alpha} \|\psi(\cdot) - \bar{\psi}(\cdot)\|_T.$$

The proof of (4.12)–(4.14) is completed, except for the replacement of the condition $\alpha < 1$ by the condition on the spectral radius. In [98] this is done by the following change of variables. Let Γ be a diagonal matrix with positive diagonal elements. Then

$$\Gamma x(t) = \Gamma \psi(t) + \left[I - \Gamma Q' \Gamma^{-1}\right] \Gamma y(t).$$

Thus, to solve (4.11) we can replace Q' by $\Gamma Q' \Gamma^{-1}$. But since the spectral radius of Q' is less than unity, there is such a Γ for which the row sums of $\Gamma Q' \Gamma^{-1}$ are all strictly less than unity [242, Lemma 3, applied to Q']. ■

Localizing Theorem 4.1 and bounded regions. Consider the state space and reflection directions in Figure 4.1. There is an obvious analogue of (4.11). One needs to formulate the relations among the reflection directions on the faces adjoining each corner analogously to the spectral radius condition of Theorem 4.1. Then a "localization or stopping time argument" can be used to get an analogue of Theorem 4.1. The result will be stated in Theorem 5.1 in the context of the Skorohod problem, a more convenient format.

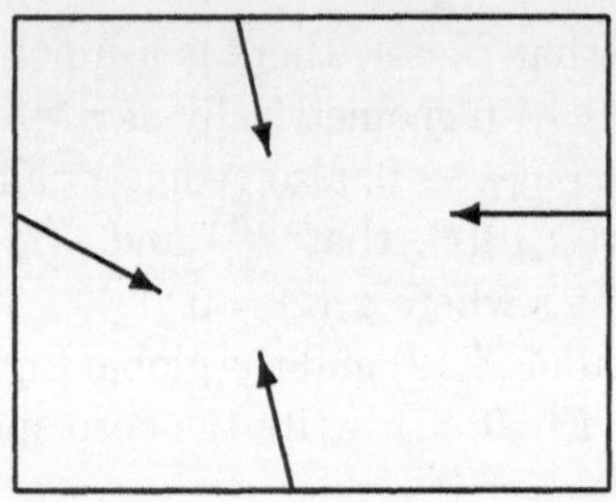

Figure 4.1. Bounded state space: constraints on all sides.

A strong-sense solution to the reflected SDE. We write the reflected SDE with an ordinary admissible control as

$$x(t) = x(0) + \int_0^t b(x(s), u(s))ds + \int_0^t \sigma(x(s))dw(s) + [I - Q'] \, y(t). \quad (4.21)$$

The form with an admissible relaxed control is written as

$$x(t) = x(0) + \int_0^t \int_{\mathcal{U}} b(x(s), \alpha) r_s(d\alpha)ds + \int_0^t \sigma(x(s))dw(s) + [I - Q'] \, y(t). \quad (4.22)$$

Here $[I - Q']y(\cdot)$ is the reflection term as in (4.11). Let $w(\cdot)$ denote an $\mathcal{F}_t$-standard Wiener process for some filtration $\{\mathcal{F}_t, t \geq 0\}$. The definitions of admissible controls or admissible pairs $(u(\cdot), w(\cdot))$ or $(r(\cdot), w(\cdot))$, and the definitions of weak and strong-sense existence and uniqueness for the reflected case are just the obvious analogues of the definitions used in Subsection 1.2 and Section 2.

Theorem 4.2. *Assume* (A1.1), (A1.2), *the conditions on* Q *in Theorem* 4.1, *and let* $(u(\cdot), w(\cdot))$ *and* $(r(\cdot), w(\cdot))$ *be admissible pairs. Then there are strong-sense unique strong-sense solutions to* (4.21) *and* (4.22) *for each initial condition* $x(0) \in G$.

Remarks on the proof. A strong-sense solution to (4.21) and (4.22) can be constructed analogously to the way that it was constructed in Theorem 1.1. An essential tool in the proof of Theorem 1.1 was the Lipschitz condition (A1.2), and it is used here in a similar way, in combination with the Lipschitz condition (4.13).

The method is the same for (4.21) and (4.22), and we work with (4.21). One constructs a sequence of approximations via the Picard iteration:

$$x^{n+1}(t) = x(0) + \int_0^t b(x^n(s), u(s))ds + \int_0^t \sigma(x^n(s))dw(s) + [I - Q'] \, y^n(t).$$

Given nonanticipative $x^n(\cdot)$, the existence of nonanticipative $y^n(\cdot)$ follows from Theorem 4.1. Then, apply (4.13) with

$$\psi^n(t) = x(0) + \int_0^t b(x^n(s), u(s))ds + \int_0^t \sigma(x^n(s))dw(s).$$

Analogously to the situation in Theorem 1.1, the Lipschitz conditions (4.13) and (A1.2) imply that there is $K_1 < \infty$ (depending only on the functions $b(\cdot), \sigma(\cdot)$ and the matrix Q) such that

$$E\|x^{n+1}(\cdot) - x^n(\cdot)\|_t^2 \leq K_1(1+t)\int_0^t E\|x^n(\cdot) - x^{n-1}(\cdot)\|_s^2 ds.$$

The procedure is then completed as indicated in Theorems 1.1 and 4.1.

A similar procedure can be applied to the equation with an admissible impulsive or singular control such as

$$x(t) = x(0) + \int_0^t b(x(s))ds + DF(t) + \int_0^t \sigma(x(s))dw(s) + [I - Q']\,y(t). \tag{4.23}$$

Unreflected jumps taking $x(\cdot)$ out of G. If there is a possibility that a jump in $F(\cdot)$ in (4.23) will take the unreflected state out of G, then we need to impose a constraint condition. Our approach is guided by what will be the heavy traffic limit in the problems of interest. Suppose that $F(\cdot)$ has a discontinuity δF at time τ that could take the (unreflected) path out of G. The value $x(\tau) - x(\tau-)$ is defined as follows. Define $\delta x(\cdot)$ and the reflection process $\delta y(\cdot)$ by $\delta y(0) = 0$ and, for $s \geq 0$,

$$\delta x(s) = x(\tau-) + s(\delta F) + [I - Q']\,\delta y(s). \tag{4.24}$$

Then set

$$x(\tau) = \delta x(1), \quad y(\tau) = y(\tau-) + \delta y(1). \tag{4.25}$$

The procedure (4.24), (4.25) is motivated by the behavior of the scaled heavy traffic physical trajectory $x^n(\cdot)$, for which there might be "fast movement" for large n, as in the $F_0^n(\cdot)$ of Section 2.6. That process $F_0^n(\cdot)$ converges to a unit jump in an obvious sense, but it is not an instantaneous jump for any particular value of $n < \infty$. In (4.24), (4.25), we are spreading the movement out over the time interval $[0, 1]$.

For a two-dimensional illustration, let $\delta F = (\rho, 0)$, where $\rho = -\rho_1 - \rho_2$, $\rho_i > 0$, and $x_1(\tau-) = \rho_1, x_2(\tau-) > \rho_2$. Let the reflection direction on the axis $\{x : x_1 = 0\}$ be $(1, -q_{12})$ for some $1 > q_{12} > 0$. Refer to Figure 4.2. The path $\delta x(\cdot)$ in (4.24) first moves to the point $(0, x_2(\tau-))$ along the indicated horizontal line. At this point, the reflection starts acting, and forces the path to move along the boundary to point b. The combination of the jump and the reflection cause a total movement $q_{12}\rho_2$ along the boundary.

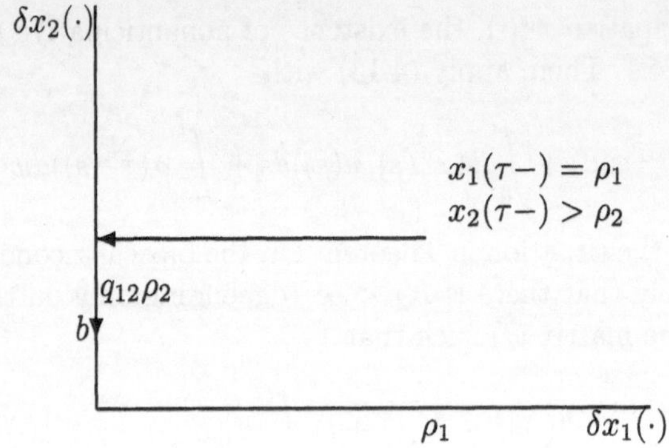

Figure 4.2. The path for the solution of (4.24).

The fact that this approach is correct can be seen from the following physical example. Consider the tandem queue of Example 1.2.1. To create an asymptotic jump in the scaled path $x^n(\cdot)$, suppose that the input to queue 1 is shut off at real time $n\tau$ for a real time of duration $O(\sqrt{n})$. Let the limit (as $n \to \infty$) of the number of missing inputs to queue 1 during this time (divided by $\sqrt{n}$)[1] converge in probability to $\rho = \rho_1 + \rho_2$. Let the value of $x^n(t)$ before shutoff be $x(\tau-)$ as above. Consider the scaled path $x^n(\cdot)$ during this shutoff time. In this scaling, time is compressed by a factor of n and size by a factor of $\sqrt{n}$. The shutoff lasts $O(1/\sqrt{n})$ units of scaled time. Thus, asymptotically, any change during this time looks like a jump. For large n, with a high probability $x^n(\cdot)$ moves very close to the trajectory in the figure. From the initial condition $x(\tau-)$, $x^n(\cdot)$ moves quickly and very close to the horizontal line until it hits the vertical axis very close to $(0, x_2(\tau-))$. Then, once the value of $x_1(t)$ is zero, there are no further inputs to queue 2, and $x_2(\cdot)$ decreases very quickly. In fact, asymptotically, it decreases by $q_{12}\rho_2$, the mean scaled number that it would have received if there were no shutdown. Thus, $x^n(\cdot)$ moves down the vertical axis an (asymptotic) distance $q_{12}\rho_2$. This yields the solution to (4.24).

Uniqueness and the Girsanov transformation. The explicit construction described in Theorem 4.2 under (A1.1) and (A1.2) implies strong-sense existence and uniqueness. Now drop the Lipschitz condition but assume the conditions used in Subsection 1.4 on $b(\cdot), \sigma(\cdot), \tilde{b}_1(\cdot)$ and that $[\sigma_1(\cdot)]^{-1}\tilde{b}_1(\cdot)$ satisfies the Novikov condition. Suppose that the uncontrolled form of (4.21) or (4.22) has a unique weak-sense solution for the initial condition of interest. Then, the Girsanov transformation can be used to construct weak-sense solutions, and the comments concerning it in Subsection 1.4 hold here as well.

---

[1]Such convergence will be shown to hold under broad conditions in Chapter 5.

Weak-sense uniqueness and the Feller and strong Markov property. Suppose that an (uncontrolled) solution exists in the weak-sense and is weak-sense unique for each initial condition. Let $b(\cdot)$ and $\sigma(\cdot)$ be continuous and suppose that for each bounded set B and $T < \infty$,

$$\lim_{K\to\infty} \sup_{x\in B} P\left\{ \sup_{s\le T} |x(s)| \ge K \right\} = 0. \tag{4.26}$$

Then, for each bounded and continuous function $f(\cdot)$ and $t \ge 0$, $E_x f(x(t))$ is a continuous function of the initial condition x. We say that $x(\cdot)$ has the *Feller* property or is a *Markov–Feller* process. This implies that $x(\cdot)$ is a *strong Markov* process. That is, for any stopping time τ with respect to the filtration engendered by $x(\cdot)$, $P\{x(\tau + t) \in \cdot | x(s), s \le \tau\} = P(x(\tau), t, \cdot))$, where $P(x, t, \cdot)$ is the transition function of the Markov process $x(\cdot)$ [68]. To prove the Feller property, let $x^n(0) \to x(0)$. By the weak-sense uniqueness and the continuity and boundedness in probability assumptions, the corresponding sequence of processes $(x^n(\cdot), z^n(\cdot))$ with initial conditions $x^n(0)$ converges weakly to the process with initial condition $x(0)$. The fact that weak-sense uniqueness implies the Feller property continues to hold for the process (5.8) defined in the next section if the coefficients $b(\cdot)$ and $\sigma(\cdot)$ are continuous and (4.26) holds. As noted in the next paragraph, it also holds for any process obtained via the Girsanov transformation, where the untransformed system is

$$dx(t) = b_0(x(t))dt + \sigma(x(t))dw(t), \tag{4.27}$$

where $b_0(\cdot)$, $\sigma(\cdot)$, and $\sigma^{-1}(\cdot)$ are bounded and continuous.

The Feller and strong Feller property and the Girsanov transformation. The Markov–Feller process $x(\cdot)$ is said to be a *strong Feller* process if $E_x f(x(t))$ is continuous for each $t > 0$ for each bounded and measurable real-valued function $f(\cdot)$. Consider the process (4.27) where the inverse $\sigma^{-1}(\cdot)$ exists and $b_0(\cdot), \sigma(\cdot)$ and $\sigma^{-1}(\cdot)$ are bounded and continuous, and there is a unique solution for each initial condition $x(0) = x$. All of the solutions are taken in the weak sense. With an appropriate choice of the probability space, this process can be obtained from the solution to

$$dx(t) = \sigma(x(t))dw(t) \tag{4.28}$$

by the Girsanov transformation method of Theorem 1.2. Let $g(\cdot)$ be bounded and measurable. Then the solution to

$$dx(t) = b_0(x)dt + g(x(t))dt + \sigma(x(t))dw(t)$$

can also be obtained from that of (4.28) by the Girsanov transformation method. Absorb $b_0(\cdot)$ into $g(\cdot)$.

The Feller and strong Feller properties will hold for the transformed processes if they hold for the original ones. An outline of the argument follows. Let $(\Omega, \mathcal{F}, P_x)$ denote a probability space with a filtration $\{\mathcal{F}_t, t \geq 0\}$ on which are defined a standard $\mathcal{F}_t$-Wiener process and a solution $x(\cdot)$ to (4.28) with $x(0) = x$. Define the mollified function

$$g_\epsilon(x) = \frac{1}{\sqrt{2\pi\epsilon}} \int e^{-|y-x|^2/2\epsilon} g(y)dy.$$

Define

$$R(g,t) = \exp\left(\int_0^t g'(x(s))\sigma^{-1}(x(s))dw(s) - \frac{1}{2}\int_0^t \left|\sigma^{-1}(x(s))g(x(s))\right|^2 ds\right). \tag{4.29}$$

Then

$$E_x \left|R(g_\epsilon, t) - R(g,t)\right| \to 0$$

uniformly in x in any bounded set as $\epsilon \to 0$. Thus, as $\epsilon \to 0$ the measures defined by the Radon–Nikodym derivatives $R(g_\epsilon, t)$ converge in variation to that defined by the Radon–Nikodym derivative $R(g_\epsilon, t)$, uniformly in x in any bounded set. This implies that if the process with new drift term $g_\epsilon(\cdot)$ is a Feller or strong Feller process for each $\epsilon > 0$, then so is the process with transformed drift $g(\cdot)$. The weak-sense uniqueness of the solution to (4.28) for each x implies that it is a Feller process.

These remarks continue to hold for the reflected stochastic differential equation which is dealt with in the next section.

3.5 The Skorohod Problem

The Skorohod problem model is perhaps the most versatile approach for the reflected diffusion type problems of concern in this book. The development relies heavily on [63, 64], which contains a comprehensive development of the theory of the Skorohod problem, particularly as it pertains to the solutions of stochastic differential equations with reflection. The following material is specialized to the cases of interest in this book.

3.5.1 Definitions and Lipschitz Conditions

Definitions: Directions of reflection $d(x)$. The following condition will be used throughout the book. Additional conditions on $d(x)$ will be given when needed. Note that the state space G is redefined.

A5.1. *The state space G is the intersection of a finite number of closed half spaces in $\mathbb{R}^r$, and is the closure of its interior (i.e., it is a closed convex* *polyhedron with an interior and planar sides). Let ∂G_i, $i = 1, \ldots,$ denote the faces of G, and n_i the interior normal to ∂G_i. Interior to ∂G_i, the reflection direction is denoted by the unit vector d_i, and $\langle d_i, n_i \rangle > 0$ for each i. The possible reflection directions at points on the intersections of the ∂G_i are in the convex hull of the directions on the adjoining faces. Let $d(x)$ denote the set of reflection directions at the point $x \in \partial G$, whether it is a singleton or not.*

By an *edge*, we mean the intersection of (more than one) boundary faces. If the intersection is a single point, then it is referred to as a *corner*.

Definition: The Skorohod problem. Let $z(\cdot)$ be an $I\!R^r$-valued function of bounded variation on each finite interval. Let $|z|(t)$ denote its total variation over $[0,t]$, and $\mu_z(\cdot)$ the measure on $[0,\infty)$ that is defined by the total variation. Let $\psi(\cdot) \in D(I\!R^r; 0, \infty)$ with $\psi(0) \in G$. Then we say that $(x(\cdot), z(\cdot))$ solves the *Skorohod problem* for $\psi(\cdot)$ if the following hold:

$$x(\cdot) = \psi(\cdot) + z(\cdot), \quad x(0) = \psi(0), \tag{5.1a}$$

$$x(t) \in G \text{ for all } t, \tag{5.1b}$$

$$|z|(t) < \infty \text{ for each } t, \tag{5.1c}$$

$$|z|(t) = \int_0^t I_{\{x(s) \in \partial G\}} d|z|(s), \tag{5.1d}$$

there exists an $I\!R^r$-valued measurable function $\gamma(\cdot)$ such that $\gamma(t)$ is in the positive convex cone $C(d(x(t)))$ generated by $d(x(t))$ for $\mu_z(\cdot)$-almost all t, and $z(t) = \int_0^t \gamma(s) d|z|(s)$. (5.1e)

Alternatively, we can write

$$z(t) = \sum_i y_i(t) d_i, \tag{5.1f}$$

where $y_i(0) = 0$, and $y_i(\cdot)$ is nondecreasing and can increase only at t where $x(t) \in \partial G_i$. The $y_i(\cdot)$ are not necessarily unique if the $d_i \in d(x)$ are not linearly independent at each x, since there might be a "choice" as to which $y_i(\cdot)$ increases on some corner or edge. Theorem 4.3.6 shows that (under a nondegeneracy condition and (A5.3)) the change in $|z|(\cdot)$ is zero during the time that $x(t)$ is on the corners or edges.

Consider the setup of Subsection 4.2, where $G = \{x : x_i \geq 0, i \leq r\}$. In the form (4.11), $z(t) = [I - Q']y(t)$. On the open face where $x_i = 0$ and $x_j > 0$, $j \neq i$, the value of $d(x)$ can be obtained from the matrix Q, by recalling that $y_i(\cdot)$ can increase only at t where $x_i(t) = 0$. Let e_i denote the unit vector in the ith coordinate direction, and q_i the vector $(q_{i1}, \ldots, q_{ir})$.

The contribution of $y_i(\cdot)$ to the reflection term $[I - Q']y(t)$ is $[e_i - q_i]y_i(t)$. Thus, $d(x)$ consists of the (unit vector normalization of the) single point $e_i - q_i$ on the considered open face. On the corners and edges, $d(x)$ is the convex combination of the values on the adjoining open faces.

Although the definition of the Skorohod problem is somewhat abstract, it provides a very useful tool for the construction and analysis of reflecting diffusions. The definition of $x(\cdot)$ can be seen as one "natural" constrained version of $\psi(\cdot)$: That is, $x(\cdot)$ is obtained from $\psi(\cdot)$ by adding a function $z(\cdot)$ that will "push" $x(\cdot)$ in the proper direction when it tries to leave G. Loosely speaking, (5.1d) and (5.1e) imply that $z(\cdot)$ can push $x(\cdot)$ only while $x(\cdot)$ is on the boundary of G, and then only in the directions that are consistent with the current position of $x(\cdot)$.

The generalized Skorohod problem. There are important applications where condition (5.1c) is violated; that is, the process $z(\cdot)$ might have an unbounded variation. This is characteristic of the so-called processor-sharing algorithms [65], where the reflection directions are as depicted in Figure 1.2.2. The main problem concerns the definition of $z(\cdot)$ in (5.1e) in terms of its variation process and the sets of allowed reflection directions. For $x \notin \partial G$, set $d(x) = \phi$, the empty set. We say that $(x(\cdot), z(\cdot))$ solves the *generalized Skorohod problem* if (5.1a) and (5.1b) hold, $\psi(\cdot)$ is continuous, and conditions (5.1c)–(5.1e) are replaced by the following: $z(\cdot)$ is continuous and for each $t \geq 0$ and $s > 0$,

$$z(t+s) - z(t) \in \text{ cone}\left\{d(x(u)) : t \leq u \leq t+s\right\}, \tag{5.2}$$

where cone is the positive cone generated by the convex hull.

A Lipschitz condition. Under an additional condition on the reflection directions, there is an analogue of (4.12) and the Lipschitz condition (4.13). Such a result also holds for the generalized Skorohod problem for some processor-sharing cases [65]. The following condition (A5.2) implies the Lipschitz condition. The condition is not always easily verifiable, but seems to be the most general condition that can be used. It was heavily used in [62, 63, 65], and is implied by the spectral radius condition as noted in the comments after (A5.4), which implies both (A5.2) and (A5.3) [63]. So, one can assume either ((A5.2) and (A5.3)) or (A5.4). The condition (A5.3) guarantees that a solution to the Skorohod problem exists and implies the bounded variation of $z(\cdot)$. Illustrations of some two-dimensional cases will be given below.

A5.2. *Let x be an arbitrary point on a corner or edge of G with reflection directions d_i, $i = 1, \ldots, k$, on the adjacent faces. Shift coordinates so that $x = 0$. For each such x, there is a compact and convex set B in $I\!R^r$ with*

$0 \in B^0$ *such that if* $\nu(\bar{x})$ *denotes the set of inward normals at* $\bar{x} \in \partial B$, *then*

$$\bar{x} \in \partial B \text{ and } |\langle \bar{x}, n_i \rangle| < 1 \text{ together imply that}$$
$$\langle \nu, d_i \rangle = 0, \text{ for all } \nu \in \nu(\bar{x}) \text{ and } i = 1, \ldots, k.$$

We will also have use for the following condition.

A5.3. *For each* $x \in \partial G$, *define the index set* $I(x) = \{i : x \in \partial G_i\}$. *Suppose that* $x \in \partial G$ *lies in the intersection of more than one boundary; that is,* $I(x)$ *has the form* $I(x) = \{i_1, \ldots, i_k\}$ *for some* $k > 1$. *Let* $N(x)$ *denote the convex hull of the interior normals* $n_{i_1}, \ldots, n_{i_k}$ *to* $\partial G_{i_1}, \ldots, \partial G_{i_k}$, *respectively, at* x. *Then, there is some vector* $v \in N(x)$ *such that* $\gamma' v > 0$ *for all* $\gamma \in d(x)$.

There is a neighborhood $N(\partial G)$ *and an extension of* $d(\cdot)$ *to* $\overline{N(\partial G)}$ *that is upper semicontinuous in the following sense: For each* $\epsilon > 0$, *there is* $\rho > 0$ *that goes to zero as* $\epsilon \to 0$ *and such that if* $x \in N(\partial G) - \partial G$ *and distance*$(x, \partial G) \leq \rho$, *then* $d(x)$ *is in the convex hull of the directions* $\{d(v); v \in \partial G, \text{distance}(x, v) \leq \epsilon\}$.

A5.4. *Either* (a) *or* (b) *holds:* (a) $d_i = n_i$ *for all* i. (b) *For an arbitrary corner or edge, let* d_i *denote the directions of reflection on the adjoining faces. Then there are constants* $a_i > 0$ *such that for all* i,

$$a_i \langle n_i, d_i \rangle > \sum_{j \neq i} a_j \left| \langle n_i, d_j \rangle \right|.$$

Comment on (A5.4). Note that the d_i need not be normalized to have unit length. Consider the model of Section 4, where $G = \{x : x_i \geq 0, i \leq r\}$ and let Q denote the matrix with entries $\{q_{ij}; i, j\}$, where $q_{ij} \geq 0$. Then $\langle n_i, d_i \rangle = 1 - q_{ii}$ and, for $j \neq i$, $\langle n_i, d_j \rangle = -q_{ji}$. Then (A5.4) reduces to finding $a_i > 0$, $i \leq r$, such that $a_i > \sum_j a_j q_{ji}$, $i \leq r$. For $a = (a_i, i \leq r)$, this is equivalent to solving $[I - Q']a = b$ for some b, where the components of a and b are positive. If $b_i > 0, i \leq r$, then $[I - Q']^{-1} b$ has positive components, by the spectral radius condition and the nonnegativity of the q_{ij}. Thus (A5.4) is an extension of the spectral radius condition to state spaces that are not necessarily orthants or hyperrectangles.

More generally, normalize the d_i such that $\langle d_i, n_i \rangle = 1 - v_{ii}$ for $0 \leq v_{ii} < 1$. For $i \neq j$, define $v_{ij} = |\langle d_i, n_j \rangle|$. Then (A5.4) holds if and only if the spectral radius of the matrix $V = \{v_{ij}; i, j\}$ is less than unity [63].

Remark on (A5.3) and the completely-S condition. Let A be a square matrix of dimension r. Suppose that for each set of indices $i_1, \ldots, i_k$, $k \leq r$, the submatrix $A(i_1, \ldots, i_k)$ obtained by deleting the rows and columns $i_1, \ldots, i_k$ has the property that there is a vector $\alpha > 0$ (i.e., each component is positive) such that $A(i_1, \ldots, i_k)\alpha > 0$. Then A is said to be *completely-S*. If A is completely-S, then so is its transpose A'. For a history of the condition and its name and other applications, see [45, 212].

When G is an orthant or hyperrectangle, the first part of (A5.3) is equivalent to the completely-S condition. Fix a corner or edge, shift the coordinates so that it contains the origin, let D denote the matrix whose *rows* are the reflection directions $\{d_{i_1}, \ldots, d_{i_k}\}$ on the adjoining faces, and let N denote the matrix whose columns are the interior normals to the adjacent faces. Then the first part of (A5.3) says that there is a vector $\alpha \geq 0$ such that $DN\alpha > 0$. It also says that the same is true if we restrict attention to any subset of the adjoining faces. Thus DN, the matrix whose (i, j)th element is $\langle d_i, n_j \rangle$, is completely-S. Thus, $N'D'$ is also completely-S. Hence there is a vector $\alpha > 0$ such that $N'D'\alpha > 0$. Locally, about the corner or edge of concern, $G = \{x : \langle n_i, x \rangle \geq 0; i = i_{i_1}, \ldots, i_{i_k}\}$. Thus, there is a nonnegative linear combination of the reflection directions that points strictly inward.

The basic definition of the Skorohod problem requires only that we specify $d(x)$ for $x \in \partial G$. In the "discrete" example of Subsection 4.1, the reflection term was a correction to the attempt of the process to leave the set G. In a sense, the process left G, but was immediately pushed back to it in the correct direction. This requires that $d(x)$ be defined and satisfy an "upper semicontinuity" condition in a small neighborhood of the boundary, which explains the last part of (A5.3). Recalling the previous paragraph, it is seen that at each point in $N(\partial G) - G$, for small enough neighborhoods there is a vector in $d(x)$ that takes x to G and the length of this vector goes to zero as the distance between x and G goes to zero. The first part of (A5.3) implies that if $c(\cdot)$ takes values on ∂G, then $z(t) = 0$ is the only solution to

$$x(t) = c(t) + z(t). \tag{5.3}$$

Examples of (A5.2). The set B is easy to construct for two-dimensional problems. Refer to Figure 5.1, where the angles between the d_i and the n_i are less than or equal to 45°. For $x = 0$, the box B is constructed so that the sides are parallel to the d_i and the box is large enough to eliminate interference between the directions. The example of Figure 5.1 satisfies the spectral radius condition if some angle is less than 45°, and then (A5.4) also holds. The example of Figure 5.2, where the angle between the d_i and n_i is 45° for both $i = 1, 2$, does not satisfy the spectral radius condition or (A5.4), which shows that (A5.2) is a strict extension of the spectral radius condition.

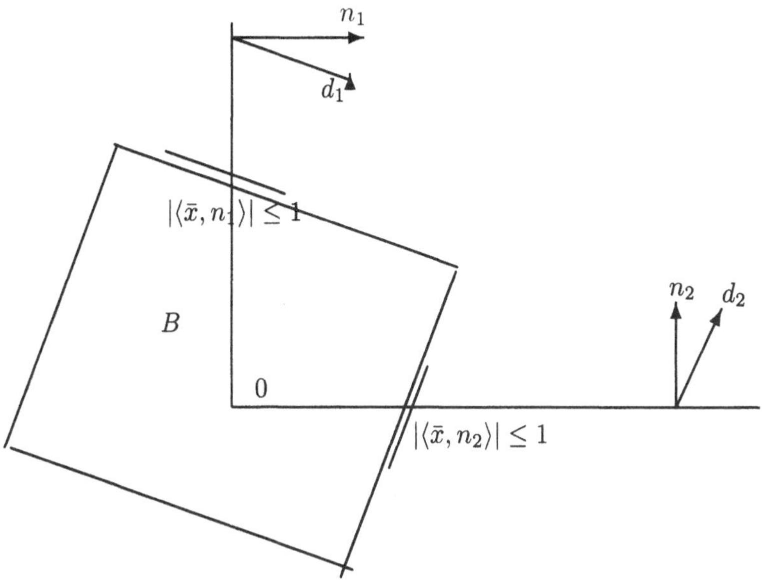

Figure 5.1. Constructing the box B in (A5.2).

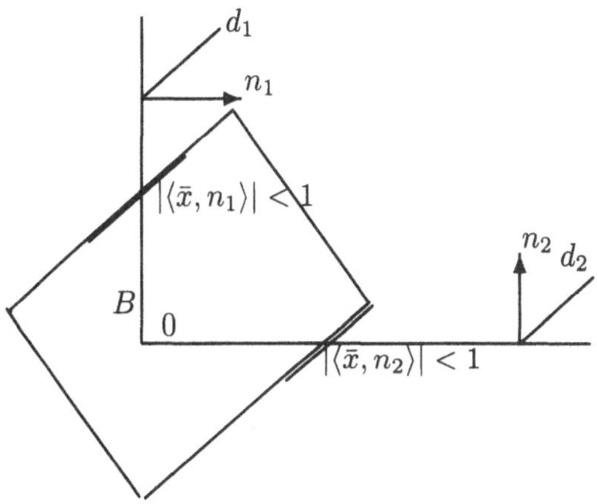

Figure 5.2. Constructing the box B in (A5.2).

The first part of the following theorem is [63, Theorem 2.2.]. The proof of the second part uses a localization of that proof and the fact that (A5.3) implies that if we start in a small neighborhood of a corner or edge then there is a constant C_1 such that $|y(t)| \leq C_1 |z(t)|$ until some new constraint (not involved in that corner or edge) appears.

Theorem 5.1. *Assume* (A5.1), (A5.2), *and that a solution to* (5.1a) *exists. Then there is* $C < \infty$ *that does not depend on* t *or on* $\psi(\cdot) \in D(\mathbb{R}^r; 0, \infty)$ *such that* (4.12) *and* (4.13) *hold, where* $y(\cdot)$ *is replaced by* $z(\cdot)$. *Now add the assumption* (A5.3). *Then a solution exists and* (4.12) *holds.*

3.5.2 Reflected Stochastic Differential Equations

Let $(\Omega, \mathcal{F}, P)$ be the probability space, $\{\mathcal{F}_t, t \geq 0\}$ a filtration, and $w(\cdot)$ an $\mathcal{F}_t$-standard vector-valued Wiener process. Let $u(\cdot)$ be an admissible control with respect to $w(\cdot)$. Then an $\mathcal{F}_t$-adapted process $x(\cdot)$ is a solution to the reflected stochastic differential equation for initial condition $x(0)$ if it solves (always, with probability one)

$$x(t) = x(0) + \int_0^t b(x(s), u(s))ds + \int_0^t \sigma(x(s))dw(s) + z(t), \; x(t) \in G, \quad (5.4)$$

where

$$|z|(t) = \int_0^t I_{\{x(s)\in\partial G\}} d|z|(s) < \infty, \quad (5.5)$$

and where there exists an $\mathcal{F}_t$-adapted process $\gamma(\cdot)$ with $\gamma(s)$ taking values in $d(x(s))$ for almost all (ω, s) and such that

$$z(t) = \int_0^t \gamma(s) d|z|(s). \quad (5.6)$$

Thus (pathwise, with probability one) $(x(\cdot), z(\cdot))$ solves the Skorohod problem for $\psi(t) = x(0) + \int_0^t b(x(s), u(s))ds + \int_0^t \sigma(x(s))dw(s)$. The relaxed control form is

$$x(t) = x(0) + \int_0^t \int_{\mathcal{U}} b(x(s), \alpha) r_s(d\alpha)ds + \int_0^t \sigma(x(s))dw(s) + z(t), \; x(t) \in G. \quad (5.7)$$

Without the control, we write simply

$$x(t) = x(0) + \int_0^t b(x(s))ds + \int_0^t \sigma(x(s))dw(s) + z(t). \quad (5.8)$$

The Lipschitz condition of Theorem 5.1 and the iterative method of Theorem 4.2 give the following result, whose proof is left to the reader. The bounded variation of $z(\cdot)$ is considered in Theorem 5.3.

Theorem 5.2. *Assume* (A1.1), (A1.2), (A5.1)–(A5.3). *Then the Skorohod problem given by* (5.4)–(5.6) *has a strong-sense unique solution for each admissible control or relaxed control and each initial condition.*

The first assertion of the next theorem is [167, Theorem 11.1.1]. The rest is a consequence of Theorem 5.1.

Theorem 5.3. *Assume* (A1.1), (A5.1), (A5.3), *and that* $b(\cdot)$ *and* $\sigma(\cdot)$ *are measurable, with* $\sigma(\cdot)$ *bounded. Let* $(r(\cdot), w(\cdot))$ *be admissible, and let* $(x(\cdot), r(\cdot), w(\cdot), z(\cdot))$ *solve* (5.7). *Let* $b(\cdot)$ *be bounded. Then, for each* $T < \infty$,

$$\sup_{x,r} E^r_{x(0)} |z|^2 (T) < \infty.$$

Now, add (A5.2), *drop the boundedness assumption on* $b(\cdot)$, *let* G_1 *be a compact set in* G, *and suppose that*

$$\sup_r \sup_{x(0)\in G_1} E^r_{x(0)} \int_0^t \int_{\mathcal{U}} |b(x(s),\alpha)|^2 r_s(d\alpha)ds = O(t).$$

Then

$$\sup_r \sup_{x(0)\in G_1} E^r_{x(0)} |z|^2 (t) = O(t^2),$$

$$\sup_r \sup_{x(0)\in G_1} E^r_{x(0)} |y(t)|^2 = O(t^2),$$

where $y(\cdot)$ *is defined in* (5.1f).

Example: Unbounded $b(\cdot)$. Let G be the nonnegative two-dimensional orthant, and consider the system with $\dot{x}_1 = x_1$, $\dot{x}_2 = -x_1$ with $x_2(0) = 0$, $x_1(0) = 1$. Let the reflection directions be the normals. Then $z_2(t) = e^t - 1$. Such a possibility is the reason for assuming the bound on the integral of square of $b(x(s),\alpha)$.

On the uniqueness of $y(\cdot)$ in the representation $z(t) = \sum_i d_i y_i(t)$. The conditions of Theorem 5.3 ensure that any process $y(\cdot)$ has the asserted moments, but it does not ensure that $y(\cdot)$ is uniquely defined by $z(\cdot)$. To see this, consider the two dimensional case where $G = \{x : x_1 \geq 0, x_2 \geq 0\}$, $dx = -dt + dz$, with $d_i = (1,1), i = 1,2$. Then there is an ambiguity at $x = 0$, since $d_1 = d_2$. Such uniqueness is sometimes important in the control problem in order to guarantee that the costs associated with a sequence of controls converge to the cost associated with the limit. If the covariance matrix $a(x)$ is nondegenerate for all $x \in G$, then, by Theorem 4.2.1, the contribution to $z(\cdot)$ during the time that more than one boundary constraint is active is zero, in which case $z(\cdot)$ yields $y(\cdot)$ uniquely w.p.1. Otherwise, an additional condition is needed.

Weak-sense existence. The next theorem is a good illustration of the use of the boundary conditions to get weak-sense existence. We drop (A5.2), and thus lose the Lipschitz condition of Theorem 5.1. But (A5.3) will still give us existence. The proof constructs a sequence of approximations, proves appropriate tightness, and then shows that the weak-sense limit processes solve (5.5)–(5.7). We work with relaxed controls, but the proof can be adapted to work with the singular or impulsive control cases as well. Two of the methods introduced in the proof will be used frequently in the sequel. The first (actually completed in Theorem 6.1) is the "proof by contradiction" of the tightness of the sequence of reflection processes. The conditions (A5.1) and (A5.3) on the boundary are used to show a contradiction if this sequence is not tight. Second, the martingale method of Theorem 2.1.3 is used to show that (for a given weakly convergent subsequence) the weak-sense limit solution is nonanticipative with respect to the weak-sense limit Wiener processes.

Theorem 5.4. *Assume* (A1.1), (A5.1), (A5.3), *and that* $b(\cdot), \sigma(\cdot)$ *are continuous, with* $\sigma(\cdot)$ *bounded. Assume either that* G *or* $b(\cdot)$ *is bounded or that* $b(\cdot, \alpha)$ *has at most linear growth in* x*, uniformly in* $\alpha \in \mathcal{U}$*. Let* $(\bar{r}(\cdot), \bar{w}(\cdot))$ *be admissible and* $x(0) \in G$*. Then there exists a weak-sense solution to* (5.5)–(5.7) *in the sense that there exists a probability space* $(\Omega, \mathcal{F}, P)$*, a filtration* $\{\mathcal{F}_t, t \geq 0\}$*, an* $\mathcal{F}_t$*-standard Wiener process* $w(\cdot)$*, and an admissible relaxed control* $r(\cdot)$*, where* $(r(\cdot), w(\cdot))$ *has the same distribution as* $(\bar{r}(\cdot), \bar{w}(\cdot))$*. Also, there are* $\mathcal{F}_t$*-adapted processes* $x(\cdot)$ *and* $z(\cdot)$*, such that* (5.5)–(5.7) *hold with initial condition* $x(0)$.

Proof. The proof uses a sequence of approximations and a weak convergence argument. For later use, we will work with a more general problem where there is a sequence $(x^n(\cdot), r^n(\cdot), w^n(\cdot))$, each $(r^n(\cdot), w^n(\cdot))$ is admissible, $x^n(0) = x(0)$, and $w^n(\cdot)$ is a standard vector-valued Wiener process. The probability spaces on which they are defined is unimportant. It will be shown that there is a weakly convergent subsequence whose weak-sense limit $(x(\cdot), r(\cdot), w(\cdot))$ satisfies (5.5)–(5.7). Since the distribution of each $(r^n(\cdot), w^n(\cdot))$ is that of $(\bar{r}(\cdot), \bar{w}(\cdot))$, we have the theorem as stated. Without loss of generality, we suppose that all of the processes are defined on the same probability space. The proof will be done for bounded G. Otherwise, the truncation method mentioned in Subsection 2.5.5 can be used. See also [157] for other uses of the truncation method in weak convergence.

For each $n < \infty$, define a sequence x_k^n, with $x_0^n = x(0)$, as the solution to the discrete-time Skorohod problem:

$$x_{k+1}^n = x_k^n + \delta\psi_k^n + \delta z_k^n, \tag{5.9}$$

where

$$\delta\psi_k^n = \int_{k/n}^{(k+1)/n} \int_{\mathcal{U}} b(x_k^n, \alpha) r^n(d\alpha, ds) + \sigma(x_k^n) \left[w^n((k+1)/n) - w^n(k/n)\right].$$

The term δz_k^n is the reflection term, and it is well-defined for large n owing to the last part of (A5.3) and the fact that for each $T, \mu > 0$,

$$\lim_{n\to\infty} P\left\{\max_{k \leq Tn} |\delta\psi_k^n| \geq \mu\right\} = 0. \tag{5.10}$$

Define $z_0^n = \psi_0^n = 0$ and for $k \geq 1$ set

$$z_k^n = \sum_{l=0}^{k-1} \delta z_l^n, \quad \psi_k^n = \sum_{l=0}^{k-1} \delta\psi_l^n.$$

Define the continuous-time interpolations $z^n(t) = z_k^n$ on $[k/n, (k+1)/n)$, and define $x^n(\cdot)$ and $\psi^n(\cdot)$ analogously. Then we can write

$$x^n(t) = x(0) + \psi^n(t) + z^n(t). \tag{5.11}$$

Owing to the properties of the stochastic integrals, the boundedness of G, and compactness of $\mathcal{U}$, the set $\{\psi^n(\cdot), r^n(\cdot), w^n(\cdot)\}$ is tight. The main problem concerns the tightness of the set of reflection terms $\{z^n(\cdot)\}$. It will be proved in Theorem 6.1 that this sequence is "asymptotically continuous," which implies the tightness. By asymptotic continuity we mean that for each $T > 0$ and $\epsilon > 0$,

$$\lim_{\delta \to 0} \limsup_n P\left\{\sup_{t \le T} \sup_{v \le \delta} |z^n(t+v) - z^n(t)| \ge \epsilon\right\} = 0. \tag{5.12}$$

The limit equality (5.12) implies the tightness (via the criterion of Theorem 2.5.6, as always) of $\{z^n(\cdot)\}$ and that each weak-sense limit process has continuous paths with probability one, although it does not imply that any weak-sense limit has bounded variation.

Suppose that (5.12) does not hold for some $T > 0$ and $\epsilon > 0$. Then there are $\delta_n \to 0$ and $\eta > 0$ such that

$$\limsup_n P\left\{\sup_{t \le T} \sup_{v \le \delta_n} |z^n(t+v) - z^n(t)| \ge \epsilon\right\} \ge \eta. \tag{5.13}$$

The impossibility of (5.13) is shown in Theorem 6.1, and is based on the following facts. By the boundedness of $b(\cdot)$ in G and the martingale properties of $w(\cdot)$, for each $\nu > 0$,

$$\lim_{\delta \to 0} \limsup_n P\left\{\sup_{t \le T} \sup_{s \le \delta} |\psi^n(t+s) - \psi^n(t)| \ge \nu\right\} = 0. \tag{5.14}$$

(That is, $\psi^n(\cdot)$ is asymptotically continuous.) Furthermore, the $|\delta z_n^n|$ are asymptotically small in the sense that, for each $T, \mu > 0$, (5.10) implies that

$$\lim_{n \to \infty} P\left\{\max_{k \le Tn} |\delta z_k^n| \ge \mu\right\} = 0. \tag{5.15}$$

To get (5.15), we have used the fact that there is $C < \infty$, which depends on the "biggest" angle of reflection, such that $|\delta z_k^n| \le C|\delta \psi_k^n|$.

Now, (5.14), (5.15), (A5.3), and Theorem 6.1 imply that $\{x^n(\cdot)\}$ and $\{z^n(\cdot)\}$ are asymptotically continuous. The next step is to take a weakly convergent subsequence of $\{x^n(\cdot), r^n(\cdot), w^n(\cdot), z^n(\cdot)\}$ and characterize the weak-sense limit as a solution to (5.7). To simplify the notation, suppose that n indexes this weakly convergent subsequence. The process $r(\cdot)$ must have continuous paths with probability one, since $r^n(t+\delta, \mathcal{U}) - r^n(t, \mathcal{U}) = \delta$.

Let $(x(\cdot), r(\cdot), w(\cdot), z(\cdot))$ denote the weak-sense limit of the selected subsequence. Let $\Delta > 0$. Write (5.11) in the form

$$\begin{aligned} x^n(t) = x(0) + &\int_0^t \int_{\mathcal{U}} b(x^n(s), \alpha) r^n(d\alpha\, ds) \\ &+ \sum_{i:i\Delta<t} \sigma(x^n(i\Delta)) \left[w^n((i\Delta + \Delta) \wedge t) - w^n(i\Delta)\right] + z^n(t) + \rho_\Delta^n(t), \end{aligned} \tag{5.16}$$

where $\rho_\Delta^n(\cdot)$ is due to the use of $x^n(\cdot)$ rather than x_k^n in the integral representing the drift term and to the error introduced by using $x^n(i\Delta)$ in lieu of x_k^n on the time intervals $[i\Delta, i\Delta + \Delta)$ in the stochastic integral. For each $\Delta > 0$, the set $\{\rho_\Delta^n(\cdot)\}$ is tight, and for any $T > 0$, $\epsilon > 0$,

$$\lim_{\Delta \to 0} \limsup_n P\left\{\sup_{t \le T} |\rho_\Delta^n(t)| \ge \epsilon\right\} = 0. \tag{5.17}$$

By the weak convergence and the continuity of $x(\cdot)$,

$$\begin{aligned} x(t) = x(0) + &\int_0^t \int_{\mathcal{U}} b(x(s), \alpha) r(d\alpha\, ds) \\ &+ \sum_{i:i\Delta<t} \sigma(x(i\Delta)) \left[w((i\Delta + \Delta) \wedge t) - w(i\Delta)\right] + z(t) + \rho_\Delta(t), \end{aligned}$$

where $\rho_\Delta(\cdot)$ is the weak-sense limit of $\rho_\Delta^n(\cdot)$. By (5.17), $\rho_\Delta(\cdot)$ goes to the "zero" process as $\Delta \to 0$. Since $w(\cdot)$ is the weak-sense limit of a sequence of standard Wiener processes, it is a standard Wiener process.

With $\Delta \to 0$ in the sum in the above expression, it would seem that the sum converges to the desired stochastic integral. But before proving this, we need to show that the other processes are nonanticipative with respect to $w(\cdot)$, which in turn implies that the sum converges to the stochastic integral. Let $\{\mathcal{F}_t, t \ge 0\}$ denote the filtration generated by $(x(\cdot), r(\cdot), w(\cdot), z(\cdot))$. Let $\{\phi_k(\cdot)\}$ be an arbitrary finite set of continuous functions on $\mathcal{U}$. Let $h(\cdot)$ be a real-valued bounded and continuous function of its arguments. Let $t > 0$, $\tau > 0$, and for an arbitrary integer p, let $0 \le s_i \le t$, $i = 1, \ldots, p$, be arbitrary. By the admissibility of $r^n(\cdot)$ and the definitions of $x^n(\cdot)$ and $z^n(\cdot)$, for $s \ge 0$ we have

$$\begin{aligned} &Eh(x^n(s_i), w^n(s_i), z^n(s_i), \langle r^n(s_i), \phi_k\rangle, i \le p, k) \\ &\quad \times \left[w^n(t + \tau + 1/n) - w^n(t + 1/n)\right] = 0. \end{aligned} \tag{5.18}$$

The weak convergence, (5.18), and the uniform integrability of $\{w^n(t+1/n), n < \infty\}$ for each t imply that

$$Eh(x(s_i), w(s_i), z(s_i), \langle r(s_i), \phi_k\rangle, i \le p, k)[w(t + \tau) - w(t)] = 0. \tag{5.19}$$

It can also be readily shown that

$$\begin{aligned} &0 = Eh\left(x^n(s_i), w^n(s_i), z^n(s_i), \langle r^n(s_i), \phi_k\rangle, i \le p, k\right) \times \\ &\left[(w^n(t + \tau + 1/n) - w^n(t + 1/n))\,(w^n(t + \tau + 1/n) - w^n(t + 1/n))' - \tau I\right]. \end{aligned} \tag{5.20}$$

By the weak convergence and the uniform integrability of $\{|w^n(t+1/n)|^2, n < \infty\}$ for each t, (5.20) holds with the n-variables dropped. Now, these limits and the arbitrariness of $h(\cdot), \phi_k(\cdot), t, \tau, s_i, p, k$, and Theorem 2.1.3 imply that $w(\cdot)$ is an $\mathcal{F}_t$-standard Wiener process and the other processes are nonanticipative with respect to it.

Finally, consider the weak-sense limit $z(\cdot)$ of the reflection terms. By the weak convergence and the fact that $z^n(\cdot)$ can change only at t such that $x^n(t)$ is arbitrarily close to ∂G (for large n), we see that $z(\cdot)$ can change only at t such that $x(t) \in \partial G$. By the weak convergence and the upper-semicontinuity property (in x) of the sets $d(x)$, for any $t, \tau \geq 0$, with probability one, we must have

$$z(t+\tau) - z(t) \in \text{cone}\left\{d(x(s)) : t \leq s \leq t+\tau\right\}, \tag{5.21}$$

where *cone* denotes the positive cone generated by the convex hull. ■

A cost function. An analogue of the discounted cost function (2.13) for the reflected SDE is

$$W_\beta(x(0), r) = E_{x(0)}^r \int_0^\infty e^{-\beta t} \left[\int_{\mathcal{U}} k(x(t), \alpha) r_t(d\alpha)dt + c'dy(t)\right], \tag{5.22}$$

where the components of the vector c are nonnegative.

3.5.3 Impulsive and Singular Control Problems

Results analogous to Theorems 5.2–5.4 also hold for the reflected SDE analogues of the impulsive and singular control problems. Then, for admissible $F(\cdot)$ the controlled reflected SDE is

$$dx(t) = b(x(t))dt + DdF(t) + \sigma(x(t))dw(t) + dz. \tag{5.23}$$

If a jump in $F(\cdot)$ can take the unreflected process out of G, then we define the value of the reflected process $x(\cdot)$ just after the jump as done in the comments below (4.23).

An analogue of (5.22) for the impulsive control problem is

$$\begin{aligned} W_\beta(x(0), F) = E_{x(0)}^F &\int_0^\infty e^{-\beta t} \left[k_0(x(t))dt + c'dy(t)\right] \\ &+ E_{x(0)}^F \sum_i e^{-\beta\tau_i} g(x(\tau_i-), \nu_i). \end{aligned} \tag{5.24}$$

The discounted cost function for the singular control problem is

$$W_\beta(x(0), F) = E_{x(0)}^F \int_0^\infty e^{-\beta t} \left[k_0(x(t))dt + \bar{c}'dF(t) + c'dy(t)\right], \tag{5.25}$$

where the components of the vector $\bar{c}$ are positive.

3.6 Tightness of a Sequence of Solutions to the Skorohod Problem

The proofs of the tightness and asymptotic continuity of the reflection terms in Theorem 5.4 were deferred until this section, where a general method will be given that will be used throughout the book.

The Skorohod problem formulation. The next theorem allows us to work with the general model of Section 5, under assumptions (A5.1) and (A5.3) on the state space G and the set of reflection directions. The system equation is

$$x^n(t) = x^n(0) + \psi^n(t) + z^n(t), \; x^n(t) \in G. \tag{6.1}$$

The following condition will be used.

A6.1. *There are real $\epsilon_n \to 0$ such that $z^n(\cdot)$ can change only at t where $x^n(t)$ is within a distance of ϵ_n from the boundary ∂G. Let $d_\epsilon(x)$ denote the (perhaps empty) set of reflection directions at all points $y \in \partial G$ within a distance of ϵ from $x \in G$. Let $z^n(0) = 0$ and suppose that for each $t \geq 0$ and $s > 0$, there are $\epsilon_n \leq \mu_n \to 0$ such that $z^n(\cdot)$ satisfies the relation*

$$z^n(t+s) - z^n(t) \in cone\left\{d_{\mu_n}(x^n(u)) : t \leq u \leq t+s\right\}, \tag{6.2a}$$

where cone is the positive cone generated by the convex hull, and for each $T < \infty$,

$$\sup_{t \leq T} |z^n(t) - z^n(t-)| \to 0 \text{ in probability.} \tag{6.2b}$$

Assumptions (A5.1), (A5.3), and (A6.1) imply that $z^n(\cdot)$ can be written in the form

$$z^n(t) = \sum_i y_i^n(t) d_i, \tag{6.3}$$

where we recall that d_i is the reflection direction on ∂G_i and points to the interior of G_i, and where $y_i^n(0) = 0$ and the $y_i^n(\cdot)$ are nondecreasing and can increase only at t where $x^n(t)$ is within a distance ϵ_n of ∂G_i. The proof of the following theorem adapts a method of [167], and will be used often in the sequel.

Theorem 6.1. *Assume the form (6.1), and that $\psi^n(\cdot)$ is asymptotically continuous in the sense that for each $\nu > 0, T > 0$,*

$$\lim_{\delta \to 0} \limsup_n P\left\{\sup_{t \leq T} \sup_{s \leq \delta} |\psi^n(t+s) - \psi^n(t)| \geq \nu\right\} = 0. \tag{6.4}$$

Assume (A5.1) and (A5.3) and that $\{x^n(0)\}$ is tight and $z^n(\cdot)$ satisfies (A6.1). Then $\{z^n(\cdot), y^n(\cdot)\}$ is tight and the limit of any weakly convergent subsequence is continuous with probability one.

Proof. By using the Skorohod representation of Theorem 2.5.5 for the $\psi^n(\cdot)$, it is sufficient to work with a deterministic sequence $\psi^n(\cdot)$ that converges uniformly to a continuous function. Assume that the asymptotic continuity assertion is false. In particular, suppose that there is an asymptotic jump (of size $\geq \nu_0 > 0$) in some $y_i^n(\cdot)$ somewhere on an interval $[0,T]$. We can shift the time origin and suppose that there are $\nu_0 > 0$ and $\delta_n \to 0$ such that $y_i^n(\delta_n) \geq \nu_0$ for large n and some nonempty set of i. Since we will work on $[0, \delta_n]$ for arbitrarily small δ_n, we can suppose that $\psi^n(t)$ converges uniformly to zero. By (6.2b) and (A6.1), the jump in $y^n(\cdot)$ cannot (for large n) take the path $x^n(\cdot)$ interior to the state space G, say more than some $\nu_n \to 0$ from the boundary. Thus, the asymptotic jump in $y^n(\cdot)$ must correspond to an asymptotic jump in $x^n(\cdot)$ arbitrarily close to the boundary (hence, along the boundary, asymptotically). By taking a subsequence if necessary, we can suppose (without loss of generality) that $x^n(0)$ actually converges to a particular point $x(0) \in \partial G$.

By using a time scale change as in Section 2.6, we will next show that there is a solution to the Skorohod problem $x(t) = x(0) + z(t)$, where $x(0) \in \partial G$, $z(t) = \sum_i y_i(t) d_i$, where $y(\cdot)$ is nondecreasing and differentiable, has zero initial condition, and can increase only when $x(t) \in \partial G_i$, and at least one of the $y_i(t)$ is positive for $t > 0$.

Define $T^n(t) = t + \sum_i y_i^n(t)$ and $\hat{T}^n(t) = \inf\{s : T^n(s) > t\}$. Define $\hat{y}^n(t) = y^n(\hat{T}^n(t))$ and similarly define $\hat{x}^n(\cdot), \hat{\psi}^n(\cdot)$ and $\hat{z}^n(\cdot)$. Then

$$\hat{x}^n(t) = x(0) + \hat{\psi}^n(t) + \sum_i \hat{y}_i^n(t) d_i,$$

where $\hat{\psi}^n(\cdot)$ can be assumed to go to zero uniformly in t, and $\hat{y}_i^n(\cdot)$ can increase only when the distance between $\hat{x}^n(t)$ and ∂G_i is no greater than ϵ_n. By construction, the $\hat{y}_i^n(t)$ are asymptotically Lipschitz continuous (Lipschitz constant no larger than unity). Take a convergent subsequence with limits satisfying

$$\hat{x}(t) = x(0) + \sum_i \hat{y}_i(t) d_i.$$

For some nonempty subset of i, $\hat{y}_i(t) > 0$ for $t > 0$.

First suppose that $x(0)$ is interior to the bounding face ∂G_i and $\hat{y}_i(t) > 0$ for $t > 0$. (The $\hat{y}_j(t), j \neq i$, must then be zero.) Then, by (A5.3) the vector d_i points inward, so that $\hat{x}(t) \in G^0$ for small $t > 0$, a contradiction.

Next, suppose that $x(0)$ lies on the edge that is the intersection of ∂G_1 and ∂G_2. Then we can suppose that one or both of $\hat{y}_1(\cdot), \hat{y}_2(\cdot)$ are nonzero, with the other $\hat{y}_i^n(\cdot)$, $i \neq 1, 2$, being zero. If only, say, $\hat{y}_1(\cdot)$ is nonzero, then by the argument of the previous paragraph, $\hat{x}^n(\cdot)$ immediately moves away from ∂G_1, which leads to a contradiction. Now, suppose that both $\hat{y}_1(\cdot), \hat{y}_2(\cdot)$ are nonzero. The condition (A5.3) implies that no positive linear combination of d_1 and d_2 can be zero. Using this, (A5.3), and the fact that $\hat{x}(0) \in \partial G$ implies that $\hat{x}(\cdot)$ must move away from the edge onto one of the adjoining faces. But then we are in the case of the first paragraph. The proof when $x(0)$ lies on the intersection of more that two faces is done in the same way. ■

Figure 6.1 is an example of the reflection directions for the model of Section 4, where the q_{ij} are transition probabilities. The example of Figure 6.2 arises in the model of Section 7.1, and Figure 6.3 presents another example that cannot be represented in terms of transition probabilities.

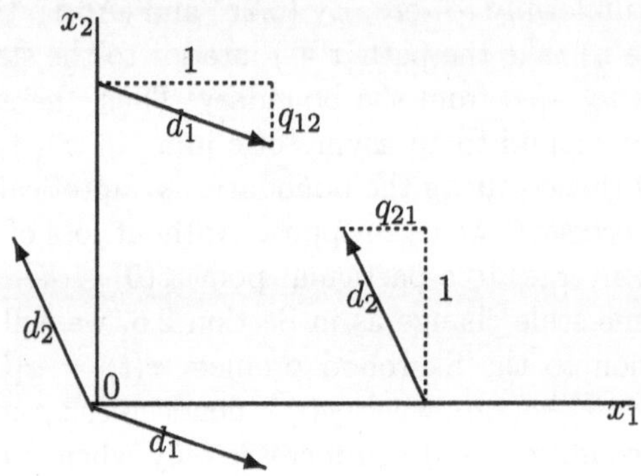

Figure 6.1. An example of reflection directions.

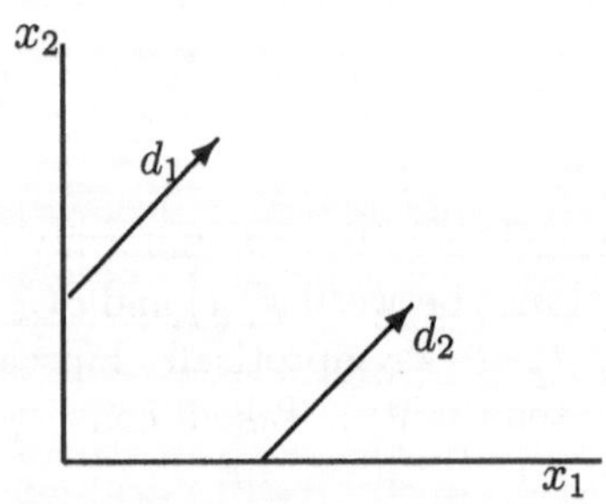

Figure 6.2. An example of reflection directions.

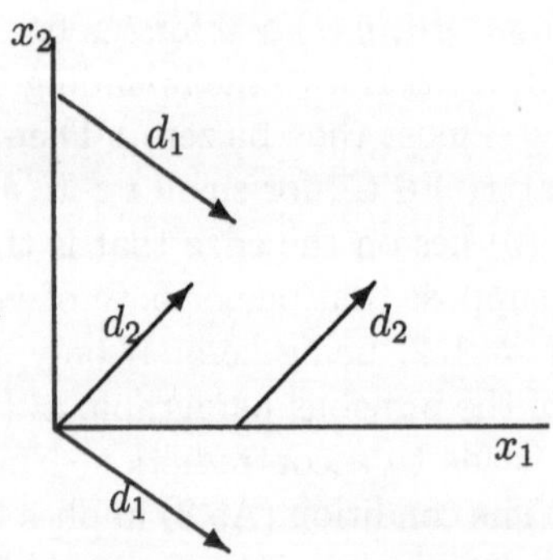

Figure 6.3. An example of reflection directions.

Tightness under the spectral radius condition. An alternative method for proving tightness of the reflection terms for the system of Section 4 uses the spectral radius condition directly, although (A6.2) implies (A5.3). Let the state space be $G = \{x : 0 \leq x_i \leq B_i, i \leq r\}$, where $B_i \leq \infty$. The system is

$$x^n(t) = x^n(0) + \psi^n(t) + [I - Q'_n]\, y^n(t) - U^n(t),\ y^n(0) = U^n(0) = 0, \quad (6.5)$$

where $y_i^n(\cdot)$ (respectively, $U_i^n(\cdot)$) can increase only at those t where $x_i^n(t) \approx 0$ (respectively, where $x_i^n(t) \approx B_i$). The next condition will be used.

A6.2. $Q_n \to Q = \{q_{ij}; i, j\}$, *where the spectral radius of the matrix* $\{|q_{ij}|; i, j\}$ *is less than unity. There are* $\epsilon_n \to 0$ *such that* $y_i^n(\cdot)$ *(respectively,* $U_i^n(\cdot)$*) can increase only at* t *where* $x_i^n(t) \leq \epsilon_n$ *(respectively, where* $x_i^n(t) \geq B_i - \epsilon_n$*).*

Theorem 6.2. *Assume the system* (6.5), *where* $\{x^n(0)\}$ *is tight,* $\psi^n(\cdot)$ *satisfies* (6.4), *and let* (A6.2) *and* (6.2b) *hold. Then* $\{y^n(\cdot), U^n(\cdot)\}$ *is tight, and any weak-sense limit process is continuous with probability one.*

Proof. Proceeding as in Theorem 6.1, if the theorem is false then there is a solution to

$$x(t) = x(0) + [I - Q']\, y(t) - U(t),\ x(t) \in \partial G,\ y(t) \neq 0,\ t > 0.$$

Since $x_i(t)$ cannot be near both zero and B_i simultaneously, it is sufficient to allow only one boundary for each i. Then, by choosing and fixing the boundaries and translating coordinates, we can suppose that the model is

$$x(t) = x(0) + [I - Q']\, y(t),\ x(0) \in \partial G,$$

where the new Q also satisfies the spectral radius condition in (A6.2), G becomes the nonnegative orthant, and $y_i(\cdot)$ can increase only at t where $x_i(t) = 0$. But (A6.2) ensures that the only solution is $x(t) = x(0), y(t) = 0$. ■

Remark on the proof. The proof is simple but of quite general applicability. To ensure understanding, it will be illustrated for the two-dimensional problem of Figure 6.1, where $x(t) = x(0) + [I - Q']y(t)$. Suppose that either one or both of the components $y_i(\cdot)$ are not zero for some small t when $x(0)$ is at the corner. Then some component $x_i(\cdot)$ must increase, taking the state away from the corner, either to the interior, where no component of $y(\cdot)$ can increase, or an adjacent edge, where only one component of $y(\cdot)$ can increase. In the latter case, an increase in that component of $y(\cdot)$ immediately forces the state to the interior where no component of $y(\cdot)$ can increase. The situation is similar in the cases illustrated in Figures 6.2 and 6.3, although not all of the values of q_{ij} are nonnegative.

3.7 Reflected Jump–Diffusion Processes

Jump–diffusion processes can occur in heavy traffic modeling if there are sudden changes due to process interruptions or large (but infrequent) bursts of arrivals. Here, a jump component is added to (4.21) or (5.4), (5.7) or (5.8). Recall the discussion of the Poisson measure in Section 2.3. Let $(\Omega, \mathcal{F}, P)$ denote the probability space, $\{\mathcal{F}_t, t \geq 0\}$ a filtration on it, $w(\cdot)$ a standard $\mathcal{F}_t$-Wiener process, $N(\cdot)$ an $\mathcal{F}_t$-Poisson measure, $p(\cdot)$ the jump process derived from it, with jump rate $\lambda < \infty$ and jump distribution $\Pi(\cdot)$.

The simplest uncontrolled jump–diffusion process (with no boundary reflections) is an $\mathcal{F}_t$-adapted process that satisfies

$$x(t) = x(0) + \int_0^t b(x(s))ds + \int_0^t \sigma(x(s))dw(s) + p(t). \tag{7.1}$$

If the jump size depends on the current state x, then we could replace $p(t)$ by

$$J(t) = \int_0^t q(x(s-))dp(s),$$

for some appropriate function $q(\cdot)$. Then the jumps $dp(t)$ in (7.1) are changed to $q(x(t-))dp(t)$. They depend on the value of the state just before the jump in $p(\cdot)$ occurs. A more useful model is in terms of the associated Poisson measure. We will always use the following assumption.

A7.1. *$q(\cdot)$ is a measurable $\mathbb{R}^r$-valued function on $\mathbb{R}^r \times \Gamma$, with $q(\cdot, \gamma)$ being continuous for each γ. There is $C < \infty$ such that $|q(x,\gamma)| \leq C[1+|x|+|\gamma|]$, λ is bounded and $\int_\Gamma |\gamma|^2 \Pi(d\gamma) < \infty$.*

The following Lipschitz condition (A7.2) is sometimes used, when one wishes to duplicate the proof of Theorem 1.1 and its dependence on the Lipschitz condition.

A7.2. *There is $C < \infty$ such that*

$$\int_\Gamma |q(x,\gamma) - q(y,\gamma)|^2 \, \Pi(d\gamma) \leq C \, |x-y|^2 .$$

Now consider the controlled equation, but still with no reflections:

$$x(t) = x(0) = \int_0^t \int_{\mathcal{U}} b(x(s), \alpha) r_s(d\alpha)ds + \int_0^t \sigma(x(s))dw(s) + J(t), \tag{7.2}$$

where

$$J(t) = \int_0^t \int_\Gamma q(x(s-), \gamma) N(ds\, d\gamma). \tag{7.3}$$

With the form (7.3), and appropriate choices of $q(\cdot)$, the jump rate and distribution can depend on the current value of the state. In fact, the jump rate at x is

$$\bar{\lambda}(x) = \lambda \int_{\{\gamma: q(x,\gamma) \neq 0\}} \Pi(d\gamma) \leq \lambda, \tag{7.4}$$

and the jump distribution at x is

$$\bar{\Pi}(x, A) = \int_{\{\gamma: q(x,\gamma) \in A\}} \Pi(d\gamma). \tag{7.5}$$

One can also start with $\bar{\lambda}(x)$ and $\Pi(x, \cdot)$ and obtain $q(\cdot)$ from them, as shown in [118]. The theory of uncontrolled and unreflected jump–diffusion processes can be found in [82, 83, 113, 117].

The reflected and controlled process. Now add the boundary G, assume (A1.1), (A5.1), (A7.1), and consider the equation

$$x(t) = x(0) = \int_0^t \int_{\mathcal{U}} b(x(s), \alpha) r_s(d\alpha) ds + \int_0^t \sigma(x(s)) dw(s) + J(t) + z(t), \tag{7.6}$$

where the jump term $J(\cdot)$ is defined by (7.3).

Under the additional conditions of any of Theorems 1.1, 1.2, 2.1, and 5.2–5.4, the conclusions of those theorems continue to hold. If a jump would take the unreflected process out of G, then we would use the method (4.24), (4.25) to get the path. The solution to (7.6) can be constructed in pieces: If τ_n denotes the time of the nth jump of $\int_0^t \int_\Gamma \gamma N(ds\, d\gamma)$, then the solution to the process without a jump term is constructed on $[0, \tau_1)$, then the jump at τ_1 is added, then with the new initial condition, the solution the process without the jump term is constructed on $[\tau_1, \tau_2)$, etc. All of the comments concerning strong and weak-sense uniqueness continue to hold. The theory follows from that for the diffusion process, since we can work from one jump to the next. If (7.6) has a unique weak (or strong) sense solution for any initial condition when the $J(\cdot)$ is dropped, then it will have a unique weak (or strong) sense solution for any initial condition when the $J(\cdot)$ is included. The only change in the definition involves the replacement of $w(\cdot)$ by the pair $(w(\cdot), N(\cdot))$. The definition of admissible control involves the same change. Analogous remarks hold for the singular and impulsive control forms.

Itô's lemma. The following formula follows directly from the definition of a solution to (7.6) and the corresponding formula for the case of diffusions with no jumps [113], provided that $z(\cdot)$ has bounded variation. See also (1.3) and (2.4.4). Let $r(\cdot)$ be admissible, $x(\cdot)$ solve (7.6), $f(\cdot) \in C^2$,

$\int_0^{\cdot} \int_{\mathcal{U}} b(x(s),\alpha)r_s(d\alpha)ds \in \mathcal{H}_1, \sigma(x(\cdot)) \in \mathcal{H}_2$, and assume (A7.1). Then

$$\begin{aligned} f(x(t)) = f(x(0)) &+ \int_0^t \mathcal{L}_s^r f(x(s))ds + \int_0^t f_x'(x(s))\sigma(x(s))dw(s) \\ &+ \int_0^t f_x'(x(s))dz(s) + J_f(t), \end{aligned} \tag{7.7}$$

where $a(x) = \sigma(x)\sigma'(x)$,

$$\begin{aligned} \mathcal{L}_s^r f(x) = f_x'(x) &\int_{\mathcal{U}} b(x,\alpha)r_s(d\alpha) + \frac{1}{2}\text{trace}\left[f_{xx}(x)a(x)\right] \\ &+ \bar{\lambda}(x)\int_{\Gamma} \left[f(x+\gamma) - f(x)\right]\bar{\Pi}(x,d\gamma), \end{aligned} \tag{7.8}$$

and

$$\begin{aligned} J_f(t) = &\sum_{0<s\leq t} \left[f(x(s)) - f(x(s-))\right] \\ &- \int_0^t \bar{\lambda}(x(s))\int_{\Gamma} \left[f(x(s)+\gamma) - f(x(s))\right]\bar{\Pi}(x(s),d\gamma)ds. \end{aligned}$$

$J_f(\cdot)$ is a local martingale.

3.8 Approximations of Optimal Controls

Definition: ϵ-optimal control. An admissible relaxed control $r(\cdot)$ is said to be ϵ-optimal for the system (5.7) or (7.6) and cost function (5.22) if $W_\beta(x(0),r) \leq V_\beta(x(0))+\epsilon$. There always exists such a control if $V_\beta(x(0)) > -\infty$. The proof that the optimal costs $V_\beta^n(x(0))$ for the heavy traffic problem converge to the optimal cost $V_\beta(x(0))$ for the limit problem requires the use of suitable approximations to the ϵ-optimal controls for the weak-sense limit models. The relevant results are in [167] for the unreflected case with bounded G. A nearly identical proof can be used for the reflected problem. The controls in the following theorems are used for *theoretical* purposes in the proofs only, and do not have any value in practice.

Theorem 8.1 shows that we can approximate the optimal control by a control that is a function of the initial condition and the driving Wiener process and Poisson measure at a finite number of time points, and that is continuous in the values of the initial condition and Wiener process at those time points.

Remark on notation. In the following theorems we use various standard Wiener processes and Poisson measures, which will be denoted by $w^\mu(\cdot)$ and $N^\mu(\cdot)$, respectively, for appropriate scalar or vector values of the superscript

μ. When a control $r^\mu(\cdot)$ has the same superscript, it is assumed to be admissible with respect to the first pair. Also, $(x^\mu(\cdot), z^\mu(\cdot))$ will denote the associated solution process. The initial condition of all $x^\mu(\cdot)$ and $x(\cdot)$ is $x(0)$. For any ordinary stochastic control $u^\mu(\cdot)$ we use $r^\mu(\cdot)$ for its relaxed control representation. The following rather weak conditions will be needed.

A8.1. *Assume the system* (7.6) *and discounted cost criterion* (5.22) *with* $c_i \geq 0$. *For each* $\epsilon_0 > 0$, *and initial condition* $x(0)$ *of interest, there is an* ϵ_0*−optimal process* $(x(\cdot), z(\cdot), r(\cdot), w(\cdot), N(\cdot))$ *satisfying* (7.6), *and the solution under this control is unique in the weak sense.*

A8.2. *Let* $u(\cdot)$ *be an admissible ordinary control. Suppose that it is piecewise constant and takes only a finite number of values. Then, for each initial condition, there is a solution to* (7.6) *where* $r(\cdot)$ *is the relaxed control representation of* $u(\cdot)$, *and the solution is unique in the weak sense.*

Theorem 8.1. [167, a slight modification of Theorem 3.1, Chapter 10, with the same proof.] *Assume the model* (5.7) *or* (7.6) *and cost* (5.22) *with* $c_i \geq 0$. *Assume* (A1.1), (A2.1), (A5.1), (A5.3), (A7.1), (A8.1), *and* (A8.2). *Let* G *and* Γ *be bounded. Fix* $\epsilon_0 > 0$, *and let* $(x(\cdot), z(\cdot), r(\cdot), w(\cdot), N(\cdot))$ *be an* ϵ_0*-optimal solution whose existence is asserted in* (A8.1). *Then, for each* $\epsilon > \epsilon_0$, *there is a* $\delta > 0$ *and a probability space on which are defined a pair* $(w^\epsilon(\cdot), N^\epsilon(\cdot))$, *an admissible control* $u^\epsilon(\cdot)$ *that takes values in a finite set* $\{\alpha_1^\epsilon, \ldots, \alpha_{p_\epsilon}^\epsilon\} = \mathcal{U}_\epsilon \subset \mathcal{U}$ *and is constant on the intervals* $[n\delta, n\delta + \delta)$, *and a solution* $x^\epsilon(\cdot)$ *such that*

$$|W_\beta(x(0), r^\epsilon) - W_\beta(x(0), r)| \leq \epsilon. \tag{8.1}$$

There is $\theta > 0$ *and a partition* $\{\Gamma_j^q, j \leq q\}$ *of* Γ *such that* $\Pi(\partial \Gamma_j^q) = 0$, *for all* j, *and the approximating* $u^\epsilon(\cdot)$ *can be chosen so that its probability law at any time* $k\delta$, *conditioned on* $\{x(0), w^\epsilon(s), N^\epsilon(s), s \leq k\delta, u^\epsilon(i\delta), i < k\}$, *depends only on the initial condition* $x(0)$ *and on*

$$\{w^\epsilon(p\theta), N^\epsilon(p\theta, \Gamma_j^q), j \leq q, p\theta < k\delta;\ u^\epsilon(i\delta), i < k\} \tag{8.2}$$

and is continuous in the $w^\epsilon(p\theta)$ *and* $x(0)$ *arguments for each value of the other arguments. More particularly, there are functions* $q_k(\cdot)$ *that are continuous in the* w *and* $x(0)$ *variables for each value of the other variables and such that*

$$\begin{aligned} &P\left\{u^\epsilon(k\delta) = \alpha \middle| x(0), u^\epsilon(i\delta), i < k,\ w^\epsilon(s), N^\epsilon(s), s \leq k\delta\right\} \\ &\quad = q_k\left(\alpha; x(0), u^\epsilon(i\delta), i < k,\ w^\epsilon(p\theta), N^\epsilon(p\theta, \Gamma_j^q), j \leq q, p\theta < k\delta\right). \end{aligned} \tag{8.3}$$

Now drop the boundedness of G and Γ. Suppose, in addition, that for each $\mu > 0$ there is $T_\mu < \infty$ such that for the initial condition of interest

$$E^r_{x(0)} \int_{T_\mu}^{\infty} e^{-\beta t} \left[\int_{\mathcal{U}} |k(x(t), \alpha)| \, r_t(d\alpha) dt + c' dy(t) \right] \leq \mu, \tag{8.4}$$

uniformly in the relaxed control $r(\cdot)$, and the set

$$\left\{ \sup_{\alpha \in \mathcal{U}} |k(x(t), \alpha)|, y(t); t \leq T, \text{ all piecewise constant controls} \right\} \tag{8.5}$$

is uniformly integrable for each T. Then the conclusions remain true.

Comment. In [167] the theorem is proved for $p\theta \leq k\delta$, but the same proof yields it for $p\theta < k\delta$. The same remark applies to the next theorem.

The impulsive and singular control problems. Theorem 8.1 has an analogue for the singular and impulsive control problems. The reflected jump–diffusion process which is analogous to (3.2) is

$$x(t) = x(0) + \int_0^t b(x(s))ds + \int_0^t \sigma(x(s))dw(s) + DF(t) + J(t) + z(t). \tag{8.6}$$

The cost function will be (5.24) for the impulsive and (5.25) for the singular controls.

A8.3. *For each $\epsilon > 0$ and initial condition of interest there is an ϵ-optimal solution $(x(\cdot), z(\cdot), F(\cdot), w(\cdot), N(\cdot))$ to (8.6), and it is unique in the weak sense.*

A8.4. *Let $F(\cdot)$ be an admissible singular control for (8.6), and suppose that $F(\cdot)$ is piecewise constant and takes only a finite number of values. Then for each initial condition there exists a weak-sense unique solution to (8.6).*

See [170] for the details of a closely related case.

Theorem 8.2. [167, a slight extension of Theorem 2.5, Chapter 11.] *Assume the model* (8.6) *and* (5.25). *Assume the conditions* (A2.1), (A5.1), (A5.3), (A7.1), (A8.3), *and* (A8.4). *Let G and Γ be bounded, $c_i \geq 0, \bar{c}_i > 0$ and $\epsilon > 0$. Then there is an ϵ-optimal solution $(x^\epsilon(\cdot), z^\epsilon(\cdot), F^\epsilon(\cdot), w^\epsilon(\cdot), N^\epsilon(\cdot))$ to* (8.6) *with the following properties:* (a) *There are $T_\epsilon < \infty$ and $\delta > 0, \theta > 0$ and sets Γ_j^ϵ satisfying the conditions in Theorem 8.1, and a finite set of real numbers $\{v_1, \ldots, v_m\}$ such that $F^\epsilon(\cdot)$ is constant on the intervals $[n\delta, n\delta + \delta)$, only one of its components can jump at a time, and the jumps take values in the discrete set $\{v_1, \ldots, v_m\}$. Also, $F^\epsilon(\cdot)$ is bounded and is constant after time T_ϵ.* (b) *The values of the control are determined* *by the conditional probability law (the expression defines the functions* $q_k(\cdot)$*)*

$$P\left\{dF^\epsilon(k\delta) = v_i \middle| x(0), F^\epsilon(i\delta), i < k,\ w^\epsilon(s), N^\epsilon(s), s \le k\delta\right\}$$
$$= q_k\left(v_i; x(0), F^\epsilon(i\delta), i < k,\ w^\epsilon(p\theta), N^\epsilon(p\theta, \Gamma_j^q), j \le q, p\theta < k\delta\right), \tag{8.7}$$

where the $q_k(\cdot)$ *are continuous in the* w *and* $x(0)$ *variables for each value of the other variables.*

Now drop the boundedness of G *and* Γ*. Suppose, in addition, that for each* $\mu > 0$ *there are* $T_\mu < \infty$, $K_\mu < \infty$, *and a* μ*-optimal admissible control* $F_\mu(\cdot)$ *that is constant after* T_μ, $|F_\mu(\infty)| \le K_\mu$, *and such that*

$$\lim_{T\to\infty} E_{x(0)}^F \int_T^\infty e^{-\beta t}\left[|k_0(x(t))|\, dt + c' dy(t)\right] = 0$$

for the initial condition of interest, and uniformly in controls $F(\cdot)$ *bounded by* K_μ *and not increasing after* T_μ*. Let*

$$\{|k_0(x(t))|, y(t); t \le T, \text{ all piecewise constant controls bounded by } K_\mu\}$$

be uniformly integrable for each T*. Then the conclusions remain true. The analogues of the above results for the impulsive control problem also hold.*

An extension of Theorems 8.2 and 8.3. In applications where we wish to adapt Theorems 8.2 or 8.3, it is not always convenient to compute a "pre-Wiener" process that will converge to the limit $w^\epsilon(\cdot)$. Then the following extension, which is stated without proof, is useful.

Theorem 8.3. *Assume the conditions of Theorem* 8.1 *for either* (5.7) *or* (7.6)*. Suppose that the equation*

$$x(t) = x(0) + \int_0^t b(x(s), u(s))ds + M(t) + J(t) + z(t) \tag{8.8}$$

has a unique solution for each piecewise constant $u(\cdot)$*, continuous function* $M(\cdot)$*, and initial condition* $x(0)$*. Then there is an* ϵ*-optimal control* $u^\epsilon(\cdot)$ *that is defined by* (8.3)*, with*

$$M^\epsilon(t) = \int_0^t \sigma(x^\epsilon(s))dw^\epsilon(s) \tag{8.9}$$

replacing $w^\epsilon(\cdot)$ *in the right side of* (8.3)*, where* $x^\epsilon(\cdot)$ *is the solution to* (5.7) *or* (7.6) *under* $u^\epsilon(\cdot)$*. The analogous result holds for the cases of Theorem* 8.2.

4

Invariant Measures and the Ergodic Problem

Chapter Outline

A major concern in heavy traffic analysis is the long-time behavior of the physical systems, whether uncontrolled, controlled or optimally controlled, as well as the long time behavior of the weak-sense limit processes. The ergodic cost (or average cost per unit time over a long time interval) problem is particularly important in applications of heavy traffic analysis to communications systems. Due to their high speed, they tend to reach their new "steady states" rapidly as the operating conditions change.

This chapter provides much of the necessary foundation. Parts are quite technical and difficult, and if the reader is not interested in the details of the proofs for the ergodic cost problems, the chapter can be skipped. Up to Section 7 we are concerned only with the limit models. Section 1 is concerned with some general results concerning the strong convergence of the transition function of a strong Feller diffusion process to the stationary measure. Section 2 concerns the reflected diffusion process. Under a nondegeneracy assumption, it is shown that negligible time is spent in an arbitrarily small neighborhood of the boundary. This is then used to prove the strong Feller property and uniform mutual absolute continuity of the transition function with respect to Lebesgue measure, as well as the fact that the process satisfies a Doeblin condition.

Section 3 uses the Girsanov transformation method to model the controlled problem in terms of the uncontrolled problem. The properties of this transformation are then used to show that the absolute continuity results proved in the previous section are all uniform in the control. In Section 4 it is shown that the ergodic cost under any feedback control and an arbitrary initial condition is equal to the ergodic cost for the stationary system under that control. It is also shown there that the invariant measure has a convenient continuity property in the control, and that any feedback control can be approximated arbitrarily well by a "smooth" feedback control. The cost function involves a "running" cost as well as a boundary reflection cost. Sections 5 and 6 develop a useful maximum principle for the ergodic cost problem. In the heavy traffic analysis, one shows that the limits of the pathwise average costs are costs for a stationary weak-sense limit system, but not necessarily under a feedback control. The maximum principle is used to prove that this cannot be less than the cost associated with a feedback control, an important step in proving its optimality and allowing attention to be restricted to feedback controls.

Finally, Section 7 introduces the powerful occupation measure method, which allows us to characterize limits of pathwise costs for the heavy traffic problems as ergodic costs for some weak-sense limit problem. In Section 8, we approximate an ϵ-optimal control so that it can be used for showing that the optimal costs for the limit and for the physical model under heavy traffic are close. Section 8 gives some approximations to the optimal control that will be used in Chapter 9 to prove the convergence of the optimal ergodic costs for the heavy traffic problem to that for the weak-sense limit model.

4.1 Convergence to Invariant Measures

Let $x(\cdot)$ be a Markov process with paths in $D(G; 0, \infty)$, where $G \subset I\!R^r$, and is the closure of its interior. The transition function is denoted by $P(x, t, \cdot)$. It is the distribution of $x(s+t)$, given $x(s) = x$, and it is stationary in that it does not depend on the initial time s. A measure μ on $(G, \mathcal{B}(G))$ is said to be *invariant* or stationary if the fact that $x(0)$ has the distribution μ implies that $x(t)$ does, for all t. In this section we establish some results concerning the convergence of $P(x, t, \cdot)$ to an invariant measure as $t \to \infty$. In the next theorem, (1.1) is a consequence of [58, Theorem 4]. The other assertions are readily proved and the details are omitted. Let $l(\cdot)$ denote Lebesgue measure and P_x the probability given that $x(0) = x$.

Theorem 1.1. *Let $P(x, t, \cdot)$ be mutually absolutely continuous with respect to Lebesgue measure for each $t > 0$ and each x, and let $\mu(\cdot)$ be an invariant probability measure for $x(\cdot)$, which we suppose exists. Then $\mu(\cdot)$ is mutually absolutely continuous with respect to Lebesgue measure and*

$$P(x, t, A) \to \mu(A) \text{ for each } x \text{ and Borel set } A. \tag{1.1}$$

Hence, $\mu(\cdot)$ is the unique invariant probability measure.

Theorem 1.2. *Let $P(x,t,\cdot)$ be mutually absolutely continuous with respect to Lebesgue measure for each $t > 0$ and each x. Suppose that, for each bounded, continuous real-valued function $f(\cdot)$ on G and $t > 0$,*

$$\int f(y)P(x,t,dy) \text{ is continuous in } x. \tag{1.2}$$

(That is, $x(\cdot)$ is a Feller process.) Then (i) : $x(\cdot)$ is a strong Feller process in that (1.2) holds for each bounded and measurable $f(\cdot)$. Also, (ii) : for any compact set $D \subset G$ and $t > 0$, $\{P(x,t,\cdot), x \in D\}$ is uniformly absolutely continuous with respect to Lebesgue measure in that for each $\epsilon > 0$ there is a $\delta > 0$ such that $l(A) \leq \delta$ implies that $P(x,t,A) \leq \epsilon, x \in D$. Assume, in addition, that for each $\epsilon > 0$ and compact set $D_1 \subset G$,

$$\lim_{t\to 0} \sup_{x \in D_1} P_x \{|x(t) - x| \geq \epsilon\} = 0. \tag{1.3}$$

Then (iii) : for any $0 < t_0 < t_1 < \infty$ the uniformity in the absolute continuity holds for $(x,t) \in D \times [t_0, t_1]$.

Proof. We work with the first two assertions, since the last assertion is proved in the same way. Fix $t > 0$ and compact D. Suppose that the assertion (ii) of the theorem is false for D. Then given $\epsilon > 0$, there are sets A^n with $l(A^n) \to 0$ and $x^n \in D$ such that $P(x^n, t, A^n) \geq \epsilon$. Thus, we need only show that any sequence $\{P(x^n, t, \cdot)\}$ is uniformly absolutely continuous with respect to Lebesgue measure, and we can suppose that there is x such that $x^n \to x$. Now, by the Vitali–Hahn–Saks theorem [61, Theorem 2, page 158], we need only show that $P(\cdot, t, A)$ is continuous at x for each Borel set A. [This implies (i).] We can suppose that A is contained in a compact set since (1.2) implies that for each $\delta > 0$ there is a compact set $D_\delta \subset G$ such that

$$\sup_{x \in D} P(x,t,G - D_\delta) \leq \delta.$$

By (1.2) and Theorem 2.5.3, $P(x^n, t, A) \to P(x,t,A)$ for each A such that $P(x,t,\partial A) = 0$. By the absolute continuity of each $P(y,t,\cdot)$ with respect to Lebesgue measure, this holds for any set A that is a finite (or countable) union of (open or closed or partly open) rectangles, or the intersection of a sequence of such unions. Now, approximate A by such sets B^n as follows. Let $\epsilon > 0$. Define the symmetric difference Δ by $A \Delta B = (A - B) \cup (B - A)$. For each n, there are sets B^n such that $l(A \Delta B^n) \leq \epsilon 2^{-n}$, $P(x^n, t, A\Delta B^n) \leq \epsilon 2^{-n}$, and $P(x,t,A\Delta B^n) \leq \epsilon 2^{-n}$. Let $B_1 = \cup_n B^n$, $B_0 = \cap_n B^n$. We have

$$P(x^n, t, A) \leq P(x^n, t, B_1) + \epsilon 2^{-n},$$

$$P(x^n, t, A) \geq P(x^n, t, B_1) - P(x^n, t, B_1 - B^n) - P(x^n, t, A\Delta B^n),$$

$$\limsup_n P(x^n, t, B_1 - B^n) \leq \limsup_n P(x^n, t, B_1 - B_0).$$

The conclusion now follows owing to the arbitrariness of ϵ and the fact that $P(x^n, t, B_1 - B_0) \to P(x, t, B_1 - B_0) \leq 2\epsilon$ and $l(B_1 - B_0) \leq 2\epsilon$. ∎

Theorem 1.3. *Assume* (1.2) *and the conditions of Theorem* 1.1. *For* α *in some index set, let* $\{q_\alpha(\cdot)\}$ *be a tight family of probability measures on* G. *Then for each Borel set* A,

$$\int q_\alpha(dx)P(x, t, A) \to \mu(A)$$

uniformly in α, *as* $t \to \infty$.

Proof. Let $s > 0$ and define $q_{\alpha,s}(\cdot) = \int q_\alpha(dx)P(x, s, \cdot)$. By (1.2), $\{P(x, s, \cdot), x \in D\}$ is tight for compact D. It follows from this, the tightness of $\{q_\alpha(\cdot)\}$, and Theorem 1.2 that $\{q_{\alpha,s}(\cdot), \alpha\}$ is tight and uniformly (in α) absolutely continuous with respect to Lebesgue measure.

Given $\epsilon > 0$, let k be such that $q_{\alpha,s}\{x : |x| \geq k\} \leq \epsilon$, for all α. Write

$$\int q_\alpha(dx)P(x, t+s, A) = \int_{|x|\leq k} q_{\alpha,s}(dx)P(x, t, A) + \int_{|x|>k} q_{\alpha,s}(dx)P(x, t, A). \tag{1.4}$$

The $\lim_{t\to\infty} \sup_\alpha$ of the second term on the right side is less than or equal to ϵ. Let $q_{\alpha,s,x}(\cdot)$ denote the derivative of $q_{\alpha,s}(\cdot)$ with respect to x. Rewrite the first term on the right of (1.4) as

$$\int_{|x|\leq k} q_{\alpha,s,x}(x)P(x, t, A)dx.$$

By the uniform (in α) absolute continuity of $q_{\alpha,s}(\cdot)$ with respect to Lebesgue measure,

$$\lim_{c\to\infty} \sup_\alpha \int_{|x|\leq k, q_{\alpha,s,x}(x)\geq c} q_{\alpha,s,x}(x)P(x, t, A)dx = 0.$$

This and (1.1) imply that the first term on the right of (1.4) minus

$$\mu(A) \int_{\{x:|x|\leq k\}} q_{\alpha,s,x}(x)dx$$

goes to zero as $t \to \infty$, uniformly in α. The theorem follows since $\epsilon > 0$ is arbitrarily small. ∎

4.2 Properties of Solutions to the Reflected Stochastic Differential Equation

In this section we establish various results on the transition function of a solution to the Skorohod problem

$$dx = b_0(x)dt + \sigma(x)dw + dz \tag{2.1}$$

or

$$dx = b_0(x)dt + b_1(x)dt + \sigma(x)dw + dz. \tag{2.2}$$

Note the boundedness condition on G in (A2.1), which will be assumed in the rest of the chapter. Assume the following.

A2.1. *$\sigma(\cdot)$ is an $r \times r$ matrix, Hölder continuous, with $\sigma^{-1}(x)$ bounded in G. The function $b_0(\cdot)$ is Hölder continuous, $b_1(\cdot)$ is measurable and there is a real B such that $\sup_x |b_1(x)| \leq B$. G is bounded and is the closure of its interior.*

A2.2. *There is a weak-sense unique solution on $[0, \infty)$ to (2.1) for each initial condition $x(0) = x \in G$.*

Assumptions (A2.1) and (A2.2) imply that the solution to (2.1) is a Markov–Feller process. Denote its transition function by $P(x, t, \cdot)$. The solution to (2.2) is obtained from that of (2.1) by the Girsanov transformation of Subsection 3.1.4 and it is also a Markov–Feller process. See the comments at the end of Section 3.4 and in Section 4 of this chapter for more detail.

The unconstrained process. The proofs of the next few theorems depend on the properties of the *unconstrained* processes

$$dx = b_0(x)dt + \sigma(x)dw \tag{2.1$'$}$$

and

$$dx = b_0(x)dt + b_1(x)dt + \sigma(x)dw, \tag{2.2$'$}$$

where $G = I\!R^r$, the $b_i(\cdot)$ are bounded, and the transition functions are both denoted by $P_0(x, t, \cdot)$. Some of these properties are listed below. Proofs for (2.1$'$) can be found in [68] and those for (2.2$'$) can be obtained via the Girsanov transformation method. Let D be an arbitrary compact set in G, and let $0 < t_0 < t_1 < \infty$ be arbitrary numbers.

1. For each $x = x(0)$, there is a weak-sense unique weak-sense solution to (2.1$'$) and (2.2$'$).

2. $P_0(x, t, \cdot)$ is absolutely continuous with respect to Lebesgue measure, uniformly in $D \times [t_0, t_1]$.

3. $P_0(x,t,\cdot)$ has a continuous density $p_0(x,t,y)$ with respect to Lebesgue measure, which is positive on $D \times [t_0, t_1] \times D$.

4. Let D_i, $i = 1,2$, be compact sets that are the closures of their interiors, have twice continuously differentiable boundaries, and satisfy $D_1 \subset D_2^0$. Let $P_{0,k}(x,t,\cdot)$ denote the transition function of the process that is killed on first hitting ∂D_2. Then $P_{0,k}(x,t,\cdot)$ has a density $p_{0,k}(x,t,y)$ with respect to Lebesgue measure, which is continuous in x, y, t for $t > 0$, and

$$\inf_{x,y \in D_1, t_0 \le t \le t_1} p_{0,k}(x,t,y) > 0.$$

The probability of being killed goes to zero (uniformly in $x \in D_1$) as $t \to 0$.

Theorem 2.1. [102, an elaboration of Lemma 7.2.] *Assume*[1] (2.2), (A3.5.1), (A3.5.3), (A2.1), *and* (A2.2). *Then*

$$E_x \int_0^\infty I_{\partial G}(x(s))ds = 0 \tag{2.3}$$

for each initial value $x \in G$. *Furthermore,*

$$P(x,t,\partial G) = 0, \quad \textit{for } t > 0 \textit{ and all } x \in G. \tag{2.4}$$

$$\lim_{\delta \to 0} \sup_{t_0 \le t \le t_1} \sup_{x \in G} P(x,t,N_\delta(\partial G)) = 0, \quad 0 < t_0 < t_1 < \infty, \tag{2.5}$$

where $N_\delta(\partial G)$ *is a* δ*-neighborhood of the boundary.*

Proof. The proof is essentially a copy of that in the reference. We will work with each boundary face separately. Consider the ith face ∂G_i. Do an affine transformation of the coordinates so that we can suppose that $x_i = 0$ on this face and it contains the origin. Let $I_{\{0\}}(u)$ denote the indicator function of the point $u = 0$. We will first show that

$$E_x \int_0^t I_{\{0\}}(n_i' x(s))ds = 0, \quad \text{for each } t < \infty \text{ and all } i, \tag{2.6}$$

where n_i is the interior normal on ∂G_i.

Let $\phi(\cdot)$ be a continuous and twice continuously differentiable nonnegative and nonincreasing real-valued function on $[0,\infty)$, with $\phi(0) = 1$,

[1] Here and in the rest of the chapter (A3.5.3) is used to ensure that $z(\cdot)$ has bounded variation (see Theorem 3.5.3), so that Itô's lemma (3.7.7) is applicable.

$\phi_y(0) = -1$, $\phi(y) = 0$ for $y \geq 1$, and whose second derivative is nonincreasing, nonnegative and equals $\phi_{yy}(y) = 1$ for $y \leq .5$. Let $\epsilon > 0$ be a rescaling factor. By Theorem 3.5.3, for each t,

$$\sup_{x \in G} E_x |z|(t) < \infty. \tag{2.7}$$

Then applying Itô's lemma, as given in equation (3.7.7), to $\phi(n_i' x(\cdot)/\epsilon)$ yields

$$\phi\left(\frac{n_i' x(t)}{\epsilon}\right) - \phi\left(\frac{n_i' x(0)}{\epsilon}\right) = \\ \frac{1}{\epsilon}\int_0^t \phi_y\left(\frac{n_i' x(s)}{\epsilon}\right) n_i' dx(s) + \frac{1}{2\epsilon^2}\int_0^t [n_i' a(x(s)) n_i]\, \phi_{yy}\left(\frac{n_i' x(s)}{\epsilon}\right) ds, \tag{2.8}$$

where $a(x) = \sigma(x)\sigma'(x)$. Now multiply all terms by ϵ^2, take expectations, use (2.7) and let $\epsilon \to 0$ to get

$$\lim_{\epsilon \to 0} \sup_{x \in G} E_x \int_0^t [n_i' a(x(s)) n_i]\, \phi_{yy}\left(\frac{n_i' x(s)}{\epsilon}\right) ds = 0.$$

Repeat the procedure for all i. Fatou's lemma and the facts that $\phi_{yy}(y) = 1$ for y near zero and the strict positivity of $n_i' a(y) n_i$ yield (2.3), and also (2.4) for almost all (x, t), say for $(x, t) \in B \subset G \times [t_0, t_1]$, where B is dense.

Now (2.4) will follow if we can show that $P(x, t, \partial G)$ is continuous in $G \times [t_0, t_1]$. It is enough to show continuity at each $(x, t) \in B$. Suppose that it is not continuous at $(x, t) \in B$. Then there are $x^n \to x$ and $t^n \to t$ and $\rho > 0$ such that $P(x^n, t^n, \partial G) \geq \rho$. But $P(x^n, t^n, \cdot) \Rightarrow P(x, t, \cdot)$. Hence, $\liminf_n P(x^n, t^n, \partial G) \leq P(x, t, \partial G) = 0$ by Theorem 2.5.3, which yields a contradiction. ■

The next theorem essentially follows from Theorems 1.2, 2.1, and the argument in [102, Lemma 9].

Theorem 2.2. *Assume the forms in* (2.2), (A3.5.1), (A3.5.3), (A2.1), *and* (A2.2). *Let* $0 < t_0 < t_1 < \infty$. *For* $t > 0$ *the transition probability* $P(x, t, \cdot)$ *is uniformly mutually absolutely continuous with respect to Lebesgue measure in the following sense. Given* $\delta > 0$ *there is* $\epsilon > 0$ *such that* $l(A) \leq \epsilon$ *implies that*

$$P(x, t, A) \leq \delta,\ x \in G,\ t_0 \leq t \leq t_1. \tag{2.9}$$

Also, for any $\epsilon > 0$ *there is* $\delta > 0$ *such that*

$$\inf_{t_0 \leq t \leq t_1} \inf_{x \in G} \inf_{\{A : l(A) \geq \epsilon\}} P(x, t, A) \geq \delta, \tag{2.10}$$

and $x(\cdot)$ *is a strong Feller process.*

Proof. By (2.5), we need only work with Borel $A \in G^0$. Let G_ϵ denote the set $\{x \in G : d(x, \partial G) \geq \epsilon\}$, where $d(\cdot)$ denotes the Euclidean distance. It is sufficient to work with $A \subset G_\epsilon$ for arbitrarily small ϵ. Define the stopping times $\sigma_1 = \min\{t : x(t) \in G_\epsilon\}$, and for $n \geq 1$,

$$\rho_n = \min\{t > \sigma_n : x(t) \in \partial G\},$$

$$\sigma_{n+1} = \min\{t > \rho_n : x(t) \in G_\epsilon\}.$$

By Theorem 2.1, we need only work with processes starting at $x(\sigma_n)$ and defined on $[\sigma_n, \rho_n]$. Properties (2.9) and (2.10) then follow from property (4) of the unconstrained process and Theorem 2.1. The strong Feller property follows from Theorem 1.2 and the mutual absolute continuity. ∎

Definition. $x(\cdot)$ is said to satisfy a Doeblin condition [196] if there is a probability measure $\nu(\cdot)$ and $t > 0$, $\epsilon < 1$, $\delta > 0$ such that $\nu(A) \geq \epsilon$ implies that $P(x, t, A) \geq \delta$ for all $x \in G$. The next theorem follows from Theorem 2.2.

Theorem 2.3. *Assume the forms in* (2.2), (A3.5.1), (A3.5.3), (A2.1),*and* (A2.2). *Then $x(\cdot)$ satisfies a Doeblin condition with respect to the measure $\nu(\cdot) = l(\cdot)/l(G)$, and this condition is uniform in the choice of $b_1(\cdot)$ in* (2.2), *provided that they are uniformly bounded.*

Remark on convergence to the invariant measure. Let $k(\cdot)$ be a bounded and measurable real-valued function on G, and assume the conditions of Theorem 2.3 so that the process satisfies a Doeblin condition and is a strong Feller process. Then there is a unique invariant measure $\mu(\cdot)$ and [196]

$$E_x k(x(t)) \to \int k(x)\mu(dx) \tag{2.11}$$

exponentially, uniformly in $x \in G$ as $t \to \infty$.

4.3 The Process Model and Approximations of Controls

This section is concerned with properties of the transition function and invariant measure, and with the convergence of the transition function to the invariant measure, when these are considered as functions of the control. The concept of relaxed feedback control due to Borkar [19], who writes the relaxed control form of Section 3.2 in feedback form, will be used. The approach to be used for the stochastic maximum principle for the ergodic cost problem is a little more efficient and self contained than the original development in [156], which adapted the methods of Bismut [16] and assumed that the set of all possible (drift, running cost rate) $(b(x,\mathcal{U}), k(x,\mathcal{U}))$ was convex. The relaxed control has a "convexifying effect."

Definition: Relaxed feedback control $m(\cdot)$ [19]. $m(x,\cdot)$ is a measure on the Borel sets of $\mathcal{U}$ for each $x \in G$ and $m(\cdot, A)$ is Borel measurable for each Borel set in $\mathcal{U}$. Let $0 \in \mathcal{U}$, corresponding to no control.

Remark. The cost criteria of interest in this chapter are the $\gamma(u)$ or $\gamma(m)$ that are introduced in Section 4. The relaxed feedback control is used for mathematical purposes only, as is the relaxed control of Section 3.2. It simplifies the development and is not intended to be a practical control. Analogously to the situation for control over a finite time interval or for the discounted cost criterion, under the assumptions used here, there might not be an optimal control of the classical feedback type. But even in the ergodic case, a relaxed feedback control can be approximated by a classical (perhaps time dependent) feedback control as in Theorem 4.5. The following conditions will be used. The function $b_c(\cdot)$ is the part of the drift term that contains the control, and we define $b(x,\alpha) = b_0(x) + b_c(x,\alpha)$.

A3.1. *$\sigma(\cdot)$ is an $r\times r$ matrix and is bounded and Hölder continuous, $\sigma^{-1}(x)$ is bounded and continuous on G, $b_c(\cdot)$ is continuous, $b_0(\cdot)$ is Hölder continuous, and $b_c(x,0) = 0$ for all x. The set $\mathcal{U}$ is compact and $0 \in \mathcal{U}$. The set G is compact and is the closure of its interior.*

A3.2. *For each initial condition $x(0) \in G$, there is a unique weak-sense solution $x(\cdot)$ to (3.1).*

For a relaxed feedback control $m(\cdot)$, define

$$b_m(x) = \int_{\mathcal{U}} b_c(x,\alpha)m(x,d\alpha).$$

Thus, if $m(x,\cdot)$ is concentrated on $\{0\}$ for all x, then $b_m(x) = 0$ for all x. The uncontrolled system will be the solution to the Skorohod problem

$$dx(t) = b_0(x(t))dt + \sigma(x(t))dw(t) + dz(t),\ x(t) \in G \subset I\!R^r, \tag{3.1}$$

where $w(\cdot)$ is a standard Wiener process.

The controlled process. Notation. The controlled system will be defined via the Girsanov transformation. In Subsection 3.1.4, the Girsanov transformation method was used to construct a weak-sense solution. Recall that the method worked by transforming the measure of the original process. The function $R(t)$ in (3.1.8) depended on both the initial condition $x(0)$ and the added drift term $\tilde{b}_1(\cdot)$ in (3.1.12). Strictly speaking, the original probability measure P should have been indexed by the initial condition $x(0)$, and the Radon–Nikodym derivative $R(t)$ should be indexed by

$x(0)$ and $\tilde{b}_1(\cdot)$. At that time we were concerned only with the solution for some given initial condition and new drift. But now we are concerned with all initial conditions and all possible added drifts $b_m(\cdot)$, and we need to use the appropriate notation.

First, consider the uncontrolled problem. When $x = x(0)$, denote the probability space by $(\Omega, P_x, \mathcal{F})$, with filtration $\{\mathcal{F}_t, t \geq 0\}$, where $\mathcal{F} = \cup_{t \geq 0} \mathcal{F}_t$ and Ω is a canonical path space for $(x(\cdot), w(\cdot), z(\cdot))$, say $D(G \times I\!R^v; 0, \infty)$, where v is the sum of the dimensions of $x(t), w(t)$, and $z(t)$. By a solution we mean that there is such a space and an $\mathcal{F}_t$-standard Wiener process $w(\cdot)$ such that the uncontrolled form of (3.1) holds, with $(x(\cdot), z(\cdot))$ nonanticipative. Strictly speaking, we should index $w(\cdot)$ by the initial condition $x = x(0)$ as well, but this only complicates the notation, without adding anything useful. We also use $(\Omega_T, P_{x,T}, \mathcal{F}_T)$ to denote the restriction to functions on the time interval $[0, T]$.

Now, define the controlled system. For a relaxed feedback control $m(\cdot)$ and $T > 0$, define

$$\zeta(T, m) = \int_0^T \left[\sigma^{-1}(x(s)) b_m(x(s))\right]' dw(s) - \frac{1}{2} \int_0^T \left|\sigma^{-1}(x(s)) b_m(x(s))\right|^2 ds, \tag{3.2}$$

and set

$$R(T, m) = e^{\zeta(T,m)}. \tag{3.3}$$

Following the method of Subsection 3.1.4, for each $(x, T, m(\cdot))$, define the measure $P^m_{x,T}$ on $(\Omega_T, \mathcal{F}_T)$ via the Radon–Nikodym derivative $R(T, m)$:

$$dP^m_{x,T} = R(T, m) dP_{x,T}. \tag{3.4}$$

For each $(x, m(\cdot))$, the family $P^m_{x,T}$ of measures, indexed by T, is consistent and can be extended uniquely to a measure P^m_x on $(\Omega, \mathcal{F})$ that is consistent with the $P^m_{x,T}$. When there is no control, we omit the superscript m. By Section 3.1.4, the process $w_m(\cdot)$ defined by

$$dw_m(t) = dw(t) - \left[\sigma^{-1}(x(s)) b_m(x(s))\right] dt \tag{3.5}$$

is an $\mathcal{F}_t$-standard Wiener process on $(\Omega, P^m_x, \mathcal{F})$. Now, rewrite (3.1) as

$$dx(t) = b_0(x(t))dt + b_m(x(t))dt + \sigma(x(t))dw_m(t) + dz(t). \tag{3.6a}$$

Let $P^m(x, t, \cdot)$ denote the transition function of the process (3.6a) and use $P(x, t, \cdot)$ for that of (3.1). The process $w_m(\cdot)$ should be indexed also by the initial condition $x = x(0)$, but we omit it for notational simplicity. If we use a relaxed, but not a relaxed feedback, control $r(\cdot)$ with derivative $r_t(\cdot)$, then define $b_r(x, t) = \int_{\mathcal{U}} b(x, \alpha) r_t(x, d\alpha)$ and $P^r_{x,T}$ by (3.4) with $b_r(x(t), t)$ and $w_r(\cdot)$ replacing $b_m(x(t))$ and $w_m(\cdot)$, respectively. Then we write

$$dx(t) = b_0(x(t))dt + b_r(x(t))dt + \sigma(x(t))dw_r(t) + dz(t). \tag{3.6b}$$

The following theorem uses nondegeneracy of the unconstrained process (that is, the fact that $\sigma(x)\sigma'(x)$ is positive definite). Such nondegeneracy is not so important for the proof of (3.7) for the unconstrained problem, and "hypoellipticity" can replace the nondegeneracy as in [151]. This is harder to do when there are boundaries.

Theorem 3.1. *Assume the model* (3.6a) *and assumptions* (A3.5.1), (A3.5.3), (A3.1), *and* (A3.2). *Let* $m^n(y,\cdot) \Rightarrow m(y,\cdot)$ *for almost all* y, *where* $m(\cdot)$ *and* $m^n(\cdot)$ *are relaxed feedback controls. Then for any Borel set* A,

$$P^{m^n}(x,t,A) \to P^m(x,t,A), \tag{3.7}$$

uniformly in $(x,t) \in G \times [t_0,t_1]$, *for any* $0 < t_0 < t_1 < \infty$. *That is, for any bounded and measurable real-valued function* $f(\cdot)$,

$$\int f(y)P^{m^n}(x,t,dy) \to \int f(y)P^m(x,t,dy) \tag{3.8}$$

uniformly in $G \times [t_0,t_1]$. *For any* $t > 0$, $P^m(x,t,\cdot)$ *is absolutely continuous with respect to Lebesgue measure, uniformly in the relaxed feedback control* $m(\cdot)$ *and in* $(x,t) \in G \times [t_0,t_1]$. *For each relaxed feedback control* $m(\cdot)$, *the process given by* (3.6a) *is a strong Feller process. The equation has a unique weak-sense solution for each* x.

Proof. We concentrate on the uniformity in x in (3.8) and leave the other details to the reader. The expression (3.8) can be written equivalently as (P_x-measure always used when writing E_x)

$$E_x f(x(t))R(t,m^n) - E_x f(x(t))R(t,m) \to 0. \tag{3.9}$$

For notational simplicity, let $\sigma(x) = I$, the identity. The details are the same for general nondegenerate $\sigma(\cdot)$. We will use the inequalities:

$$\left|e^a - e^b\right| \le |a-b| \left|e^a + e^b\right|, \tag{3.10}$$

$$\begin{aligned} E_x &\left| \int_0^t b_m'(x(s))dw(s) - \int_0^t b_{m^n}'(x(s))dw(s) \right|^2, \\ &\le E_x \int_0^t \left|b_m(x(s)) - b_{m^n}(x(s))\right|^2 ds. \end{aligned} \tag{3.11}$$

By the continuity and uniform boundedness of $b_c(\cdot)$ and the weak convergence assumption on the $m^n(y,\cdot)$, we have

$$b_{m^n}(y) = \int_{\mathcal{U}} b_c(y,\alpha)m^n(y,d\alpha) \to b_m(y) = \int_{\mathcal{U}} b_c(y,\alpha)m(y,d\alpha) \tag{3.12}$$

for almost all y (Lebesgue measure, as usual). Define

$$\tilde{b}_n(y) = \left|b_m(y) - b_{m^n}(y)\right|^2.$$

By Egoroff's theorem [61, Theorem 12, page 149], for each $\epsilon > 0$, there is a measurable set A_ϵ with $l(A_\epsilon) \leq \epsilon$ such that $\tilde{b}_n(y) \to 0$ uniformly in $y \notin A_\epsilon$. Furthermore, $P(x,t,\cdot)$ is absolutely continuous with respect to Lebesgue measure for each x and $t > 0$ (and uniformly in $(x,t) \in G \times [t_0, t_1]$ for any $0 < t_0 < t_1 < \infty$). These facts imply that

$$\int_0^t E_x \tilde{b}_n(x(s))ds \to 0,$$

uniformly in $x \in G$. The last expression together with (3.10)–(3.11) implies (3.9) uniformly in $x \in G$. ■

Theorem 3.2. *Assume the conditions of Theorem 3.1. For each relaxed feedback control $m(\cdot)$ there is a unique invariant measure $\mu_m(\cdot)$, and the set $\{\mu_m(\cdot)\}$ over all $m(\cdot)$ is tight. The $\mu_m(\cdot)$ are absolutely continuous with respect to Lebesgue measure, uniformly in $m(\cdot)$.*

Proof. Since $x(\cdot)$ is a Markov–Feller process for each $m(\cdot)$ and G is bounded, there is at least one invariant measure [11, 12]. The set of invariant measures is tight, since G is compact. By the Markov property and the Girsanov transformation,

$$\mu_m(A) = \int \mu_m(dx)P^m(x,t,A) = \int \mu_m(dx)E_x I_A(x(t))R(t,m). \quad (3.13)$$

There is a real K such that for all $x \in G$ and all $m(\cdot)$, $E_x I_A(x(t))R(t,m) \leq E_x^{1/2} I_A(x(t)) E_x^{1/2} R^2(t,m) \leq K E_x^{1/2} I_A(x(t))$. Now, by the absolute continuity (uniformly in $x \in G$ for each $t > 0$) of $P(x,t,\cdot)$ with respect to Lebesgue measure, for any $\delta > 0$ there is $\epsilon > 0$ such that $l(A) \leq \epsilon$ implies that for all $m(\cdot)$ and all $x \in G$, $E_x I_A(x(t))R(t,m) \leq \delta$, which implies the theorem. ■

The next theorem is a partial converse to the absolute continuity assertions of Theorems 3.1 and 3.2.

Theorem 3.3. *Assume the conditions of Theorem 3.1. Let $0 < t_0 < t_1 < \infty$. For each $\epsilon > 0$ there is $\delta > 0$ such that*

$$\inf_{t_0 \leq t \leq t_1} \inf_{x \in G} \inf_m \inf_{\{A: l(A) \geq \epsilon\}} P^m(x,t,A) \geq \delta, \quad (3.14)$$

$$\inf_{t_0 \leq t \leq t_1} \inf_m \inf_{\{A: l(A) \geq \epsilon\}} \mu_m(A) \geq \delta. \quad (3.15)$$

Proof. For positive k, define the set $B_k^m = \{\inf_{t_0 \leq t \leq t_1} \zeta(t,m) \geq -k\}$, where $\zeta(t,m)$ was defined by (3.2). Then

$$E_x R(t,m) I_A(x(t)) \geq e^{-k} E_x I_{B_k^m} I_A(x(t)). \quad (3.16)$$

Define

$$\delta_1 = \inf_{x \in G} \inf_{t_0 \le t \le t_1} \inf_{\{A: l(A) \ge \epsilon\}} P(x,t,A).$$

By Theorem 2.2, $\delta_1 > 0$ if $\epsilon > 0$. Choose k such that $\inf_{x \in G, m} P_x\{B_k^m\} \ge 1 - \delta_1/2$. Now,

$$\inf_{t_0 \le t \le t_1} \inf_{x \in G, m} E_x I_{B_k^m} I_A(x(t)) \ge \delta_1/2,$$

which yields (3.14) for $\delta = e^{-k}\delta_1/2$. The relation (3.15) is proved in a similar manner. It is sufficient to work with a sequence $m^n(\cdot)$. For $t > 0$, write

$$\mu_{m^n}(A) = \int \mu_{m^n}(dy) P^{m^n}(y,t,A) \ge \int \mu_{m^n}(dy) g(y),$$

where $g(y) = \inf_n \inf_{\{A: l(A) \ge \epsilon\}} P^{m^n}(y,t,A)$. By (3.14), $g(y) \ge \delta > 0$. Thus,

$$\mu_{m^n}(A) \ge \delta \int \mu_{m^n}(dy) \ge \delta,$$

which yields (3.15). ∎

Smoothing and discretizing a relaxed feedback control. The following approximations will be useful in the proofs of convergence of the ergodic costs for the heavy traffic problems. Extend the definition of $m(y, \cdot)$ so that it is defined for y in some small neighborhood of G. For an integer n and relaxed feedback control $m(\cdot)$, define the smoothed control

$$m_\epsilon(x, \cdot) = \frac{1}{\sqrt{2\pi\epsilon}} \int_{I\!R^r} e^{-|y-x|^2/2\epsilon} m(y, \cdot) dy, \quad x \in G. \tag{3.17}$$

Then it is easy to verify the next theorem, whose details are left to the reader.

Theorem 3.4. *Assume the conditions of Theorem* 3.1. *Then $m_\epsilon(\cdot)$ is a relaxed feedback control and $m_\epsilon(x, \cdot) \Rightarrow m(x, \cdot)$ for almost all x. The function $b_{m_\epsilon}(\cdot)$ is continuous for each ϵ, and $b_{m_\epsilon}(\cdot)$ converges almost everywhere to $b_m(\cdot)$.*

Let $m_\epsilon(\cdot)$ be obtained from some relaxed feedback control $m(\cdot)$ by smoothing as in (3.17). Given $\rho > 0$, let $\{U_i^\rho\}$ be a finite partition of $\mathcal{U}$, where each U_i^ρ has diameter less than ρ, and let $u_i^\rho \in U_i^\rho$. Define the relaxed feedback control $m_{\epsilon,\rho}(\cdot)$, concentrated on the set $\mathcal{U}_\rho = \{u_i^\rho\}$, by $m_{\epsilon,\rho}(x, u_i^\rho) = m_\epsilon(x, U_i^\rho)$. The functions $p_i(x) \equiv m_{\epsilon,\rho}(x, u_i^\rho)$ can be assumed to be arbitrarily smooth in x. The proof of the following theorem is straightforward, and the details are omitted.

Theorem 3.5. *Let G and $\mathcal{U}$ be compact, and G the closure of its interior. Suppose that $b_c(\cdot)$ is continuous. Then as $\rho \to 0$, for each x the sequence*

$m_{\epsilon,\rho}(x,\cdot)$ converges weakly to $m_\epsilon(x,\cdot)$ and, uniformly in $x \in G$,

$$\int_{\mathcal{U}} b_c(x,\alpha) m_{\epsilon,\rho}(x,d\alpha) \to \int_{\mathcal{U}} b_c(x,\alpha) m_\epsilon(x,d\alpha).$$

Time on the intersection of two faces. The following theorem asserts that the time spent near the intersection of two or more boundary faces has little effect on the reflection process in (2.2).

Theorem 3.6. *Assume the conditions of Theorem 3.1, but where G is not necessarily compact and $b_0(\cdot)$ and $b_m(\cdot)$ are bounded. Suppose that no more than r constraints (that are taken to be linearly independent) are active at any boundary point. Then for any pair of intersecting boundary faces ∂G_i and $\partial G_j, j \neq i$, and each initial condition and control,*

$$\int_0^\infty I_{\{x(s)\in \partial G_i \cap \partial G_j\}} d|z|(s) = 0. \tag{3.18}$$

Proof. If G were the orthant $[\mathbb{R}^+]^r = \{x : x_i \geq 0, i \leq r\}$, $b_0(\cdot)$ a constant, and there were no control, then the proof is in [212]. Suppose that for some $k < r$, G was actually the wedge $\{x : x_i \geq 0, i \leq k\}$. The proof in [212] can be modified by restricting the test functions to depend only on the relevant first k coordinates. The proof in [212] actually covers the case where $b_0(\cdot) + b_m(\cdot)$ is bounded and measurable but not necessarily a constant, since we still have a strong Markov process, and the Girsanov transformation technique of the reference still applies.

A linear transformation of coordinates can be used[2] to extend the proof to cover the general form of G that is defined by (A3.5.1) and (A3.5.3). Since it is sufficient to restrict attention to a neighborhood of any corner or edge, without loss of generality suppose that G is an infinite cone or wedge defined by $G = \{x : N'x \geq 0\}$, where N is the matrix whose columns are the interior normals to the boundary faces. We will restrict attention to the case where G is determined by r constraints, so that $N = \{n_1, \ldots, n_r\}$, where n_i is the (column vector) interior normal to ∂G_i. Let d_i denote the reflection direction on ∂G_i.

Define $\Gamma = N^{-1}$. Inequalities between vectors are defined component by component. Then

$$\begin{aligned}[\mathbb{R}^+]^r &= \{x : Ix = N'\Gamma' x \geq 0\} = \{x : N'v \geq 0 \text{ for } v = \Gamma' x\} \\ &= \{x : x = [\Gamma']^{-1} v \text{ for } N'v \geq 0\} = [\Gamma']^{-1} G = N'G.\end{aligned}$$

Thus $G = \Gamma'[\mathbb{R}^+]^r$ and G and $[\mathbb{R}^+]^r$ are related by invertible linear transformations.

---

[2]The proof of this extension is due to R. Williams.

We need only show that the condition (A3.5.3) on G continues to hold after the linear transformation. Condition (A3.5.3) is equivalent to the existence of a vector $\alpha = (\alpha_1, \ldots, \alpha_r)$, with $\alpha_i \geq 0$ and such that

$$\left\langle \sum_i \alpha_i d_i, n_j \right\rangle = \sum_i \alpha_i \langle d_i, n_j \rangle > 0, \quad \text{for all } j.$$

Let e_i denote the unit vector in the ith coordinate direction, which is the interior normal to the transformed face $\{x : x_i = 0, x_j > 0, \ j \neq i\}$, and note that $N'd_i$ is the transformed reflection direction on this face. Then

$$\langle d_i, n_j \rangle = \langle d_i, Ne_j \rangle = \langle N'd_i, e_j \rangle.$$

Thus, $\langle \sum_i \alpha_i N'd_i, e_j \rangle > 0$ for all j, which implies that (A3.5.3) continues to hold under the transformation. ■

4.4 Optimal Feedback Controls

This section demonstrates several important facts concerning optimal controls for ergodic cost functions. First, it is shown that whatever the initial condition, the actual limit cost per unit time is the same as the cost for the stationary system. Then it is shown that the invariant measure is "continuous" in the control in a specific sense and that an optimal relaxed feedback control exists. Finally, approximations to the optimal control are given, and the special case where the control appears linearly in the dynamics and cost is exhibited.

Assumptions on the cost function.

A4.1. *Let $k(x, \alpha) = k_0(x) + k_c(x, \alpha)$, where all functions are continuous and real-valued and let c_i be nonnegative real numbers.*

Recall the representation $z(t) = \sum_i y_i(t) d_i$ of Section 3.5. Define $y(\cdot)$ to be the vector with components $y_i(\cdot)$, and define c analogously. For an admissible relaxed control $r(\cdot)$ with derivative $r_t(\cdot)$, define $k_r(x,t) = \int k_c(x,\alpha) r_t(d\alpha)$ and

$$\gamma_T(x, r) = \frac{1}{T} E_x^r \int_0^T \left[k_0(x(s)) + k_r(x(s), s) \right] ds + \frac{1}{T} E_x^r c' y(T).$$

Then the cost function of interest is

$$\gamma(x, r) = \limsup_T \gamma_T(x, r). \tag{4.1}$$

For a relaxed feedback control $m(\cdot)$, define $k_m(x) = \int_{\mathcal{U}} k_c(x,\alpha)m(x,d\alpha)$, with cost function $\gamma(m)$, where

$$\gamma_T(x,m) = E_x^m \frac{1}{T}\int_0^T [k_0(x(s)) + k_m(x(s))]\,ds + \frac{1}{T}E_x^m c'y(T), \qquad (4.2a)$$

$$\gamma(m) = \limsup_T \gamma_T(x,m). \qquad (4.2b)$$

We omit the $x = x(0)$ from the argument of $\gamma(m)$, since it will not depend on the initial condition under our conditions. Define $\bar{\gamma} = \inf_m \gamma(m)$.

Representation of the cost in terms of a stationary system. Let $m(\cdot)$ be a relaxed feedback control. The system (3.6a) starts with an arbitrary initial condition, that does not necessarily have the stationary distribution. It turns out that the limit (4.2b) is the same as if the initial condition were distributed as $\mu_m(\cdot)$. This is the assertion of the next theorem.

Theorem 4.1. *Assume the model* (3.6a) *and assumptions* (A3.5.1), (A3.5.3), (A3.1), (A3.2), *and* (A4.1), *and let* $m(\cdot)$ *be a relaxed feedback control. Then the* $E_x^m y_i(1)$ *are continuous functions of* x *and*

$$\limsup_T \gamma_T(x,m) = \lim_T \gamma_T(x,m) = \gamma(m)$$
$$= \int [k_0(x) + k_m(x)]\,\mu_m(dx) + \int E_x^m\,[c'\bar{y}(1)]\,\mu_m(dx).$$

Proof. Theorem 3.5.3 implies that $\sup_m \sup_{x\in G} E_x^m \sum_i y_i^2(1) < \infty$. Hence the expectations in the cost functional are well defined. The proof of continuity in x uses a weak convergence argument and the weak-sense uniqueness of the solution to (3.6a) like that outlined at the end of Section 3.4 and the details are omitted.

Let N denote the integer part of T and define

$$Q^{m,T}(x,\cdot) = \frac{1}{T}\int_0^T P^m(x,t,\cdot)dt, \quad Q^{m,N}(x,\cdot) = \frac{1}{N}\sum_{n=1}^{N} P^m(x,n,\cdot).$$

Then the cost equals (modulo an asymptotically negligible error)

$$\gamma_T(x,m) = \int [k_0(y) + k_m(y)]\,Q^{m,T}(x,dy) + \int E_y^m\,[c'y(1)]\,Q^{m,N}(x,dy).$$

By the fact that $x(\cdot)$ under $m(\cdot)$ is a Markov–Feller process with a unique invariant measure $\mu_m(\cdot)$, $Q^{m,T}(x,\cdot)$ and $Q^{m,N}(x,\cdot)$ converge weakly to $\mu_m(\cdot)$ as $T \to \infty$ [11, 12]. In fact, Theorem 1.1 implies that $Q^{m,T}(x,A) \to \mu_m(A)$

and $Q^{m,N}(x,A) \to \mu_m(A)$ for each x and Borel set A. The theorem follows from this. ∎

Continuity of the invariant measure in the control.

Theorem 4.2. *Assume the conditions of Theorem 4.1. Then $\mu_m(\cdot)$ is continuous in the control in that if $m^n(x,\cdot) \Rightarrow m(x,\cdot)$ for almost all x, then for each Borel set A,*

$$\mu_{m^n}(A) \to \mu_m(A). \tag{4.3}$$

Proof. By extracting a weakly convergent subsequence of $\{\mu_{m^n}(\cdot)\}$ and relabeling, we can suppose that there is $\mu(\cdot)$ such that $\mu_{m^n}(\cdot) \Rightarrow \mu(\cdot)$. Let $f(\cdot)$ be a bounded and continuous real-valued function on G. Let $t > 0$. Then using the definition of an invariant measure,

$$\int \mu_{m^n}(dx) \int f(y) P^{m^n}(x,t,dy) = \int f(x)\mu_{m^n}(dx) \to \int f(x)\mu(dx). \tag{4.4}$$

By (3.7), $\int f(y)P^{m^n}(x,t,dy) \to \int f(y)P^m(x,t,dy)$, uniformly in $x \in G$, and the limit is continuous in $x \in G$ since $x(\cdot)$ is a strong Feller process under the relaxed feedback control $m(\cdot)$. Thus, (4.4) implies that

$$\begin{aligned}\int f(x)\mu(dx) &= \lim_n \int \mu_{m^n}(dx) \int f(y)P^m(x,t,dy) \\ &= \int \mu(dx) \int f(y)P^m(x,t,dy).\end{aligned} \tag{4.5}$$

The expression (4.5), the arbitrariness of $f(\cdot)$, and the fact that the invariant measure is unique (Theorem 1.1) imply that $\mu(\cdot) = \mu_m(\cdot)$. Now, for a Borel set A, let $f(x) = I_A(x)$. Then the two right-hand terms in (4.5) are still equal, and they equal the limit of the left-hand term in (4.4), which proves (4.3). ∎

Theorem 4.3. *Assume the conditions of Theorem 4.1. Then in the class of relaxed feedback controls, there exists an optimal control.*

Proof. We will use a modification of a technique of Borkar [19]. Let $m^n(\cdot)$ be a minimizing sequence of relaxed feedback controls with associated standard Wiener processes $w^n(\cdot)$, and define the measure $q^n(\cdot)$ on the Borel sets of $G \times \mathcal{U}$ by $q^n(dx\,d\alpha) = m^n(x,d\alpha)dx$. The compactness of $\mathcal{U}$ and G implies that the set $\{q^n(\cdot)\}$ is tight. Suppose that it converges weakly (extract a subsequence, if necessary) to the measure $q(\cdot)$. Factor $q(\cdot)$ as $q(dx,d\alpha) = m(x,d\alpha)dx$. In addition, by extracting a further subsequence if necessary, let $\mu_{m^n}(\cdot)$ converge weakly to a measure $\mu(\cdot)$. We need to show that $m(\cdot)$ is a relaxed feedback control and $\mu(\cdot)$ is the associated invariant measure. By its definition, $m(\cdot,A)$ is measurable for each Borel set A, and

$m(x,\cdot)$ is a measure on the Borel sets of $\mathcal{U}$ for almost all x. Define $m(x,\cdot)$ on the exceptional x-set so that it is a relaxed feedback control. For notational simplicity, absorb $b_0(\cdot)$ into $b_c(\cdot)$. Define $\psi^n(t) = \int_0^t b_{m^n}(x^n(s))ds$, where $(x^n(\cdot), w^n(\cdot), z^n(\cdot))$ denotes the processes associated with $m^n(\cdot)$ and initial measure $\mu_{m^n}(\cdot)$, and $x^n(\cdot)$ is stationary.

The sequence of processes $\{x^n(\cdot), w^n(\cdot), \psi^n(\cdot), z^n(\cdot)\}$ is tight. By extracting a further subsequence, if necessary, suppose that it converges weakly to $(x(\cdot), w(\cdot), \psi(\cdot), z(\cdot))$. Then $x(\cdot)$ is stationary, in that $x(t)$ has the distribution $\mu(\cdot)$ for all $t \geq 0$. In addition,

$$x(t) = x(0) + \psi(t) + \int_0^t \sigma(x(s))dw(s) + z(t), \tag{4.6}$$

where $z(\cdot)$ is the reflection term and $(x(\cdot), \psi(\cdot), z(\cdot))$ are nonanticipative with respect to the standard vector-valued Wiener process $w(\cdot)$. (This can be shown by an adaptation of the proof of Theorem 3.5.4.) Suppose that we can write

$$\psi(t) = \int_0^t b_m(x(s))ds. \tag{4.7}$$

Then we will have shown that $x(\cdot)$ is the stationary process under the relaxed feedback control $m(\cdot)$.

Let $h(\cdot)$ be a bounded continuous function of its arguments, p an arbitrary integer, $0 \leq s_i \leq t \leq t+\tau$, $i \leq p$, arbitrary, and let $f(\cdot)$ be an arbitrary twice continuously differentiable real-valued function with compact support. Write $\Psi(t) = (x(t), w(t), \psi(t), z(t))$ and define $\Psi^n(t)$ analogously for the processes under $m^n(\cdot)$. We will show that for each small $\delta > 0$ and where $\mathcal{L}^0$ denotes the differential generator of the *unreflected and uncontrolled process*,

$$Eh(\Psi(s_i), i \leq p)\left[f(x(t+\tau)) - f(x(t+\delta)) - F(t+\delta, t+\tau)\right] = 0, \tag{4.8}$$

where

$$\begin{aligned} &F(t, t+\tau) \\ &= \int_t^{t+\tau} \mathcal{L}^0 f(x(s))ds + \int_t^{t+\tau} f_x'(x(s))dz(s) + \int_t^{t+\tau} f_x'(x(s))b_m(x(s))ds. \end{aligned}$$

Equation (4.8) holds if $x^n(\cdot)$, $z^n(\cdot)$, and $m^n(\cdot)$ are used. It is not hard to show that the limits of the parts of the expression (4.8) that are due to the components of $F(t+\delta, t+\tau)$ are the analogous expressions for the limit process, except possibly for the one due to $b_m(\cdot)$, and we will concentrate on this component. Write

$$\begin{aligned} &E_{x^n(t)}^{m^n} \int_{t+\delta}^{t+\tau} f_x'(x^n(s))b_{m^n}(x^n(s))ds \\ &= \int_\delta^\tau ds \int P^{m^n}(x^n(t), s, dy) \int_{\mathcal{U}} f_x'(y)b_c(y,\alpha)m^n(y, d\alpha). \end{aligned} \tag{4.9}$$

Write $P^{m^n}(x,s,dy)$ in terms of its density as $p^n(x,s,y)dy$. By the uniform absolute continuity in Theorem 3.1, the set of functions $\{p^n(x,s,\cdot); x \in G, \delta \le s \le \tau\}$ is uniformly integrable. Hence, for our purposes, it can be supposed to be bounded. Let $\epsilon > 0$. For each n, x, s, define $p^n_\epsilon(x,s,\cdot)$ by smoothing $p^n(x,s,\cdot)$ analogously to what was done in (3.17). Then for small ϵ, the value of (4.9) changes arbitrarily little, uniformly in n, if $p^n_\epsilon(x,s,y)dy$ is used for $P^{m^n}(x,s,dy)$. Thus, the approximation to (4.9) is now

$$\int_\delta^t ds \int dy \int_{\mathcal{U}} p^n_\epsilon(x^n(t),s,y) f'_x(y) b_c(y,\alpha) m^n(y,d\alpha). \tag{4.10a}$$

Owing to the smoothing, $\{p^n_\epsilon(x,s,\cdot); x \in G, \delta \le s \le \tau\}$ is equicontinuous. By the equicontinuity, we can replace $m^n(y,d\alpha)dy$ in (4.10a) by $m(y,d\alpha)dy$, while changing the value of the expression by an arbitrarily small amount for large n. The approximation to (4.9) is now

$$\int_\delta^\tau ds \int p^n_\epsilon(x^n(t),s,y) f'_x(y) b_m(y) dy, \tag{4.10b}$$

whose values differ from those of (4.9) by an arbitrarily small amount for large n and small ϵ. Now, for small $\rho > 0$ approximate $m(\cdot)$ in (4.10b) by the smoothed version $m_\rho(x,\cdot)$ obtained via (3.17). This changes the values of (4.10b) by arbitrarily little, uniformly in n and ϵ. Again, we change the values arbitrarily little, uniformly in n and ρ, by letting $\epsilon \to 0$. This final step leads to the approximating expression

$$E^{m^n}_{x^n(t)} \int_{t+\delta}^{t+\tau} f'_x(x^n(s)) b_{m_\rho}(x^n(s)) ds.$$

Thus, we need to evaluate the limit (as $n \to \infty$) of

$$Eh(\Psi^n(s_i), i \le p) \int_{t+\delta}^{t+\tau} f'_x(x^n(s)) b_{m_\rho}(x^n(s)) ds.$$

But the limit is just

$$Eh(\Psi(s_i), i \le p) \int_{t+\delta}^{t+\tau} f'_x(x(s)) b_{m_\rho}(x(s)) ds. \tag{4.11}$$

Finally, since ρ was arbitrarily small, the value of (4.11) changes arbitrarily little if $\rho \to 0$.

We can conclude that (4.8) holds. Hence, $x(\cdot)$ is a stationary process with relaxed feedback control $m(\cdot)$ and invariant measure $\mu(\cdot) = \mu_m(\cdot)$. Now, (4.8) and Theorem 2.2.1 imply that there is a Wiener process $\bar{w}(\cdot)$ with respect to which $(x(\cdot), z(\cdot))$ is nonanticipative and

$$dx(t) = b_m(x(t))dt + \sigma(x(t))d\bar{w}(t) + dz(t).$$

One can show that $w(\cdot) = \bar{w}(\cdot)$, but the details are omitted.

It remains to show that $\gamma(m^n) \to \gamma(m)$. A sequence of approximations similar to those used above yields that

$$\begin{aligned} &\int [k_0(x) + k_c(x,\alpha)]\, m^n(x,d\alpha)\mu_{m^n}(dx) \\ &\qquad \to \int [k_0(x) + k_c(x,\alpha)]\, m(x,d\alpha)\mu(dx). \end{aligned} \tag{4.12}$$

The sequence $E_x^{m^n} y_i^n(1)$ converges to a continuous limit $E_x^m y_i(1)$, uniformly in $x \in G$. Hence

$$\int \mu_{m^n}(dx) E_x^{m^n} y_i^n(1) \to \int \mu_m(dx) E_x^m y_i(1),$$

which concludes the proof. ∎

Smooth ϵ-optimal relaxed feedback controls. Since Theorem 4.3 implies that $\gamma(m^n) \to \gamma(m)$ if $m^n(x,\cdot)$ converges weakly to $m(x,\cdot)$ for almost all x, Theorem 4.3 and Theorems 3.4 and 3.5 yield the next theorem.

Theorem 4.4. *Under the conditions of Theorem 4.1, for each $\epsilon > 0$ there is an ϵ-optimal feedback relaxed control $m_\epsilon(\cdot)$, where $m_\epsilon(x, A)$ is arbitrarily smooth in x, uniformly in the Borel set A.*

Given $\epsilon > 0$, there is an integer ρ and a subset $\mathcal{U}_\epsilon = \{u_i, i \le \rho\} \subset \mathcal{U}$ and infinitely differentiable functions $p_i(\cdot), i \le \rho$, such that $\sum_i p_i(x) = 1$ and the relaxed control defined by

$$m_\epsilon(x, B) = \sum_i p_i(x) I_{\{u_i \in B\}} \tag{4.13}$$

is ϵ-optimal.

Controls appearing linearly. Suppose that the controls appear linearly in that $b_c(x,\alpha) = D\alpha$ and $k_c(x,\alpha) = \bar{c}'\alpha$. Let $\mathcal{U}$ be convex. Then the relaxed feedback controls are equivalent to ordinary feedback controls, since we can write $\int_{\mathcal{U}} \alpha m(x, d\alpha) = u(x)$, which is an ordinary control. The analogous result holds for relaxed controls that are not feedback. Then Theorem 4.4 simplifies to the following. See [155, Chapter 9] and [192] for the second part.

Theorem 4.5. *Under the conditions of Theorem 4.1, but with convex $\mathcal{U}$, $b_c(x,\alpha) = D\alpha$ and $k_c(x,\alpha) = \bar{c}'\alpha$, for each $\epsilon > 0$ there is an arbitrarily smooth ϵ-optimal feedback control $u(\cdot)$. More generally, keep the conditions of Theorem 4.1, but replace the linearity by the condition that the set $(b_c(x,\mathcal{U}),\ k_c(x,\mathcal{U}))$ is convex for each x. Then for each relaxed feedback* *control $m(\cdot)$ there is an equivalent ordinary feedback control $u(\cdot)$, in that $u(\cdot)$ is measurable and $\mathcal{U}$-valued with the same dynamics and cost. For any $\epsilon > 0$, there is an ϵ-optimal arbitrarily smooth ordinary feedback control $u(\cdot)$.*

4.5 A Maximum Principle

The results in this and in the next two sections will be used in the heavy traffic analysis in Chapter 9 to verify that the optimal cost per unit time for the physical model converges to the optimal ergodic cost for the limit model, and that there are good controls for the limit problem that give nearly optimal results for the physical problem under heavy traffic. Such results help justify the use of the heavy traffic limit for the analysis of optimization problems for the physical system. They are an extension and simplification of the results in [156]. The reference [57] concerns a similar maximum principle, but for finite time problems and with (ω, t)-dependent controls, and not feedback controls.

The Bellman equation. For the moment, suppose that G has a "smooth boundary" with (unique) reflection direction $d(x)$ at each $x \in \partial G$, where $d(\cdot)$ is "smooth." From a purely formal point of view, the Bellman equation for the minimal cost can be written as

$$h(x) + \bar{\gamma} = \min_{\alpha \in \mathcal{U}} \left[\mathcal{L}^{\alpha} h(x) + k(x, \alpha)\right], \tag{5.1}$$

where $\bar{\gamma}$ is the infimal ergodic cost and $h(\cdot)$ is an auxiliary function that satisfies the boundary condition $h'_x(x)d(x) =$ (boundary cost rate at $x \in \partial G$) [78]. When (5.1) is "well posed," the minimizer of the right-hand side is a candidate for the optimal control $\bar{u}(\cdot)$. To show that it is an optimal control, one uses a "verification theorem" [78]. Some additional work is often needed to ensure that there is a well-defined solution to the stochastic differential equation under this control. For the problem of interest here, the sets G do not have smooth boundaries, and little is known about (5.1). In order to avoid problems with the Bellman equation, we take a slightly indirect approach. Following the development in [156], we obtain a convenient maximum principle, that gives a necessary and sufficient condition for optimality and allows us to draw the desired conclusions.

The probability space. In this section the sample space is $\Omega = \Omega_1 \times \Omega_2 = D[G; 0, \infty) \times D[\mathbb{R}^r; 0, \infty)$. Let the canonical paths be denoted by $\omega = (\omega_1, \omega_2) = (p_1(\omega_1, \cdot), p_2(\omega_2, \cdot))$. Ω_1 contains the paths of the $x(\cdot)$ process and Ω_2 those for the $\mathbb{R}^r$-valued standard Wiener process $w(\cdot)$. Let $\mathcal{F}_t$ denote the σ-algebra induced by the coordinate projections to time t. Define the left shift operator θ_t on Ω_1 by $\theta_t \omega_1 = p_1(\omega_1, t + \cdot)$.

Definition: Homogeneous additive functionals of $x(\cdot)$. Suppose that the transition function of $x(\cdot)$ does not depend on the initial time (as would be the case under a relaxed feedback control, since $b_0(\cdot)$, $b_c(\cdot)$, and $\sigma(\cdot)$ do not depend explicitly on time). Then $A(\cdot)$ is said to be a homogeneous additive functional of $x(\cdot)$ if for each $t, s \geq 0$ [195, page 119],

$$A(\omega, t+s) = A(\omega, t) + A(\theta_t\omega, s),$$

where θ_t is the left shift operator on the path space of the $x(\cdot)$ process that was defined above. An example is the integral defined by $A(t) = \int_0^t f(x(v))dv$.

The following assumptions will be used.

A5.1. (A3.5.1)–(A3.5.3) *hold with G bounded. The functions $\sigma(\cdot)$ and $b_0(\cdot)$ are Lipschitz continuous, $\sigma^{-1}(\cdot)$ exists and is bounded and continuous, $b_c(\cdot)$ is continuous and $b_c(\cdot, \alpha)$ is Lipschitz continuous uniformly in $\alpha \in \mathcal{U}$, where $\mathcal{U}$ is compact. Also, $b_c(x, 0) = 0$ and $0 \in \mathcal{U}$.*

A5.2. *No more than r constraints, which are taken to be linearly independent, are active at any boundary point.*

Remark on (A5.1.) Under (A5.1) there is a strong-sense unique solution to (3.6a) when either $b_c(\cdot)$ is deleted or if $m(\cdot)$ is a smoothed relaxed feedback control of the type $m_\epsilon(\cdot)$ constructed in (3.17). Then $(x(\cdot), z(\cdot))$ will be a functional of the Wiener process $w_m(\cdot)$. This fact will be important for the martingale representation Theorem 5.3.

A representation of the reflection process in terms of local time. Under (A5.1) and (A5.2), Theorem 3.6 says that the contribution to the $z(\cdot) = \sum_i d_i y_i(\cdot)$ process due to time spent in an arbitrarily small neighborhood of the edges and corners is negligible. Thus, the reflection process is a "local time process, " as seen in [114, Chapter 3, Section 4.2, Chapter 4, Section 7] and $y(\cdot)$ is a homogeneous additive functional of $x(\cdot)$.

Write the uncontrolled process as

$$dx(t) = b_0(x(t))dt + \sigma(x(t))dw(t) + dz(t). \tag{5.2}$$

Let $m(\cdot)$ be a relaxed feedback control and write the associated controlled process as (3.6a), with the standard Wiener process $w_m(\cdot)$ (under P_x^m). It is convenient to split the cost into its "running" and "boundary" components as

$$\gamma(m) = \int [k_0(x) + k_m(x)]\mu_m(dx) + \int E_x^m\left[c'y(1)\right]\mu_m(dx) = \gamma_k(m) + \gamma_z(m).$$

Define $\hat{k}_m(x) = k_0(x) + k_m(x) - \gamma_k(m)$, $A_m(x) = \lim_t [E_x^m c'y(t) - \gamma_z(m)t]$ and

$$H_m(x) = \int_0^\infty E_x^m \hat{k}_m(x(s))ds + A_m(x).$$

The following result will play a fundamental role in the development.

Theorem 5.1. *Assume* (A5.1) *and* (A5.2). *Then* $E_x^m \hat{k}_m(x(t))$ *and the limit defining* $A_m(x)$ *converge exponentially, and uniformly in* $x \in G$ *and in* $m(\cdot)$. *The two components of* $H_m(\cdot)$ *are continuous. The process defined by*

$$M_m(t) = H_m(x(t)) - H_m(x) + \int_0^t \hat{k}(x(s))ds + [c'y(t) - \gamma_z(m)t] \tag{5.3}$$

is a continuous P_x^m*-martingale for each* x *and* $m(\cdot)$.

Proof. The convergence follows from the remark after Theorem 2.3 together with the fact that $E_x^m y_i(t)$ is continuous in x for each $m(\cdot)$ and $t > 0$. The martingale property is verified by a direct demonstration that the process satisfies the definition, and we omit the details. ■

Later we will use the important fact that $M_m(t)$ is a homogeneous additive functional of $x(\cdot)$. To see this in a simple case, drop the $y_i(\cdot)$ term in the cost and note that, pathwise,

$$M_m(\theta_t\omega, s) = H_m(p_1(\omega_1, t+s)) - H_m(p_1(\omega_1, t)) + \int_t^{t+s} \hat{k}(p_1(\omega_1, u))du.$$

Thus,

$$M_m(t)+M_m(\theta_t\omega, s) = H_m(x(t+s))-H_m(x)+\int_0^{t+s} \hat{k}(x(u))du = M_m(t+s),$$

as required. A similar computation can be used for the $y(\cdot)$ term. See [68] for an extensive use of this "shift" notation.

Theorem 5.3 will show that for any relaxed feedback control $m(\cdot)$ there is an $\mathbb{R}^r$-valued Borel function $\psi_m(\cdot)$ such that $\sup_{x\in G} E_x^m \int_0^t |\psi_m(x(s))|^2 ds < \infty$ for each $t < \infty$ and

$$M_m(t) = \int_0^t \psi_m'(x(s))\sigma(x(s))dw_m(s). \tag{5.4}$$

This function will be used in the next theorem.

Theorem 5.2. [156, an extension of part of Theorem 6.1.] *Assume* (A5.1) *and* (A5.2). *Given relaxed feedback controls* $m(\cdot)$ *and* $v(\cdot)$, *define*

$$e^{m,v}(x) = [k_m(x) - k_v(x)] + \psi_m'(x)[b_m(x) - b_v(x)].$$

If $e^{m,v}(x) > 0$ on a set of positive Lebesgue measure, then there is a relaxed feedback control $\hat{v}(\cdot)$ such that $\gamma(\hat{v}) < \gamma(m)$. The condition $e^{m,v}(x) \leq 0$ for almost all x for each relaxed feedback control $v(\cdot)$ is necessary and sufficient for the optimality of $m(\cdot)$ within the class of relaxed feedback controls.

Proof. We can write

$$dx(t) = [b_0(x(t)) + b_v(x(t))]\,dt + \sigma(x(s))dw_v + dz(t),$$

where

$$\begin{aligned} dw_v(t) &= dw(t) - \sigma^{-1}(x(t))b_v(x(t))dt \\ &= dw_m(t) + \sigma^{-1}(x(t))\,[b_m(x(t)) - b_v(x(t))]\,dt. \end{aligned} \tag{5.5}$$

Since the P_x^v and P_x^m are mutually absolutely continuous for each fixed x, a w.p.1 assertion with respect to one is also a w.p.1 assertion with respect to the other.

By Theorems 3.4 and 4.2, without loss of generality, we can suppose that $m(\cdot)$ is of the smoothed type of (3.17). Then $b_m(x)$ is Lipschitz continuous and there is a unique strong-sense solution to (3.6a). We first derive the expression

$$\int \mu_v(dx)\,[e^{m,v}(x) + \gamma(v) - \gamma(m)] = 0. \tag{5.6}$$

The definitions (5.3)–(5.5) yield

$$\begin{aligned} 0 = H_m(x(t)) - H_m(x) &+ \int_0^t \hat{k}_m(x(s))ds + \int_0^t [c'dy(s) - \gamma_z(m)ds] \\ &- \int_0^t \psi_m'(x(s))\sigma(x(s))\left[dw_v(s) - \sigma^{-1}(x(s))\,[b_m(x(s)) - b_v(x(s))]\,ds\right]. \end{aligned} \tag{5.7}$$

The second line is just the value of $M_m(t)$ as defined by (5.4), with $w_m(\cdot)$ obtained from (5.5).

The invariance of $\mu_v(\cdot)$ under the relaxed feedback control $v(\cdot)$ implies that

$$\int \mu_v(dx)\,[E_x^v H_m(x(t)) - H_m(x)] = 0. \tag{5.8}$$

The next step is to take the expectations of the terms in (5.7) with respect to P_x^v, and use the fact that $w_v(\cdot)$ is a standard Wiener process under the measure P_x^v to eliminate the stochastic integral with respect to $w_v(\cdot)$. While $\psi(x(\cdot))$ is square integrable on any finite time interval with respect to P_x^m measure, it might not be with respect to P_x^v measure. But a stopping time argument can be used. There is a sequence of stopping times τ^n such that $E_x^v \int_0^{t\wedge\tau^n} |\psi_m(x(s))|^2 ds < \infty$. Using this, taking limits as $n \to \infty$, and then integrating with respect to $\mu_v(\cdot)$ and using (5.8) yields

$$\begin{aligned}\int_0^t ds \int \mu_v(dx) E_x^v \left[\hat{k}_m(x(s)) + \psi_m'(x(s)) \left[b_m(x(s)) - b_v(x(s))\right]\right] \\ + \int \mu_v(dx) E_x^v \int_0^t \left[c'dy(s) - \gamma_z(m)ds\right] = 0.\end{aligned} \tag{5.9}$$

Now, note that for each t,

$$\int \mu_v(dx)\hat{k}_v(x) + \int \mu_v(dx) E_x^v \int_0^t \left[c'dy(s) - \gamma_z(v)ds\right] = 0,$$

subtract this "zero" quantity from (5.9), and use the invariance of $\mu_v(\cdot)$ under $v(\cdot)$ again to get

$$\begin{aligned}t \int \mu_v(dx) \left[\left(\hat{k}_m(x) - \hat{k}_v(x)\right) + \psi_m'(x)\left[b_m(x) - b_v(x)\right]\right] + t\left[\gamma_z(v) - \gamma_z(m)\right] \\ = 0,\end{aligned}$$

which yields (5.6).

Next, define $A = \{x : e^{m,v}(x) > 0\}$, and suppose that $l(A) > 0$. Define the relaxed control $\hat{v}(\cdot)$ by

$$\hat{v}(x,\cdot) = \begin{cases} m(x,\cdot), & \text{for } x \notin A, \\ v(x,\cdot), & \text{for } x \in A. \end{cases}$$

Then $e^{m,\hat{v}}(x) = 0$ for $x \notin A$, and $e^{m,\hat{v}}(x) > 0$ for $x \in A$. Since (5.6) holds for $\hat{v}(\cdot)$ replacing $v(\cdot)$, we have

$$\int \mu_{\hat{v}}(dx)\left[e^{m,\hat{v}}(x) + \gamma(\hat{v}) - \gamma(m)\right] = 0. \tag{5.10}$$

Theorems 3.2 and 3.3 imply that $\mu_{\hat{v}}(\cdot)$ and Lebesgue measure are mutually absolutely continuous. Since $e^{m,\hat{v}}(x) \geq 0$ and $e^{m,\hat{v}}(x) > 0$ on A, the fact that $l(A) > 0$ implies that $\mu_{\hat{v}}(A) > 0$. Then (5.10) implies that $\gamma(\hat{v}) < \gamma(m)$, which proves the first assertion of the theorem. The second is proved by a similar argument. ∎

Theorem 5.3. *Assume* (A5.1) and (A5.2), *and let* $m(\cdot)$ *be smoothed as in* (3.17). *Then* $M_m(\cdot)$ *has the representation* (5.4), *where for each* t,

$$\sup_{x\in G} E_x^m \int_0^t |\psi_m(x(s))|^2 < \infty.$$

Proof. Owing to the Lipschitz condition, $x(\cdot)$ and $z(\cdot)$ are adapted to the filtration $\{\mathcal{F}_t, t \geq 0\}$ engendered by $w_m(\cdot)$. Then by a martingale representation theorem ([57, Theorem 2.3] or [69, Theorem 16.22]), and the square integrability of $M_m(t)$ for all t, there is an adapted process $\xi_m(\cdot)$ such that $E_x^m \int_0^t |\xi_m(s)|^2 ds < \infty$ for each t, x (in fact, this expectation is bounded uniformly in $x \in G$ for each t) and such that

$$M_m(t) = \int_0^t \xi_m'(s)\sigma(x(s))dw_m(s), \quad \text{w.p.1 under } P_x^m. \tag{5.11}$$

Let $\mathcal{N}_m$ denote the class of $I\!R^r$-valued continuous random processes that are square integrable $\mathcal{F}_t$-martingales under P_x^m for each x, and are also homogeneous additive functionals of the process $x(\cdot)$. If $B(\cdot) \in \mathcal{N}_m$, then the quadratic variation (that is, the Doob–Meyer) process $\langle B\rangle(t)$ has a representation that is a homogeneous additive and nondecreasing function of $x(\cdot)$ and does not otherwise depend on $x(0) = x$. This follows from [195, Theorem 3, page 126] and is also implied by the development in [146, Appendix]. The process $M_m(\cdot)$ is in $\mathcal{N}_m$. The representation

$$w_m(t) = \int_0^t \sigma^{-1}(x(s))\,[dx(s) - [b_0(x(s)) + b_m(x(s))]\,ds - dz(s)]$$

shows that $w_m(\cdot)$ is in $\mathcal{N}_m$. Since both of $w_m(\cdot) \pm M_m(\cdot)$ are in $\mathcal{N}_m$, the form

$$\langle w_m + M_m, w_m + M_m\rangle(t) - \langle w_m - M_m, w_m - M_m\rangle(t) = 4\langle w_m, M_m\rangle(t)$$

implies that $\langle w_m, M_m\rangle(\cdot)$ has a representation that is a homogeneous additive and nondecreasing function of $x(\cdot)$ and does not otherwise depend on $x(0) = x$.

Thus, using the form (5.11), we have

$$\langle w_m, M_m\rangle(t) = \int_0^t \sigma'(x(s))\xi_m(s)ds,$$

which must have a representation as a homogeneous additive functional of $x(\cdot)$ and not otherwise depend on $x(0) = x$. It follows that there is a Borel function $\psi_m(\cdot)$ that does not depend on $x(0) = x$ and such that

$$\xi_m'(t)\sigma(x(t)) = \psi_m'(x(t))\sigma(x(t)) \tag{5.12}$$

for almost all t (w.p.1 under P_x^m, for all x). ■

4.6 Optimality Over All Admissible Controls

Theorem 5.2 gives a necessary and sufficient condition for optimality in the class of (time-independent) relaxed feedback controls. In applications it is often necessary to show that it is sufficient to restrict attention to such feedback controls, and Theorem 5.2 is easily extended to do this. For initial condition $x = x(0)$, let $r(\cdot)$ be an admissible relaxed (not a feedback) control with derivative $r_t(\cdot)$, and define the cost $\gamma(x, r)$ by (4.1).

Theorem 6.1. *Assume* (A5.1) *and* (A5.2), *and let* $m(\cdot)$ *be an optimal relaxed feedback control and* $r(\cdot)$ *an admissible relaxed control. Then for each* x,

$$\gamma(x, r) \geq \gamma(m). \tag{6.1}$$

Proof. For notational simplicity in the proof only, suppose that there are no boundary costs. The proof for the general case is similar. The proof is a minor modification of that of Theorem 5.2. See also [163, Theorem 10.6], on which the modification is based. Let $T_n \to \infty$ be such that the limsup in (4.1) is realized along that sequence for control $r(\cdot)$. The Girsanov transformation method remains valid for $r(\cdot)$: Define $b_r(x, t)$ as above Theorem 3.1, and use it to define $w_r(\cdot)$, which is a standard $\mathcal{F}_t$-Wiener process under P_x^r.

Replace $v(\cdot)$ by $r(\cdot)$ in (5.7). Dividing (5.7) by T_n and taking expectations with respect to P_x^r (using the stopping time argument of Theorem 5.2) yields

$$\begin{aligned} 0 = \frac{E_x^r H_m(x(T_n))}{T_n} - \frac{H_m(x)}{T_n} + \frac{1}{T_n}\int_0^{T_n} E_x^r \hat{k}_m(x(s))ds \\ + \frac{1}{T_n} E_x^r \int_0^{T_n} \psi_m'(x(s)) \left[b_m(x(s)) - b_r(x(s), s)\right] ds. \end{aligned} \tag{6.2}$$

Letting $n \to \infty$ yields

$$\lim_n \frac{1}{T_n}\int_0^{T_n} E_x^r \Big[\hat{k}_m(x(s)) + \psi_m'(x(s)) \left[b_m(x(s)) - b_r(x(s), s)\right]\Big] ds = 0. \tag{6.3}$$

Now, subtract the centered quantity

$$\hat{k}_r(x(s), s) = k_0(x(s)) + k_r(x(s), s) - \gamma(x, r)$$

from the integrand in (6.3), where $k_r(\cdot)$ is defined above (4.1). Owing to the definitions of $\gamma(x, r)$ and T_n, this does not affect the value of the limit. This subtraction yields

$$\begin{aligned} \lim_n \frac{1}{T_n}\int_0^{T_n} E_x^r \big[\hat{k}_m(x(s)) - \hat{k}_r(x(s), s) \\ + \psi_m'(x(s)) \left[b_m(x(s)) - b_r(x(s), s)\right]\big] ds = 0. \end{aligned} \tag{6.4}$$

By methods similar to those used in Section 4 it can be shown that for each $t > 0$ the distribution of $x(t)$ is mutually absolutely continuous with respect to Lebesgue measure under the relaxed control $r(\cdot)$. For almost all

$t > 0$ there is a relaxed feedback control $\bar{r}^t(\cdot)$ such that with probability one (P_x^r),

$$E_x^r \left[r_t(d\alpha) | x(t) \right] = \bar{r}^t(x(t), d\alpha).$$

Define $\bar{r}^t(x, \cdot)$ on the exceptional null t-set to be any probability measure on $\mathcal{B}(\mathcal{U})$. Then for each t, $\bar{r}^t(y, \cdot)$ is well-defined for almost all $y \in G$ (Lebesgue measure). Define

$$b_{\bar{r}^s}(x(s)) = \int b_c(x(s), \alpha) \bar{r}^s(x(s), d\alpha)$$

and define $k_{\bar{r}^s}(x(s))$ analogously. By the definition of $e^{m,v}$ in Theorem 5.2,

$$\begin{aligned} &E_x^r e^{m,\bar{r}^s}(x(s)) \\ &= E_x^r \left[k_m(x(s)) - k_{\bar{r}^s}(x(s)) \right] + E_x^r \psi_m'(x(s)) \left[b_m(x(s)) - b_{\bar{r}^s}(x(s)) \right]. \end{aligned} \tag{6.5}$$

Note that owing to the use of the expectation E_x^r in (6.5), the value of the expression is the same as if the derivative $r_t(\cdot)$ of the original relaxed control $r(\cdot)$ were used in the integration. This and (6.4) yield

$$\lim_n \frac{1}{T_n} \int_0^{T_n} E_x^r \left[e^{m,\bar{r}^s}(x(s)) + \gamma(x, r) - \gamma(m) \right] ds = 0. \tag{6.6}$$

By Theorem 5.2, $e^{m,\bar{r}^t}(y) \le 0$ for almost all $y \in G$ (Lebesgue measure) for each t. But this implies that $e^{m,\bar{r}^t}(x(t)) \le 0$ with probability one (P_x^r) for each t. This and (6.6) imply (6.1) ■

Discounted costs. For $\beta > 0$ and relaxed feedback control $m(\cdot)$, define the cost

$$W_\beta(x, m) = E_x^m \int_0^\infty e^{-\beta t} \left[[k_0(x(t)) + k_m(x(t))] dt + c' dy(t) \right],$$

with the analogous definition for an admissible relaxed (nonfeedback) control $r(\cdot)$. Define $V_\beta(x) = \inf_{m(\cdot)} W_\beta(x, m)$. The methods of Theorems 5.2 and 5.3 can be adapted to show that

$$\lim_{\beta \to 0} \beta V_\beta(x) \to \bar{\gamma}.$$

There is an analogous extension to discounted costs for the results of the next section.

4.7 Functional Occupation Measures

Consider the average cost per unit time problem for a controlled network in heavy traffic. Recall the costs defined by (4.1). Let n denote the heavy traffic parameter, and $x^n(\cdot)$ the scaled queue length with $x^n(0) = x$. Suppose that the cost of interest for an admissible relaxed control $r(\cdot)$ is

$$\gamma^n(x,r) = \limsup_T \gamma_T^n(x,r), \tag{7.1}$$

where

$$\gamma_T^n(x,r) = \frac{1}{T} E_x^r \int_0^T [k_0(x^n(s)) + k_r(x^n(s),s))]\, ds + \frac{1}{T} E_x^r c' y^n(T). \tag{7.2}$$

If $r^n(\cdot)$ is an optimal or ϵ-optimal relaxed control for the physical system, then we would like to know that $\gamma_T^n(x,r^n)$ converges to the optimal (or ϵ-optimal) ergodic cost for the stationary weak-sense limit process defined by, say, (3.6a) or (3.6b) *no matter how* $n, T \to \infty$. It is important that the limit does not depend on how the parameters n and T reach their limits: If larger n requires larger T to attain a good approximation, then the result would not be useful in applications. Furthermore, it is desired to show that good controls for the weak-sense limit process are also good for the physical system under heavy traffic.

The first step in proving such results is to show that $\limsup \gamma_T^n(x,r^n)$ is also the cost for a stationary weak-sense limit system, under some relaxed (but not necessarily feedback) control. Then the results of Sections 4 and 5 are used to show first that this limit cost is no greater than the optimal cost for the limit system, and then to get a nice nearly optimal feedback control that will be nearly optimal for the physical system under heavy traffic. See Chapter 9. The remainder of this section will be concerned with the first step. To simplify the development in this section, it will be supposed that $\sigma(x)\sigma'(x) = \Sigma$, not depending on x.

In fact, Theorem 7.1 will imply the stronger result that the pathwise average (without the expectation in (7.2)) costs converge in probability to the mean ergodic cost for some stationary limit system. See Theorem 9.2.1 for an important application of this fact. The results in Section 2.7 concerning measure-valued random variables will be used.

A general model. Assume (A5.1) and (A5.2). Suppose that $x^n(\cdot)$ satisfies

$$\begin{aligned} x^n(t) = x^n(0) + \int_0^t \int_{\mathcal{U}} [b_0(x^n(s)) + b_c(x^n(s),\alpha) r_s^n(d\alpha)]\, ds \\ + M^n(t) + z^n(t) + \epsilon^n(t), \end{aligned} \tag{7.3}$$

where $\epsilon^n(\cdot)$ converges weakly to the "zero" process, and $z^n(\cdot)$ has the representation

$$z^n(t) = \sum_{i=1}^q d_i y_i^n(t), \tag{7.4}$$

where q is the number of faces of G, and $y_i^n(\cdot)$ can increase only at t where $x^n(t)$ is arbitrarily close (distance less than $\epsilon_n \to 0$) to ∂G_i, and its discontinuities converge to zero. The paths of $x^n(\cdot)$ and $M^n(\cdot)$ are in $D(G;0,\infty)$ and $D(\mathbb{R}^r;0,\infty)$, respectively, and $r^n(\cdot)$ is a relaxed control. The sequence $M^n(\cdot)$ will converge to a Wiener process. It is convenient to represent the paths of $z^n(\cdot)$ as in (7.4), in terms of the q individual boundary processes. Define $\Psi^n(\cdot) = (x^n(\cdot), r^n(\cdot), y^n(\cdot), M^n(\cdot))$.

For any function $\phi(\cdot)$, define the shifted function $\phi_t(\cdot) = \phi(t+\cdot)$ and the shifted and centered function $\Delta_t\phi(\cdot) = \phi(t+\cdot) - \phi(t)$. In this notation, $(\Delta_t r^n)_u(d\alpha) = r^n_{t+u}(d\alpha)$. Then for any $t \geq 0$, (7.3) can be rewritten as

$$\begin{aligned} x^n_t(s) = x^n(t) + \int_0^s du \int_{\mathcal{U}} \left[b_0(x^n_t(u)) + b_c(x^n_t(u),\alpha)(\Delta_t r^n)_u(d\alpha)\right] \\ + \Delta_t M^n(s) + \Delta_t z^n(s) + \Delta_t \epsilon^n(s). \end{aligned} \tag{7.5}$$

Stationary solution. For the processes defined in (3.6a) or (3.6b), if the distribution of $(x_t(\cdot), \Delta_t r(\cdot), \Delta_t y(\cdot), \Delta_t w_r(\cdot))$ does not depend on t, then the solution is said to be *stationary*.

Representing the cost in terms of an occupation measure. Define $\Psi^n_t(\cdot) = (x^n_t(\cdot), \Delta_t r^n(\cdot), \Delta_t y^n(\cdot), \Delta_t M^n(\cdot))$, and the path space

$$S = D(G;0,\infty)) \times \mathcal{R}(\mathcal{U} \times [0,\infty)) \times D(\mathbb{R}^q;0,\infty) \times D(\mathbb{R}^r;0,\infty),$$

the path space for $\Psi^n(\cdot)$ and $\Psi^n_t(\cdot)$. Let $\bar{F}(\cdot)$ be a measurable real-valued function on S, depending on the first three components only. Let $x^n(0)$ be the initial condition. Consider the *pathwise average* cost function over $[0,T]$,

$$C^n_T(x^n(0), r^n) = \frac{1}{T}\int_0^T \bar{F}(x^n_t(\cdot), \Delta_t r^n(\cdot), \Delta_t y^n(\cdot))dt. \tag{7.6}$$

Define the *occupation measures* $\mathcal{Q}^{t,n}$ and $\mathcal{Q}^n_T$ by

$$\begin{aligned} \mathcal{Q}^{t,n}(C) &= I_C(\Psi^n_t(\cdot)), \quad C \in \mathcal{B}(S), \\ \mathcal{Q}^n_T(C) &= \frac{1}{T}\int_0^T \mathcal{Q}^{t,n}(C)dt. \end{aligned} \tag{7.7}$$

Let $\tilde{\psi}(\cdot) = (\tilde{x}(\cdot), \tilde{r}(\cdot), \tilde{y}(\cdot), \tilde{m}(\cdot))$ denote the canonical variables on the associated path spaces of the components of $\Psi^n(\cdot)$ or $\Psi^n_t(\cdot)$. The notation

$$\tilde{\psi}_t(\cdot) = (\tilde{x}(t+\cdot), \tilde{r}(t+\cdot) - \tilde{r}(t), \tilde{y}(t+\cdot) - \tilde{y}(t), \tilde{m}(t+\cdot) - \tilde{m}(t))$$

will also be used. The cost (7.6) can be written as

$$C^n_T(x^n(0), r^n) = \int \bar{F}(\tilde{\psi}(\cdot)) d\mathcal{Q}^n_T(\tilde{\psi}(\cdot)). \tag{7.8}$$

Now, Q_T^n is a measure-valued random variable. The limits of the costs $C_T^n(x^n(0), r^n)$ as $n, T \to \infty$ depend on the weak-sense limits of the Q_T^n, and these limits might be either random or deterministic. Recall the discussion and definitions of Section 2.7. Let Q denote a weak-sense limit of $\{Q_T^n; n, T\}$ as both $T \to \infty$ and $n \to \infty$ in some way, and let $\mathcal{P}(S)$ denote the space of probability measures on $(S, \mathcal{B}(S))$. Then Q is a $\mathcal{P}(S)$-valued random variable. Let the probability space on which Q is defined be denoted by $(\Omega', \mathcal{F}', P')$. For each $\omega' \in \Omega'$, the sample value $Q^{\omega'}$ is a measure on the path space S; hence it induces a random process $\Psi^{\omega'}(\cdot) = (x^{\omega'}(\cdot), r^{\omega'}(\cdot), y^{\omega'}(\cdot), M^{\omega'}(\cdot))$. By the weak convergence and under the uniform integrability and almost everywhere continuity assumption (A7.4) below,

$$C_T^n(x^n(0), r^n) \to \int \bar{F}(\psi(\cdot))dQ(\psi(\cdot)) \tag{7.9}$$

in the sense of convergence in distribution.

Several questions arise. Do the $\Psi^{\omega'}(\cdot)$ satisfy the desired limit equation (3.6b) for almost all ω'? Is $r^{\omega'}(\cdot)$ admissible? Are the limit processes stationary, and is the stationary solution unique? Such questions will now be answered.

Example of a cost function. Cost functions of the type (7.2) can readily be put into or approximated by the form (7.6), for appropriate $\bar{F}(\cdot)$. Actually, the only problem might concern the boundary costs such as $y_i^n(T)/T$. This is dealt with as follows:

$$\frac{1}{T}\int_0^T \Delta_t y_i^n(1)dt = \frac{1}{T}\int_0^T \left[y_i^n(t+1) - y_i^n(t)\right]dt = \frac{1}{T}y_i^n(T) + \delta_T^n,$$

where

$$\delta_T^n = \frac{1}{T}\left[\int_T^{T+1} \left[y_i^n(t) - y_i^n(T)\right]dt - \int_0^1 y_i^n(t)dt\right].$$

Theorem 3.5.3 implies that $\lim_T \sup_n E\delta_T^n = 0$. Thus, the boundary costs can also be put into the form (7.6).

One can do a similar treatment for the running cost. For $s \in [0,1]$ and a bounded integrable function $h(\cdot)$,

$$\int_0^T h(t+s)dt = \int_0^T h(t)dt - \int_0^s h(t)dt + \int_T^{T+s} h(t)dt.$$

Thus,

$$\frac{1}{T}\int_0^T h(t)dt = \frac{1}{T}\int_0^T \left[\int_0^1 h(t+s)ds\right]dt + \text{ small error},$$

where the small error goes to zero as $T \to \infty$. Thus, for a continuous function $k(\cdot)$, modulo an asymptotically negligible error, and writing $k_{r_s^n}(x^n(s)) =$

$\int_{\mathcal{U}} k_c(x^n(s), \alpha) r_s^n(d\alpha)$,

$$\frac{1}{T}\int_0^T [k_0(x^n(s)) + k_{r_s^n}(x^n(s))]\, ds$$
$$= \frac{1}{T}\int_0^T \left[\int_0^1 \left[k_0(x^n(t+s)) + k_{r_{t+s}^n}(x^n(t+s))\right] dt\right] ds.$$

Hence, modulo an asymptotically negligible term, we can write

$$C_T^n(x^n(0), r^n) = \int \bar{F}(\tilde{\psi}) dQ_T^n(\tilde{\psi}),$$

where

$$\bar{F}(\tilde{\psi}) = \int_0^1 [k_0(\tilde{x}(s)) + k_{\tilde{r}_s}(\tilde{x}(s))]\, ds + c'\tilde{y}(1). \tag{7.10}$$

To see the last expression, note that by the definitions

$$\int \tilde{y}_i(1) dQ_T^n(\tilde{\psi}) = \frac{1}{T}\int_0^T \tilde{y}_i(1) dQ^{t,n}(\tilde{\psi}) = \frac{1}{T}\int_0^T [y_i^n(t+1) - y_i^n(t)]\, dt.$$

Note that the $\bar{F}(\cdot)$ defined by (7.10), as a real-valued function on S, is not necessarily continuous in the Skorohod topology. See the comments concerning continuous functions on $D(\mathbb{R}; 0, \infty)$ in Section 2.5. But $\bar{F}(\cdot)$ is continuous at each $\tilde{\psi}(\cdot)$ for which $\tilde{y}(\cdot)$ is a continuous function. The limit processes will all have continuous paths with probability one. The integral $\int_0^1 \cdot ds$ was introduced so that the function $\bar{F}(\tilde{\psi}(\cdot))$ in (7.10) would be continuous (w.p.1) in the relaxed control, as well as in the paths of the other processes.

Assumptions. Assumption (A7.3) might look formidable, but it is essentially the criterion of Theorem 2.1.3 for a Wiener process, applied to $M^n(t + \alpha + \cdot) - M^n(t + \alpha)$ for large t, α, where the expectation in (2.1.9) is replaced by the conditional expectation, given the data to time t. For a large "gap" α between the time t of the conditioning data and the initial time $t + \alpha$ of the process, the conditional expectation approximates the expectation under broad "mixing" conditions.

A7.1. $\Delta_t \epsilon^n(\cdot)$ *converges weakly to the "zero" process as* $n \to \infty$, *uniformly in* t*: That is, for any sequence* t_n, *bounded or not,* $\Delta_{t_n}\epsilon^n(\cdot)$ *converges to the "zero" process as* $n \to \infty$.

A7.2. $z^n(t) = \sum_i y_i^n(t) d_i$. *There are real* $\epsilon_n \to 0$ *such that* $y_i^n(\cdot)$ *can change only at* t *where* $x^n(t)$ *is within a distance of* ϵ_n *from the boundary face* ∂G_i. *Also, for each* $T > 0$,

$$\lim_{\epsilon \to 0} \limsup_n \sup_t P\left\{\sup_{s \le T} |z^n(t+s) - z^n((t+s)-)| \ge \epsilon\right\} = 0.$$

A7.3. *$\Delta_t M^n(\cdot)$ is asymptotically continuous as $n \to \infty$, and t denotes any sequence. Let Σ be a positive definite and symmetric matrix. For any integer p, arbitrary $s \geq 0$, $s_i \leq s$, $i \leq p$, $\tau > 0$, any bounded and continuous real-valued function $h(\cdot)$, and as $n, t \to \infty$ in any way at all,*

$$E_t^n h(\Psi_{t+\alpha}^n(s_i), i \leq p) \times \Big[f(\Delta_{t+\alpha} M^n(s+\tau)) - f(\Delta_{t+\alpha} M^n(s)) \\ - \frac{1}{2} \int_s^{s+\tau} \text{trace}[f_{ww}(\Delta_{t+\alpha} M^n(u))\Sigma] du \Big] \to 0$$

in the sense of probability as $n, \alpha, t \to \infty$, where E_t^n denotes the expectation given $\{\Psi^n(s), s \leq t\}$ and $f(\cdot)$ is an arbitrary real-valued function with compact support that is continuous together with its partial derivatives up to second order.

A7.4. *The set $\{\bar{F}(\Psi_t^n(\cdot)); n, t\}$ is uniformly integrable. Also, the function $\bar{F}(\cdot)$ is continuous w.p.1 with respect to the measure of each stationary solution to (3.6b) with $\sigma(x)\sigma'(x) = \Sigma$ and where we identify $(\tilde{x}(\cdot), \tilde{r}(\cdot), \tilde{y}(\cdot))$ as the canonical path of $(x(\cdot), r(\cdot), y(\cdot)$.*

Theorem 7.1. *Assume the model (7.3), cost (7.1), (A5.1), (A5.2), and (A7.1)–(A7.4). Let $\{x^n(0)\}$ be tight. Then for any sequence $T \to \infty$ and $n \to \infty$, $\{\mathcal{Q}_T^n; n, T\}$ is tight as a sequence with values in $\mathcal{P}(S)$. Let $\mathcal{Q}$ denote the weak-sense limit of a weakly convergent subsequence, indexed by $n, T_n \to \infty$. Let ω' denote the canonical point in the probability space on which $\mathcal{Q}$ is defined, and let $\mathcal{Q}^{\omega'}$ denote the sample values. Then for almost all ω', there is a stationary process $\Psi^{\omega'} = (x^{\omega'}(\cdot), r^{\omega'}(\cdot), y^{\omega'}(\cdot), M^{\omega'}(\cdot))$ satisfying the Skorohod problem*

$$x^{\omega'}(t) = x^{\omega'}(0) + \int_0^t ds \int_{\mathcal{U}} \left[b_0(x^{\omega'}(s)) + b_c(x^{\omega'}(s), \alpha) r_s^{\omega'}(d\alpha) \right] \\ + M^{\omega'}(t) + z^{\omega'}(t), \tag{7.11}$$

where $z^{\omega'}(\cdot) = \sum_i d_i y^{\omega'}(\cdot)$, $M^{\omega'}(\cdot)$ is a Wiener process with covariance matrix Σ, $r^{\omega'}(\cdot)$ is an admissible relaxed control, $z^{\omega'}(\cdot)$ is the reflection term, and $(x^{\omega'}(\cdot), r^{\omega'}(\cdot), y^{\omega'}(\cdot))$ are nonanticipative with respect to $M^{\omega'}(\cdot)$. Also,

$$C_{T_n}^n(x^n(0), r^n) \Rightarrow \int \bar{F}(\tilde{\psi}(\cdot)) d\mathcal{Q}(\tilde{\psi}(\cdot)). \tag{7.12}$$

Suppose that the Skorohod representation (Theorem 2.5.5) is used with $(\Omega', \mathcal{F}', P')$ still denoting the common probability space, and $\pi(\cdot)$ the Prohorov metric. Then $\pi(\mathcal{Q}_{T_n}^{n,\omega'}, \mathcal{Q}^{\omega'}) \to 0$ for almost all ω'. Also, if $\rho(\cdot)$ denotes the metric on the path space S, then for almost all ω',

$$\rho(\Psi_{T_n}^{n,\omega'}(\cdot), \Psi^{\omega'}(\cdot)) \to 0,$$

with probability one.

Proof. The set $\{\Delta_t z^n(\cdot); n, t\}$ is tight and has continuous weak-sense limits by Theorem 3.6.1. Hence $\{\Psi_t^n(\cdot); n, t\}$ is tight, and has continuous weak-sense limits. By (A5.2), the same assertion can be made for the $y_i^n(\cdot)$. For each n, t, the expectation $EQ^{n,t}$ of the sample occupation measure is the probability measure of the process $\Psi_t^n(\cdot)$. Hence the set of measures $\{EQ^{n,t}; n.t\}$ is tight. Thus, by Theorem 2.7.1, the set of measure-valued random variables $\{\mathcal{Q}_T^n; n, T\}$ is tight. For notational simplicity, let $n, T_n \to \infty$ index a weakly convergent subsequence of $\mathcal{Q}_T^n$, with weak-sense limit $\mathcal{Q}$. The convergence (7.12) follows by the weak convergence and (A7.4), provided that (7.11) holds.

We need to identify the processes induced by the $\mathcal{Q}^{\omega'}$ and demonstrate (7.11). The Skorohod representation for $\{\mathcal{Q}_{T_n}^n, \mathcal{Q}\}$ will be used in the rest of the proof. The stationarity will be proved next. Let $C \in \mathcal{B}(S)$, and define its left shift

$$C_c = \left\{\tilde{\psi}(\cdot) : \tilde{\psi}_c(\cdot) \in C\right\}.$$

Then

$$\mathcal{Q}_{T_n}^n(C_c) = \frac{1}{T_n}\int_0^{T_n} I_{C_c}(\Psi_t^n(\cdot))dt = \frac{1}{T_n}\int_0^{T_n} I_C(\Psi_{t+c}^n(\cdot))dt,$$

$$\mathcal{Q}_{T_n}^n(C_c) - \mathcal{Q}_{T_n}^n(C) = \frac{1}{T_n}\int_{T_n}^{T_n+c} I_C(\Psi_t^n(\cdot))dt - \frac{1}{T_n}\int_0^c I_C(\Psi_t^n(\cdot))dt.$$

Thus $\mathcal{Q}_{T_n}^{n,\omega'}(C) - \mathcal{Q}_{T_n}^{n,\omega'}(C_c) \to 0$ as $n \to \infty$ for all ω' and C. This and the weak convergence imply that for any bounded and continuous function $\phi(\cdot)$

$$\int \phi(\tilde{\psi}(\cdot))d\mathcal{Q}(\tilde{\psi}(\cdot)) = \int \phi(\psi_c(\cdot))d\mathcal{Q}(\tilde{\psi}(\cdot))$$

w.p.1, which implies the stationarity of $\Psi^{\omega'}(\cdot)$ for almost all ω'.

Let $C_\epsilon \in \mathcal{B}(S)$ denote the set on which $\psi(\cdot)$ has a discontinuity of size bigger than $\epsilon > 0$ on the arbitrary interval $[0, T_0]$. By the definition of $\mathcal{Q}_T^n$,

$$E\mathcal{Q}_T^n(C_\epsilon) = \frac{1}{T}E\int_0^T I_{\{\Psi_t^n(\cdot)\in C_\epsilon\}}dt.$$

But this goes to zero uniformly in T as $n \to \infty$, since any weak-sense limit of $\Psi_t^n(\cdot)$ (with $n \to \infty$ and t being any sequence) has continuous paths, which implies that the limit processes associated with the $\mathcal{Q}^{\omega'}$ are continuous w.p.1 for almost all ω'.

The form (7.11) will be proved next. Define

$$f(\tilde{\psi}(\cdot)) =$$

$$\left[\tilde{x}(t) - \tilde{x}(0) - \int_0^t\int_{\mathcal{U}} [b_0(\tilde{x}(s)) + b_c(\tilde{x}(s), \alpha)\tilde{r}_s(d\alpha)]\, ds - \tilde{m}(t) - \tilde{z}(t)\right]^2 \wedge 1.$$

It will be shown that

$$E \int f(\tilde{\psi}(\cdot)) dQ^{\omega'}(\tilde{\psi}(\cdot)) = 0.$$

The function $f(\cdot)$ is bounded, and it is continuous w.p.1 $(Q^{\omega'})$ for almost all ω'. We need only show that

$$E \int f(\tilde{\psi}(\cdot)) dQ^{n,\omega'}_{T_n}(\tilde{\psi}(\cdot)) \to 0.$$

The last expression is

$$\frac{1}{T_n} E \int_0^{T_n} f(\Psi_t^n(\cdot)) dt,$$

and this goes to zero by virtue of the weak convergence of $\Delta_t \epsilon^n(\cdot)$ to the "zero" process.

To show that the $M^{\omega'}(\cdot)$ are the desired Wiener processes, and to prove nonanticipativity, by Theorem 2.1.4 it is sufficient to prove that for almost all ω', any bounded and continuous $h(\cdot)$, any $s > 0, \tau > 0$, integer p any $s_i \leq s, i \leq p$, and where $f(\cdot)$ is as in (A7.3),

$$\int h(\tilde{\psi}(s_i), i \leq p) F_0(\tilde{m}(\cdot)) dQ^{\omega'}(\tilde{\psi}(\cdot)) = 0, \tag{7.13}$$

where

$$F_0(\tilde{m}(\cdot)) = \left[f(\tilde{m}(s+\tau)) - f(\tilde{m}(s)) - \frac{1}{2} \int_s^{s+\tau} \text{trace} f_{ww}(\tilde{m}(u)) \Sigma du \right] = 0. \tag{7.14}$$

The relationship (7.13) holds if

$$\frac{1}{T^2} E \left[\int_0^T h(\Psi_t^n(s_i), i \leq p) F_0(\Delta_t M^n(\cdot)) dt \right]^2 \to 0 \tag{7.15}$$

as $n \to \infty$ and $T \to \infty$. The limit (7.15) holds by (A7.3).

It remains only to prove that $z^{\omega'}(\cdot) = \sum_i d_i y_i^{\omega'}(\cdot)$ is the reflection process w.p.1, for almost all ω'. Fix i. For arbitrary $T > 0$, let $A \in \mathcal{B}(S)$ denote the set of paths such that $\tilde{y}_i(\cdot)$ changes at some time $s \leq T$ where $\tilde{x}(s)$ is not on ∂G_i. It needs to be shown that $E \int I_A dQ^{\omega'}(\tilde{\psi}(\cdot)) = 0$. For each t and n let I_t^n denote the indicator of the event that $y_i^n(t+s) - y_i^n(t)$ changes at some $s \leq T$ while $x^n(t+s)$ is greater than $\epsilon > 0$ from ∂G_i. Then $\lim_n \sup_t E I_t^n = 0$. This implies that

$$\lim_n \sup_T \frac{1}{T} E \int_0^T I_t^n dt = 0,$$

which, since ϵ is arbitrary, implies that $Q^{\omega'}(A) = 0$ for almost all ω'. ■

Note on state dependence. The assumption (A7.3) ensures that the martingales $M^{\omega'}(\cdot)$ in (7.11) are Wiener processes with covariance matrix Σ. The more general stochastic integral form can also be treated. Let $\sigma(\cdot)$ be bounded and continuous, define $a(x) = \sigma(x)\sigma'(x)$, and suppose that it is uniformly positive definite. In (A7.3), replace Σ by $a(x^n_{t+\alpha}(u))$. The theorem continues to hold, but with $M^{\omega'}(\cdot)$ replaced by $\int_0^t \sigma(x^{\omega'}(s))dw^{\omega'}(s)$, where $w^{\omega'}(\cdot)$ is a standard Wiener process and the nonanticipativity is with respect to $w^{\omega'}(\cdot)$.

4.8 Approximating Controls

An approximation by an ordinary control for the nonlinear problem. For use in applications to the actual heavy-traffic models, we will need to have nearly optimal (say, 2ϵ-optimal) ordinary (rather than relaxed) controls. If the control appears linearly in the dynamics and cost function, then Theorem 4.5 can be used. Otherwise, we need to approximate some smooth nearly optimal relaxed control. This was done in Section 3.8 for problems that are of interest over a finite time or for the discounted cost function.

The approximation for the ergodic cost function is more subtle, and will be based on the smooth relaxed feedback control $m_\epsilon(\cdot)$ defined in (4.13), that was concentrated on some finite set $\mathcal{U}_\epsilon = \{u_i, i \le \rho\}$ for some integer ρ. Recall the definition of $p_i(\cdot)$ there. The construction to follow is for theoretical purposes only, and is not intended for practical use. Given $\delta > 0$, an approximation $u^\delta(\cdot)$ to $m_\epsilon(\cdot)$ will be constructed recursively, on each interval $[\nu\delta, \nu\delta + \delta)$, $\nu = 0, 1, \ldots$, in sequence. Let $x^\delta(\cdot)$ denote the solution under $u^\delta(\cdot)$. Subdivide each interval $[\nu\delta, \nu\delta + \delta)$ into successive subintervals with lengths $p_i(x^\delta(\nu\delta))\delta$, $i = 1, \ldots, \rho$, respectively. Define $u^\delta(\cdot)$ on $[\nu\delta, \nu\delta + \delta)$ by setting $u^\delta(t) = u_i$ on the ith subinterval of $[\nu\delta, \nu\delta+\delta)$. The proof of the following theorem is omitted, but it uses the fact that the "smoothness" of the paths does not depend on the control in the sense that for any $T, \epsilon > 0$,

$$\sup_t P\left\{ \sup_{\tau \le T} \sup_{s \le \mu} |x^\delta(t + \tau + s) - x^\delta(t + \tau)| \ge \epsilon \right\}$$

goes to zero as $\mu \to 0$, uniformly in the control.

In essence, the control $u^\delta(\cdot)$ that was just constructed approximates the effect of the relaxed control $m_\epsilon(\cdot)$ of (4.13) by using the values u_i on small time intervals of the appropriate lengths, in a delayed "piecewise constant" feedback manner. The proof uses occupation measure arguments similar to those in Section 7, where $x^\delta(\cdot)$ and $u^\delta(\cdot)$ are the process and the control, but the mean (rather than the sample value) value of the occupation measure is used. The fact that the limit (as $\delta \to 0$) of the ergodic costs is characterized in terms of a stationary process is shown as in Section 7, and one need only characterize the control for that process. The control turns out to be just $m_\epsilon(\cdot)$. The discussion is summarized in the next theorem.

Let $r^\delta(\cdot)$ denote the relaxed control representation of $u^\delta(\cdot)$. For the associated solution $x^\delta(\cdot)$ and reflection term $z^\delta(\cdot)$, define $\Psi^\delta(\cdot) = (x^\delta(\cdot), w(\cdot), r^\delta(\cdot), z^\delta(\cdot))$.

Theorem 8.1. *Assume the conditions of Theorem 4.1, and define $u^\delta(\cdot)$ as above the theorem. Let the $p_i(\cdot)$ in (4.13) be continuously differentiable and define the mean occupation measure*

$$\bar{Q}_T^\delta(C) = \frac{1}{T}\int_0^T P\left\{\Psi_t^\delta(\cdot) \in C\right\} dt.$$

Let $\bar{\Psi}^\delta(\cdot)$ denote the process associated with some weak-sense limit of the $\{\bar{Q}_T^\delta(\cdot);\, T\}$ as $T \to \infty$, and let $\gamma(x, u^\delta)$ denote the associated ergodic cost for initial condition $x^\delta(0) = x$. Then

$$\lim_{\delta \to 0} \gamma(x, u^\delta) = \gamma(m_\epsilon).$$

Application of Theorem 8.1 to (7.5). In the next theorem, the control $u^\delta(\cdot)$ of Theorem 8.1 is adapted for use on the system (7.5). The direct adaptation is simple. One just uses the path values of (7.5) in the recursive construction. In applications, it is not always possible to apply any desired control. It can happen that the mechanism that realizes the actual control introduces a small perturbation or error, that goes to zero as $n \to \infty$. This issue is discussed in detail, and examples given in Chapter 9. In preparation for the applications, we proceed as follows.

Let $u^{\delta,n}(\cdot)$ denote the actual control that is applied. Its values at time t depend on the data to time $t-$. Let $x^{\delta,n}(\cdot)$ denote the corresponding process (7.5). Given the path $x^{\delta,n}(\cdot)$, let $\bar{u}^{\delta,n}(\cdot)$ be constructed as $u^\delta(\cdot)$ was, but with $x^{\delta,n}(\nu\delta)$ replacing $x^\delta(\nu\delta)$. It is assumed that the errors or perturbations are small in the sense that applied control $u^{\delta,n}(\cdot)$ and desired control $\bar{u}^{\delta,n}(\cdot)$ are close in the sense of (8.1) below. Under $u^{\delta,n}(\cdot)$, write $\Psi_t^{\delta,n}(\cdot)$ instead of $\Psi_t^n(\cdot)$, that was defined above (7.6). For the present case, write the occupation measure in (7.7) as $Q_T^{\delta,n}(\cdot)$. The proof of the following theorem is similar to what would be used for Theorem 8.1.

Theorem 8.2. *Return to the assumptions and setup of Theorem 7.1 and suppose that, for the controls discussed above,*

$$\limsup_n \sup_t E \int_t^{t+1} \left|\bar{u}^{\delta,n}(s) - u^{\delta,n}(s)\right| ds = 0. \tag{8.1}$$

Let $Q^\delta(\cdot)$ denote a weak-sense limit of the sequence of random measures $Q_T^{\delta,n}(\cdot)$ as $n \to \infty$ and $T \to \infty$. Let $\gamma^{\delta,\omega'}$ denote the ergodic cost for the *stationary process engendered by the measure-valued sample* $Q^{\delta,\omega'}(\cdot)$ *of the random measure* $Q^{\delta}(\cdot)$. *Then* $\gamma^{\delta} \to \gamma(m_{\epsilon})$ *in probability as* $\delta \to 0$.

5

The Single-Processor Problem

Chapter Outline

In this chapter we begin the systematic formulation and development of the heavy traffic models. We start with the uncontrolled case of a single processor or queue, where the concepts are simplest and can be handled with minimal notation. The results serve as a foundation for the subsequent work on networks and controls and are a simpler venue in which to develop an intuitive feeling for the methods. Many of the basic ideas in the formulation, scaling, and proofs carry over to the network case, with only minor modifications.

Many of the queueing problems of interest are difficult enough so that approximation methods are required for their analysis. As noted in Chapter 1, the heavy traffic assumption allows us to approximate the (appropriately) scaled queue by a reflected diffusion process, which can be used to obtain approximations to quantities of interest for the original physical problem. Quantities of interest might include the mean or variance of the length or waiting time, or the stationary distribution of a continuous nonlinear function of the length or waiting time, the probability of being empty, the probability of buffer overflow, the probability that the supremum of the length will exceed some value on a given time interval, etc. These might be the values for the stationary process or the values at some particular time or even an integral of the values over some time interval.

By heavy traffic we mean that the rate at which the processor can work is very close to the rate of arrival of work; equivalently, that the fraction of time that the processor is idle is small. In all cases in this book, unless explicitly noted otherwise (as might be the case when controls are used), it is assumed that the processor continues to work if there is work to do. Also, unless mentioned otherwise, we assume a first come first served (FCFS) discipline, perhaps within priority classes, if appropriate. Many of the results are insensitive to the service discipline, since they do not distinguish between queued jobs.

Let $Q(t)$ denote the size of the physical queue at real time t. As noted in Chapter 1, under conditions of heavy traffic and to allow the exploitation of the methods of weak convergence theory, it is natural to scale the queue size (and perhaps time as well) and to work with the scaled model. To formalize this, one embeds the physical queue in a sequence parameterized by n, where the traffic intensity goes to unity as $n \to \infty$, and $Q^n(t)$ denotes the size of the nth member of this sequence at real time t. Roughly speaking, there are two forms that the scaling can take when the limit is to be a reflected diffusion, the first being more prevalent in classical queues and manufacturing systems, and the second being more prevalent in communications systems. For the first form, the convergence to zero of the difference of the rates is made precise in (A1.2), the so-called heavy traffic assumption. The scaling used in the state equation (1.1) is one of the standard ones in practice. It was used in Sections 1.1 and 1.2. Owing to the small difference between the arrival and service rates, the queue builds up over time. If $b < 0$ in (A1.2), so that the queues are (barely) stable, then for any moderate initial condition the size will build up slowly to an asymptotic average of $O(\sqrt{n})$. If $b > 0$, so that the queue is (barely) unstable, then its size goes to infinity, but slowly, so that at time nt there are $O(\sqrt{n})$ queued. These comments corroborate the scaling. In order to relate the asymptotic results to a particular queue, one needs to select n and the parameter b in (A1.2). This is usually done, as noted in Sections 1.1 and 1.2, by letting n be a large integer so that the b^n in (A1.2) is of moderate size and defining it as b. The chapter is concerned with the convergence of the sequence of scaled queues.

In many problems arising in applications to communication systems, the jobs are "packets," "cells," or similar units of data partition or organization. Then the basic arrival and service rates are usually very high, and it is more appropriate to scale the number queued, but not time. Then the parameter n denotes the basic size or speed of the system, and the difference between arrival and service rates in real time is $O(\sqrt{n})$. This alternative, and equally important, scaling will be briefly discussed in Section 7. In a sense that will be seen, a limit result for any one of the scalings immediately transforms into a result for the other scaling. This scaling was discussed at the end of Subsection 1.1.1 and was used for the examples in Sections 1.4 and 1.5.

The above scalings are similar to what one uses for the central limit theorem. Alternatively, it is often the case that a scaling reminiscent of the law of large numbers is most appropriate. For example, if the limit

b in (A1.2) is $\pm\infty$, then the limit approximating diffusion process will either explode or reduce to zero, and an alternative scaling that reduces the problem to "mean flows" might yield more information. This leads to the so-called *fluid* approximation, which will be discussed in the next chapter, since it is trivial in the context of the single-processor problems of this introductory chapter.

Despite the seemingly special nature of the scaling, the limits have been very useful in shedding light on quite complicated problems, as seen from the examples in Chapter 1 and the references cited there. The problems of this chapter are fundamental, but classical. Although some of the techniques and assumptions might be more efficient and general, many of the problem formulations were dealt with in the original works [21, 110, 111].

The basic single processor problem is introduced in Subsection 1.1. This provides the foundation for all subsequent results. The proof of tightness of the reflection terms is deferred to Section 2, where it is dealt with by two different methods, The first uses the so-called reflection mapping. The second, which will be much more useful, uses a method by contradiction, showing that if the reflection terms are not tight, then something impossible happens. The reflection term arises from the idle time of the processor. Two ways of dealing with the representation of this idle time are discussed. The method of Subsection 1.3 introduces so-called "fictitious services," while that of Subsection 1.1 uses the idle time directly. While both methods have their place, the one of Subsection 1.1 is the more versatile. Subsection 1.2 extends the result of Theorem 1.1 to the case with multiple arrival streams.

Up to now, we have worked with the scaled queue size. Section 3 deals with the formulation in terms of the scaled workload, which is simple in the context of this chapter and will be very important later on. The limit equations are derived in two ways. The first way is direct, via a copy of the method of Theorem 1.1. The second way is the consequence of an important result (a type of law of large numbers) concerning the asymptotic linear relationship between the scaled queued size and workload.

The versatility of the methods is illustrated by the extensions to batch arrivals and services and a multiprocessor case in Section 4. There are applications where the arrival or service rates depend on some external or environmental modulating influence. Section 5 shows how to use some of the powerful results of Section 2.8 to identify the Wiener process in the heavy traffic limit for a class of such problems.

The processing can be interrupted, either due to the arrival of priority work or by processor breakdown, etc. The necessary modifications are in Section 6. The interruptions change the limit by either the addition of additional randomness, an increase in the drift parameter, or by the addition of jump terms.

5.1 A Canonical Single-Server Problem

5.1.1 The Basic Model

In this section we will work with a canonical model for the first type of scaling for a sequence of single processor queues $Q^n(\cdot)$ with a single arrival stream. The model is canonical in that it sets up the basic structures and issues that will have to be dealt with in more complex problems. Define the scaled queue size

$$x^n(t) = Q^n(nt)/\sqrt{n}.$$

Let $A^n(t)$ (respectively, $D^n(t)$) denote $1/\sqrt{n}$ times the number of arrivals (respectively, departures) by real time nt. Then the state equation is

$$x^n(t) = x^n(0) + A^n(t) - D^n(t). \tag{1.1}$$

The arrival and service rates for $Q^n(\cdot)$ depend on n and as $n \to \infty$, their difference goes to zero. Recall the definitions of $\Delta_l^{a,n}$ and $\Delta_l^{d,n}$ as the lth interarrival and service interval, respectively, for $Q^n(\cdot)$. For some centering constants $\bar{\Delta}^{a,n}$ and $\bar{\Delta}^{d,n}$, to be specified further in (A1.1), define the processes

$$w^{a,n}(t) = \frac{1}{\sqrt{n}} \sum_{l=1}^{nt} \left[1 - \frac{\Delta_l^{a,n}}{\bar{\Delta}^{a,n}}\right], \tag{1.2a}$$

$$w^{d,n}(t) = \frac{1}{\sqrt{n}} \sum_{l=1}^{nt} \left[1 - \frac{\Delta_l^{d,n}}{\bar{\Delta}^{d,n}}\right]. \tag{1.2b}$$

Whenever a real variable, such as the nt in (1.2a,b), is used as a limit of summation, we use the *integer part. Throughout the book, even if not explicitly noted, all of the $\bar{\Delta}$ terms are positive.* The term *driving random variables* refers to the initial condition and the set of arrival times and service intervals. More generally, it will include the initial condition and any influences on the system that occur after the initial time; for example, routing or control decisions. The following assumptions will be used.

A1.0. *The initial condition $x^n(0)$ is independent of all other driving random variables.*

A1.1. *The centering constants $(\bar{\Delta}^{a,n}, \bar{\Delta}^{d,n})$ converge to positive $(\bar{\Delta}^a, \bar{\Delta}^d)$ as $n \to \infty$, and there is a nonnegative definite and symmetric matrix Σ such that the pair $(w^{\alpha,n}(\cdot), \alpha = a, d)$ defined in* (1.2a) *and* (1.2b) *converges weakly to a Wiener process with covariance matrix Σ.*

For notational simplicity, we will often write $1/\bar{\Delta}^\alpha = \bar{\lambda}^\alpha$, $\alpha = a, d$, which defines the "rates" $\bar{\lambda}^\alpha$ as the inverses of the mean interval lengths.

A1.2. *The heavy traffic condition. For some real number b,*

$$\sqrt{n}\left[\frac{1}{\bar{\Delta}^{a,n}} - \frac{1}{\bar{\Delta}^{d,n}}\right] = b^n \to b, \quad \text{as } n \to \infty. \tag{1.3}$$

Define the traffic intensity $\rho^n = \bar{\Delta}^{d,n}/\bar{\Delta}^{a,n}$. Then (1.3) can be written as

$$\frac{\sqrt{n}}{\bar{\Delta}^d}\left[\rho^n - 1\right] = b^n \to b. \tag{1.4}$$

As noted in the Chapter Outline and in Chapter 1, in applications to a single queue, one defines b from (1.4), by choosing n large but keeping b^n moderate, and then defining it as b. A simpler alternative to (A1.1) uses (A1.3) and (A1.4):

A1.3. $\left\{|\Delta_l^{a,n}|^2, |\Delta_l^{d,n}|^2; l, n\right\}$ *is uniformly integrable.*

A1.4. *For each n, the members of the set of all interarrival and service times are mutually independent. There are constants $\bar{\Delta}^{\alpha,n}, \bar{\Delta}^{\alpha}, \sigma_{\alpha}^2$, $\alpha = a, d$, such that for $\alpha = a, d$,*

$$E\Delta_l^{\alpha,n} = \bar{\Delta}^{\alpha,n} \to \bar{\Delta}^{\alpha}, \tag{1.5}$$

$$\lim_{n\to\infty} E\left[1 - \frac{\Delta_l^{\alpha,n}}{\bar{\Delta}^{\alpha,n}}\right]^2 \to \sigma_{\alpha}^2. \tag{1.6}$$

The limits in (1.5) *and* (1.6) *are taken uniformly in l.*

Condition (A1.1) allows dependence among the individual service times and among the individual interarrival times, as well as between the service and interarrival intervals. Such dependence might occur, for example, if successive customers arriving closer together required less work than successive customers arriving far apart. Condition (A1.2) implies that $\bar{\Delta}^a = \bar{\Delta}^d$. The condition (A1.3) implies the standard Lindeberg condition (2.8.3) used in proofs of the central limit theorem.

Idle time and the reflection term. A main problem in heavy traffic modeling concerns the treatment of the idle time of the processors. In the models of this and the next chapter, the idle time is the source of the reflection term in the heavy traffic limit process. As in the comparative treatment of Examples 1.1.1 and 1.1.2, two methods will be used. In Theorem 1.1, we introduce the idle time directly in expanding the state equation (1.1). This is the method of Example 1.1.2, and is very close to that employed in [213]. It is generally simpler than the following alternative, which was employed in the earliest work on the subject [21, 25, 110, 111]. The alternative uses the idea of "fictitious services" during the idle time, as in Example 1.1.1. The reflection term then arises as the correction for the fictitious services. Although this alternative method (dealt with in Subsection 1.3) can be employed in most cases of interest, it can get complicated when the service times are correlated. Both methods are presented here, since each has a role to play in modern developments, although the bulk of the sequel uses the first method. When both can be used, they yield the same results.

Notation. Define $S^{d,n}(t)$ (respectively, $S^{a,n}(t)$) to be $1/n$ times the number of services (respectively, arrivals) completed by real time nt. Let $N^{d,n}(t)$ (resp, $N^{a,n}(t)$) denote $(1/n)$ times the number of service completions (respectively, arrivals) needed to cover the real time interval $[0, nt]$. In particular,

$$\begin{aligned} S^{a,n}(t) &= \frac{1}{n} \max \left\{ m : \sum_{l=1}^{m} \Delta_l^{a,n} \le nt \right\}, \\ N^{a,n}(t) &= \frac{1}{n} \min \left\{ m : \sum_{l=1}^{m} \Delta_l^{a,n} \ge nt \right\}. \end{aligned} \tag{1.7}$$

Define

$$\mathcal{T}^{\alpha,n}(t) = \frac{1}{n} \sum_{l=1}^{nt} \Delta_l^{\alpha,n}. \tag{1.8}$$

For this problem, $S^{a,n}(t) = A^n(t)/\sqrt{n}$, and analogously for the departures. This redundant notation is useful for later generalizations. If there is an arrival at time nt, then $N^{a,n}(t) = S^{a,n}(t)$. Otherwise, $N^{a,n}(t)$ is $1/n$ larger. Similarly, $N^{d,n}(t)$ and $S^{d,n}(t)$ differ by at most $1/n$. For any $t \ge 0$, the random variables $N^{\alpha,n}(t)$, $\alpha = a, d$, will be useful, since the $nN^{\alpha,n}(t)$ are stopping times for the filtrations $\{\mathcal{F}_l^{\alpha,n}, l \ge 0\}$ defined above (A1.5). The functions $N^{a,n}(\cdot)$ (or $S^{a,n}(\cdot)$) and $\mathcal{T}^{a,n}(\cdot)$ are inverses in that (modulo an error that is no larger than $1/n$ for (1.9b) and less than $1/n$ times the interval $\Delta_l^{a,n}$ covering nt for (1.9a))

$$\mathcal{T}^{a,n}(N^{a,n}(t)) = t, \quad \mathcal{T}^{a,n}(S^{a,n}(t)) = t. \tag{1.9a}$$

$$N^{a,n}(\mathcal{T}^{a,n}(t)) = t, \quad S^{a,n}(\mathcal{T}^{a,n}(t)) = t. \tag{1.9b}$$

Given the current real time nt, the real time since the current service started or has to go, and the real time since or until the next arrival, are called *residual times*. We define a *residual-time error term* to be a random process that is bounded by (constant/$\sqrt{n}$)×[constant plus a finite sum of such residual-time terms]. Such residual-time error terms will occur frequently in the various representations of the scaled queue process, and will usually be denoted by $\epsilon^n(t)$ or with an additional sub or superscript. Successive uses of the symbol $\epsilon^n(\cdot)$ might refer to different residual-time error terms, but they are always defined as above. The conditions that will be used guarantee that $\epsilon^n(\cdot)$ converges weakly to the "zero" process in all cases, whether noted explicitly or not.

Define the scaled idle time

$$T^{d,n}(t) = \frac{1}{\sqrt{n}}\left[nt - \sum_{l=1}^{nS^{d,n}(t)} \Delta_l^{d,n} - \tilde{\Delta}^{d,n}(t)\right], \tag{1.10}$$

where $\tilde{\Delta}^{d,n}(t)$ is the time that any customer being served at nt has been in service. We have $S^{d,n}(t) \leq N^{a,n}(t) + x^n(0)/\sqrt{n}$ and (possibly modulo a residual-time error term, that is just $1/n$ times the time that any customer being served at real time nt has been in service)

$$\frac{1}{\sqrt{n}}T^{d,n}(t) + \mathcal{T}^{d,n}(S^{d,n}(t)) = t. \tag{1.11}$$

Also (possibly modulo $1/n$)

$$S^{d,n}(\mathcal{T}^{d,n}(t)) \leq t. \tag{1.12}$$

In (1.12), the argument $\mathcal{T}^{d,n}(t)$ is $1/n$ times the time taken by the first (the integer part) nt services, and $S^{d,n}(\mathcal{T}^{d,n}(t))$ is $1/n$ times the number of actual services that are completed by that time. Define

$$z^n(t) = T^{d,n}(t)/\bar{\Delta}^{d,n}. \tag{1.13}$$

The basic convergence theorem. Now we are prepared to state and prove the first theorem. Many of the techniques used in the proof will be used repeatedly in the sequel. The main point of the theorem is the weak convergence of the sequence of scaled queues $x^n(\cdot)$ to the reflected diffusion process $x(\cdot)$ that is defined by (1.14). Additionally, it asserts the convergence of the other processes of interest; namely, the normed arrival and service processes, the processes that keep track of the randomness in the arrivals and services, and the reflection (scaled idle time) process.

Theorem 1.1. *Assume* (A1.0), (A1.2), *and either* (A1.1) *or* (A1.3) *and* (A1.4). *Let* $x^n(0) \Rightarrow x(0)$. *Define*

$$\Psi^n(\cdot) = \left(x^n(\cdot), S^{a,n}(\cdot), S^{d,n}(\cdot), w^{a,n}(\cdot), w^{d,n}(\cdot), z^n(\cdot)\right).$$

Then $\{\Psi^n(\cdot)\}$ *is tight and converges weakly to a set*

$$\left(x(\cdot), S^a(\cdot), S^d(\cdot), w^a(\cdot), w^d(\cdot), z(\cdot)\right)$$

satisfying

$$S^a(t) = S^d(t) = t/\bar{\Delta}^a = \bar{\lambda}^a t$$

and

$$x(t) = x(0) + w(t) + bt + z(t), \; x(t) \geq 0, \tag{1.14}$$

where

$$w(t) = w^a(S^a(t)) - w^d(S^d(t)) = w^a(\bar{\lambda}^a t) - w^d(\bar{\lambda}^a t).$$

Let $\mathcal{F}_t$ denote the minimal σ-algebra that measures $\{x(0), w^\alpha(\bar{\lambda}^a s), z(s), s \leq t, \alpha = a, d\}$. Then $(w^\alpha(\bar{\lambda}^a \cdot), \alpha = a, d)$ is an $\mathcal{F}_t$-Wiener process with covariance matrix $\bar{\lambda}^a \Sigma$. Under (A1.3) *and* (A1.4), *Σ is diagonal with diagonal entries σ_a^2, σ_d^2. The function $z(\cdot)$, called the reflection term, is continuous and nondecreasing, $z(0) = 0$, and it can increase only at t where $x(t) = 0$. It satisfies*

$$z(t) = \max\left\{0, -\min_{s \leq t}\left[x(0) + bs + w(s)\right]\right\}. \tag{1.15}$$

The $(x(\cdot), z(\cdot))$ are nonanticipative with respect to $(w^a(\bar{\lambda}^a \cdot), w^d(\bar{\lambda}^a \cdot))$.

Remark on the proof. The proof is typical of most others in the rest of this chapter and in Chapters 6 and 7 in that it starts by expanding and reorganizing the terms in (1.1) into expressions that have a formal resemblance to the drift, the Wiener process, and the reflection terms in the "limit process," analogously to the method used in the examples in Chapter 1. This resemblance gives us confidence in the approach, and gives an additional intuitive feeling for the relations between the physical process and the weak-sense limit process. With such a representation available, the weak convergence is then proved. The quantity b^n in (1.3) appears in the expansion. This term is the difference between the two large (in scaled time and space) mean rates of arrival and of service. This heavy traffic assumption (A1.2) ensures that this term is of the proper magnitude. In the reflected diffusion limit process, the reflection term $z(\cdot)$ plays the role of a constraint. It keeps the state (that represents the weak-sense limit of the scaled physical queue lengths) from becoming negative. Alternatively put, it represents the weak-sense limit of the scaled cumulative idle time processes of the processor, divided by $\bar{\Delta}^d$. In the heavy traffic limit, the idle time does not affect the drift term bt or the noise term $w(t)$, which would be the same if the processor were never idle. The Wiener process terms represent the weak-sense limit of the randomness in the arrival and service processes.

Proof. Let us start by assuming (A1.1). Repeating the expansions used in Sections 1.1 and 1.2, write (1.1) as

$$x^n(t) = x^n(0) + \frac{1}{\sqrt{n}} \sum_{l=1}^{nS^{a,n}(t)} 1 - \frac{1}{\sqrt{n}} \sum_{l=1}^{nS^{d,n}(t)} 1. \tag{1.16}$$

Then separate the "noise" from the "drift" by factoring the sums as

$$
\begin{aligned}
x^n(t) = x^n(0) &+ \frac{1}{\sqrt{n}} \sum_{l=1}^{nS^{a,n}(t)} \left[1 - \frac{\Delta_l^{a,n}}{\bar{\Delta}^{a,n}}\right] - \frac{1}{\sqrt{n}} \sum_{l=1}^{nS^{d,n}(t)} \left[1 - \frac{\Delta_l^{d,n}}{\bar{\Delta}^{d,n}}\right] \\
&+ \frac{1}{\sqrt{n}\bar{\Delta}^{a,n}} \sum_{l=1}^{nS^{a,n}(t)} \Delta_l^{a,n} - \frac{1}{\sqrt{n}\bar{\Delta}^{d,n}} \sum_{l=1}^{nS^{d,n}(t)} \Delta_l^{d,n}.
\end{aligned}
\tag{1.17}
$$

The second and third terms on the right side of (1.17) are $w^{a,n}(S^{a,n}(t))$ and $-w^{d,n}(S^{d,n}(t))$, respectively. The third *sum* is nt plus an "error," where the error is $[nt$ minus the time of the last arrival before or at $nt]$. Thus, $1/(\sqrt{n}\bar{\Delta}^{a,n})$ times this error is a residual-time error term. The last term of (1.17) can be written as

$$
-\frac{1}{\bar{\Delta}^{d,n}} \left[\sqrt{n}t - T^{d,n}(t) - \frac{1}{\sqrt{n}}\tilde{\Delta}^{d,n}(t)\right], \tag{1.18}
$$

where $\tilde{\Delta}^{d,n}(t)/\sqrt{n}$ is defined below (1.10) and is a residual-time error term. Let $\epsilon^n(t)$ denote the sum of the residual-time error terms. We can now write

$$
\begin{aligned}
x^n(t) = x^n(0) &+ \sqrt{n}\left[\frac{1}{\bar{\Delta}^{a,n}} - \frac{1}{\bar{\Delta}^{d,n}}\right] t \\
&+ w^{a,n}(S^{a,n}(t)) - w^{d,n}(S^{d,n}(t)) + z^n(t) + \epsilon^n(t).
\end{aligned}
\tag{1.19}
$$

The scaled idle time process $z^n(\cdot)$ is nondecreasing and can increase only at t where $x^n(t) = 0$. By (A1.2), the second term on the right of (1.19) converges to bt.

Next, prove the weak convergence of $\mathcal{T}^{a,n}(\cdot)$ to the process with values $\bar{\Delta}^a t$. Factor $\mathcal{T}^{a,n}(t)$ as

$$
\mathcal{T}^{a,n}(t) = \frac{1}{n}\sum_{l=1}^{nt} \bar{\Delta}^{a,n} + \frac{1}{n}\sum_{l=1}^{nt} \left[\Delta_l^{a,n} - \bar{\Delta}^{a,n}\right]. \tag{1.20}
$$

By the weak convergence (A1.1) of $w^{a,n}(\cdot)$ to a process with continuous paths, for each $t < \infty$,

$$
\lim_n P\left\{\sup_{s\le t} \frac{1}{n}\left|\sum_{l=1}^{ns}[\Delta_l^{a,n} - \bar{\Delta}^{a,n}]\right| \ge \delta\right\} = 0. \tag{1.21}
$$

This implies that the process defined by the second sum in (1.20) converges weakly to the "zero" process. Combining this with the fact that $\bar{\Delta}^{a,n} \to \bar{\Delta}^a$ implies that $\mathcal{T}^{a,n}(\cdot)$ converges weakly to the process with values $\bar{\Delta}^a t$. Now, the inverse relation (1.9b) implies that $S^{a,n}(\cdot)$ converges weakly to the inverse function, that with values $t/\bar{\Delta}^a = \bar{\lambda}^a t$. A similar argument shows that $\mathcal{T}^{d,n}(\cdot)$ converges weakly to the process with values $\bar{\Delta}^d t$. Since the paths of $w^a(\cdot)$ are continuous, the weak convergence implies that $w^{a,n}(S^{a,n}(\cdot))$ converges weakly to the process with values $w^a(\bar{\lambda}^a t)$.

The analysis of $S^{d,n}(\cdot)$ is more subtle due to the idle time. But a similar argument can be used. Equation (1.12) and the weak convergence of $\mathcal{T}^{d,n}(\cdot)$ implies that for any $\epsilon > 0$ and $t > 0$,

$$\lim_n P\left\{\sup_{s\le t} S^{d,n}(s) \le \bar{\lambda}^d t + \epsilon\right\} = 1. \tag{1.22}$$

The residual-time error process $\epsilon^n(t)$ is bounded by a constant times the maximum of the $\Delta_l^{a,n}/\sqrt{n}$ up to the first arrival after (or at) real time nt, plus an analogous term for the departures. The fact that the residual-time error terms converge to the "zero" process is implied by the weak convergence in (A1.1); in particular by the facts that the $w^{\alpha,n}(\cdot), \alpha = a, d$, and the $S^{a,n}(\cdot)$ converge weakly to continuous processes. This implies [1] that the maximum value of the $\Delta_l^{\alpha,n}/\sqrt{n}, \alpha = a, d$, on any real time interval $[0, nt]$ goes to zero in probability as $n \to \infty$.

Equation (1.11) implies that for any $\tau > 0$ (and modulo a residual-time error term, which is bounded by $1/\sqrt{n}$ times the sum of the times that any customers being served at real times nt and $nt + n\tau$ have been in service)

$$\mathcal{T}^{d,n}(S^{d,n}(t+\tau)) - \mathcal{T}^{d,n}(S^{d,n}(t)) = \tau - \frac{1}{\sqrt{n}}\left[T^{d,n}(t+\tau) - T^{d,n}(t)\right] \le \tau. \tag{1.23}$$

The weak convergence of $\mathcal{T}^{d,n}(\cdot)$, as proved, and (1.22), (1.23) imply that $\{S^{d,n}(\cdot)\}$ is tight and that any weak-sense limit process has Lipschitz continuous paths with Lipschitz constant no greater than $\bar{\lambda}^d$.

By the reflection mapping (3.4.6),

$$z^n(t) = \max\left\{0, -\min_{s\le t}\left[x^n(0) + b^n s + w^{a,n}(S^{a,n}(s)) - w^{d,n}(S^{d,n}(s)) + \epsilon^n(s)\right]\right\}. \tag{1.24}$$

This and the weak convergence of the processes on the right side of (1.24) to continuous weak-sense limit processes imply the weak convergence of $z^n(\cdot)$ to some continuous weak-sense limit process $z(\cdot)$. By the weak convergence, it is clear that $z(0) = 0$, $z(\cdot)$ is nondecreasing, and that it can increase only at t such that $x(t) = 0$.

To complete the proof of the weak convergence of $S^{d,n}(\cdot)$, return to (1.11). By the tightness of the reflection terms $z^n(\cdot)$, the idle time processes $T^{d,n}(\cdot)/\sqrt{n}$ converge weakly to the "zero" process. This and (1.11) imply the asserted convergence of $S^{d,n}(\cdot)$. By (A1.1), $(w^\alpha(\bar{\lambda}^\alpha \cdot), \alpha = a, d)$

---

[1] All of the residual-time error terms in the rest of the book will converge to the zero process for the same reasons.

is a Wiener process. The fact that it is an $\mathcal{F}_t$-Wiener process follows from the independence of $x(0)$ and $(w^\alpha(\bar{\lambda}^a \cdot), \alpha = a, d)$ and the representation of $z(\cdot)$ via the reflection map

$$z(t) = \max\left\{0, -\min_{s \le t}\left[x(0) + bs + w^a(\bar{\lambda}^a s) - w^d(\bar{\lambda}^a s)\right]\right\}.$$

Replacing (A1.1) by (A1.3) and (A1.4). All that needs to be done is to show that $(w^{\alpha,n}(\cdot), \alpha = a, d)$ converges to a pair of real-valued and mutually independent Wiener processes with variances σ_a^2, σ_d^2, respectively. But this follows from Theorem 2.8.8. [2] ■

A martingale difference condition on the intervals. Assumption (A1.4) can be modified so that the condition that the intervals be independent is replaced by the condition that the centered intervals be "martingale differences." A similar condition can be used in all the theorems of the sequel. In preparation for this, the following definitions are needed.

Let $\mathcal{F}_l^{a,n}$ denote the minimal σ-algebra that measures $x^n(0)$, $\Delta_j^{a,n}, j \le l$, and all the other (service time in this case) intervals that begin at or before the arrival of the lth exogenous customer. Let $\mathcal{F}_l^{d,n}$ denote the minimal σ-algebra that measures $x^n(0)$, $\Delta_j^{d,n}, j \le l$, and all the other (interarrival, in this case) intervals that begin at or before the lth service completion. Let $E_l^{a,n}$ (respectively, $E_l^{d,n}$) denote the associated conditional expectations.

A1.5. *There are constants* $\bar{\Delta}^{\alpha,n}, \bar{\Delta}^\alpha, \sigma_\alpha^2$, $\alpha = a, d$, *such that*

$$E_l^{a,n}\Delta_{l+1}^{a,n} = \bar{\Delta}^{a,n} \to \bar{\Delta}^a, \quad E_l^{d,n}\Delta_{l+1}^{a,n} = \bar{\Delta}^{d,n} \to \bar{\Delta}^d, \tag{1.25}$$

$$\lim_{n\to\infty} E_l^{a,n}\left[1 - \frac{\Delta_{l+1}^{a,n}}{\bar{\Delta}^{a,n}}\right]^2 \to \sigma_a^2, \quad \lim_{n\to\infty} E_l^{d,n}\left[1 - \frac{\Delta_{l+1}^{d,n}}{\bar{\Delta}^{a,n}}\right]^2 \to \sigma_d^2. \tag{1.26}$$

The limits in (1.25) *and* (1.26) *are in the mean and are uniform in* l.

Theorem 1.2. *Assumptions* (A1.3) *and* (A1.5) *imply that* $w^{a,n}(\cdot), w^{d,n}(\cdot)$ *converge weakly to mutually independent Wiener processes* $w^a(\cdot), w^d(\cdot)$, *and Theorem* 1.1 *holds with* Σ *being diagonal with entries* σ_a^2, σ_d^2.

Comments on the proof. Theorem 2.8.8 shows that each $w^{\alpha,n}(\cdot)$ converges individually to a Wiener process $w^\alpha(\cdot)$ with variance σ_α^2, $\alpha = a, d$,

---

[2] Recall how uniform integrability such as (A1.3) was used in Theorem 2.8.8 to prove that the maximum of the intervals $\Delta_l^{\alpha,n}/\sqrt{n}$ for $l \le nt$ goes to zero in probability as $n \to \infty$.

but care must be exercised in using Theorem 2.8.8 to get the correlation between $w^\alpha(\cdot), \alpha = a, d$, since the conditioning σ-algebras $\mathcal{F}_l^{a,n}$ and $\mathcal{F}_l^{d,n}$ depend on $\alpha = a, d$. The lth arrival might occur before, after, or at the same time as the lth service completion.

To show independence, it is sufficient to show that

$$\prod_i Ef_i(w^a(t_i))g_i(w^d(s_i)) = \prod_i Ef_i(w^a(t_i))Eg_i(w^d(s_i))$$

for each finite set of bounded continuous, and twice continuously differentiable real-valued functions $f_i(\cdot), g_i(\cdot)$ with compact support and any t_i, s_i. It is more natural to show that

$$\prod_i Ef_i(w^a(\bar{\lambda}^a t_i))g_i(w^d(\bar{\lambda}^a s_i)) = \prod_i Ef_i(w^a(\bar{\lambda}^a t_i))Eg_i(w^d(\bar{\lambda}^a s_i)),$$

since then the natural order of time is introduced. By the weak convergence and continuity of the weak-sense limit processes, it is only necessary to show that

$$\prod_i Ef_i(w^{a,n}(S^{a,n}(t_i)))g_i(w^{d,n}(S^{a,n}(s_i)))$$
$$- \prod_i Ef_i(w^{a,n}((S^{a,n}(t_i)))Eg_i(w^{d,n}(S^{a,n}(s_i)))$$

converges to zero. This can be done by a Taylor expansion as is the proof of Theorem 2.8.8, and the use of (A1.3) and (A1.5). In the evaluation, without loss of generality, the $S^{a,n}(t)$ (respectively, $S^{d,n}(s)$) would be replaced by $N^{a,n}(t)$ (respectively, by $N^{d,n}(s)$) since it is an $\mathcal{F}_l^{a,n}$-stopping time (respectively, an $\mathcal{F}_l^{d,n}$-stopping time). An analogous, but more general, result for networks is proved in Theorem 6.1.2, and we defer to that.

5.1.2 Multiple Arrival Streams of Different Rates

We now consider the model where there are a finite number of independent input streams, called $A^{k,n}$, $1 \leq k \leq \kappa$, replacing the single stream A^n in Theorem 1.1. They are served FCFS, and if more than one job arrives at the same time, order them in any way at all. It is assumed that the service time distributions do not depend on the arrival class. (See Subsection 3.2 if they do.) The discussion is divided according to whether the arrivals in the additional streams are "frequent," "infrequent," or "moderately frequent," all of which occur in applications. Let the kth arrival stream have interarrival times $\{\Delta_l^{a,k,n}, l < \infty\}$.

Frequent arrivals. For some centering constants $\bar{\Delta}^{a,k,n}$ that converge to $\bar{\Delta}^{a,k} \equiv 1/\bar{\lambda}^{a,k}$, $k \leq \kappa$, define

$$w^{a,k,n}(t) = \frac{1}{\sqrt{n}} \sum_{l=1}^{nt} \left[1 - \frac{\Delta_l^{a,k,n}}{\bar{\Delta}^{a,k,n}} \right]. \tag{1.27}$$

Define $A^{k,n}(\cdot)$ and $S^{a,k,n}(\cdot)$ analogously to $A^n(\cdot)$ and $S^{a,n}(\cdot)$, but for the kth arrival stream. We will use the following conditions.

A1.6. *For some real number* b,

$$\sqrt{n}\left[\sum_k \frac{1}{\bar{\Delta}^{a,k,n}} - \frac{1}{\bar{\Delta}^{d,n}}\right] = b^n \to b. \tag{1.28}$$

A1.7. $(w^{a,k,n}(\cdot),\ k \le \kappa, w^{d,n}(\cdot))$ *converges weakly to mutually independent Wiener processes with variances* $\sigma^2_{a,k}, k \le \kappa,$ *and* σ^2_d, *respectively.*

The proof of the next theorem is essentially that of Theorem 1.1 but where we use the representations

$$x^n(t) = x^n(0) + \sum_k A^{k,n}(t) - D^n(t) \tag{1.29}$$

and

$$\begin{aligned} x^n(t) = x^n(0) &+ \frac{1}{\sqrt{n}}\sum_k \sum_{l=1}^{nS^{a,k,n}(t)}\left[1 - \frac{\Delta_l^{a,k,n}}{\bar{\Delta}^{a,k,n}}\right] - \frac{1}{\sqrt{n}}\sum_{l=1}^{nS^{d,n}(t)}\left[1 - \frac{\Delta_l^{d,n}}{\bar{\Delta}^{d,n}}\right] \\ &+ \sum_k \frac{1}{\sqrt{n}\bar{\Delta}^{a,k,n}} \sum_{l=1}^{nS^{a,k,n}(t)} \Delta_l^{a,k,n} - \frac{1}{\sqrt{n}\bar{\Delta}^{d,n}} \sum_{l=1}^{nS^{d,n}(t)} \Delta_l^{d,n}. \end{aligned} \tag{1.30}$$

Theorem 1.3. *Let* $x^n(0)$ *converge weakly to* $x(0)$ *and assume* (A1.0), (A1.6), *and* (A1.7). *Then the conclusions of Theorem* 1.1 *continue to hold, but where* $S^{a,k,n}(\cdot)$ *converges weakly to the process with values* $\bar{\lambda}^{a,k}t$, $w(\cdot)$ *is*

$$w(t) = \sum_k w^{a,k}(\bar{\lambda}^{a,k}\cdot) - w^d(\bar{\lambda}^d t),$$

where $(w^{a,k}(\cdot),\ k \le \kappa, w^d(\cdot))$ *are mutually independent Wiener processes with variances* $\sigma^2_{a,k}, k \le \kappa,$ *and* σ^2_d, *respectively.*

If we wish to use the analogue of (A1.3) and (A1.4) to imply (A1.7), then suppose that the members of the set of all intervals are mutually independent, replace (1.5) and (1.6) by

$$\begin{aligned} &E\Delta_l^{a,k,n} = \bar{\Delta}^{a,k,n} \to \bar{\Delta}^{a,k}, \quad E\Delta_l^{d,n} = \bar{\Delta}^{d,n} \to \bar{\Delta}^d, \\ &E\left[1 - \frac{\Delta_l^{a,k,n}}{\bar{\Delta}^{a,k,n}}\right]^2 \to \sigma^2_{a,k}, \quad E\left[1 - \frac{\Delta_l^{d,n}}{\bar{\Delta}^{d,n}}\right]^2 \to \sigma^2_d. \end{aligned} \tag{1.31}$$

and assume that

$$\left\{|\Delta_l^{a,k,n}|^2, |\Delta_l^{d,n}|^2; k, l, n\right\} \text{ is uniformly integrable.} \tag{1.32}$$

The analogue of the martingale conditions of Theorem 1.2 can be used in lieu of (1.31).

Infrequent arrivals. Sometimes, the arrivals of some input stream occur far less frequently than the arrivals of the others. Then the following trivial extension of Theorem 1.1 is useful. Use the setup of Theorem 1.3, but let $A^{0,n}$ denote another arrival stream. Let $S^{a,0,n}(\cdot)$ denote $1/n$ times the number of arrivals from stream $A^{0,n}$ by real time nt.

Theorem 1.4. *Assume the conditions of Theorem 1.3, but suppose that $\sqrt{n}S^{a,0,n}(\cdot)$ converges weakly to the process with values $\bar{\lambda}^{a,0}t$. Then the conclusions of Theorem 1.3 hold, but with $\bar{\lambda}^{a,0}$ added to b.*

The input steam $A^{0,n}$ is called *marginal* in that it affects the drift of the weak-sense limit, but not the diffusion part. Although the arrivals from $A^{0,n}$ occur much less frequently than those from the other streams, they can still have a substantial effect on the weak-sense limit process, under heavy traffic, via the change in the drift.

Moderately frequent arrivals. In the case of Theorem 1.4, $1/\sqrt{n}$ times the mean number of arrivals of $A^{0,n}$ by real time nt converged to $\bar{\lambda}^{a,0}t$. The original heavy traffic condition (1.28) still worked, and the only change in the result concerned the increased value of the drift term in the weak-sense limit process. If the mean number of arrivals from $A^{0,n}$ in real time $[0, nt]$ rises to the order of n, then we are back to the setup in Theorem 1.3.

Intermediate cases are possible, where $1/n$ times the mean number of arrivals on $[0, nt]$ goes to zero, but $1/\sqrt{n}$ times this number goes to infinity. Then, the service rate must be increased to accommodate this new stream $A^{0,n}$, but the limit Wiener processes are not affected. Return to the setup in Theorem 1.3. Suppose that in addition to the input streams $A^{k,n}$, $k \geq 1$, of Theorem 1.3, there is an additional input stream $A^{0,n}$ with intervals $\{\Delta_l^{a,0,n}, l < \infty\}$. For some centering constant $\bar{\Delta}^{a,0,n}$ that goes to infinity as $n \to \infty$, define

$$w^{a,0,n}(t) = \frac{1}{\sqrt{n}} \sum_{l=1}^{nt} \left[1 - \frac{\Delta_l^{a,0,n}}{\bar{\Delta}^{a,0,n}} \right]. \tag{1.33}$$

Suppose that $\{w^{a,0,n}(\cdot)\}$ converges weakly to some continuous process. The new heavy traffic condition is

$$\lim_n \sqrt{n} \left[\sum_{k=1}^{K} \frac{1}{\bar{\Delta}^{a,k,n}} + \frac{1}{\bar{\Delta}^{a,0,n}} - \frac{1}{\bar{\Delta}^{d,n}} \right] = b. \tag{1.34}$$

We can now state the following theorem, of which Theorem 1.4 is a special case.

Theorem 1.5. *Assume the conditions of Theorem* 1.3, *but with the changes in the above paragraph. Then the conclusions of Theorem* 1.3 *hold.*

Remark on the proof. The proof is nearly identical to those of Theorems 1.1 and 1.3. Define $S^{a,0,n}(t)$ to be the (scaled by $1/n$) number of arrivals from $A^{0,n}$ by real time nt. One does the same expansions, except with the new input stream added, as was done to get (1.30). This yields a new drift term $\sqrt{n}t/\bar{\Delta}^{a,0,n}$, which is taken care of by the new heavy traffic condition. We also have the new term $w^{a,0,n}(S^{a,0,n}(t))$. The only new problem concerns the asymptotic behavior of $S^{a,0,n}(t)$. This converges weakly to the "zero" process. To show this, we modify the proof in Theorem 1.1 slightly. Let $\bar{S}^{a,0,n}(t)$ denote $1/n$ times the number of arrivals from $A^{0,n}$ by real time $nt\bar{\Delta}^{a,0,n}$. Define

$$\bar{\mathcal{T}}^{a,0,n}(t) = \frac{1}{n\bar{\Delta}^{a,0,n}} \sum_{l=1}^{nt} \Delta_l^{a,0,n}. \tag{1.35}$$

Relations similar to (1.9) hold for these new definitions: In particular (modulo an error that is no larger than $1/n$),

$$\bar{S}^{a,0,n}(\bar{\Delta}^{a,0,n}\mathcal{T}^{a,0,n}(t)) = t. \tag{1.36}$$

By the assumption concerning the weak convergence of the processes defined by (1.33), $\bar{\mathcal{T}}^{a,0,n}(\cdot)$ converges weakly to the process with values t. This, (1.36), and the fact that $\bar{\Delta}^{a,0,n} \to \infty$ imply that $S^{a,0,n}(\cdot)$ converges weakly to the "zero" process. Hence $w^{a,0,n}(S^{a,0,n}(\cdot))$ converges weakly to the "zero" process.

5.1.3 The Fictitious Services Model

In the representation of the queue-length process (1.16) in the form (1.19), the reflection term $z^n(\cdot)$ appeared directly as a way of accounting for the effect of the idle time. There is an alternative approach to dealing with the idle time in the expansion of (1.16), that has been of great value historically [21, 110], and was illustrated in Example 1.1.1. It is still of importance, although the method used in Theorem 1.1 is simpler for the bulk of problems of interest in this book. Theorem 1.1 will be proved again with this alternative approach. *Unless explicitly noted, in the rest of the book the idle time will be handled by the method of Theorem* 1.1.

A convention concerning service times: The residual service time assumption. As noted before the proof of Theorem 1.1, the proof of the weak convergence of $x^n(\cdot)$ to the reflected diffusion "heavy traffic limit" proceeds by expanding and reorganizing the terms in (1.1) into expressions that have a "resemblance" to the drift, Wiener process, and reflection terms in the "limit process." In Theorem 1.1, the reflection term was represented directly as a function of the cumulative idle time of the processor.

Let us change the model slightly by assuming that the processor continues to serve even when there are no customers to be served. It acts as though there were customers present at all times. This procedure generates "fictitious" outputs, and these must be corrected for. It is the correction to the fictitious outputs that creates the reflection term. The reflection term keeps the scaled queue length from becoming negative due to the fictitious outputs. A problem will arise with this formulation only when there is an arrival of a customer in the midst of a fictitious service interval. This is handled by a slight modification of the model. Suppose that a customer does arrive in the midst of a fictitious service interval. Then modify the model by supposing that there is no fictitious departure but that the service time of that newly arrived customer is modified to be the remaining (residual) service time of the current fictitious service interval. With this modification, that customer departs at the time that the fictitious service interval would end. With this general scheme, time is divided into intervals of lengths $\Delta_l^{d,n}$, which we assume satisfy the conditions of Theorem 1.1. Owing to the residual service-time assumption, the queue size for the modified problem might be different from that in the actual physical problem. It turns out that the modified queue differs from the original queue by at most one customer [110]. A general proof of the asymptotic irrelevance of such a "residual service time" modification will be given in Section 2, and a similar form can be used for the network problem.

Let $D_f^n(t)$ denote $1/\sqrt{n}$ times the number of customers (either real or fictitious) served on the real time interval nt, and define $z_f^n(t)$ to be $1/\sqrt{n}$ times the number of fictitious departures on the real time interval nt. Define $x_f^n(t)$ to be $1/\sqrt{n}$ times the queue size at real time nt in the system modified with the "residual service time" assumption. Then the state equation for the modified system is

$$x_f^n(t) = x^n(0) + A^n(t) - D_f^n(t) + z_f^n(t). \tag{1.37}$$

The compensation term $z_f^n(\cdot)$ acts as the constraint. As noted above, it corrects for the fictitious departures. We can see the relation between the definition of $z_f^n(t)$ given above and the $z^n(t)$ defined by (1.13) by noting that the division of the idle time by $\bar{\Delta}^{d,n}$ in (1.13) gives the "mean value" of the services that could have been done during the idle time.

Theorem 1.6. *Assume the conditions of Theorem* 1.1, *except for the use of* $(x_f^n(\cdot), z_f^n(\cdot))$ *in lieu of* $(x^n(\cdot), z^n(\cdot))$. *Then the conclusions of Theorem* 1.1 *hold.*

Remark. The theorem shows that the measures of $x^n(\cdot)$ and $x_f^n(\cdot)$ are asymptotically equivalent in the sense of weak convergence. Thus the fictitious service and residual time service model give the same limit distributions. This, in itself, does not imply that $x^n(\cdot) - x^n_f(\cdot)$ converges weakly to the "zero" process. This latter result is implied by Theorem 2.4.

Proof. The proof is essentially that of Theorem 1.1, and we will comment on the differences only. For concreteness, suppose (A1.1). Using the fictitious services and residual service time convention, define

$$S_f^{d,n}(t) = \frac{1}{n} \times \max\left\{ m : \sum_{l=1}^{m} \Delta_l^{d,n} \le nt \right\}. \tag{1.38}$$

We now have, possibly modulo a residual-time error term,

$$S_f^{d,n}(\mathcal{T}^{d,n}(t)) = t. \tag{1.39}$$

Rewrite (1.37), analogously to what was done in Theorem 1.1, as

$$x_f^n(t) = x^n(0) + \frac{1}{\sqrt{n}} \sum_{l=1}^{nS^{a,n}(t)} 1 - \frac{1}{\sqrt{n}} \sum_{l=1}^{nS_f^{d,n}(t)} 1 + z_f^n(t). \tag{1.40}$$

Factor the sums in (1.40) to yield

$$\begin{aligned} x_f^n(t) = x^n(0) &+ \frac{1}{\sqrt{n}} \sum_{l=1}^{nS^{a,n}(t)} \left[1 - \frac{\Delta_l^{a,n}}{\bar{\Delta}^{a,n}}\right] - \frac{1}{\sqrt{n}} \sum_{l=1}^{nS_f^{d,n}(t)} \left[1 - \frac{\Delta_l^{d,n}}{\bar{\Delta}^{d,n}}\right] \\ &+ \frac{1}{\sqrt{n}\bar{\Delta}^{a,n}} \sum_{l=1}^{nS^{a,n}(t)} \Delta_l^{a,n} - \frac{1}{\sqrt{n}\bar{\Delta}^{d,n}} \sum_{l=1}^{nS_f^{d,n}(t)} \Delta_l^{d,n} + z_f^n(t). \end{aligned} \tag{1.41}$$

Noting that the fourth and fifth terms of (1.41) are (modulo a residual-time error term) $\sqrt{n}t/\bar{\Delta}^{a,n}$ and $-\sqrt{n}t/\bar{\Delta}^{d,n}$, respectively, and using the heavy traffic condition (A1.2) as in Theorem 1.1, equation (1.41) can be rewritten as

$$x_f^n(t) = x^n(0) + b^n t + w^{a,n}(S^{a,n}(t)) - w^{d,n}(S_f^{d,n}(t)) + z_f^n(t) + \epsilon^n(t), \tag{1.42}$$

where $\epsilon^n(\cdot)$ is a residual-time error term.

From this point on, the proof of all but the tightness of $\{z_f^n(\cdot)\}$ is exactly as in Theorem 1.1, and this tightness proof is deferred to Theorem 2.4, since it is an application of the method introduced in Section 2. ■

5.2 Tightness of the Reflection Processes: A General Approach

Finite buffers. The method for showing the tightness will be developed for a slightly more complicated model than that used in Section 1. The methods are actually special cases of what was done in Section 3.6. But it is worthwhile specializing the development for the simple cases of this chapter. Up to this point, we supposed that there was an unlimited waiting room for the customers awaiting service. We will now consider the case where there is a constant B such that the queue is limited to at most $B\sqrt{n}$, the buffer size. Arrivals to a full buffer are assumed to be lost. This buffer size is chosen because it is the only one that gives results that are not either of the form of Theorem 1.1 or trivial. Indeed, if the buffer size is $O(n^\alpha)$ where $\alpha > 0.5$, then the effect of the buffer does not appear in the limit. If $\alpha < 0.5$, then the limit process is "zero" and there is a substantial buffer overflow, since $O(n^\alpha)/\sqrt{n} \to 0$. The critical scaling factor $\sqrt{n}$ tells us what the appropriate "dimensioning" of the system should be, when operating in the heavy traffic regime. Owing to the upper bound, there are two reflection processes, one occurring at the origin and the other at the full buffer level.

The proof of weak convergence for the problem with a buffer of the size $O(\sqrt{n})$ is essentially that of Theorem 1.1. The major difference concerns the proof of tightness of the two reflection processes, since they depend on each other. Two ways of handing this problem will be given. The first uses the reflection mapping (3.4.6), (1.15), or (1.24) "in pieces," by dividing time into intervals such that in each one there can be a reflection from only one of the ends. An alternative method specializes the approach of Theorem 3.6.1, the proof by contradiction, by showing that some consequence of the lack of tightness (actually, more strongly, lack of asymptotic continuity) is not possible. Both methods of proof will be given, since they can be used on more general problems. The second method is more versatile and will be used throughout the rest of the book. It can be used in a much wider class of problems, including the workload formulation, and to show the asymptotic unimportance of the "residual service time" assumption that was used in Section 1.3 concerning the service time of a customer who arrives when the system is empty.

For the finite buffer problem, the state equation (1.1) needs to be corrected by subtracting the scaled number of customers lost due to the buffer being full on their arrival. Define $U^n(t)$ to be $1/\sqrt{n}$ times the number of customers who arrived by real nt when the buffer was full. The state equation (1.1) is replaced by

$$x^n(t) = x^n(0) + A^n(t) - D^n(t) - U^n(t). \tag{2.1}$$

The terms $z^n(t)$ and $z(t)$ will be redefined to be the total reflection process, whatever the boundary. Define $L^n(t) = T^{d,n}(t)/\bar{\Delta}^{d,n}$, the reflection process at the lower boundary, as in (1.13), and $z^n(t) = L^n(t) - U^n(t)$.

The expansions used up to (1.19) hold here also, and we can write

$$x^n(t) = x^n(0) + w^{a,n}(S^{a,n}(t)) - w^{d,n}(S^{d,n}(t)) + b^n t + L^n(t) - U^n(t) + \epsilon^n(t), \tag{2.2}$$

where $\epsilon^n(\cdot)$ is a residual-time error term.

Theorem 2.1. *Assume the conditions of Theorem 1.1, but with the upper bound $B\sqrt{n}$ added. Then the conclusions of that theorem continue to hold with*

$$x(t) = x(0) + w(t) + bt + z(t),\ x(t) \in [0, B], \tag{2.3}$$

where $z(\cdot) = L(\cdot) - U(\cdot)$ and $L(0) = U(0) = 0$. The $L(\cdot)$ (respectively, $U(\cdot)$) is continuous and can increase only at t where $x(t) = 0$ (respectively, where $x(t) = B$).

Proof. The proof is identical to that of Theorem 1.1, except for the proof of tightness and asymptotic continuity of $\{L^n(\cdot), U^n(\cdot)\}$. Two useful proofs are given below in Theorems 2.2 and 2.3. ■

First proof of tightness of $\{L^n(\cdot), U^n(\cdot)\}$. Write (2.2) as

$$x^n(t) = h^n(t) + z^n(t), \quad x^n(t) \in [0, B]. \tag{2.4}$$

Theorem 2.2. *Suppose that $\{h^n(\cdot)\}$ is tight, and that the weak-sense limit of any weakly convergent subsequence has continuous paths with probability one. Then $\{L^n(\cdot), U^n(\cdot)\}$ is tight and the weak-sense limit process of any weakly convergent subsequence has continuous paths with probability one.*

Remark. The tightness and asymptotic continuity assumption is equivalent to the following. For each $T > 0$ and $\mu > 0$,

$$\lim_{\delta\to 0} \limsup_{n} P\left\{ \sup_{t\le T} \sup_{s\le\delta} |h^n(t+s) - h^n(t)| \ge \mu \right\} = 0. \tag{2.5}$$

Proof. This proof is based on the use of the reflection map (1.15) or (1.24) on appropriate subintervals. Let $0 < B_1 < B_2 < B$. Define the stopping times $\tau_0^n = 0$, and for $k > 0$, define, recursively,

$$\begin{aligned} \sigma_k^n &= \min\left\{t > \tau_{k-1}^n : x^n(t) \ge B_2\right\}, \\ \tau_k^n &= \min\left\{t > \sigma_k^n : x^n(t) \le B_1\right\}. \end{aligned} \tag{2.6}$$

By the assumption on $\{h^n(\cdot)\}$, we can suppose, without loss of generality, that $x^n(\cdot)$ never jumps from $[0, B_1]$ to $[B_2, B]$ in one step and that it always takes more than one step to go from $[B_1, B_2]$ to B, and similarly in the reverse direction.

Divide time into the intervals $[\tau_k^n, \sigma_{k+1}^n)$ and $[\sigma_{k+1}^n, \tau_{k+1}^n)$, $k = 0, 1, \ldots$. At most one component of $z^n(\cdot)$ can increase on each interval, the $L^n(\cdot)$

on the $[\tau_k^n, \sigma_{k+1}^n)$ and the $U^n(\cdot)$ on the $[\sigma_{k+1}^n, \tau_{k+1}^n)$. The application of the reflection map (1.24) yields that for $t < \sigma_{k+1}^n - \tau_k^n$,

$$\begin{aligned} &L^n(\tau_k^n + t) - L^n(\tau_k^n) \\ &\quad = \max\left\{0, -\min_{s\leq t}\left[x^n(\tau_k^n) + h^n(\tau_k^n + s) - h^n(\tau_k^n)\right]\right\}. \end{aligned} \tag{2.7}$$

Analogously, for $t < \tau_{k+1}^n - \sigma_{k+1}^n$,

$$\begin{aligned} &U^n(\sigma_{k+1}^n + t) - U^n(\sigma_{k+1}^n) \\ &\quad = \max\left\{0, -\min_{s\leq t}\left[(B - x^n(\sigma_{k+1}^n)) - (h^n(\sigma_{k+1}^n + s) - h^n(\sigma_{k+1}^n))\right]\right\}. \end{aligned} \tag{2.8}$$

Let k be an integer. Since $\{h^n(\cdot)\}$ is tight with continuous weak-sense limit processes, (2.7) and (2.8) imply that $\{L^n(\tau_k^n \wedge \cdot), U^n(\tau_k^n \wedge \cdot), n < \infty\}$ is tight for each k and the weak-sense limit processes have continuous paths with probability one. To complete the proof we need only ascertain that for each $t > 0$,

$$\lim_{k\to\infty} \liminf_n P\{\tau_k^n \geq t\} = 1. \tag{2.9}$$

But (2.9) holds owing to the fact that $\{h^n(\cdot)\}$ is tight and all weak-sense limit processes have continuous paths with probability one. ■

A second proof of weak convergence of the reflection terms. An alternative and more powerful method uses the argument by contradiction of Theorem 3.6.1.

Theorem 2.3. *Suppose that $x^n(\cdot)$ has the representation* (2.4) *and that* (2.5) *holds. Let $L^n(\cdot)$ and $U^n(\cdot)$ be nondecreasing processes with zero initial values and the following properties. As $n \to \infty$, $L^n(\cdot)$ can increase only when $x^n(\cdot)$ is arbitrarily close to the boundary $x = 0$, and $U^n(\cdot)$ can increase only when $x^n(\cdot)$ is arbitrarily close to the boundary $x = B$. For each $T < \infty$, let the maximum values of the discontinuities in $L^n(\cdot)$ and $U^n(\cdot)$ on $[0, T]$ go to zero in probability as $n \to \infty$. Then $\{L^n(\cdot), U^n(\cdot)\}$ is tight and asymptotically continuous. Condition* (2.5) *holds under the conditions of Theorem* 1.1, *with or without the upper bound B added.*

Proof. We will use a direct argument by contradiction (instead of the reflection map) by showing that $z^n(\cdot)$ is asymptotically continuous. In fact, specializing the proof of Theorem 3.6.1 (equivalently, adapting a method of [167]) it will be shown that for each $T > 0$ and $\nu > 0$,

$$\lim_{\delta\to 0} \limsup_n P\left\{\sup_{t\leq T}\sup_{s\leq\delta} |z^n(t+s) - z^n(t)| \geq \nu\right\} = 0. \tag{2.10}$$

The limit equality (2.10) implies the tightness and that each weak-sense limit process has continuous paths with probability one.

It is clear that (2.5) holds under the conditions of Theorem 1.1, with or without the upper bound added. Suppose that (2.10) does not hold for some $T > 0$ and $\nu > 0$. Then there are $0 < \delta_n \to 0$ and $\eta > 0$ such that

$$\limsup_n P\left\{\sup_{t \leq T} |z^n(t+\delta_n) - z^n(t)| \geq \nu\right\} \geq \eta. \tag{2.11}$$

Now, (2.11) implies that there is a subsequence $z^{n_k}(\cdot)$ and ω sets Ω_k satisfying $P\{\Omega_k\} \geq \eta/2$, and such that for $\omega \in \Omega_k$ and large k, $|z^{n_k}(\cdot)|$ has at least one increase of magnitude at least $\nu/2 > 0$ on an arbitrarily small time interval in $[0, T]$. Such a jump must also be an (asymptotically and on the same subsequence and ω-sets) jump of the path $x^{n_k}(\cdot)$, and it must take this path (asymptotically and on the same subsequence and ω-sets) from some point arbitrarily close to the boundary points 0 or B to some point at least a distance of $\nu/4$ from the boundary. But this possibility is contradicted by the hypotheses. ■

On the "residual service time" assumption in Theorem 1.6. The argument of Theorem 2.3 can be used to validate the "residual service time" assumption made for the arrivals to an empty queue in Theorem 1.6. The proof will be given for a more general one-dimensional problem. In Theorem 1.6, $x^n_f(\cdot)$ refers to the physical process, where an arrival during a "fictitious service" (or, equivalently, when the processor is idle) has a service time that is just the residual service time for the current fictitious service interval. Let $x^{+,n}(\cdot)$ be the scaled queue-length process for the system that is modified so that a fictitious customer is immediately added to the queue whenever it becomes empty. Define $z^{+,n}(\cdot)$ to be the scaled process defined by the number of fictitious customers added, and $x^n(\cdot)$ the scaled true queue-length process. Let $D^{+,n}(t)$ denote $1/\sqrt{n}$ times the number of departures by nt, with this modification. Then,

$$x^{+,n}(t) = x^n(0) + A^n(t) - D^{+,n}(t) + z^{+,n}(t)$$

and

$$x^{+,n}(t) \geq x^n_f(t) \geq x^n(t).$$

Note the important fact that $x^{+,n}(\cdot)$ can be viewed as the original scaled queue-length process, but with reflection at $x = 1/\sqrt{n}$ instead of at $x = 0$.

Theorem 2.4. *Assume the conditions of Theorem 1.1 with or without the upper bound B. Then $\{x^n(\cdot), x^{+,n}(\cdot)\}$ is tight and any weak sense limit $(x(\cdot), x^+(\cdot))$ satisfies (2.3) with the same $w(\cdot)$. Hence $x^n_f(\cdot) - x^n(\cdot)$ converges weakly to the "zero" process.*

Comments on the proof. Theorem 2.3 and the expansions used in Theorems 1.1 and 1.6 imply that $\{z^{+,n}(\cdot), z^n(\cdot)\}$ is asymptotically continuous and can increase only when the respective state is arbitrarily close to the boundary. The $w^{d,n}(\cdot)$–processes in (1.17) and (1.42) are asymptotically equivalent, owing to the convergence of $S^{d,n}(\cdot)$ and $S_f^{d,n}(\cdot)$ to the same limit. Then the proofs of Theorem 1.1 and 1.6 imply the tightness of $\{x^{+,n}(\cdot), x^n(\cdot)\}$ and that any limit satisfies (2.3) with the same Wiener process. But (2.3) has a unique strong-sense solution ■

5.3 The Workload Processes

In Theorem 1.1 the process of interest was the scaled queue size $x^n(\cdot)$. The (scaled) work queued is also important, since the waiting time for a customer is the queued work on its arrival. In Section 6.4 and Chapters 11 and 12 it will be seen that the formulation in terms of the scaled work can greatly simplify the treatment of multiclass systems. Indeed, the workload formulation is essential for the treatment of problems where some processor must handle several different classes of jobs. For the simple single processor and single job class setup of Theorem 1.1, there are two ways of getting the limit workload equation. It can be derived directly, via a copy of the proof of Theorem 1.1, or it can be obtained from (1.14) or (2.3) by a simple rescaling and a law of large numbers. This latter method also works in reverse in that one can obtain the results of Theorem 1.1 for the limit of the scaled queue sizes from the equation for the limit of the scaled workload. Although the latter method is both more insightful and quicker, both methods will be developed. If the workload is to be controlled and the cost function involves the expectation of an unbounded function of the state, then the proof of convergence of the costs often requires an estimate of some moment of the workload. Getting such an estimate usually requires that we write out the workload equations for the physical problem in full, analogously to what was done in (1.17) and (1.41), so that all terms are exhibited and the appropriate moment estimated before the limit is taken. The next two subsections use a copy of the method of Theorem 1.1, and the rescaling method will be given in Subsection 3.3.

5.3.1 The Workload Equation for the Basic Model

Define the workload $WL^n(t)$, to be $1/\sqrt{n}$ times the real time that the processor must work to complete service on all of the jobs that are in the system at real time nt. *Redefine*

$$z^n(t) = T^{d,n}(t), \tag{3.1}$$

where we recall that $T^{d,n}(t)$ is the scaled (by $1/\sqrt{n}$, as usual) cumulative idle time by real time nt. The scaled (by $1/\sqrt{n}$) work that has been done by real time nt is $t\sqrt{n} - z^n(t)$. The (scaled) work that has arrived by time

nt can be written as

$$\frac{1}{\sqrt{n}} \sum_{l=1}^{nS^{a,n}(t)} \Delta_l^{d,n}. \tag{3.2}$$

We can now write the (scaled by $1/\sqrt{n}$) current work queued as

$$WL^n(t) = WL^n(0) + \frac{1}{\sqrt{n}} \sum_{l=1}^{nS^{a,n}(t)} \Delta_l^{d,n} - t\sqrt{n} + z^n(t). \tag{3.3}$$

A minor modification of the proof of Theorem 1.1 yields the following.

Theorem 3.1. *Assume* (A1.1)–(A1.2), *let* $WL^n(0) \Rightarrow WL(0)$, *and suppose that* (A1.0) *holds for* $WL^n(0)$. *Redefine*

$$\Psi^n(\cdot) = \left(WL^n(\cdot), S^{a,n}(\cdot), w^{a,n}(\cdot), w^{d,n}(\cdot), z^n(\cdot)\right).$$

Then $\{\Psi^n(\cdot)\}$ *is tight and converges weakly to a set*

$$\left(WL(\cdot), S^a(\cdot), w^a(\cdot), w^d(\cdot), z(\cdot)\right),$$

where

$$WL(t) = WL(0) + \bar{\Delta}^d bt + \bar{\Delta}^d w(t) + z(t), \tag{3.4}$$

where $w(\cdot), w^\alpha(\cdot)$, $\alpha = a, d$, *and* $S^a(\cdot)$ *are as in Theorem* 1.1, $z(\cdot)$ *is the reflection term, and* $WL(\cdot)$ *and* $z(\cdot)$ *are nonanticipative with respect to* $(w^\alpha(\bar{\lambda}^a\cdot), \alpha = a, d)$ *analogously to the case in Theorem* 1.1.

Proof. The proof is essentially a copy of that of Theorem 1.1. Expand the sum in (3.3) as

$$\frac{1}{\sqrt{n}} \sum_{l=1}^{nS^{a,n}(t)} \left[\Delta_l^{d,n} - \bar{\Delta}^{d,n}\right] + \frac{1}{\sqrt{n}} \sum_{l=1}^{nS^{a,n}(t)} \bar{\Delta}^{d,n}. \tag{3.5}$$

The first term of (3.5) is $-\bar{\Delta}^{d,n} w^{d,n}(S^{a,n}(t))$. Expand the last term in (3.5) as

$$\frac{1}{\sqrt{n}} \bar{\Delta}^{d,n} \sum_{l=1}^{nS^{a,n}(t)} \left[1 - \frac{\Delta_l^{a,n}}{\bar{\Delta}^{a,n}}\right] + \frac{1}{\sqrt{n}} \frac{\bar{\Delta}^{d,n}}{\bar{\Delta}^{a,n}} \sum_{l=1}^{nS^{a,n}(t)} \Delta_l^{a,n}. \tag{3.6}$$

The right-hand *sum* in (3.6) equals nt minus the time between nt and the last arrival at or before real time nt. Rewriting and using (A1.2) yields

$$\begin{aligned} WL^n(t) &= WL^n(0) + \bar{\Delta}^{d,n} w^{a,n}(S^{a,n}(t)) - \bar{\Delta}^{d,n} w^{d,n}(S^{a,n}(t)) \\ &+ \sqrt{n}\bar{\Delta}^{d,n} \left[\frac{1}{\bar{\Delta}^{a,n}} - \frac{1}{\bar{\Delta}^{d,n}}\right] t + z^n(t) + \epsilon^n(t), \end{aligned} \tag{3.7}$$

where $\epsilon^n(\cdot)$ is a residual-time error term. From here on, the proof is the same as that of Theorem 1.1. ■

Remark on the scale factor $\bar{\Delta}^d$ in (3.4) and (3.7). Note that (3.4) differs from (1.14) (and (3.7) from (1.19)) only in the scale factor $\bar{\Delta}^d$. Thus, the (scaled and in the limit) queued work equals the (scaled and in the limit) queued number of customers times the average work per customer. Such a relationship generally holds under broad conditions, a fact probably noted first in [213] and thoroughly treated in [239]. The result, that is of considerable importance, is a consequence of the assumption (A1.1), as seen in Subsection 3.3.

5.3.2 Several Input Streams: Different Work Requirements

Now let us return to the situation in Theorems 1.3 to 1.5 where there are several input streams $A^{k,n}$, still only one processor, but now we allow the work requirements to depend on the particular stream. Owing to this last complication, it is hard to develop the limit problem unless the workload formulation is used. Continue to assume the FCFS service discipline. We consider the analogue of Theorem 1.3. The analogues of Theorems 1.4 and 1.5 should then be obvious. Let $\{\Delta_l^{d,k,n}, l < \infty\}$ be the service time requirements for the members of input stream $A^{k,n}$. For some centering constants $\bar{\Delta}^{d,k,n}$, which converge to $\bar{\Delta}^{d,k}$, define

$$w^{d,k,n}(t) = \frac{1}{\sqrt{n}} \sum_{l=1}^{nt} \left[1 - \frac{\Delta_l^{d,k,n}}{\bar{\Delta}^{d,k,n}}\right]. \tag{3.8}$$

Define $S^{d,k,n}(\cdot)$ and $N^{d,k,n}(\cdot)$ analogously to $S^{d,n}(\cdot)$ and $N^{d,n}(\cdot)$, but for the kth arrival stream. We will use the following assumptions.

A3.1. *There is a real b such that*

$$\sqrt{n}\left[\sum_k \frac{\bar{\Delta}^{d,k,n}}{\bar{\Delta}^{a,k,n}} - 1\right] = b^n \to b.$$

A3.2. $w^{a,k,n}(\cdot), w^{d,k,n}(\cdot)$, $k \le \kappa$, *converge weakly to mutually independent Wiener processes with variances* $\sigma_{a,k}^2, \sigma_{d,k}^2$, $k \le \kappa$.

Theorem 3.2. *Let $WL^n(0)$ converge weakly to $WL(0)$, and assume* (A3.1), (A3.2), *and that* (A1.0) *holds for $WL^n(0)$. Then the conclusions of Theorem* 3.1 *hold, but where the weak-sense limit system is*

$$WL(t) = WL(0) + bt + w(t) + z(t), \tag{3.9a}$$

where b is defined in (A3.1) *and*

$$w(t) = \sum_k \bar{\Delta}^{d,k}\left[w^{a,k}(\bar{\lambda}^{a,k}t) - w^{d,k}(\bar{\lambda}^{a,k}t)\right]. \tag{3.9b}$$

Proof. In analogy to (3.3), the scaled workload queued can be written as

$$WL^n(t) = WL^n(0) + \sum_k \frac{1}{\sqrt{n}} \sum_{l=1}^{nS^{a,k,n}(t)} \Delta_l^{d,k,n} - t\sqrt{n} + z^n(t). \tag{3.10}$$

Expand the sum in (3.10) as in (3.5). The analogue of the first term of (3.5) is

$$-\sum_k \bar{\Delta}^{d,k,n} w^{d,k,n}(S^{a,k,n}(t)).$$

The analogue of the right side of (3.6) is

$$\sum_k \bar{\Delta}^{d,k,n} w^{a,k,n}(S^{a,k,n}(t)) + \frac{1}{\sqrt{n}} \sum_k \frac{\bar{\Delta}^{d,k,n}}{\bar{\Delta}^{a,k,n}} \sum_{l=1}^{nS^{a,k,n}(t)} \Delta_l^{a,k,n}.$$

From this point on, the development is the same as for Theorem 1.1 or 3.1. ■

5.3.3 *Asymptotic Relations Between the Queue Size and Workload*

Theorem 3.3 establishes a linear relationship between the workload and the scaled queue size, and provides a general way of getting (3.4) from (1.14) and vice versa. It can be used for the network case as well. Relationships between the queue sizes for the individual classes and the workload for feedforward networks can also be found in [206].

Theorem 3.3. *Consider the system model used in Theorem 1.1. Suppose that $w^{d,n}(\cdot)$ converges weakly to a continuous process, $\bar{\Delta}^{d,n} \to \bar{\Delta}^d > 0$, and that $x^n(\cdot)$ is bounded in probability in that for each $T > 0$,*

$$\lim_N \limsup_n P\left\{\sup_{t \le T} |x^n(t)| \ge N\right\} = 0. \tag{3.11}$$

Then, the difference

$$WL^n(\cdot) - \bar{\Delta}^{d,n} x^n(\cdot) \tag{3.12}$$

converges weakly to the "zero" process. Now, drop the assumption (3.11), but suppose that it holds for $WL^n(\cdot)$ replacing $x^n(\cdot)$. Suppose that, asymptotically, with an arbitrarily high probability and on any bounded time interval, $S^{a,n}(\cdot)$ is bounded by a function with a linear growth plus a constant and that $x^n(0)/\sqrt{n}$ converges weakly to zero. Then (3.11) holds.

Proof. Define $I^{d,n}(t) = nS^{d,n}(t)$, the index of the last customer to complete service by real time nt. Then

$$\frac{1}{\sqrt{n}} \sum_{l=I^{d,n}(t)+2}^{I^{d,n}(t)+\sqrt{n}x^n(t)} \Delta_l^{d,n} \le WL^n(t) \le \frac{1}{\sqrt{n}} \sum_{l=I^{d,n}(t)+1}^{I^{d,n}(t)+\sqrt{n}x^n(t)} \Delta_l^{d,n}. \tag{3.13}$$

Asymptotically, with an arbitrarily high probability and on any bounded time interval $[0,T]$, $S^{d,n}(\cdot)$ is bounded by the function with values $\bar{\lambda}^d t+1$. This follows from (1.12) and the proof of convergence of $\mathcal{T}^{d,n}(\cdot)$ in Theorem 1.1. Now, the fact that $1/n$ times the upper index of the summations in (3.13) is bounded with a high probability on any interval $[0,T]$ together with the weak convergence of $w^{d,n}(\cdot)$ to some continuous process implies that the difference between the expressions in the brackets is a residual-time error term and converges weakly to the "zero" process. Thus, we can work with the right-hand expression in (3.13).

Write $\Delta_l^{d,n} = [\Delta_l^{d,n} - \bar{\Delta}_l^{d,n}] + \bar{\Delta}_l^{d,n}$. Then, modulo a residual-time error term, the right-hand expression in (3.13) is the sum of

$$\bar{\Delta}^{d,n} x^n(t) \tag{3.14}$$

and

$$\frac{1}{\sqrt{n}} \sum_{l=I^{d,n}(t)+1}^{I^{d,n}(t)+\sqrt{n}x^n(t)} \left[\Delta_l^{d,n} - \bar{\Delta}_l^{d,n}\right]. \tag{3.15}$$

The process defined by (3.15) converges weakly to the "zero" process due to (3.11), the convergence assumption on $w^{d,n}(\cdot)$, and the bound on $S^{d,n}(\cdot)$. Thus, (3.12) holds.

Now suppose that (3.11) holds for the $WL^n(\cdot)$ processes, but not necessarily for the $x^n(\cdot)$. The expansion of the right hand expression in (3.13) into (3.14) and (3.15) still holds. The upper limit of summation in (3.15), divided by n, is bounded by $x^n(0)/\sqrt{n} + S^{a,n}(t)$. But $x^n(0)/\sqrt{n}$ converges weakly to zero, and $S^{a,n}(\cdot)$ is asymptotically bounded on any $[0,T]$ by an affine function of t. These facts and the weak convergence of $w^{d,n}(\cdot)$ imply that the processes defined by (3.15) satisfy (3.11). Thus, so must the process defined by (3.14). ■

Getting the queue size from the workload: Several arrival streams. Return to the model used in Theorem 3.2, where each arriving stream has a different work distribution. There is still only one processor. Even when it is most convenient to prove the heavy traffic limit theorems for the workload process, the queue lengths are still of interest. The following generalization of Theorem 3.3 illustrates a method of broad applicability. Let $x^{k,n}(t)$ equal $1/\sqrt{n}$ times the number of members of the kth arrival stream $A^{k,n}$ that are in the queue at real time nt. Let $n\tau^n(t)$ denote the real time of arrival of the customer being served at real time nt, or the real time of arrival of the next customer if the queue is empty at nt. Order simultaneous arrivals in some way. Then, analogously to (3.10),

$$WL^n(t) = \sum_k \left[\frac{1}{\sqrt{n}} \sum_{l=nS^{a,k,n}(\tau^n(t))+1}^{nS^{a,k,n}(\tau^n(t))+\sqrt{n}x^{k,n}(t)} \Delta_l^{d,k,n}\right] + \epsilon^n(t), \tag{3.16}$$

where $\epsilon^n(\cdot)$ is a residual-time error term. It is bounded by $1/\sqrt{n}$ times the work requirements of the job currently being served plus $1/\sqrt{n}$ times the work in the last arrival by real time nt from the stream currently being served.

Theorem 3.4. *Suppose that $WL^n(\cdot)$ is asymptotically continuous, $\{W^n(0)\}$ tight, $\{w^{\alpha,k,n}(\cdot);\ \alpha = a,d;k\}$ converges weakly to continuous processes and $\bar{\Delta}^{d,k,n} \to \bar{\Delta}^{d,k} = 1/\bar{\lambda}^{d,k} > 0$. Let $x^{k,n}(0)/\sqrt{n}$ converge weakly to zero. Suppose that $\bar{\Delta}^{a,k,n} \to \bar{\Delta}^{a,k} = 1/\bar{\lambda}^{a,k} > 0$. Then the difference*

$$WL^n(\cdot) - \sum_k \bar{\Delta}^{d,k,n} x^{k,n}(\cdot) \tag{3.17}$$

converges weakly to the "zero" process. The process $x^{k,n}(\cdot)/x^{j,n}(\cdot)$ converges weakly to the process with constant values $\bar{\lambda}^{a,k}/\bar{\lambda}^{a,j}$ in the following sense: For any $\delta > 0$,

$$\frac{x^{k,n}(\cdot)}{\delta + x^{j,n}(\cdot)} - \frac{WL^n(\cdot)\bar{\lambda}^{a,k}}{\delta + WL^n(\cdot)\bar{\lambda}^{a,j}} \tag{3.18}$$

converges weakly to the zero process. For each k, the difference

$$x^{k,n}(\cdot) - c^k WL^n(\cdot), \quad c^k = \frac{\bar{\lambda}^{a,k}}{\sum_i \bar{\lambda}^{a,i}\bar{\Delta}^{d,i}}, \tag{3.19}$$

converges weakly to the "zero" process.

Remark. The use of $\delta > 0$ in (3.18) is solely to deal with the situation where $WL(t) = 0$.

Proof. The last assertion is a consequence of the previous assertions. An argument like that used in Theorem 3.3 for the residual-time error term can be used to show that $\epsilon^n(\cdot)$ in (3.16) is a residual-time error term and converges weakly to the "zero" process. Factor the bracketed term in (3.16) as

$$\begin{aligned} &\frac{1}{\sqrt{n}} \sum_{l=nS^{a,k,n}(\tau^n(t))+1}^{nS^{a,k,n}(\tau^n(t))+\sqrt{n}x^{k,n}(t)} \bar{\Delta}^{d,k,n} \\ &\qquad + \frac{1}{\sqrt{n}} \sum_{l=nS^{a,k,n}(\tau^n(t))+1}^{nS^{a,k,n}(\tau^n(t))+\sqrt{n}x^{k,n}(t)} \left[\Delta_l^{d,k,n} - \bar{\Delta}^{d,k,n}\right]. \end{aligned} \tag{3.20}$$

The right hand term of (3.20) equals

$$-\bar{\Delta}^{d,k,n}\left[w^{d,k,n}(S^{a,k,n}(\tau^n(t)) + x^{k,n}(t)/\sqrt{n}) - w^{d,k,n}(S^{a,k,n}(\tau^n(t)))\right]. \tag{3.21}$$

Note that (modulo $1/n$)

$$S^{a,k,n}(\tau^n(t)) + x^{k,n}(t)/\sqrt{n} = S^{a,k,n}(t). \tag{3.22}$$

Thus, by the weak convergence (Theorem 1.1) of $S^{a,k,n}(\cdot)$ to the process with values $\bar{\lambda}^{a,k}t$ and the weak convergence of $w^{d,k,n}(\cdot)$, (3.21) satisfies (3.11).

Now (3.16) can be written as (modulo a residual-time error term)

$$WL^n(t) = \sum_k \bar{\Delta}^{d,k,n} x^{k,n}(t) + \sum_k \text{expression (3.21)}. \tag{3.23}$$

If some $x^{k,n}(\cdot)$ does not satisfy (3.11), then (3.23) implies that $WL^n(\cdot)$ does not, a contradiction to the first hypothesis of the theorem. Thus, all $x^{k,n}(\cdot)$ satisfy (3.11), which implies that the process (3.21) converges to the "zero" process, which in turn implies that the process (3.17) does also.

In view of (3.22), to prove the convergence assertion for (3.18), we need to show that the process defined by

$$\frac{\sqrt{n}\left[S^{a,k,n}(t) - S^{a,k,n}(\tau^n(t))\right]}{\delta + \sqrt{n}\left[S^{a,j,n}(t) - S^{a,j,n}(\tau^n(t))\right]} \tag{3.24}$$

is asymptotically equivalent to the process in the second term of (3.18). The expression (3.24) can be written as the ratio $F^{k,n}(t)/[\delta + F^{j,n}(t)]$, where

$$\begin{aligned} F^{i,n}(t) &= \frac{1}{\sqrt{n}} \sum_{l=nS^{a,i,n}(\tau^n(t))+1}^{nS^{a,i,n}(t)} 1 \\ &= \frac{1}{\sqrt{n}} \sum_{l=nS^{a,i,n}(\tau^n(t))+1}^{nS^{a,i,n}(t)} \left[1 - \frac{\Delta_l^{a,i,n}}{\bar{\Delta}^{a,i,n}}\right] + \frac{1}{\sqrt{n}} \sum_{l=nS^{a,i,n}(\tau^n(t))+1}^{nS^{a,i,,n}(t)} \frac{\Delta_l^{a,i,n}}{\bar{\Delta}^{a,i,n}}. \end{aligned}$$

This equals

$$\left[w^{a,i,n}(S^{a,i,n}(t)) - w^{a,i,n}(S^{a,i,n}(\tau^n(t)))\right] + \frac{\sqrt{n}(t-\tau^n(t))}{\bar{\Delta}^{a,i,n}} + \epsilon^{i,n}(t), \tag{3.25}$$

where $\epsilon^{i,n}(\cdot)$ is a residual-time error term and converges weakly to the "zero" process.

The *numerator* of the second component of (3.25) can be rewritten as

$$WL^n(\tau^n(t)) + \tilde{\epsilon}^{i,n}(t), \tag{3.26}$$

where $\tilde{\epsilon}^{i,n}(\cdot)$ is a residual-time error term. The relationship (3.22), the asymptotic continuity of the $S^{a,k,n}(\cdot)$ and $w^{a,k,n}(\cdot)$, and the fact that $x^{k,n}(\cdot)/\sqrt{n}$ converges weakly to the "zero" process imply that the first component of (3.25) converges weakly to the "zero" process and so does $WL^n(\cdot) - WL^n(\tau^n(\cdot))$. Thus, the assertion concerning (3.18) holds. ■

5.4 Batch Arrivals or Services and Many Servers

The proof of Theorem 1.1 illustrated some of the basic methods. These methods are readily extendable to many models with realistic complications, a couple of which will be considered in this section. Each of the extensions is typical of a class of problems, and introduces and illustrates a useful method of analysis. We will work with the scaled queue size. The development in terms of queued workload combines the methods used here with the representations used in Section 3.

Suppose that the arrivals occur in groups, or "batches," with the size of the batch being random. Continue to let $x^n(t)$ denote $1/\sqrt{n}$ times the number of individual customers in the system at real time nt. The development differs from that in Theorem 1.1 in essentially notational details, although the extra randomness due to the random batch size leads to an additional Wiener process in the weak-sense limit equation. The service discipline might also be in batches. For concreteness, we will suppose that there is a number $\bar{S}$, the service batch size, such that when the processor becomes available, $\bar{S}$ new customers enter service and are served simultaneously. If there are fewer than $\bar{S}$ customers available, then all available enter service, with the same service time distribution. The limit equations are the same for the alternative formulation where the processor cannot start service unless there are $\bar{S}$ customers to be served.

In Subsection 4.2 we consider the case where there are $\bar{s} > 1$ processors. The development is also quite close to that for Theorem 1.1, except for an additional complication concerning the reflection term. This complication arises, since if fewer than $\bar{s}$ processors are serving actual customers at some time, then there will be idle time, even though the system is not empty. This problem is also readily taken care of by the method of Theorem 2.3.

5.4.1 Batch Arrivals and Services

Let $\Delta_{b,l}^{a,n}$ and $\Delta_{b,l}^{d,n}$ denote the times between arrivals of (respectively, service times of) the *batches*. Let $\beta_l^{a,n}$ denote the size of the lth arriving batch. Service occurs in batches of size $\bar{S}$. If there are fewer than $\bar{S}$ remaining in the queue at the end of a service interval, then all remaining customers are served as a batch. Define $S_b^{a,n}(t)$ (respectively, $S_b^{d,n}(t)$) to be $1/n$ times the number of batches that have arrived (respectively, been served) by real time nt. For some centering constants $\bar{\beta}^{a,n}$, $\bar{\Delta}_b^{\alpha,n}$, $\alpha = a, d$, define

$$
\begin{aligned}
w_b^n(t) &= \frac{1}{\sqrt{n}} \sum_{l=1}^{nt} \left[\beta_l^{a,n} - \bar{\beta}^{a,n}\right], \\
w_b^{\alpha,n}(t) &= \frac{1}{\sqrt{n}} \sum_{l=1}^{nt} \left[1 - \frac{\Delta_{b,l}^{\alpha,n}}{\bar{\Delta}_b^{\alpha,n}}\right].
\end{aligned}
\tag{4.10}
$$

We will need the following assumptions.

A4.1. For some real number b, $\sqrt{n}\left[\frac{\bar{\beta}^{a,n}}{\bar{\Delta}_b^{a,n}} - \frac{\bar{S}}{\bar{\Delta}_b^{d,n}}\right] = b^n \to b.$

A4.2. *There are* $\bar{\beta}^a$, $\bar{\Delta}_b^\alpha$, $\alpha = a, d$, *such that* $\bar{\beta}^{a,n} \to \bar{\beta}^a$ *and* $\bar{\Delta}_b^{\alpha,n} \to \bar{\Delta}_b^\alpha = 1/\bar{\lambda}_b^\alpha$. *Also,* $w_b^{a,n}(\cdot), w_b^{d,n}(\cdot), w_b^n(\cdot)$ *converge weakly to mutually independent Wiener processes* $w_b^a(\cdot), w_b^d(\cdot), w_b(\cdot)$, *with variances* $\sigma_{b,a}^2, \sigma_{b,d}^2, \sigma_b^2$, *respectively.*

Let $z_0^n(t)$ denote $1/\sqrt{n}$ times the difference between [$\bar{S}$ times the number of completed batch services by real time nt] and [the number of individual customers that have completed service by real time nt]. The process $z_0^n(\cdot)$ can increase at t only if a batch service with fewer than $\bar{S}$ customers is completed at nt, and then it increases by [$\bar{S}$ – number served]/$\sqrt{n}$. Recall that $T^{d,n}(t)$ is the scaled idle time defined by (1.10). *Redefine*

$$z^n(t) = z_0^n(t) + \frac{\bar{S}}{\bar{\Delta}_b^{d,n}} T^{d,n}(t), \tag{4.2}$$

which will converge to the reflection term in the limit, even though it might increase when $x^n(t) > 0$.

Theorem 4.1. *Assume* (A1.0), (A4.1), (A4.2), *and let* $x^n(0) \Rightarrow x(0)$. *Then*

$$\{x^n(\cdot), S_b^{\alpha,n}(\cdot), w_b^{\alpha,n}(\cdot), w_b^n(\cdot), z^n(\cdot), \alpha = a, d\}$$

is tight and converges weakly to a set

$$\{x(\cdot), S_b^\alpha(\cdot), w_b^\alpha(\cdot), w_b(\cdot), z(\cdot), \alpha = a, d\}.$$

The conclusions of Theorem 1.1 *hold, but where*

$$w(t) = \bar{\beta}^a w_b^a(\bar{\lambda}_b^a t) - \bar{S} w_b^d(\bar{\lambda}_b^d t) + w_b(\bar{\lambda}_b^a t),$$

where the new value of b is used and the Wiener processes are mutually independent.

Proof. The proof follows closely the lines of that of Theorem 1.1. We can write

$$x^n(t) = x^n(0) + \frac{1}{\sqrt{n}} \sum_{l=1}^{nS_b^{a,n}(t)} \beta_l^{a,n} - \frac{\bar{S}}{\sqrt{n}} \sum_{l=1}^{nS_b^{d,n}(t)} 1 + z_0^n(t). \tag{4.3}$$

The third term on the right of (4.3) assumes that all services are in groups of $\bar{S}$ customers, even if there are fewer than $\bar{S}$ there when service started. This is corrected for by the term $z_0^n(\cdot)$, which can only increase at the end of a service interval if the (unscaled) number in the system is between 1 and $\bar{S}-1$ when that service started, equivalently, when $1/n \leq x^n(t) \leq (\bar{S}-1)/\sqrt{n}$ at that starting time.

Write the second term on the right of (4.3) as

$$\frac{1}{\sqrt{n}} \sum_{l=1}^{nS_b^{a,n}(t)} \left[\beta_l^{a,n} - \bar{\beta}^{a,n}\right] + \frac{1}{\sqrt{n}} \sum_{l=1}^{nS_b^{a,n}(t)} \bar{\beta}^{a,n}$$

that equals

$$w_b^n(S_b^{a,n}(t)) + \frac{\bar{\beta}^{a,n}}{\sqrt{n}} \sum_{l=1}^{nS_b^{a,n}(t)} \left[1 - \frac{\Delta_{b,l}^{a,n}}{\bar{\Delta}_b^{a,n}}\right] + \frac{\bar{\beta}^{a,n}}{\sqrt{n}\bar{\Delta}_b^{a,n}} \sum_{l=1}^{nS_b^{a,n}(t)} \Delta_{b,l}^{a,n}.$$

The second term of the last expression is $\bar{\beta}^{a,n} w_b^{a,n}(S_b^{a,n}(t))$.

Repeat the procedure for the third term on the right of (4.3) to get (see Theorem 1.1 for an analogous computation)

$$\frac{\bar{S}}{\sqrt{n}} \sum_{l=1}^{nS_b^{d,n}(t)} 1 = \frac{\bar{S}}{\bar{\Delta}_b^{d,n}} \left[\sqrt{n}t - T^{d,n}(t)\right] + \bar{S} w_b^{d,n}(S_b^{d,n}(t)) + \epsilon^n(t),$$

where $\epsilon^n(t)$ is a residual-time error term. Putting the pieces together yields

$$\begin{aligned} x^n(t) &= x^n(0) + \bar{\beta}^{a,n} w_b^{a,n}(S_b^{a,n}(t)) - \bar{S} w_b^{d,n}(S_b^{d,n}(t)) + w_b^n(S_b^{a,n}(t)) \\ &\quad + \sqrt{n}t \left[\frac{\bar{\beta}^{a,n}}{\bar{\Delta}_b^{a,n}} - \frac{\bar{S}}{\bar{\Delta}_b^{d,n}}\right] + z^n(t) + \epsilon^n(t), \end{aligned}$$

where $\epsilon^n(t)$ is a residual-time error term. The proof is completed as for Theorem 1.1. Note that Theorem 2.3 needs to be used to get the tightness and asymptotic continuity of $z^n(\cdot)$ since, strictly speaking, it is not a reflection term. Theorem 2.3 can be used, since for any $\epsilon > 0, T > 0$,

$$\lim_n P\left\{\frac{1}{\sqrt{n}} \text{max. \# of arrivals during a service interval on } [0, nT] \geq \epsilon\right\} = 0.$$

■

Comment on the effect of grouping into batches. The results of Theorems 1.1 and 4.1 are hard to compare, without making specific assumptions on the constants. Suppose that $\sigma_\alpha^2 = \sigma_{b,\alpha}^2$, $\alpha = a, d$, and that the mean arrival and service rates for individual customers is the same in both cases. Thus, we set $\bar{\beta}^{a,n}/\bar{\Delta}_b^{a,n} = 1/\bar{\Delta}^{a,n}$, and $\bar{S}/\bar{\Delta}_b^{d,n} = 1/\bar{\Delta}^{d,n}$. With this scheme, when customers are grouped, the interarrival or service times are added, except that we do not change the variance of the interarrival times and service intervals, a concession to the "possible savings" in dealing with several customers at a time.

With the above suppositions, the value of b is the same in both Theorems 1.1 and 4.1. The variances of the limit Wiener processes are

$$[\sigma_{b,a}^2 \bar{\beta}^a \bar{\lambda}_b^a]\bar{\beta}^a + [\sigma_{b,d}^2 \bar{S} \bar{\lambda}_b^d]\bar{S} + \sigma_b^2 \bar{\lambda}_b^a, \text{ Theorem 4.1,}$$

$$[\sigma_{b,a}^2 \bar{\beta}^a \bar{\lambda}_b^a] + [\sigma_{b,d}^2 \bar{S} \bar{\lambda}_b^d], \text{ Theorem 1.1.}$$

Clearly, with the above assignments, the variance of the $w(\cdot)$ is smaller without the grouping into batches.

The variance of the "arrival" noise is multiplied by $\bar{\beta}^a$ and the variance of the "service-time noise" is multiplied by $\bar{S}$ in going from Theorem 1.1 to Theorem 4.1. This is in addition to the effect of the "batch noise" $w_b(\cdot)$. One would hope, however, that there would also be a reduction in the mean times with batch service, yielding a smaller value of b in Theorem 4.1.

5.4.2 Many Servers

Return to the problem of Theorem 1.1, but suppose that the processor is composed of $\bar{s} < \infty$ processors that are not necessarily identical. When one of them completes a service, it takes the next customer in the system, if any, that is not being served. Various terms will be redefined. Let $\Delta_l^{d,i,n}, l < \infty$, be the service times of processor i. Redefine $S^{d,i,n}(t)$ to be $1/n$ times the number of services completed by real time nt by processor $i \leq \bar{s}$. For some centering constants $\bar{\Delta}^{d,i,n}$ that converge to $\bar{\Delta}^{d,i} \equiv 1/\bar{\lambda}^{d,i}$, define

$$w^{d,i,n}(t) = \frac{1}{\sqrt{n}} \sum_{l=1}^{nS^{d,i,n}(t)} \left[1 - \frac{\Delta_l^{d,i,n}}{\bar{\Delta}^{d,i,n}} \right]. \tag{4.4}$$

An infinite processor (self service) model is treated in Chapter 8. The following conditions will be needed.

A4.3. *There is a real b such that*

$$\sqrt{n} \left[\frac{1}{\bar{\Delta}^{a,n}} - \sum_i \frac{1}{\bar{\Delta}^{d,i,n}} \right] = b^n \to b.$$

A4.4. *The processes $w^{a,n}(\cdot)$, $w^{d,i,n}(\cdot)$, $i \leq \bar{s}$, converge weakly to mutually independent Wiener processes with variances $\sigma_a^2, \sigma_{d,i}^2, i \leq \bar{s}$, respectively.*

Define the scaled idle time for processor i:

$$T^{d,i,n}(t) = \frac{1}{\sqrt{n}} \left[nt - \sum_{l=1}^{nS^{d,i,n}(t)} \Delta_l^{d,i,n} - \tilde{\Delta}^{d,i,n}(t) \right], \tag{4.5}$$

where $\tilde{\Delta}^{d,i,n}(t)$ is the time that any customer being served at real time nt at processor i has been in service. The $\tilde{\Delta}^{d,i,n}(\cdot)/\sqrt{n}$ is a residual-time error term. Define the scaled idle time

$$z^n(t) = \sum_i T^{d,i,n}(t)/\bar{\Delta}^{d,i,n}, \tag{4.6}$$

which will be the reflection term in the limit.

Theorem 4.2. *Assume* (A1.0), (A4.3)–(A4.4), *and let* $x^n(0)$ *converge weakly to* $x(0)$. *Then the conclusions of Theorem* 1.1 *hold, with weak-sense limit*

$$x(t) = x(0) + bt + w(t) + z(t), \tag{4.7}$$

where

$$w(t) = w^a(\bar{\lambda}^a t) - \sum_i w^{d,i}(\bar{\lambda}^{d,i} t). \tag{4.8}$$

Comments on the proof. The proof follows that of Theorem 1.1 closely, and only a few remarks will be made. Equation (1.16) is replaced by

$$x^n(t) = x^n(0) + \frac{1}{\sqrt{n}} \sum_{l=1}^{nS^{a,n}(t)} 1 - \frac{1}{\sqrt{n}} \sum_i \sum_{l=1}^{nS^{d,i,n}(t)} 1.$$

Proceeding as below (1.16), rewrite the last term on the right as

$$-\sum_i w^{d,i,n}(S^{d,i,n}(t)) - \frac{1}{\sqrt{n}} \sum_i \frac{1}{\bar{\Delta}^{d,i,n}} \sum_{l=1}^{nS^{d,i,n}(t)} \Delta_l^{d,i,n}.$$

The right- hand term of the last expression can be written as (modulo a residual-time error term)

$$-\sum_i \frac{1}{\bar{\Delta}^{d,i,n}} \left[\sqrt{n}t - T^{d,i,n}(t)\right].$$

From this point on, the development is the same as that for Theorem 1.1, except for a slight difference concerning the treatment of the reflection term $z^n(\cdot)$. The term $z^n(\cdot)$ can increase at t even if $x^n(t) > 0$, provided that some processor is not working. Note that for $z^n(\cdot)$ to increase at interpolated time t, there can be no more than $\bar{s} - 1$ customers in the queue at that time; equivalently, $x^n(t) \leq (\bar{s} - 1)/\sqrt{n}$. With this, the proof in Theorem 2.3 can be used without any change to get the asymptotic continuity of the $z^n(\cdot)$.

5.5 Correlated Processes: Bursty Arrivals

Assumption (A1.1) in Theorem 1.1 simply asserted that $(w^{a,n}(\cdot), w^{d,n}(\cdot))$ converges to a Wiener process. In applications where the interarrival or service intervals are correlated, the correlation usually arises from some specific aspect of the physical model, and not in the form of a process where the convergence to a Wiener process is a priori obvious. This is the reason for the various criteria in Section 2.8 for the verification that a process is indeed a Wiener process. This section will deal with one such application where the rate of arrivals and the required service times are "bursty." They vary in time, depending on the state of the "environment." For example, the arrival frequencies and/or required service times might depend on the time of day, the sudden appearance of an external problem that causes a burst (up or down) in the customer creation rate, and so forth. The general scheme that is to be used is intended to be illustrative of a method for handling a broad range of external "modulating processes" that control the arrival or service rates. It is also an illustration of the use of Theorem 2.8.9.

The modulating state process and assumptions. To keep the development relatively simple, we will work with a Markov "modulating" process to model the effects of the changing external environment, and there is a single exogenous arrival stream, called A^n. For each n, let $B^n(\cdot)$ be an irreducible stationary Markov chain in continuous time and on a finite state space $\{1, \ldots, M\}$ with uniformly (in n) bounded transition rates b^n_{km}, where k and m denote the states. The $B^n(\cdot)$ need not actually depend on n. In what follows, the process $B^n(\cdot)$ modulates the arrival stream A^n, although the same scheme can be applied to the service process, as seen in [166]. If the same process $B^n(\cdot)$ modulates both the arrival and departure processes, then the development is (notationally) more complicated and the $w^a(\cdot)$ and $w^d(\cdot)$ will be correlated. But the general method can be adapted.

Assumptions and notation.

A5.1. *There are real numbers b_{km} such that $b^n_{km} \to b_{km}$, as $n \to \infty$. There is a unique stationary distribution $\{\pi^n_k\}$ for each n, and it converges to limit probabilities that are positive for each state.*

The interarrival intervals $\Delta^{a,n}_l$ will depend on the current value of the modulating Markov state, and we now define the relationships. Let B^n_l denote the value of the modulating state at the beginning of the lth interarrival interval.

A5.2. *The modulating process and the set of interarrival intervals are independent of the service process. Conditioned on the modulating states*

$\{B_l^n, l < \infty\}$, the random variables in the set $\{\Delta_l^{a,n}, l < \infty\}$ are mutually independent. The process defined by

$$w^{d,n}(t) = \frac{1}{\sqrt{n}} \sum_{l=1}^{nt} \left[1 - \frac{\Delta_l^{d,n}}{\bar{\Delta}^{d,n}} \right] \tag{5.1}$$

converges weakly to a Wiener process $w^d(\cdot)$ with variance σ_d^2. Let the state of the modulating process be k at the beginning of the lth interarrival interval. Then, there are $\bar{\Delta}^a(k)$ such that for all k,

$$\bar{\Delta}^{a,n}(k) \equiv E\left[\Delta_l^{a,n} \middle| B_m^n, m < l, B_l^n = k\right] = E\left[\Delta_l^{a,n} \middle| B_l^n = k\right] \to \bar{\Delta}^a(k) > 0,$$

$$\bar{\Delta}^{d,n} \to \bar{\Delta}^d > 0. \tag{5.2}$$

The variances of the $\Delta_l^{a,n}$ are bounded uniformly in n, l.

We need some condition that guarantees that the embedded modulating process $\{B_l^n, l < \infty\}$ converges weakly as $n \to \infty$. This is not implied by the current conditions on the intervals $\Delta_l^{a,n}$. Thus, we make the following assumption.

A5.3. *The conditional distribution of the random variable $\Delta_l^{a,n}$, given the modulating state $B_l^n = k$ and the other past data, does not depend on l or on the other past data and converges weakly as $n \to \infty$, for each value k of the modulating state.*

A5.4. $\left\{ |\Delta_l^{a,n}|^2, |\Delta_l^{d,n}|^2; n, l \right\}$ is uniformly integrable.

Under (A5.3), for each n the embedded process $\{B_l^n, l < \infty\}$ is also Markov, with transition probabilities that do not depend on time. Let us denote its steady state probabilities by π_k^n.

Now, the sequence of processes $\{B^n(\cdot), B_l^n, \Delta_l^{a,n}, l < \infty\}$ converges weakly as $n \to 0$. Let $\{B(\cdot), B_l, \Delta_l^a, l < \infty\}$ denote the weak-sense limit process. Then $B(\cdot)$ is a stationary continuous-time Markov chain. The $\{\Delta_l^a, l < \infty\}$ can be considered to be interarrival intervals and B_l the state of the (stationary) chain $B(\cdot)$ at the beginning of the lth such interval. Then $\{B_l\}$ will have positive steady state probabilities π_k that satisfy $\pi_k = \lim_n \pi_k^n$. Also,

$$\bar{\Delta}^a(k) = E[\Delta_l^a | B_l = k], \quad \text{for all } l. \tag{5.3}$$

Define or redefine

$$\begin{aligned} \bar{\Delta}^{a,n} &= \sum_k \bar{\Delta}^{a,n}(k) \pi_k^n, \\ \bar{\Delta}^a &= \lim_n \bar{\Delta}^{a,n} = \sum_k \bar{\Delta}^a(k) \pi_k, \\ \bar{\lambda}^a &= 1/\bar{\Delta}^a. \end{aligned} \tag{5.4}$$

A5.5. *With the new definition of $\bar{\Delta}^{a,n}$ and where the expectation uses the stationary distribution of B_l^n, there is σ_a^2 such that*

$$E\left[1-\frac{\Delta_l^{a,n}}{\bar{\Delta}^{a,n}}\right]^2 \to \sigma_a^2. \tag{5.5}$$

A5.6. There is $\bar{b}$ such that $\sqrt{n}\left[\frac{1}{\bar{\Delta}^{a,n}}-\frac{1}{\bar{\Delta}^{d,n}}\right]=\bar{b}^n \to \bar{b}.$

We continue to use the definition

$$w^{a,n}(t)=\frac{1}{\sqrt{n}}\sum_{l=1}^{nt}\left[1-\frac{\Delta_l^{a,n}}{\bar{\Delta}^{a,n}}\right],$$

but now $\bar{\Delta}^{a,n}$ is defined by (5.4).

Terminology for the limit Wiener processes. The following quantities are needed to describe the limit Wiener processes. Define, for $l \neq u$,

$$\begin{aligned} R_a^{k,m}(u-l) &= \frac{\bar{\Delta}^a(m)\bar{\Delta}^a(k)}{[\bar{\Delta}^a]^2} E\left[\pi_k - I_{\{B_l=k\}}\right]\left[\pi_m - I_{\{B_u=m\}}\right], \\ R_a(u-l) &= \sum_{k,m} R_a^{k,m}(u-l), \end{aligned}$$

and

$$R_a = \sigma_a^2 + 2\sum_{l=1}^{\infty} R_a(l). \tag{5.6}$$

Theorem 5.1. *Assume that $x^n(0)$ converges weakly to $x(0)$, and that (A1.0) and (A5.1)–(A5.6) hold. Then the conclusions of Theorem 1.1 hold, but where $\bar{b}$ replaces b and the new definitions of $\bar{\Delta}^a = 1/\bar{\lambda}^a$ and σ_a^2 are used. The $w^a(\cdot)$ and $w^d(\cdot)$ are mutually independent Wiener processes with variances, respectively*

$$E[w^a(1)]^2 = R_a,\ E[w^d(1)]^2 = \sigma_d^2. \tag{5.7}$$

Remark on a control problem. Suppose that there are several exogenous input streams and the control problem involves assigning on arrival to a queue, while minimizing, say, a discounted cost as in Chapter 10. The controller would be able to use all of the available information in making the assignment. This information would not include the current or past values of the modulating states, which are assumed to be unobservable. Since the current value of the modulating state would be important in making the assignment, an optimal controller would contain, either explicitly or implicitly, a filter to estimate it. Theorem 5.1 and the heavy traffic results for the control problem in Chapters 10 and 11 imply that in the heavy traffic limit, and when only the arrival processes are modulated, one need only use the average statistics, with little loss.

Proof. Expand exactly as in Theorem 1.1 to get (1.19), with the new definition of $\bar{\Delta}^{a,n}$. Since the interarrival and modulating processes are independent of the service process, any weak-sense limit of $w^{a,n}(\cdot)$ is independent of any weak-sense limit of $w^{d,n}(\cdot)$. It will be shown that $\{w^{a,n}(\cdot)\}$ is tight and that the weak-sense limits are continuous. Given this, the tightness and asymptotic continuity of the $\Psi^n(\cdot)$ (defined in Theorem 1.1) follow, as in Theorem 1.1. Consequently, the processes $w^{d,n}(\cdot)$ and $S^{d,n}(\cdot)$ converge to the same limits as in Theorem 1.1. The theorem then follows.

Thus, we need only prove the tightness and characterize the weak-sense limits of $w^{a,n}(\cdot)$. The main complication is that these processes are not martingales (when sampled at l/n, $l = 1, \ldots$), since the summands are "coupled" by the effects of the modulating process. The method of Theorem 2.8.9 will be used with $\xi_l^n = (1 - \Delta_l^{a,n}/\bar{\Delta}^{a,n})$.

The uniform integrability condition (2.8.15) holds by (A5.4). Next, consider (2.8.16). Let $E_i^{a,n}$ denote the expectation given all system data to the start of the ith interarrival interval. Fix $T > 0$, and evaluate

$$\sum_{l=i+1}^{nT} E_i^{a,n} \xi_l^n. \tag{5.8}$$

For $l \geq i$,

$$E_i^{a,n} \Delta_l^{a,n} = \sum_k \bar{\Delta}^{a,n}(k) P\left\{B_l^n = k | B_i^n\right\}, \tag{5.9}$$

and it converges to the mean value $\bar{\Delta}^{a,n}$ geometrically as $l - i \to \infty$, uniformly in n, l, i. Thus, the expression (5.8) is bounded (uniformly in n, i, T) by a constant. Hence the set (2.8.16) is uniformly integrable. Consequently, Theorem 2.8.9 implies the tightness and the continuity of the weak-sense limits of $w^{a,n}(\cdot)$.

We need only check conditions (2.8.17)–(2.8.19) to get the Wiener property. We will evaluate the conditional variances and limit variances, and the computations will yield the desired results with variance parameters $\Sigma_0 = \sigma_a^2$ and $\Sigma_1 = \sum_{l=1}^{\infty} R_a(l)$. Below, we use the representation

$$\xi_l^n = \frac{1}{\bar{\Delta}^{a,n}} \sum_k \left[\bar{\Delta}^{a,n}(k)\pi_k^n - \Delta_l^{a,n} I_{\{B_l^n = k\}}\right]$$

either explicitly or implicitly. Using the definition of $\bar{\Delta}^{a,n}$ of (5.4), rewrite $w^{a,n}(\cdot)$ as

$$w^{a,n}(t) = \frac{1}{\sqrt{n}\bar{\Delta}^{a,n}} \sum_k \sum_{l=1}^{nt} \left[\bar{\Delta}^{a,n}(k)\pi_k^n - \Delta_l^{a,n} I_{\{B_l^n=k\}}\right]. \tag{5.10}$$

For $u \neq l$ and any modulating states k, m, define the (stationary) expectations of the products of two arbitrary summands in (5.10):

$$\begin{aligned} &R_a^{k,m,n}(u-l) \\ &\quad \equiv \frac{1}{[\bar{\Delta}^{a,n}]^2} E\left[\bar{\Delta}^{a,n}(k)\pi_k^n - \Delta_l^{a,n} I_{\{B_l^n=k\}}\right]\left[\bar{\Delta}^{a,n}(m)\pi_m^n - \Delta_u^{a,n} I_{\{B_u^n=m\}}\right]. \end{aligned} \tag{5.11}$$

By first conditioning on the modulating states, and then using (5.3) and the conditional independence properties when the modulating states are given, this can be shown to equal

$$\frac{\bar{\Delta}^{a,n}(m)\bar{\Delta}^{a,n}(k)}{[\bar{\Delta}^{a,n}]^2} E\left[\pi_k^n - I_{\{B_l^n=k\}}\right]\left[\pi_m^n - I_{\{B_u^n=m\}}\right], \tag{5.12}$$

which can be evaluated from the properties of the stationary modulating process $\{B_l^n, l < \infty\}$. Note that, for $u \neq l$

$$R_a^n(u-l) \equiv E\left[1 - \frac{\Delta_l^{a,n}}{\bar{\Delta}^{a,n}}\right]\left[1 - \frac{\Delta_u^{a,n}}{\bar{\Delta}^{a,n}}\right] = \sum_{k,m} R_a^{k,m,n}(u-l).$$

Also,

$$E\left[1 - \frac{\Delta_l^{a,n}}{\bar{\Delta}^{a,n}}\right]^2 \equiv [\sigma_a^n]^2 = \frac{1}{[\bar{\Delta}^{a,n}]^2} E\left[\sum_k \left[\bar{\Delta}^{a,n}(k)\pi_k^n - \Delta_l^{a,n} I_{\{B_l^n=k\}}\right]\right]^2, \tag{5.13}$$

which can be evaluated by first conditioning on the modulating state. For $u \neq l$, the convergence

$$E I_{\{B_l^n=k\}} I_{\{B_u^n=m\}} \to \pi_k^n \pi_m^n,$$

as $|u-l| \to \infty$, is geometric, uniformly in n, for each k and m. A straightforward calculation shows that

$$\lim_n E[w^{a,n}(1)]^2 = \lim_n \left[[\sigma_a^n]^2 + 2\sum_{l=1}^{\infty} R_a^n(l)\right] \tag{5.14}$$

and that the right-hand term equals R_a, defined in (5.6).

We need to evaluate

$$E_i^{a,n}\left[1 - \frac{\Delta_l^{a,n}}{\bar{\Delta}^{a,n}}\right]\left[1 - \frac{\Delta_u^{a,n}}{\bar{\Delta}^{a,n}}\right], \tag{5.15}$$

where $u - l > 0$ is fixed but $i \to \infty, l - i \to \infty$. This is done just as the expectations were computed, but using the conditional probabilities

$$E_i^{a,n} I_{\{B_l^n = k\}} I_{\{B_u^n = m\}} = \sum_{k,m} P\{B_u^n = m | B_l^n = k\} P\{B_l^n = k | B_i^n\}.$$

Using the geometric convergence of the p-step transition probabilities to the stationary values as $p \to \infty$, uniformly in n and initial state and time, yields that (5.15) is bounded by a constant times μ^{u-l}, where $0 \le \mu < 1$. It also converges to its stationary value geometrically as $l - i \to \infty$. Then do the analogous computation for $u = l$. These computations imply (2.8.17) to (2.8.19), as claimed. ■

5.6 Processor Interruptions, Priorities, and Vacations

In this section we will consider the problem where the processor is not always available for serving the queued customers. This might occur if high priority jobs suddenly arrive or if the processors break down or need to be serviced. Such periods of unavailability have been called "vacations," and we will use that term [38, 123]. A few of the standard models will be dealt with. The vacations can be either short or long, and they can occur frequently or infrequently. If a vacation starts while a job is being processed, we must decide what happens to that job. Depending on the reason for the vacation, one might delay the vacation until the job is completed, or the vacation might preempt the job. If the work on a job is stopped, then we need to define the time needed to complete its service once service on it resumes. If this time is simply the residual time at the moment of stopping, then we say that the discipline is preempt–resume. We will either allow the job to be completed or else use the preempt–resume discipline. If the service intervals are identically and exponentially distributed for each n, then the results are the same as if the work was restarted from scratch after the vacation ends. The vacation problem is more interesting with networks, owing to the effects of a vacation of one processor on the queues for the others.

5.6.1 Short and Frequent Vacations

By "short and frequent" we mean that the length of the vacation and intervacation intervals are $O(1)$, the order of the service and interarrival times. Except for the vacations, the setup is as in Theorem 1.1.

Preempt–resume discipline. Time is divided into successive intervals where the processor is available and where it is not. Let $\Delta_l^{v,n}$ denote the duration (real time) of the lth vacation, and $\Delta_l^{s,n}$ the duration of the lth processing interval. Suppose, for notational specificity, that the processor is operating at time 0. We will need the following conditions. For some centering constants $\bar{\Delta}^{\alpha,n}, \alpha = v, s$, define the processes

$$w^{v,n}(t) = \frac{1}{\sqrt{n}} \sum_{l=1}^{nt} \left(\Delta_l^{v,n} - \bar{\Delta}^{v,n}\right), \tag{6.1}$$

$$w^{s,n}(t) = \frac{1}{\sqrt{n}} \sum_{l=1}^{nt} \left[1 - \frac{\Delta_l^{v,n} + \Delta_l^{s,n}}{\bar{\Delta}^{v,n} + \bar{\Delta}^{s,n}}\right]. \tag{6.2}$$

A6.1. *There are constants $\bar{\Delta}^\alpha$, $\alpha = v, s$, such that $\bar{\Delta}^{\alpha,n} \to \bar{\Delta}^\alpha \equiv 1/\bar{\lambda}^\alpha > 0$, $\alpha = v, s$, and $(w^{v,n}(\cdot), w^{s,n}(\cdot))$ converges weakly to a Wiener process with covariance matrix Σ_v. For each n, the intervals $\{\Delta_l^{v,n}, \Delta_l^{s,n}\}$ are independent of $\{\Delta_l^{a,n}, \Delta_l^{d,n}\}$.*

A6.2. *For some real number b,*

$$\sqrt{n}\left[\frac{1}{\bar{\Delta}^{a,n}} - \frac{1}{\bar{\Delta}^{d,n}} \frac{\bar{\Delta}^{s,n}}{\bar{\Delta}^{v,n} + \bar{\Delta}^{s,n}}\right] = b^n \to b.$$

Note that in (A6.2), $1/\bar{\Delta}^{d,n}$ is multiplied by the mean fraction of time that the processor is available. Define

$$\bar{\lambda}^{vac} = \frac{1}{\bar{\Delta}^v + \bar{\Delta}^s}, \quad \bar{\lambda}^{dv} = \frac{\bar{\lambda}^d \bar{\Delta}^s}{\bar{\Delta}^v + \bar{\Delta}^s}. \tag{6.3}$$

Theorem 6.1. *Assume the preempt–resume discipline,* (A1.0), (A1.1), *and* (A6.1)–(A6.2). *Then the conclusions of Theorem* 1.1 *hold, but with the new definitions of b and $w(\cdot)$:*

$$w(t) = w^a(\bar{\lambda}^a t) - w^d(\bar{\lambda}^{dv} t) + \bar{\lambda}^d \left[w^v(\bar{\lambda}^{vac} t) + \bar{\Delta}^v w^s(\bar{\lambda}^{vac} t)\right], \tag{6.4}$$

where $(w^v(\cdot), w^s(\cdot))$ is the weak-sense limit of $(w^{v,n}(\cdot), w^{s,n}(\cdot))$. Let $\{\mathcal{F}_t, t < \infty\}$ denote the filtration engendered by

$$\left\{x(\cdot), z(\cdot), w^a(\bar{\lambda}^a \cdot), w^d(\bar{\lambda}^{dv} \cdot), w^v(\bar{\lambda}^{vac} \cdot), w^s(\bar{\lambda}^{vac} \cdot)\right\}. \tag{6.5}$$

Then the w-processes in (6.5) *are all $\mathcal{F}_t$-Wiener processes. The first two are independent of the last two. The first two are mutually independent, and have variances $\sigma_a^2 \bar{\lambda}^a$ and $\sigma_d^2 \bar{\lambda}^{dv}$, the second two have covariance matrix $\bar{\lambda}^{vac} \Sigma_v$.*

Proof. The proof is a modification of that of Theorem 1.1. Let $S^{v,n}(t)$ denote $1/n$ times the number of vacation periods that have been completed by real time nt. Redefine $T^{d,n}(t)$ to be $1/\sqrt{n}$ times the idle time by real time nt that is not in a vacation interval. Equation (1.17) still holds, but its last term no longer equals (1.18), owing to the effects of the vacations.

Analogously to what was done in Theorem 1.1, we can write

$$-\frac{1}{\sqrt{n}} \sum_{l=1}^{nS^{d,n}(t)} \Delta_l^{d,n}$$
$$= -\sqrt{n}t + T^{d,n}(t) + \frac{1}{\sqrt{n}}\tilde{\Delta}^{d,n}(t) + \frac{1}{\sqrt{n}} \sum_{l=1}^{nS^{v,n}(t)} \Delta_l^{v,n} + \frac{1}{\sqrt{n}}\tilde{\Delta}^{v,n}(t), \tag{6.6}$$

where $\tilde{\Delta}^{v,n}(t)$ is the time that a vacation (if any) that is active at real time nt has been in progress. The term $\tilde{\Delta}^{v,n}(t)/\sqrt{n}$ is a residual-time error term and converges weakly to the "zero" process. Rewrite the fourth term on the right-hand side of (6.6) as

$$\frac{1}{\sqrt{n}} \sum_{l=1}^{nS^{v,n}(t)} \left(\Delta_l^{v,n} - \bar{\Delta}^{v,n}\right) + \frac{1}{\sqrt{n}}\bar{\Delta}^{v,n} \sum_{l=1}^{nS^{v,n}(t)} \left[1 - \frac{\Delta_l^{v,n} + \Delta_l^{s,n}}{\bar{\Delta}^{v,n} + \bar{\Delta}^{s,n}}\right]$$
$$+ \frac{1}{\sqrt{n}} \frac{\bar{\Delta}^{v,n}}{\bar{\Delta}^{v,n} + \bar{\Delta}^{s,n}} \sum_{l=1}^{nS^{v,n}(t)} \left(\Delta_l^{v,n} + \Delta_l^{s,n}\right). \tag{6.7}$$

The last term on the right of (6.7) is (modulo another residual-time error term)

$$\sqrt{n}\frac{\bar{\Delta}^{v,n}}{\bar{\Delta}^{v,n} + \bar{\Delta}^{s,n}} t. \tag{6.8}$$

Dividing (6.6)–(6.8) by $\bar{\Delta}^{d,n}$, putting the results into (1.17), and using the heavy traffic condition (A6.2) yields the following replacement for (1.19):

$$x^n(t) = x^n(0) + b^n t + w^{a,n}(S^{a,n}(t)) - w^{d,n}(S^{d,n}(t))$$
$$+ \frac{1}{\bar{\Delta}^{d,n}} \left[w^{v,n}(S^{v,n}(t)) + \bar{\Delta}^{v,n} w^{s,n}(S^{v,n}(t))\right] + z^n(t) + \epsilon^n(t). \tag{6.9}$$

The weak convergence of $S^{a,n}(\cdot)$ to the process with values $\bar{\lambda}^a t$ and of $S^{v,n}(\cdot)$ to the process with values $\bar{\lambda}^{vac} t$ is proved as the convergence of $S^{a,n}(\cdot)$ was proved in Theorem 1.1. An argument similar to that used in Theorem 1.1 shows that $S^{d,n}(\cdot)$ is tight and that any weak-sense limit is Lipschitz continuous with Lipschitz constant no greater than $\bar{\lambda}^d$. Then, as in Theorem 2.3, we can conclude that $\{z^n(\cdot)\}$ is tight and that any weak-sense limit process is continuous and nondecreasing. This implies that $\{x^n(\cdot)\}$ is tight and that any weak-sense limit is continuous. Let $(x(\cdot), z(\cdot))$ denote a weak-sense limit. Then the properties of $z^n(\cdot)$ and the weak convergence imply that $z(\cdot)$ can increase only at t where $x(t) = 0$.

An analogue of (1.11) is (modulo a residual-time error term)

$$\mathcal{T}^{d,n}(S^{d,n}(t)) = t - \frac{T^{d,n}(t)}{\sqrt{n}} - \frac{1}{n} \sum_{l=1}^{nS^{v,n}(t)} \Delta^{v,n}(t). \tag{6.10}$$

The process defined by the right-hand term (without the minus sign) converges weakly to the process with values $\bar{\Delta}^v/[\bar{\Delta}^v + \bar{\Delta}^s]t$. As in Theorem 1.1, $\mathcal{T}^{d,n}(\cdot)$ converges weakly to the process with values $\bar{\Delta}^d t$. These facts, the convergence of the idle-time term $T^{d,n}(\cdot)/\sqrt{n}$ in (6.10) to the zero process, and (6.10) imply that $S^{d,n}(\cdot)$ converges weakly to the process with values

$$\frac{\bar{\lambda}^d \bar{\Delta}^s}{\bar{\Delta}^v + \bar{\Delta}^s} t = \bar{\lambda}^{dv} t. \tag{6.11}$$

The rest of the details are as in Theorem 1.1 and are omitted. ∎

5.6.2 Priorities

The method of Theorem 6.1 can be adapted to the case where there are several input streams and priorities, since the vacations play the role of priorities. Suppose that there are two input streams, denoted by A_k^n, $k = 1, 2$, and let the stream A_1^n have priority. A high-priority input can either interrupt any low-priority job that is being served, with a preempt–resume discipline, or wait until that job is completed. The asymptotic results are the same. The fact that the scaled queue of the priority classes disappears in the limit was noted by [89, 206, 249].

Let $\Delta_{1k,l}^{a,n}$ denote the interarrival intervals and $\Delta_{1k,l}^{d,n}$ the service intervals for stream k. The subscript 1, which stands for processor 1, is not needed, but it is used to coordinate the terminology with that in Section 6.4. For some centering constants $\bar{\Delta}_{1k}^{\alpha,n}$, $\alpha = a, d$, define

$$w_{1k}^{a,n}(t) = \frac{1}{\sqrt{n}} \sum_{l=1}^{nt} \left[1 - \frac{\Delta_{1k,l}^{a,n}}{\bar{\Delta}_{1k}^{a,n}} \right],$$

$$w_{1k}^{d,n}(t) = \frac{1}{\sqrt{n}} \sum_{l=1}^{nt} \left[1 - \frac{\Delta_{1k,l}^{d,n}}{\bar{\Delta}_{1k}^{d,n}} \right].$$

The following assumptions will be used.

A6.3. $\bar{\Delta}_{1k}^{\alpha,n} \to \bar{\Delta}_{1k}^{\alpha} = 1/\bar{\lambda}_{1k}^{\alpha}$, $\alpha = a, d$. *For each k, the processes $w_{1k}^{\alpha,n}(\cdot)$, $\alpha = a, d$, converge weakly to mutually independent Wiener processes with variances $\sigma_{\alpha,1}^2$, $\alpha = a, d$, respectively. The interarrival and service intervals for A_1^n are independent of those for A_2^n.*

A6.4. *There is a real b such that*

$$\frac{\sqrt{n}}{\bar{\Delta}_{12}^{d,n}}\left[\frac{\bar{\Delta}_{12}^{d,n}}{\bar{\Delta}_{12}^{a,n}}+\frac{\bar{\Delta}_{11}^{d,n}}{\bar{\Delta}_{11}^{a,n}}-1\right]=b^n\to b.$$

Note that (A6.4) implies that

$$\bar{\lambda}_{12}^{a}=\bar{\lambda}_{12}^{d}\left[1-\frac{\bar{\lambda}_{11}^{a}}{\bar{\lambda}_{11}^{d}}\right]. \tag{6.12}$$

Define $x_{11}^n(t)$ (respectively, $x_{12}^n(\cdot)$) to be $1/\sqrt{n}$ times the number of high (respectively, low) priority jobs in the system at real time nt.

Theorem 6.2. *Assume that A_1^n has priority in either of the senses used above and that $x_{11}^n(0)$ converges weakly to zero. Assume* (A1.0), (A1.1), (A6.3), *and* (A6.4). *Then the conclusions of Theorem* 1.1 *hold, but with the new value of b, and $w(\cdot)$ taking the form*

$$w(t)=w_{12}^{a}(\bar{\lambda}_{12}^{a}t)-w_{12}^{d}(\bar{\lambda}_{12}^{a}t)+\frac{\bar{\Delta}_{11}^{d}}{\bar{\Delta}_{12}^{d}}\left[w_{11}^{a}(\bar{\lambda}_{11}^{a}t)-w_{11}^{d}(\bar{\lambda}_{11}^{a}t)\right]. \tag{6.13}$$

Remark on the initial condition. If $x_{11}^n(0)$ did not converge weakly to zero, then the server would work on its queue until $x_{11}^n(t)=0$, which would take an interpolated time of order $O(1/\sqrt{n})$. Thus, in the limit there would be a discontinuity in $x^n(\cdot)$ at $t=0$, and we would have to work on $[\delta,\infty)$ for any $\delta>0$.

Proof. One form of the proof follows the lines of Theorem 1.1, with the changes used in Theorem 6.1 to account for the fact that the processor might not always be available for the A_2^n stream. We will use a simpler alternative, which exploits Theorem 3.2. Assume that a low-priority job in process cannot be interrupted. The proof would be simpler under the preempt–resume policy. Under the hypotheses and Theorem 3.2, $WL^n(\cdot)$ satisfies (3.9). If we can show that asymptotically, $x_{11}^n(\cdot)$ converges weakly to the "zero" process, then (3.17) implies that $x^n(\cdot)-WL^n(\cdot)/\bar{\Delta}_{12}^d$ converges weakly to the zero process, and the theorem is proved. Thus, we need only show that $x_{11}^n(\cdot)$ converges weakly to the "zero" process.

Fix $T>0$. It will next be shown that the maximum (scaled) number of arrivals from A_1^n during any low-priority service period by real time nT goes to zero in probability. Let $n\tau\le nT$ be a stopping time that is the start of a low-priority service period of length $\Delta_{12,q}^{d,n}$. Let $S_{1k}^{a,n}(t)$ denote $1/n$ times the number of arrivals from input stream A_k^n by real time nt. The scaled number of high-priority arrivals during this interval is, modulo

$2/\sqrt{n}$,

$$\frac{1}{\sqrt{n}} \sum_{l=nS_{11}^{a,n}(\tau)+2}^{nS_{11}^{a,n}(\tau+\Delta_{12,q}^{d,n}/n)-1} 1. \tag{6.14a}$$

This equals

$$\frac{1}{\sqrt{n}} \sum_{l=nS_{11}^{a,n}(\tau)+2}^{nS_{11}^{a,n}(\tau+\Delta_{12,q}^{d,n}/n)-1} \left[1 - \frac{\Delta_{11,l}^{a,n}}{\bar{\Delta}_{11}^{a,n}}\right] + \frac{1}{\sqrt{n}\bar{\Delta}_{11}^{a,n}} \sum_{l=nS_{11}^{a,n}(\tau)+2}^{nS_{11}^{a,n}(\tau+\Delta_{12,q}^{d,n}/n)-1} \Delta_{11,l}^{a,n}. \tag{6.14b}$$

The sup over $\tau \leq T$ of the first term goes to zero by (A6.3). Since the first and last arrivals from A_{11}^n during the low priority service interval in question are excluded from the right-hand sum in (6.14b), the second term is bounded by $\Delta_{12,q}^{d,n}/[\bar{\Delta}_{11}^{a,n}\sqrt{n}]$, which is a residual-time error term.

Finally, prove that $x_{11}^n(\cdot)$ converges weakly to the "zero" process. Write

$$x_{11}^n(t) = x_{11}^n(0) + \frac{1}{\sqrt{n}} \sum_{l=1}^{nS_{11}^{a,n}(t)} 1 - \frac{1}{\sqrt{n}} \sum_{l=1}^{nS_{11}^{d,n}(t)} 1,$$

which equals (modulo a residual-time error term)

$$x_{11}^n(0) + \sqrt{n}t\left[\frac{1}{\bar{\Delta}_{11}^{a,n}} - \frac{1}{\bar{\Delta}_{11}^{d,n}}\right] + \left[w_{11}^{a,n}(S_{11}^{a,n}(t)) - w_{11}^{d,n}(S_{11}^{d,n}(t))\right]$$
$$+ \frac{\text{“waiting time” by } nt}{\sqrt{n}} + \frac{\text{“idle time” by } nt}{\sqrt{n}}.$$

Here, "waiting time" means the time spent waiting for a low-priority job to be completed, and "idle time" is the actual idle time plus the rest of the time spent on the low-priority class. Both of the "time" expressions can increase only when $x_{11}^n(t)$ is arbitrarily close to zero, by the previous paragraph. For the purposes of this proof, it does not matter what these "time" functions are. The rest of the proof uses the fact that the first term in the above expression goes to zero and the second to minus infinity (by (A6.4)), and the details are omitted. ■

5.6.3 Short and Infrequent Vacations

In Theorem 6.1 the order of the duration of the vacations, as well as of the time between the vacations, was $O(1)$, and did not depend on n. Sometimes, we wish to model the situation where the vacations are rare, but short in the sense that their duration is still of order $O(1)$ as in Theorem 6.1. A convenient way of covering many such cases is to use the following assumption.

A6.5. *There are constants $\bar{\Delta}^v$ and $\bar{\lambda}^s$ such that the process defined by*

$$\frac{1}{\sqrt{n}} \sum_{l=1}^{nS^{v,n}(t)} \Delta_l^{v,n}$$

converges weakly to the process with values $\bar{\Delta}^v \bar{\lambda}^s t$.

The parameters $\bar{\Delta}^v$ and $\bar{\lambda}^s$ in (A6.5) are used to reflect the usual situation where $\bar{\lambda}^s$ is the limit of a scaled "rate of vacations" and the limit of the mean duration is $\bar{\Delta}^v$. Condition (6.5) implies that the real time between vacations is of order $O(\sqrt{n})$. Theorem 6.3 can be modified, analogously to what was done in Theorem 1.5, to allow "moderately frequent" vacations, where the $1/\sqrt{n}$ in (A6.5) is replaced by $1/n^\gamma$, $\gamma < 1$.

Theorem 6.3. *Assume the conditions of Theorem* 1.1, *but add the vacations,* (A6.5), *and either use the preempt–resume discipline or wait for the current job to be completed before the vacation begins. Then Theorem* 1.1 *continues to hold, but b is replaced by*

$$b + \frac{\bar{\Delta}^v \bar{\lambda}^s}{\bar{\Delta}^d}. \tag{6.15}$$

Comment on the proof. The proof differs only slightly from that of Theorem 6.1. Condition (A6.5) is used directly in (6.6) and (6.10). The following is a sufficient condition for (A6.5). For each n, let $\{\Delta_l^{v,n}, l < \infty\}$ be identically distributed and mutually independent, with mean converging to $\bar{\Delta}^v$, and with a variance that is bounded uniformly in n. Suppose that $\{\Delta_l^{v,n}, l < \infty\}$ are independent of $\{\Delta_l^{s,n}, l < \infty\}$ and that $\sqrt{n}S^{s,n}(\cdot)$ converges weakly to the process with values $\bar{\lambda}^s t$.

5.6.4 Long but Infrequent Vacations

Now we discuss the case where the duration of the vacation intervals is of order $O(\sqrt{n})$ and the intervacation intervals are of order $O(n)$ in real time. The orders here and in Theorems 6.1 and 6.3 were chosen because they lead to well-defined limit processes. If the duration of the vacations used below is of order larger than $O(\sqrt{n})$, then the scaled system will explode. The formulations in Theorems 6.1, 6.3, and 6.4 do not exhaust the possibilities, and various intermediate cases are possible. The given results illustrate the general methods.

A6.6. *For each n, the intervals between the end of a vacation and the start of the next one are denoted by $n\tau_k^{s,n}$, $k = 1, \ldots$. The $\tau_k^{s,n}$ are mutually independent, exponentially distributed, independent of all the other "driving"* *random variables, and have rate $\bar{\lambda}^{s,n}$, where $\bar{\lambda}^{s,n}$ converges to $\bar{\lambda}^s > 0$ as $n \to \infty$.*

A6.7. *For each n, there are mutually independent and identically distributed random variables $\tau_k^{v,n}$, $k = 1, \ldots,$ such that the duration of the kth vacation interval is $\sqrt{n}\tau_k^{v,n}$. Also, $\tau_k^{v,n}$ converges weakly to random variables τ_k^v as $n \to \infty$. The $\tau_k^{v,n}$, $k = 1, \ldots,$ are independent of all other "driving" random variables. For simplicity, suppose that the processor is not on vacation at $t = 0$. Set $\tau_0^{v,n} = 0$.*

For $k > 0$, define

$$\nu_k^n = \sum_{l=1}^{k} \left[\tau_l^{s,n} + \tau_{l-1}^{v,n}/\sqrt{n}\right], \tag{6.16}$$

which is $1/n$ times the real time of the start of the kth vacation. The kth vacation and intervacation intervals in scaled time are the half open sets $[\nu_k^n, \nu_k^n + \tau_k^{v,n}/\sqrt{n})$ and $[\nu_k^{v,n} - \tau_k^{s,n}, \nu_k^{v,n})$, respectively.

Note on a fixed queue. Suppose that we are concerned with a fixed queue and not a sequence. Then estimate the drift parameter b and the scale parameter n as in Sections 1.1–1.3 or in the outline of this chapter. Given n, get the distributions of the τ_k^v and τ_k^s. If the mean of τ_k^s is too small or that of τ_k^v too large, then heavy traffic analysis might not be appropriate for the problem at hand.

The paths on the intervacation and vacation intervals. Define the *kth section* of $x^n(\cdot)$, the path segment between the end of the $(k-1)$st and the beginning of the kth vacation, as the process with values

$$x^n\left((\nu_k^n - \tau_k^{s,n} + t) \wedge \nu_k^n\right), \quad t \geq 0. \tag{6.17}$$

Under the conditions of Theorem 1.1, the first section of $x^n(\cdot)$ is tight and its weak-sense limit satisfies (1.14) over an interval that is exponentially distributed with rate $\bar{\lambda}^s$ and is independent of the $w(\cdot)$ and $x(0)$ in (1.14). The analogous statement can be said of the kth section of $x^n(\cdot)$ for any integer k, if the set of initial states of that section is tight.

In scaled time, the kth vacation lasts $\tau_k^{v,n}/\sqrt{n}$ units of time, which goes to zero in probability as $n \to \infty$. During this time there are inputs, but no outputs. Loosely speaking, the inputs arrive at a "rate" of $n/\bar{\Delta}^{a,n}$ in scaled time, which implies that $x^n(\cdot)$ increases at a rate $\sqrt{n}/\bar{\Delta}^{a,n}$ during the vacation. Thus, over the vacation interval of scaled duration $\tau_k^{v,n}/\sqrt{n}$, there would be an asymptotic "jump" in $x^n(\cdot)$ of $\tau_k^{v,n}/\bar{\Delta}^{a,n}$. For $n < \infty$, this "jump" is approximated by discrete steps. The situation is illustrated in Figure 6.1.

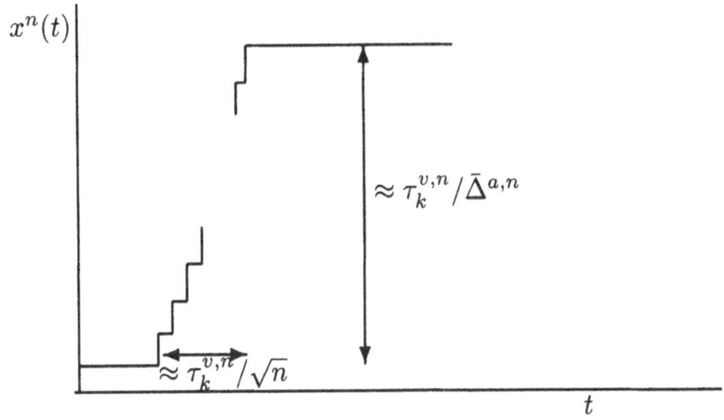

Figure 6.1. Behavior of $x^n(\cdot)$ during a vacation, in scaled time.

The quick change in the value of $x^n(\cdot)$ over a time interval which goes to zero as $n \to \infty$ is nearly a jump discontinuity in an obvious sense. Nevertheless, the sequence depicted in Figure 6.1 does not converge to the jump in the Skorohod topology as $n \to \infty$. Hence, one cannot assert weak convergence in that topology. For most applications this is merely an annoying technicality. Since the problem concerns the values taken by $x^n(\cdot)$ near the jump times, one expects that the distribution of any almost everywhere (with respect to the measure of $x(\cdot)$ in (6.19)) continuous functional $f(x(\cdot))$ that has a "negligible" dependence on the values of $x(\cdot)$ in an arbitrarily small neighborhood of the jump times will converge in distribution. That is, for the present case, this includes any continuous function of $x(t)$ for any t, integrals $\int_0^T k(x(s))ds$ for continuous $k(\cdot)$, etc.

One way of dealing with this lack of tightness is to revise the topology so that tightness is guaranteed. This is essentially a restriction of the class of functionals $f(\cdot)$ such that $f(x^n(\cdot))$ converges in distribution to $f(x(\cdot))$, as $n \to \infty$. The time transformation method outlined in Section 2.6 is ideal for this purpose. But owing to the simplicity of the problem, we will take a more direct approach and work with the sections and vacations separately.

Let $A_k^{v,n}(t)$ denote $1/\sqrt{n}$ times the number of arrivals on the real time interval $[n\nu_k^n, n\nu_k^n + (\tau_k^{v,n} \wedge t)\sqrt{n})$. Modulo $3/\sqrt{n}$, this equals

$$\frac{1}{\sqrt{n}} \sum_{l=nS^{a,n}(\nu_k^n)+2}^{nS^{a,n}(\nu_k^n+(\tau_k^{v,n}\wedge t)/\sqrt{n})-1} 1. \tag{6.18}$$

Thus, $A_k^{v,n}(\infty)$ is $1/\sqrt{n}$ times the number arriving during this vacation. A crucial result of Theorem 6.4 is that $A_k^{v,n}(\cdot)$ converges weakly (as $n \to \infty$) to the process with values $\bar{\lambda}^a(t \wedge \tau_k^v)$, where τ_k^v is the weak-sense limit of $\tau_k^{v,n}$. Thus, during the vacation, asymptotically and in the "stretched out" time scale used in (6.18), the queue size increases linearly at the mean rate $\bar{\lambda}^d$, which justifies the "rate" terminology used in the discussion concerning

Figure 6.1. Such results will play a crucial role in later chapters, particularly in Chapter 11.

The "stretched out" time scale used in (6.18) and in the definition of $A_k^{v,n}(\cdot)$ will be called the "local-fluid time scale." It is useful for studying detailed behavior during intervals of scaled time duration $O(1/\sqrt{n})$.

Remark. The term "concatenation" in the theorem means simply that the various sections and the effects of the vacations are put together in sequence to construct our limit process, as noted above. The first section is defined on $[0, \nu_1^n)$. There is a well defined limit interval $[0, \nu_1)$, and a limit value, called $x(\nu_1-)$. Then the well defined limit of the effects of the first vacation are added. This is the limit $A_1^v(\infty)$ of $A_1^{v,n}(\infty)$. Continuing, the initial condition of the second section of the limit process is $x(\nu_1-)+A_1^v(\infty)$, and so forth.

Theorem 6.4. *Assume the conditions of Theorem* 1.1, *but with vacations added, where the vacation intervals satisfy* (A6.6) *and* (A6.7). *Then the sections defined in* (6.17) *are tight, as are the* $A_k^{v,n}(\infty)$, *for each* k. *Also,* $A_k^{v,n}(\cdot)$ *converges weakly to the process* $A_k^v(\cdot)$ *with values* $\bar{\lambda}^a[t \wedge \tau_k^v]$. *The* ν_k^n *converge weakly to* $\nu_k = \sum_{l=1}^k \tau_l^s$, *where* τ_l^s *are exponentially distributed with rate* $\bar{\lambda}^s$. *The* τ_k^v, τ_k^s, $k = 1, \ldots, w^\alpha(\cdot), \alpha = a, d$, *are mutually independent. The concatenation of the weak-sense limits of the sections satisfies*

$$x(t) = x(0) + bt + w(t) + J^v(t) + z(t), \tag{6.19}$$

where

$$J^v(t) = \sum_{k:\nu_k^s \le t} A_k^v(\infty),$$

$z(\cdot)$ *is the reflection term, and* $w(\cdot)$ *is defined in Theorem* 1.1.

Proof. We will prove the tightness and characterize the weak-sense limit of $A_k^{v,n}(\cdot)$. The rest of the proof follows from that of Theorem 1.1, the independence of the $\tau_k^{s,n}, \tau_k^{v,n}, k = 1, \ldots,$ for each n, and the other parts of (A6.6) and (A6.7). Rewrite (6.18) as

$$\begin{aligned} &\frac{1}{\sqrt{n}} \sum_{l=nS^{a,n}(\nu_k^n)+2}^{nS^{a,n}(\nu_k^n+(\tau_k^{v,n}\wedge t)/\sqrt{n})-1} \left[1 - \frac{\Delta_l^{a,n}}{\bar{\Delta}^{a,n}}\right] \\ &\qquad + \frac{1}{\sqrt{n}\bar{\Delta}^{a,n}} \sum_{l=nS^{a,n}(\nu_k^n)+2}^{nS^{a,n}(\nu_k^n+(\tau_k^{v,n}\wedge t)/\sqrt{n})-1} \Delta_l^{a,n}. \end{aligned} \tag{6.20}$$

The first term converges weakly to zero by the fact that $w^{a,n}(\cdot)$ converges weakly to a continuous weak-sense limit process (see (A1.1)), together with the tightness of $\{\nu_k^n, n < \infty\}$ for each k , the fact that $\tau_k^{v,n}/\sqrt{n}$ converges weakly to zero for each k, and the weak convergence of $S^{a,n}(\cdot)$. The second term in (6.20) is $[\tau_k^{v,n} \wedge t]/\bar{\Delta}^{a,n}$, modulo a residual-time error term. The asymptotic characterization of the $A_k^{v,n}(\cdot)$ follows from this. ■

5.7 An Alternative Scaling: Fast Arrivals and Services

In the examples of the previous sections, time is scaled by n. Hence real time nt was scaled time t. Then the interarrival and service times depend only slightly on the scale factor n and are $O(1)$ in magnitude. In applications to communication theory, the interarrival and service intervals are usually very short. For example, if the arrivals are measured in cells or packets of a fixed number of bytes, then the service times (that is, the time required for the transmission of that cell over a high-bandwidth channel) is quite small. In such cases, the model is embedded in a sequence of models parameterized by n, where the interarrival and service times are of order $O(1/n)$. Then time is not rescaled at all, and n parameterizes the physical scale or speed of the system. The methods for treating this alternative scaling differ little from the methods used so far. This will be illustrated by a simple example.

Denote the actual interarrival and service times by $\hat{\Delta}_l^{a,n}$ and by $\hat{\Delta}_l^{d,n}$, respectively. Define them in terms of random variables $\Delta_l^{a,n}$ and $\Delta_l^{d,n}$ by

$$\hat{\Delta}_l^{a,n} = \frac{1}{n}\Delta_l^{a,n}, \quad \hat{\Delta}_l^{d,n} = \frac{1}{n}\Delta_l^{d,n}. \tag{7.1}$$

Redefine $x^n(t)$ to be $1/\sqrt{n}$ times the number of customers (jobs) in the queue at real time t. Also, redefine $S^{a,n}(t)$ and $S^{d,n}(t)$ to be $1/n$ times the number of arrivals and departures, respectively, by real time t. If there are vacations, then rescale the intervacation intervals and the vacation durations in the same way. Then, by working with the rescaled intervals $\Delta_l^{\alpha,n}$ and their centering constants $\bar{\Delta}^{\alpha}$, all of the results in the previous sections for the queued number of jobs continue to hold.

The results for the queued workload also continue to hold, but with a different scaling. The quantity $\hat{\Delta}_l^{d,n}$ is the time required for the lth service, and this is $O(1/n)$. For such systems, the time required to complete all of the currently queued jobs will be short. Continue to suppose that the processor works at unit rate, and define $WL^n(t)$ to be $\sqrt{n}$ times the time that is required to complete the jobs queued at real time t. Then

$$WL^n(t) = WL^n(0) + \sqrt{n} \sum_{l=1}^{nS^{a,n}(t)} \hat{\Delta}_l^{d,n} - \sqrt{n}t + z^n(t), \tag{7.2}$$

where $z^n(t)$ denotes $\sqrt{n}$ times the idle time on the real time interval $[0,t]$. Now, replace $\hat{\Delta}_l^{\alpha,n}$ by $\Delta_l^{\alpha,n}/n$, and continue as previously.

6 Uncontrolled Networks

Chapter Outline

This is the first chapter that deals with networks, and it will introduce many of the techniques to be used on such systems in the sequel. As in Chapter 5, unless otherwise mentioned the processors schedule FCFS and work when there is work to be done. Most of the development is for the scaling of Theorem 5.1.1. The analogue of the model of Section 5.7, where the arrival and service rates are "fast," is noted in Section 6, and all of the results can be carried over, with only a simple rescaling. As in Chapter 5, we are concerned with a sequence of queues indexed by n, where the traffic intensity at the processors goes to unity as $n \to \infty$. To adapt the procedure to approximate for some given queue, proceed as noted in the outline of Chapter 5 or in Sections 1.1 and 1.2 to compute n as a large number such that the b^n in (A1.3) is "moderate," and define it as b.

Many of the ideas are simple or natural extensions of the methods used in Chapter 5. There are new issues in dealing with the reflection terms, but these are similar to what was done in Section 3.6, where a general method for proving tightness of the set of reflections for a sequence of approximations to a Skorohod problem was given. The canonical network of [213] is dealt with in Section 1. The method of Theorem 5.1.1 is used to get the needed representation of the state in terms of a drift term, a "pre-Wiener" process, and a reflection term. The drift term converges by the heavy traffic condition, and the method of either Theorem 3.6.1 or 3.6.2 is used to prove that the sequence of reflection processes converges to the desired limit. In this case, the reflection terms arise due the idle time of the processors. Since the idle time of any processor affects the input process of all other processors to which it is immediately connected, this coupling determines the reflection direction for this type of network. The reflection processes can be determined by factors other than idle time, and some detailed examples will be developed in the next chapter.

The assumptions on the service and interarrival intervals are in two parts. The first is the heavy traffic condition, which ensures that the processors will be idle for a negligible percentage of time as $n \to \infty$. The second assumption resembles a central limit theorem. It holds under the classical mutual independence if the set of squares of the intervals is uniformly integrable, but it also allows the intervals to be correlated in various ways. In Theorem 1.2, it is explicitly assumed that the intervals are martingale differences. This raises some questions concerning the characterization of the weak-sense limits of the "pre-Wiener" processes and nonanticipativity that are resolved via an adaptation of the method of proof of Theorem 2.8.8. This method will be useful later when state dependence and controls are involved. In Theorem 1.1, the routing is purely random. The discussion in Subsection 1.2 shows that this can be relaxed in various useful ways.

The model used so far, while broadly applicable, is based on the explicit structure of the network used in Theorem 1.1. The initial part of the proof of that theorem manipulates the model into a form that "looks like" the desired limit. An analogous procedure can be carried out under broader conditions, where the problem is put into a form that "looks like" the Skorohod problem model of Section 3.5. For such models the terms that will be the reflection process in the limit might change when the state is only very close to the boundary (distance going to zero as $n \to \infty$), and not necessarily right on it. The general tightness proofs of Section 3.6 are well suited to deal with this. A statement of a typical result is in Theorem 1.4.

Sections 2 and 3 concern a variety of extensions, each typical of a large class of applications. Closed networks arise when the same job can circulate indefinitely and there are no exogenous inputs. While seemingly not practical, it is in fact a very useful model of many situations where the physics of the problem give rise to models that are equivalent to it [253]. In Section 1, all processors operated in the heavy traffic regime. Suppose that some do but that others have lighter traffic (for example, they had a service rate that was much more than adequate for their needs). The nonbottleneck processors can be eliminated. Batches in both arrivals and service can also be handled. The environmental modulation ideas of Section 5.5 also carry over. The approach is a natural extension of what was done there, and the reader should have little difficulty in filling in the details, although it is not dealt with. In communications systems the same message might be sent to many destinations. This so-called multicast is a simple extension of what has already been done. In Theorem 1.1 jobs remained intact. In manufacturing models one might have several processes that need to be done on a job; some can be done in parallel, while others must wait until the completion of one or more of the other processes are completed. This gives rise to the so-called fork-join queue.

Section 3 deals with the problem of blocking. A processor is blocked if the buffer to which it is to send its latest completed job is full, and it must hold that job in a way that prevents it from working on anything else until the block is cleared. The reflection directions are still determined by idle time, but now there is also an idle-time component that is forced by the blocking.

The workload concept introduced in Section 5.3 was particularly useful when there are many classes. This is even more important in the network case, where the alternatives can be very cumbersome. The simplest definition of workload simply looks at the work queued at each processor, and this is dealt with in Subsection 4.1. A more useful form views the work associated with any processor as the work that it will eventually have to do on all of the jobs that are currently in the system. The basic ideas are introduced in Subsection 4.2 for a feedforward system. The discussion before Theorem 4.2 shows how one can read off the weak-sense limit form from the physical structure, and the intuitive nature of the limit form. The weak-sense limit equations do not depend on the relative priorities of the classes, although the idle-time term for a processor might possibly increase even when the workload for that processor is not zero. Given priorities, as in Section 5.6, asymptotically, there are a negligible number of nonlowest-priority jobs in the system, and one can represent the scaled numbers of the low-priority jobs in terms of the workloads. Since the idle-time term at a processor can increase only when the lowest-priority queue there is empty, this procedure is a very convenient way of getting the limit equations for the set of lowest-priority queues. This topic is continued in Chapter 12. The fluid scaling case for the single and multiclass feedforward systems are dealt with in Section 5.

6.1 The Heavy Traffic Limit Theorem

6.1.1 Independent Routing

Notation and assumptions. This section deals with a fundamental canonical form of a single class network of the type introduced and developed by Reiman [213]. The notation and techniques that are required will form the basis of much of the subsequent material on networks. The proofs adapt the methods of Section 5.1. There are K processing stations, called P_i, $i \leq K$, each of which can receive inputs from the exterior (exogenous arrivals). Their outputs can either leave the system, be routed to some other processor, or even go directly back to the same processor if desired.

The buffer of P_i is $B_i\sqrt{n}$, where $B_i \le \infty$. If a customer arrives at a full finite buffer, then that customer is assumed to be lost. Let $\Delta_{i,l}^{a,n}$ denote the lth interarrival interval for exogenous arrivals to P_i. Thus, the first exogenous arrival at P_i appears at time $\Delta_{i,1}^{a,n}$. Let $\Delta_{i,l}^{d,n}$ denote the lth service interval for P_i. Define $A_i^n(t) = 1/\sqrt{n}$ times the number of exogenous arrivals to P_i by real time nt, and define $D_i^n(t)$ analogously for the service completions at P_i. For some centering constants $\bar{\Delta}_i^{\alpha,n}$, $\alpha = a, d$, $i \le K$, define the processes

$$w_i^{\alpha,n}(t) = \frac{1}{\sqrt{n}} \sum_{l=1}^{nt} \left[1 - \frac{\Delta_{i,l}^{\alpha,n}}{\bar{\Delta}_i^{\alpha,n}} \right]. \tag{1.1}$$

Let $I_{ij,l}^n$ denote the indicator function of the event that the lth departure from P_i goes to P_j. Define the "routing" vector $I_{i,l}^n = (I_{ij,l}^n,\ j \le K)$. The following conditions will be used.

A1.0. *The initial condition $x^n(0)$ is independent of all future arrival and service times and of all routing decisions, and other future driving random variables.*

A1.1. *The set of service and interarrival intervals is independent of the set of routing decisions. There are constants $\bar{\Delta}_i^{\alpha}$, $\alpha = a, d$, $i \le K$, such that $\bar{\Delta}_i^{\alpha,n} \to \bar{\Delta}_i^{\alpha}$. The set $w_i^{\alpha,n}(\cdot)$, $\alpha = a, d$, $i \le K$, defined in (1.1) converges weakly to a set of mutually independent Wiener processes $w_i^{\alpha}(\cdot)$, $\alpha = a, d$, $i \le K$, with variances $\sigma_{\alpha,i}^2$, respectively*

A1.2. *The $\{I_{i,l}^n; i, l\}$ are mutually independent for each n. There are constants q_{ij}^n and q_{ij} such that $q_{ij}^n \to q_{ij}$ and $P\{I_{ij,l}^n = 1\} = q_{ij}^n$. The spectral radius of the matrix $Q = \{q_{ij}; i, j\}$ is less than unity. Write $Q_n = \{q_{ij}^n; i, j\}$.*

The spectral radius condition in (A1.2) implies that the matrix Q, when considered as the transition matrix of a Markov chain, is substochastic, and that all states are transient; equivalently, each customer will eventually leave the system, and the number of services of each customer is stochastically bounded by a geometrically distributed random variable. For notational simplicity we will often write $\bar{\lambda}_i^{\alpha} = 1/\bar{\Delta}_i^{\alpha}$, $\alpha = a, d$. It will also be convenient to use the symbol $\bar{\lambda}_i^r = \bar{\lambda}_i^d$. The superscript r will be used to denote quantities associated with the random routing.

A1.3. *There are b_i such that*

$$\sqrt{n}\left[\frac{1}{\bar{\Delta}_i^{a,n}} + \sum_j \frac{q_{ji}^n}{\bar{\Delta}_j^{d,n}} - \frac{1}{\bar{\Delta}_i^{d,n}}\right] = b_i^n \to b_i. \tag{1.2}$$

Note that (1.2) implies that

$$\frac{1}{\bar{\Delta}_i^a} = \frac{1}{\bar{\Delta}_i^d} - \sum_j \frac{q_{ji}}{\bar{\Delta}_j^d},$$

or, equivalently,

$$\bar{\lambda}_i^a = \bar{\lambda}_i^d - \sum_j q_{ji}\bar{\lambda}_j^d.$$

Analogously to the usage in Section 5.1, let $S_i^{a,n}(t)$ (respectively, $S_i^{d,n}(t)$) denote $1/n$ times the number of exogenous arrivals (respectively, jobs completely served) at P_i by real time nt. Let $N_i^{a,n}(t)$ (respectively, $N_i^{d,n}(t)$) denote $1/n$ times the number of exogenous arrivals (respectively, completely served jobs) at P_i that are required to bring us to at least real time nt. In particular,

$$\begin{aligned} S_i^{a,n}(t) &= \frac{1}{n} \times \max\left\{m : \sum_{l=1}^m \Delta_{i,l}^{a,n} \le nt\right\}, \\ N_i^{a,n}(t) &= \frac{1}{n} \times \min\left\{m : \sum_{l=1}^m \Delta_{i,l}^{a,n} \ge nt\right\}. \end{aligned} \tag{1.3}$$

Define

$$\mathcal{T}_i^{\alpha,n}(t) = \frac{1}{n}\sum_{l=1}^{nt} \Delta_{i,l}^{\alpha,n}, \quad \alpha = a, d, \tag{1.4}$$

and $A_i^n(t) = \sqrt{n}S_i^{a,n}(t)$, $D_i^n(t) = \sqrt{n}S_i^{d,n}(t)$. The $N_i^{\alpha,n}(t)$ and $S_i^{\alpha,n}(t)$ might differ by at most $1/n$.

The analogue of (5.1.9a) is (modulo a residual-time error term)

$$\mathcal{T}_i^{a,n}(S_i^{a,n}(t)) = t, \tag{1.5a}$$

and (possibly modulo $1/n$)

$$S_i^{a,n}(\mathcal{T}_i^{a,n}(t)) = t. \tag{1.5b}$$

Define the scaled idle time at P_i:

$$T_i^{d,n}(t) = \frac{1}{\sqrt{n}}\left[nt - \sum_{l=1}^{nS_i^{d,n}(t)} \Delta_{i,l}^{d,n} - \tilde{\Delta}_i^{d,n}(t)\right], \tag{1.6}$$

where $\tilde{\Delta}_i^{d,n}(t)$ is the time that any customer being served, if any, at P_i at real time nt has been in service. Define

$$y_i^n(t) = T_i^{d,n}(t)/\bar{\Delta}_i^{d,n}. \tag{1.7}$$

The analogues of (5.1.11) and (5.1.12) are (modulo a residual-time error term)

$$\frac{1}{\sqrt{n}} T_i^{d,n}(t) + \mathcal{T}_i^{d,n}(S_i^{d,n}(t)) = t, \tag{1.8}$$

$$S_i^{d,n}(\mathcal{T}_i^{d,n}(t)) \le t. \tag{1.9}$$

In (1.9) the argument $\mathcal{T}_i^{d,n}(t)$ is $1/n$ times the actual service time devoted to the first nt (the integer part) services at P_i, and $S_i^{d,n}(\mathcal{T}_i^{d,n}(t))$ is $1/n$ times the number of actual services that are completed by that time.

Define

$$w_{ij}^{d,n}(t) = q_{ij}^n w_i^{d,n}(t), \tag{1.10}$$

$$w_{ij}^{r,n}(t) = \frac{1}{\sqrt{n}} \sum_{l=1}^{nt} \left[I_{ij,l}^n - q_{ij}^n \right], \quad w_i^{r,n}(t) = (w_{ij}^{r,n}(t), j \le K), \tag{1.11}$$

and

$$D_{ij}^n(t) = \frac{1}{\sqrt{n}} \sum_{l=1}^{nS_i^{d,n}(t)} I_{ij,l}^n, \tag{1.12}$$

which is $1/\sqrt{n}$ times the number of outputs of P_i that go to P_j by real time nt. Define $N_i^{r,n}(\cdot) = N_i^{d,n}(\cdot)$.

The system equation. We can now write, for $i = 1, \ldots, K$,

$$x_i^n(t) = x_i^n(0) + A_i^n(t) - D_i^n(t) + \sum_j D_{ji}^n(t) - U_i^n(t), \tag{1.13}$$

where $U_i^n(t)$ is $1/\sqrt{n}$ times the number of arrivals to P_i by real time nt that are lost due to a full buffer. For later use, note that

$$\lim_n E w_i^{r,n}(1)[w_i^{r,n}(1)]' \equiv \Sigma_{r,i}$$

$$= \begin{bmatrix} (1-q_{i1})q_{i1} & -q_{i1}q_{i2} & \cdots & -q_{i1}q_{iK} \\ -q_{i2}q_{i1} & (1-q_{i2})q_{i2} & \cdots & -q_{i2}q_{iK} \\ & & \vdots & \\ -q_{iK}q_{i1} & -q_{iK}q_{i2} & \cdots & (1-q_{iK})q_{iK} \end{bmatrix}. \tag{1.14}$$

The basic convergence theorem. Now we are prepared to state and prove the first theorem. The main point of the theorem is the weak convergence of the sequence of scaled queues $x^n(\cdot)$ to the reflected diffusion process $x(\cdot)$ that is defined by (1.16), (1.17), where the driving Wiener process is defined by (1.18a) or (in vector notation) (1.18b). Additionally, it asserts the convergence of the other processes of interest, namely, the normed arrival and service processes, the processes that keep track of the randomness in the arrivals and services, and the reflection (scaled idle-time) process. The process $w_i^a(\bar{\lambda}_i^a \cdot)$ represents the randomness in the exogenous arrival process at P_i, $w_i^d(\bar{\lambda}_i^d \cdot)$ represents the randomness in the service process at P_i, and $w_{ij}^r(\bar{\lambda}_i^d \cdot)$ represents the effect on P_j of the randomness in the routing from P_i.

Theorem 1.1. *Let $x^n(0)$ converge weakly to $x(0)$ and assume* (A1.0)–(A1.3). *Define*

$$\Psi^n(\cdot) = \left(x^n(\cdot), w_i^{\alpha,n}(\cdot), w_i^{r,n}(\cdot), S_i^{\alpha,n}(\cdot), y^n(\cdot), U^n(\cdot);\ \alpha = a, d,\ i \le K\right). \tag{1.15}$$

Then $\{\Psi^n(\cdot)\}$ is tight. Let

$$\Psi(\cdot) = (x(\cdot), w_i^\alpha(\cdot), w_i^r(\cdot), S_i^\alpha(\cdot), y(\cdot), U(\cdot);\ \alpha = a, d,\ i \le K)$$

denote a weak-sense limit. Let $(w_{ij}^r(\cdot),\ j \le K)$ denote the real-valued components of $w_i^r(\cdot)$. Then the weak-sense limit satisfies

$$x(t) = x(0) + bt + w(t) + z(t), \tag{1.16}$$

where

$$z(t) = [I - Q']\, y(t) - U(t), \quad y(0) = U(0) = 0. \tag{1.17}$$

The components of $w(\cdot)$ in (1.16) can be represented as

$$w_i(t) = w_i^a(\bar{\lambda}_i^a t) + \sum_j q_{ji} w_j^d(\bar{\lambda}_j^d t) - w_i^d(\bar{\lambda}_i^d t) + \sum_j w_{ji}^r(\bar{\lambda}_j^d t). \tag{1.18a}$$

Define $\bar{w}_i^\alpha(t) = w_i^\alpha(\bar{\lambda}_i^\alpha t)$, $\bar{w}^\alpha(t) = (\bar{w}_i^\alpha(t),\ i \le K)$, $\alpha = a, d$, $\bar{w}_i^r(t) = (w_{ij}^r(\bar{\lambda}_i^d t),\ j \le K)$, and $\bar{w}^r(t) = \sum_i \bar{w}_i^r(t)$. Then alternatively written,

$$w(t) = \bar{w}^a(t) - [I - Q']\, \bar{w}^d(t) + \bar{w}^r(t). \tag{1.18b}$$

The $y_i(\cdot)$ (respectively, $U_i(\cdot)$) are nondecreasing, continuous, and can increase only at t where $x_i(t) = 0$ (respectively, where $x_i(t) = B_i$). Let $\mathcal{F}_t$ denote the minimal σ-algebra that measures $\{x(0), \bar{w}_i^\alpha(s), y(s), U(s), s \le t$, $\alpha = a, d, r,\ i \le K\}$. The $\bar{w}_i^\alpha(\cdot)$, $\alpha = a, d, r$, $i \le K$, are mutually independent $\mathcal{F}_t$-Wiener processes. The $w_i^\alpha(\cdot)$, $\alpha = a, d$, have variance $\sigma_{\alpha,i}^2$, and $w_i^r(\cdot)$ has covariance $\Sigma_{r,i}$.

Proof. To the extent possible, the development will be parallel to that of Theorem 5.1.1. By the definitions, we can write

$$A_i^n(t) = \frac{1}{\sqrt{n}} \sum_{l=1}^{nS_i^{a,n}(t)} 1 = \frac{1}{\sqrt{n}} \sum_{l=1}^{nS_i^{a,n}(t)} \left[1 - \frac{\Delta_{i,l}^{a,n}}{\bar{\Delta}_i^{a,n}}\right] + \frac{1}{\sqrt{n}} \sum_{l=1}^{nS_i^{a,n}(t)} \frac{\Delta_{i,l}^{a,n}}{\bar{\Delta}_i^{a,n}}, \tag{1.19}$$

$$D_i^n(t) = \frac{1}{\sqrt{n}} \sum_{l=1}^{nS_i^{d,n}(t)} 1 = \frac{1}{\sqrt{n}} \sum_{l=1}^{nS_i^{d,n}(t)} \left[1 - \frac{\Delta_{i,l}^{d,n}}{\bar{\Delta}_i^{d,n}}\right] + \frac{1}{\sqrt{n}} \sum_{l=1}^{nS_i^{d,n}(t)} \frac{\Delta_{i,l}^{d,n}}{\bar{\Delta}_i^{d,n}}, \tag{1.20}$$

$$\begin{aligned} D_{ij}^n(t) &= \frac{1}{\sqrt{n}} \sum_{l=1}^{nS_i^{d,n}(t)} I_{ij,l}^n \\ &= \frac{1}{\sqrt{n}} \sum_{l=1}^{nS_i^{d,n}(t)} \left[I_{ij,l}^n - q_{ij}^n\right] + \frac{q_{ij}^n}{\sqrt{n}} \sum_{l=1}^{nS_i^{d,n}(t)} \left[1 - \frac{\Delta_{i,l}^{d,n}}{\bar{\Delta}_i^{d,n}}\right] + \frac{q_{ij}^n}{\sqrt{n}} \sum_{l=1}^{nS_i^{d,n}(t)} \frac{\Delta_{i,l}^{d,n}}{\bar{\Delta}_i^{d,n}}. \end{aligned} \tag{1.21}$$

The first term on the right of (1.21) represents the randomness in the routing, and the second the randomness in the service intervals for the mean fraction going from P_i to P_j. As in Theorem 5.1.1, the right-hand term of (1.19) is $[\sqrt{n}/\bar{\Delta}_i^{a,n}]t$, modulo a residual-time error term that is accounted for in the $\epsilon_i^n(\cdot)$ in (1.25).

The right-hand term in (1.21) can be written as

$$\frac{q_{ij}^n}{\bar{\Delta}_i^{d,n}} \left[\sqrt{n}t - T_i^{d,n}(t) - \frac{1}{\sqrt{n}} \tilde{\Delta}_i^{d,n}(t)\right], \tag{1.22}$$

where $\tilde{\Delta}_i^{d,n}(t)$ is the time that a customer, if any, at P_i at real time nt has been in service. This will also be accounted for by the $\epsilon_i^n(\cdot)$. Define

$$z_i^n(t) = y_i^n(t) - \sum_j q_{ji}^n y_j^n(t) - U_i^n(t), \tag{1.23}$$

or in vector form,

$$z^n(t) = [I - Q_n'] \, y^n(t) - U^n(t). \tag{1.24}$$

Putting the above details together yields the system equation

$$\begin{aligned} x_i^n(t) = x_i^n(0) &+ \sqrt{n}\left[\frac{1}{\bar{\Delta}_i^{a,n}} + \sum_j \frac{q_{ji}^n}{\bar{\Delta}_j^{d,n}} - \frac{1}{\bar{\Delta}_i^{d,n}}\right] t + w_i^{a,n}(S_i^{a,n}(t)) \\ &- w_i^{d,n}(S_i^{d,n}(t)) + \sum_j q_{ji}^n w_j^{d,n}(S_j^{d,n}(t)) + \sum_j w_{ji}^{r,n}(S_j^{d,n}(t)) \\ &+ z_i^n(t) + \epsilon_i^n(t). \end{aligned} \tag{1.25}$$

By the heavy traffic condition (A1.3), the second term on the right of (1.25) converges to $b_i t$. The set $w_j^{r,n}(\cdot)$, $j \le K$, converges weakly to mutually independent Wiener processes with covariance matrices $\Sigma_{r,j}$, $j \le K$, by either Theorem 2.8.2 or 2.8.8 and the independence of the routing variables in (A1.2). Furthermore, the independence of the routing variables on the other driving variables implies the independence of these limits with those of the $w_i^{\alpha,n}(\cdot)$, $\alpha = a, d$, $i \le K$.

The fact that $\mathcal{T}_i^{\alpha,n}(\cdot)$, $\alpha = a, d$, converges weakly to the process with values $\bar{\Delta}_i^{\alpha} t$ and $S_i^{a,n}(\cdot)$ to the process with values $\bar{\lambda}_i^a t$, is proved as in Theorem 5.1.1, as are the tightness of $S_i^{d,n}(\cdot)$, and the fact that any weak-sense limit of $S_i^{d,n}(\cdot)$ is Lipschitz continuous with Lipschitz constant no greater than $\bar{\lambda}_i^d$. Thus, the $w_i^{\alpha,n}(S^{\alpha,n}(\cdot))$, $\alpha = a, d$, and $\{w_i^{r,n}(S^{d,n}(\cdot))\}$ are tight and have continuous weak-sense limits.

Now, (1.25) can be written in vector form as

$$x^n(t) = h^n(t) + [I - Q_n'] \, y^n(t) - U^n(t), \tag{1.26}$$

where $h^n(\cdot)$ is asymptotically continuous, in the sense of (3.6.4). Also, the $y_i^n(\cdot)$ (respectively, $U_i^n(\cdot)$) can increase only at t where $x_i^n(t) = 0$ (respectively, where $x_i^n(\cdot) = B_i$). Then the tightness and asymptotic continuity of the set $\{y^n(\cdot), U^n(\cdot)\}$ follow from Theorem 3.6.1 and the spectral radius condition on Q. By the tightness of $\{y^n(\cdot)\}$, $T^{d,n}(\cdot)/\sqrt{n}$ converges weakly to the "zero" process. Hence (1.8) and the weak convergence of $\mathcal{T}^{d,n}(\cdot)$ to the process with values $\bar{\Delta}_i^d t$ imply that $S_i^{d,n}(\cdot)$ converges weakly to the process with values $\bar{\lambda}_i^d t$.

Let n index a weakly convergent subsequence of $\Psi^n(\cdot)$, with weak-sense limit denoted by $\Psi(\cdot)$. The weak convergence and the properties of $y_i^n(\cdot)$ imply that $y_i(0) = 0$ and that $y_i(\cdot)$ is nondecreasing and can increase only at t where $x_i(t) = 0$, and analogously for $U_i(\cdot)$. The independence of the Wiener processes follows from the weak convergence and the independence assumptions in (A1.0)–(A1.2). Then the assertion that they are $\mathcal{F}_t$-Wiener processes follows by the method which was used for construction of the solution in Theorem 3.4.2 or 3.5.2. ∎

6.1.2 Martingale Difference Intervals

Theorem 5.1.2 differed from Theorem 5.1.1 only in that it used a martingale difference condition on the intervals in lieu of the assumption that the $w^{\alpha,n}(\cdot)$ converge weakly to mutually independent Wiener processes. The martingale difference conditions (A5.1.5) together with (A5.1.3) implied (A5.1.1). A similar result will now be given for the network case.

Condition (A1.4)–(A1.6) will be shown to imply (A1.1) and the convergence to the mutually independent Wiener processes of Theorem 1.1. The conditions are a special instance of a sufficient condition for (A1.1). In addition, the techniques used in the proof are of broader applicability, in particular for the state-dependent case of Chapter 8. They use an adaptation of the method of proof of Theorem 2.8.8 for showing convergence to a Wiener process and the nonanticipativity of $(x(\cdot), z(\cdot))$.

Let $\mathcal{F}_{i,l}^{a,n}$ denote the minimal σ-algebra that measures $x^n(0), \Delta_{i,k}^{a,n}, k \le l$, and all of the routings and intervals, other than $\Delta_{i,l+1}^{a,n}$, that begin before or at the time of the lth exogenous arrival at P_i. Let $\mathcal{F}_{i,l}^{d,n}$ denote the minimal

σ-algebra that measures $x^n(0), \Delta_{i,k}^{d,n}, k \le l$, and all of the routings and intervals, other than $\Delta_{i,l+1}^{d,n}$, that begin at or before the end of the lth service at P_i. Let $\mathcal{F}_{i,l}^{r,n}$ be $\mathcal{F}_{i,l}^{d,n}$, but excluding the routing of the lth completed service at P_i. Let $E_{i,l}^{\alpha,n}$, $\alpha = a, d, r$, denote the associated conditional expectations. Let $\mathcal{F}_t^n$ denote the minimal σ-algebra that measures all of the system data up to and including real time nt, and use E_t^n for the associated conditional expectation. The following conditions will be used.

A1.4. *There are constants* $\bar{\Delta}_i^{\alpha,n}, \bar{\Delta}_i^{\alpha}, \sigma_{\alpha,i}^2$ such that for $\alpha = a, d, i \le K$,

$$\bar{\Delta}_i^{\alpha,n} = E_{i,l}^{\alpha,n} \Delta_{i,l+1}^{\alpha,n} \to \bar{\Delta}_i^{\alpha}, \tag{1.27}$$

$$\lim_{n\to\infty, l\to\infty} E_{i,l}^{\alpha,n} \left[1 - \frac{\Delta_{i,l+1}^{\alpha,n}}{\bar{\Delta}_i^{\alpha,n}} \right]^2 = \sigma_{\alpha,i}^2, \tag{1.28}$$

where the convergence is in the sense of the mean.

A1.5. $\left\{ |\Delta_{i,l}^{a,n}|^2, |\Delta_{i,l}^{d,n}|^2; i, l, n \right\}$ is uniformly integrable.

A1.6.

$$P_{\mathcal{F}_{i,l}^{r,n}} \left\{ I_{ij,l}^n = 1 \right\} = q_{ij}^n \to q_{ij}, \tag{1.29}$$

where the spectral radius of $Q = \{q_{ij}; i, j\}$ *is less than unity.*

The Wiener and nonanticipativity properties.

Theorem 1.2. *Let* $x^n(0)$ *converge weakly to* $x(0)$ *and assume* (A1.3)–(A1.6). *Then the conclusions of Theorem* 1.1 *hold.*

Proof. Theorem 2.8.8 implies that $w_i^{\alpha,n}(\cdot)$ converges weakly to a Wiener process $w_i^{\alpha}(\cdot)$ for each $\alpha = a, d, r$ and $i \le K$, but as noted below Theorem 5.1.2, it cannot be used directly to get the correlations among the weak-sense limits or to show that the set of limits is a Wiener process, since the conditioning σ-algebras $\mathcal{F}_{i,l}^{\alpha,n}$ that are used in (A1.4) and (A1.6) depend on α and i, and the number of summands in the $w_i^{\alpha,n}(S^{\alpha,n}(t))$ are different and random. We will follow the idea of the proof of Theorem 2.8.8 to verify the conditions of Theorem 2.8.2. This will yield the independence of the Wiener processes and the covariance of each as well as the nonanticipativity. Actually, once the independence is proved, the nonanticipativity property follows automatically by Theorem 3.4.2 or 3.5.2, both of which prove that $x(\cdot)$ and $z(\cdot)$ are nonanticipative functions of $w(\cdot)$. However, the more general result to be obtained will be useful in Chapter 8 and elsewhere, for the state-dependent or controlled case.

By the fact that the $w_i^{\alpha,n}(\cdot)$, $\alpha = a,d,r$, $i \le K$, are asymptotically continuous and the proof of Theorem 1.1, the set $\Psi^n(\cdot)$ of (1.15) is tight, and all weak-sense limit processes are continuous, the $S_i^{\alpha,n}(\cdot)$ converge as in Theorem 1.1, and (1.16) holds, where $w(\cdot)$ satisfies (1.18). The weak-sense limits $w_i^{\alpha,n}(\bar\lambda_i^\alpha\cdot)$ are individually Wiener processes. The main problem concerns showing that the full set $(w^\alpha(\bar\lambda_i^\alpha\cdot),\ \alpha = a,d,r,\ i\le K)$ is a Wiener process with respect to the appropriate filtration. For $\alpha = a,d,r$, define $M_i^{\alpha,n}(\cdot) = w_i^{\alpha,n}(N_i^{\alpha,n}(\cdot))$, $M^{\alpha,n}(\cdot) = (M_i^{\alpha,n}(\cdot),\ i \le K)$, and

$$\bar M^n(\cdot) = \left(M^{a,n}(\cdot), M^{d,n}(\cdot), M^{r,n}(\cdot)\right). \tag{1.30}$$

Keep in mind that $N_i^{\alpha,n}(\cdot)$, $\alpha = a,d$, is left continuous, although its discontinuities go to zero. At this point it is convenient to use the left continuity. An advantage of the left continuity is that $w_i^{\alpha,n}(N_i^{\alpha,n}(\cdot))$, $\alpha = a,d$, jumps by $(1-\Delta_{i,l}^{\alpha,n}/\bar\Delta_i^{\alpha,n})$ "just after" the *beginning* of the lth interval, which simplifies the notation in the functional expansions below. With the usual abuse of terminology, let n index a weakly convergent subsequence of $\Psi^n(\cdot)$.

Let $h(\cdot)$ be a bounded and continuous real–valued function of its arguments, $t \ge 0, \tau > 0$, p an arbitrary integer, $s_i \le t$, $i \le p$, arbitrary positive numbers, and $F(\cdot)$ a bounded real–valued continuous function of its arguments with compact support, whose partial derivatives up to second order are bounded and continuous. A direct application of the method of Theorem 2.8.8 would seek to evaluate

$$\begin{aligned}
&Eh\left(\bar M^n(s_k), x^n(s_k), y^n(s_k), U^n(s_k), k\le p\right)\left[F(\bar M^n(t+\tau)) - F(\bar M^n(t))\right]\\
&\quad = Eh\left(\bar M^n(s_k), x^n(s_k), y^n(s_k), U^n(s_k), k\le p\right)\\
&\qquad \times \left[E_t^n F(\bar M^n(t+\tau)) - F(\bar M^n(t))\right].
\end{aligned} \tag{1.31}$$

Note that $N_i^{\alpha,n}(t)$ is $\mathcal{F}_t^n$-measurable. A difficulty in the analysis is caused by the fact that the number of terms in the different components of $\bar M^n(t+\tau) - \bar M^n(t)$ are different and random, and they change values at different times. This makes the type of expansion used in Theorem 2.8.8 a little awkward. Nevertheless, the method of proof of Theorem 2.8.8 is readily adapted.

Guided by the fact that $1/n$ times the number of jumps in $w_i^{\alpha,n}(N_i^{\alpha,n}(\cdot))$, $\alpha = a,d,r$, over any interval of length τ converges to $\bar\lambda_i^\alpha\tau$ in probability, without loss of generality we can replace the τ in $w_i^{\alpha,n}(N_i^{\alpha,n}(t+\tau))$ in (1.31) by

$$\tau_i^{\alpha,n} = \min\left\{s : \#\text{ of jumps of } w_i^{\alpha,n}(N_i^{\alpha,n}(\cdot)) \text{ on } (t,t+s] \ge n\bar\lambda_i^\alpha\tau\right\}. \tag{1.32}$$

Thus, we replace the sums that define

$$w_i^{\alpha,n}(N_i^{\alpha,n}(t+\tau)) - w_i^{\alpha,n}(N_i^{\alpha,n}(t)),\ \alpha = a,d,r,$$

in the argument of $F(\cdot)$ by sums containing exactly $n\bar{\lambda}_i^\alpha \tau$ (or the next largest integer) summands. For notational simplicity, and without loss of generality, we suppose that the $n\bar{\lambda}_i^\alpha \tau$ are integers.

For $\alpha = a, d$, and $l \geq 1$, define

$$\begin{aligned} \xi_{i,l}^{\alpha,n} &= (1 - \Delta_{i,l}^{\alpha,n}/\bar{\Delta}_i^{\alpha,n}), & \xi_{i,l}^{r,n} &= \left((I_{ij,l}^n - q_{ij}^n), j \leq K\right), \\ \tilde{\xi}_{i,l}^{\alpha,n} &= \xi_{i,(N_i^{\alpha,n}(t)+l)}^{\alpha,n}, & \tilde{\xi}_{i,l}^{r,n} &= \xi_{i,(N_i^{d,n}(t)+l)}^{r,n}. \end{aligned} \tag{1.33}$$

Note that $\tilde{\xi}_{i,l}^{\alpha,n}/\sqrt{n}$ is the value of the lth jump in $M_i^{\alpha,n}(\cdot)$ at or after real time nt. Define $s_{i,l}^{\alpha,n}$ by letting $t+s_{i,l}^{\alpha,n}$ denote the (scaled) time of this jump. To simplify the algebra, and without loss of generality, we suppose that these times are distinct. We now expand the replacement for the bracketed term in the right side of (1.31) up to second order.

The first-order terms of the expansion are, where the subscripts denote the derivatives or gradients according to the case,

$$\frac{1}{\sqrt{n}} \sum_i E_t^n \sum_{l=1}^{n\bar{\lambda}_i^a \tau} F_{m_i^a}(\bar{M}^n(t + s_{i,l}^{a,n}))\tilde{\xi}_{i,l}^{a,n}, \tag{1.34}$$

$$\frac{1}{\sqrt{n}} \sum_i E_t^n \sum_{l=1}^{n\bar{\lambda}_i^d \tau} F_{m_i^d}(\bar{M}^n(t + s_{i,l}^{d,n}))\tilde{\xi}_{i,l}^{d,n}, \tag{1.35}$$

$$\frac{1}{\sqrt{n}} \sum_i E_t^n \sum_{l=1}^{n\bar{\lambda}_i^d \tau} \left[F_{m_i^r}(\bar{M}^n(t + s_{i,l}^{d,n}))\right]' \tilde{\xi}_{i,l}^{r,n}. \tag{1.36}$$

The second-order terms are

$$\frac{1}{2n} \sum_i E_t^n \sum_{l=1}^{n\bar{\lambda}_i^a \tau} F_{m_i^a m_i^a}(\bar{M}^n(t + s_{i,l}^{a,n})) \left[\tilde{\xi}_{i,l}^{a,n}\right]^2, \tag{1.37}$$

$$\frac{1}{2n} \sum_i E_t^n \sum_{l=1}^{n\bar{\lambda}_i^d \tau} F_{m_i^d m_i^d}(\bar{M}^n(t + s_{i,l}^{d,n})) \left[\tilde{\xi}_{i,l}^{d,n}\right]^2, \tag{1.38}$$

$$\frac{1}{2n} \sum_i E_t^n \sum_{l=1}^{n\bar{\lambda}_i^d \tau} \left[\tilde{\xi}_{i,l}^{r,n}\right]' F_{m_i^r m_i^r}(\bar{M}^n(t + s_{i,l}^{d,n}))\tilde{\xi}_{i,l}^{r,n}, \tag{1.39}$$

$$\frac{1}{n} \sum_i E_t^n \sum_{l=1}^{n\bar{\lambda}_i^d \tau} \left[\tilde{\xi}_{i,l}^{d,n}\right]' \left[F_{m_i^d m_i^r}(\bar{M}^n(t + s_{i,l}^{d,n}))\right] \tilde{\xi}_{i,l}^{r,n}. \tag{1.40}$$

The term $F_{m_i^d m_i^r}(\cdot)$ is the gradient with respect to m_i^r of the derivative $F_{m_i^d}(\cdot)$, and $F_{m_i^r m_i^r}(\cdot)$ is the Hessian matrix with respect to the vector variable m_i^r. The second-order terms containing products of $\tilde{\xi}_{i,l}^{a,n}$ and either $\tilde{\xi}_{i,l}^{d,n}$ or $\tilde{\xi}_{i,l}^{r,n}$ do not occur, since the jumps of $N_i^{a,n}(\cdot)$ and $N_i^{d,n}(\cdot)$ are assumed to occur at different times, and analogously for the cross terms for different i values. The error terms in the second-order expansion (see (2.8.14)) are evaluated analogously, and can be shown to converge to zero in mean by use of the uniform integrability (A1.5) and the continuity of the weak-sense limit processes.

The conditional expectations in (1.34)–(1.40) are evaluated as usual. One first takes conditional expectations of each summand separately given the data up to the starting time of the appropriate interval and uses (A1.4)–(A1.6). This procedure yields that the first-order terms and (1.40) equal zero. Define

$$K^{\alpha,n}(\tau) = \frac{1}{2n}\sum_i E_t^n \sum_{l=1}^{n\bar{\lambda}_i^\alpha \tau} F_{m_i^\alpha,m_i^\alpha}(\bar{M}^n(t+s_{i,l}^{\alpha,n}))\sigma_{\alpha,i}^2, \quad \alpha = a,d, \tag{1.41a}$$

and

$$K^{r,n}(\tau) = \frac{1}{2n}\sum_i E_t^n \sum_{l=1}^{n\bar{\lambda}_i^d \tau} \text{trace}\left[F_{m_i^r,m_i^r}(\bar{M}^n(t+s_{i,l}^{d,n}))\Sigma_{r,i}\right]. \tag{1.41b}$$

By (A1.4)–(A1.6), the difference between the sum of (1.37)–(1.39) and

$$K^{a,n}(\tau) + K^{d,n}(\tau) + K^{r,n}(\tau)$$

goes to zero in mean as $n \to \infty$.

By (A1.4) and (A1.5), the expression (1.41a) is asymptotically equivalent to

$$\sum_i \frac{\sigma_{\alpha,i}^2}{2n\bar{\Delta}_i^{\alpha,n}} E_t^n \sum_{l=nS_i^{\alpha,n}(t)+1}^{nS_i^{\alpha,n}(t+\tau)} F_{m_i^\alpha,m_i^\alpha}(\bar{M}^n(t+s_{i,l}^{\alpha,n}))\Delta_{i,l}^{\alpha,n}. \tag{1.42}$$

Now, using the asymptotic continuity of $\bar{M}^n(\cdot)$, we see that the difference between (1.42) and

$$\int_t^{t+\tau} \sum_i \frac{\bar{\lambda}_i^\alpha \sigma_{\alpha,i}^2}{2} E_t^n F_{m_i^\alpha,m_i^\alpha}(\bar{M}^n(s))ds \tag{1.43}$$

goes to zero in mean. A similar analysis shows that the difference between $K^{r,n}(\tau)$ and

$$\int_t^{t+\tau} \sum_i \frac{\bar{\lambda}_i^d}{2} E_t^n \text{trace}\left[F_{m_i^r,m_i^r}(\bar{M}^n(s))\Sigma_{r,i}\right] ds \tag{1.44}$$

goes to zero in mean. The desired result now follows from the weak convergence and Theorem 2.8.2. ■

6.1.3 Correlated Routing

The routing assumptions (A1.2) and (A1.6) can be generalized in many ways. It is necessary only that the processes $w_{ij}^{r,n}(\cdot)$ converge weakly to appropriate weak-sense Wiener process limits and that the relations between these limits and the other $w_i^\alpha(\cdot)$ can be computed. Some simple cases will be considered here. The set of examples is merely illustrative of the possibilities.

Random, nonindependent routing. Return to Theorem 1.1, drop the independence in assumption (A1.2), but suppose that there are centering constants q_{ij}^n such that $w_i^{r,n}(\cdot)$ converges weakly to a Wiener process with covariance $\Sigma_{r,i}$. Continue to suppose that the routing is independent of the other driving variables and is independent from processor to processor. Then Theorem 1.1 continues to hold.

Next, suppose that for each $i = 1, \ldots, K$, there are $p_{i,jk}^n, p_{i,jk}; j, k = 0, \ldots, K$ (where the state zero denotes the exterior) such that

$$P\left\{I_{ij,l+1}^n = 1 \middle| I_{ik,l}^n = 1, \text{ other past data}\right\} = p_{i,kj}^n \to p_{i,kj}.$$

That is, the routing process from each P_i is a first order Markov process, which we suppose is a periodic. For each j, k, let either $p_{i,jk}^n = 0$ for all n or else $\inf_n p_{i,jk}^n > 0$. Then the centering constant is $q_{ij}^n = EI_{ij,l}^n$, the steady state value. For this case, Theorem 2.8.9 can be used to evaluate the $\Sigma_{r,i}$ and we leave the details to the reader.

Deterministic routing. Suppose that the aim is to route so that the proportion of completed jobs sent from P_i to P_j is kept as close to q_{ij}^n as possible. For example, suppose that customers leaving P_i can be routed only to $P_{j_1(i)}$ and $P_{j_2(i)}$, with $q_{i,j_1(i)}^n = 0.5$. Then every second departure from P_i will be sent to $P_{j_1(i)}$, and the others to $P_{j_2(i)}$. In this case it is simple to verify that $w_{ij}^{r,n}(\cdot)$ converges weakly to the zero process. There is no "routing noise." An analogous result holds for arbitrary q_{ij}^n.

Random batch routing. An interesting situation arises when a set of successive departures are identically routed. For example, the jobs might arrive in batches (say, groups of "cells" in a communications system) with the members of the same batch being processed individually but routed in the same way. For simplicity of development we will assume independence of the routing of successive batches. More general cases can be handled via the perturbed test function method (see Theorem 2.8.9 and [157, 158]).

A1.7. *Suppose that for each i and n there are uniformly (in (l, n)) bounded random variables $k_{i,l}^n$ such that the first $k_{i,1}^n$ jobs from P_i are routed to the same destination, the next group of $k_{i,2}^n$ jobs are routed to the same destination, etc. The probability, conditioned on the "past data," that any* *group goes to* P_j *is* q^n_{ij}. *The routing is done as the individual jobs are completed and is independent of all other driving variables. Let* $\{k^n_{i,l}, l\}$ *be mutually independent and identically distributed for each* i *and* n, *and independent in* i. *Suppose that* $\{k^n_{i,l}; i, l\}$ *are independent of the routings and of the other driving variables and that*

$$Ek^n_{i,l} = \bar{k}^n_i \to \bar{k}_i, \quad E\left[k^n_{i,l}\right]^2 \to v^2_i,$$

where the convergence is as $(l, n) \to \infty$. *The spectral radius of* Q *is less than unity.*

Theorem 1.3. *Assume the conditions of Theorem* 1.1 *or* 1.2, *but with* (A1.2) *or* (A1.6) *replaced by* (A1.7). *Then the conclusions of those theorems hold, but with* $w^r_i(\bar{\lambda}^d_i t)$ *being replaced by* $v_i w^r_i(\bar{\lambda}^d_i t/\bar{k}_i)$ *with covariance matrix* $v^2_i \bar{\lambda}^d_i \Sigma_{r,i}/\bar{k}_i$.

Proof. The proof will be given under the conditions of Theorem 1.1, but with (A1.7) replacing (A1.2). Define $w^{r,n}_{ij}(\cdot)$ as in (1.11). Their summands are correlated owing to the grouping, but (1.25) still holds. The only new issue in the proof concerns the weak convergence and characterization of the weak-sense limit of the $(w^{r,n}_i(S^{d,n}_i(\cdot)), i \leq K)$. Define

$$\hat{w}^{r,n}_{ij}(t) = \frac{1}{\sqrt{n}} \sum_{l=1}^{nt} k^n_{i,l} \left[\hat{I}^n_{ij,l} - q^n_{ij}\right],$$

where $\hat{I}^n_{ij,l}$ is the indicator of the event that the lth *group* from P_i goes to P_j. The sequence $(\hat{w}^{r,n}_{ij}(\cdot), j \leq K)$ converges weakly to a vector-valued Wiener process $\hat{w}^r_i(\cdot)$ with covariance matrix $v^2_i \Sigma_{r,i}$, and this is independent of the weak-sense limits of the other $w^n(\cdot)$-processes. By (A1.7), the weak-sense limits are independent in i.

Let $\hat{S}^{d,n}_i(t)$ denote $1/n$ times the number of "groups" from P_i that are completed by real time nt. Define $\hat{T}^{d,n}_i(t)$ to be $1/n$ times the real time that is required for P_i to process the first (integer part) nt groups. Then (1.8) and (1.9) continue to hold (modulo a residual-time error term) for these definitions. Furthermore, analogously to the case in Theorem 5.1.1, $\hat{T}^{d,n}_i(\cdot)$ converges weakly to the process with values $(\bar{\Delta}^d_i \bar{k}_i)t$. Then following the method of Theorem 5.1.1, it can be shown that $\hat{S}^{d,n}_i(\cdot)$ converges weakly to the process with values $(\bar{\lambda}^d_i/\bar{k}_i)t$. Modulo a residual-time error term, we can write

$$w^{r,n}_{ij}(t) = \frac{1}{\sqrt{n}} \sum_{l=1}^{n\hat{S}^{d,n}_i(t)} k^n_{i,l} \left[\hat{I}^n_{ij,l} - q^n_{ij}\right].$$

Using the last representation and the weak convergence of $\hat{S}^{d,n}_i(\cdot)$ and of $\hat{w}^{r,n}_i(\cdot)$ yields the desired characterization of $w^r(\cdot)$. ■

6.1.4 A General Skorohod Problem Model

The form of the network model in Theorem 1.1 is quite common. It is also concrete due to the specific physical structure and routing rules. The initial part of the analysis in the proof of Theorem 1.1 manipulated the problem into the form (1.25), where the right-hand side consisted of a process that is asymptotically continuous plus a "reflection" term. Once the reflection term was known to be tight and asymptotically continuous, then the characterization of the other process could be completed. Forms similar to (1.25) arise even when the problem does not have the basic form of the model of Theorem 1.1; for example, see the manufacturing models of Section 7.1. Consider the following general form. Let the $I\!R^K$-valued system satisfy the equation

$$x^n(t) = x^n(0) + \int_0^t b(x^n(s))ds + w^n(t) + z^n(t) + \epsilon^n(t), \; x^n(t) \in G, \quad (1.45)$$

for some suitable function $b(\cdot)$, and where $\epsilon^n(\cdot)$ is asymptotically negligible. This is a generalization of (1.25). A primary part of heavy traffic analysis for any specific model is to work it into such a form. The terms $z^n(\cdot)$ are "reflection" terms in the sense of (A1.10). Such a model fits the general concept of a Skorohod problem, and under suitable conditions we expect that any weak-sense limit would satisfy a Skorohod problem. We will use the following conditions.

Recall that (see (5.2.5)) a sequence $h^n(\cdot)$ is said to be asymptotically continuous if

$$\lim_{\delta\to 0} \limsup_n P\left\{\sup_{t\le T}\sup_{s\le\delta} |h^n(t+s) - h^n(t)| \ge \mu\right\} = 0, \text{ for each } T, \mu > 0. \quad (1.46)$$

A1.8. *The processes $w^n(\cdot)$ are asymptotically continuous. If the set $\{z^n(\cdot)\}$ is tight, then $w^n(\cdot)$ converges weakly to a Wiener process $w(\cdot)$.*

A1.9. *The function $b(\cdot)$ satisfies a uniform Lipschitz condition.*

A1.10. *There are d_i, the reflection direction on ∂G_i, and $\epsilon_n \to 0$ such that*

$$z^n(t) = \sum_i y_i^n(t) d_i, \quad (1.47)$$

where $y_i^n(0) = 0$, and $y_i^n(\cdot)$ is nondecreasing and can increase only at t where $x^n(t)$ is within ϵ_n of ∂G_i. For each T, the maximal discontinuity of $z^n(\cdot)$ on $[0,T]$ goes to zero in probability as $n \to \infty$.

A1.11. *$\epsilon^n(\cdot)$ converges weakly to the "zero" process.*

Theorem 1.4. *Assume the form* (1.45) *and assumptions* (A1.0), (A1.8)–(A1.11), *and* (A3.5.1)–(A3.5.3). *Let* $x^n(0)$ *converge weakly to* $x(0)$. *Then* $x^n(\cdot), w^n(\cdot)$, *and* $z^n(\cdot)$ *are asymptotically continuous, and converge weakly to the unique solution to the Skorohod problem*

$$x(t) = x(0) + \int_0^t b(x(s))ds + w(t) + z(t), \; x(t) \in G \tag{1.48}$$

of Subsection 3.5.2, *where* $w(\cdot)$ *is a Wiener process and* $z(\cdot)$ *is the reflection term. The* $(x(\cdot), z(\cdot))$ *are nonanticipative with respect to* $w(\cdot)$.

Proof. If $b(\cdot)$ were bounded or $\{x^n(\cdot)\}$ bounded in probability on any finite interval $[0, T]$, then the proof would be relatively straightforward. The asymptotic continuity of $z^n(\cdot)$ would follow from Theorem 3.6.1. Then $w^n(\cdot)$ would converge weakly to a Wiener process, and for any weak-sense limit $(x(\cdot), w(\cdot), z(\cdot))$, $z(\cdot)$ would be the reflection process. In fact, it would satisfy (3.5.1c)–(3.5.1e). The uniqueness of the limit process, as well as the nonanticipativity, is a consequence of Theorem 3.5.2.

The unboundedness of $b(\cdot)$ is dealt with via the truncation method of Subsection 2.5.5. First, replace $b(\cdot)$ by $b(x)q_K(x)$, where $q_K(\cdot)$ has compact support and is a smooth truncation function. As in Section 2.5.5, let $(\bar{x}^{K,n}(\cdot), \bar{z}^{K,n}(\cdot))$ denote the solution to the truncated problem:

$$\bar{x}^{K,n}(t) = x^n(0) + \int_0^t b(\bar{x}^{K,n}(s))q_K(x^{K,n}(s))ds + w^n(t) + \bar{z}^{K,n}(t) + \epsilon^n(t).$$

One can modify the process in any convenient way, after the first time that $|x^{K,n}(t)|$ exceeds K. The asymptotic continuity of $\bar{z}^{K,n}(\cdot)$ follows from Theorem 3.6.1, and we can suppose that $w^n(\cdot)$ converges weakly to a Wiener process. Any weak-sense limit satisfies

$$\bar{x}^K(t) = x(0) + \int_0^t b(\bar{x}^K(s))q_K(\bar{x}^K(s))ds + w(t) + \bar{z}^K(t), \; x(t) \in G,$$

where $\bar{z}^K(\cdot)$ is the reflection term. Owing to the Lipschitz condition and Theorem 3.5.2, (2.5.19) holds. Hence, $\{x^n(\cdot)\}$ is bounded in probability on any finite time interval. ■

6.2 Extensions: Closed Networks, Nonbottleneck Processors, and Batches

6.2.1 Closed Networks

Recall the examples of closed networks in Subsection 1.2.2. Other examples are in [37, 40, 50, 85, 101, 104, 137], [138, 139, 140, 141, 191, 207, 253].

Return to the setup in Theorem 1.1, but with $B_i = \infty$. Suppose that there are no exogenous arrivals A_i^n and that $\sum_j q_{ij}^n = 1$ for all i, so that the jobs circulate, without ever leaving the system. In particular, assume the following.

A2.1. *Let $q_{ij}^n \geq 0$, $\sum_j q_{ij}^n = 1$ for all i, with $Q_n = \{q_{ij}^n; i, j\} \to Q$, an irreducible stochastic matrix. Let $A_i^n(t) = 0$ for all i, so that there are no exogenous inputs, and let $\sum_i x_i^n(0) = C$. The routing variables (see (A1.2)) are mutually independent, independent of the service intervals and initial condition, and $P\{I_{ij,l}^n = 1\} = q_{ij}^n$.*

A2.2. *There are real b_i such that*

$$\sqrt{n}\left[\sum_j \frac{q_{ji}^n}{\bar{\Delta}_j^{d,n}} - \frac{1}{\bar{\Delta}_i^{d,n}}\right] = b_i^n \to b_i.$$

The state space is now $\{x : \sum_{i=1}^K x_i = C, x_i \geq 0\}$. Thus, we can work with a $(K-1)$-dimensional problem with reduced state space $\{x : \sum_{i=1}^{K-1} x_i \leq C, x_i \geq 0\}$. See Figure 2.1 for an illustration for $K = 3$.

Theorem 2.1. *Let $x^n(0)$ converge weakly to $x(0)$. Assume* (A1.0), (A1.1) *(for the service processes only),* (A2.1), *and* (A2.2). *Then*

$$\left(x_i^n(\cdot),\, i \leq K-1;\, w_i^{d,n}(\cdot), w_i^{r,n}(\cdot),\, y_i^n(\cdot),\, i \leq K\right)$$

is tight and converges weakly to the process

$$\left(x_i(\cdot),\, i \leq K-1;\, w_i^d(\cdot),\, w_i^r(\cdot),\, y_i(\cdot),\, i \leq K\right),$$

where $w_i^d(\cdot), w_i^r(\cdot), y_i(\cdot)$, $i \leq K$, are as in Theorem 1.1. In particular, $y_K(\cdot)$ can increase only at t such that $\sum_{i=1}^{K-1} x_i(\cdot) = C$. The weak-sense limits satisfy

$$x_i(t) = x_i(0) + b_i t + w_i(t) + z_i(t),\ i \leq K-1, \tag{2.1}$$

where $w_i(\cdot), z_i(\cdot)$, $i \leq K-1$, are as in Theorem 1.1. The solution to (2.1) *is unique in the strong-sense.*

Examples. Let $K = 2$, with $q_{ii} = 0$, $i = 1, 2$. Then the limit problem is one-dimensional and can be written as

$$x_1(t) = x_1(0) + b_1 t - w_1^d(\bar{\lambda}_1^d t) + w_2^d(\bar{\lambda}_2^d t) + y_1(t) - y_2(t), \quad x_1(t) \in [0, C],$$

where $y_1(\cdot)$ is the reflection term at the lower boundary and $y_2(\cdot)$ is the reflection term at the upper boundary and $\bar{\lambda}_i^d = \bar{\lambda}_i^a$.

The case $K = 3$ is illustrated in Figure 2.1.

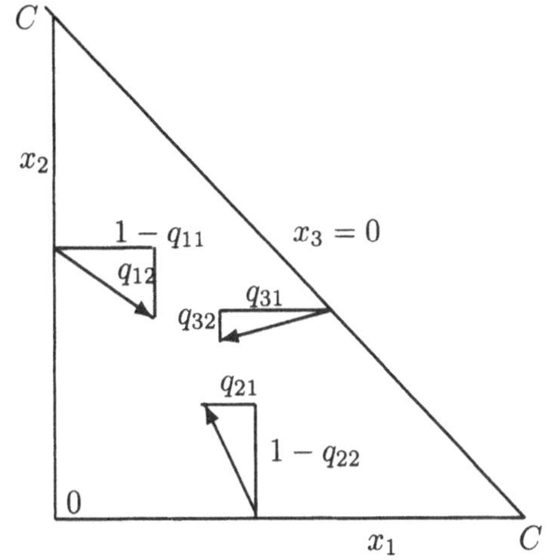

Figure 2.1. The state space for a closed three-dimensional network.

Comments on the proof. For the present problem, (1.25) reduces to the form, for $i \leq K-1$,

$$x_i^n(t) = x_i^n(0) + \sqrt{n}\left[\sum_j \frac{q_{ji}^n}{\bar{\Delta}_j^{d,n}} - \frac{1}{\bar{\Delta}_i^{d,n}}\right] t - w_i^{d,n}(S_i^{d,n}(t))$$
$$+ \sum_j q_{ji}^n w_j^{d,n}(S_j^{d,n}(t)) + \sum_j w_{ji}^{r,n}(S_j^{d,n}(t)) + z_i^n(t) + \epsilon_i^n(t), \tag{2.2}$$

where $z_i^n(\cdot)$ satisfies (1.23) (without the $U^n(\cdot)$) and $y_i^n(\cdot)$, $i \leq K$, are as in Theorem 1.1. By the heavy traffic condition (A2.2), the second term on the right of (2.2) converges to $b_i t$ as $n \to \infty$. The argument that was used to prove tightness and asymptotic continuity of the reflection terms $z^n(\cdot)$ in Theorem 3.6.1 holds here (and implies that the $y_i^n(\cdot)$ are asymptotically continuous), since (A3.5.1) and (A3.5.4) hold (see below).[1] The rest of the details concerning the characterization of the weak-sense limits of any weakly convergent subsequence are as in Theorem 1.1. The strong-sense uniqueness of the solution to (2.2) follows from Theorem 3.5.2, since (A3.5.1) and (A3.5.4) hold.

The idea of the proof of (A3.5.4) will be illustrated via the special case of Figure 2.1, where $K = 3$. The method is the same for arbitrary K. By the comment below (A3.5.4), the condition (A3.5.4) holds in the lower left-hand corner. We will verify it at the upper corner. In the upper corner, queues 1 and 3 are empty, and the problem reduces to one in two dimensions.

[1] Alternatively, one can work in sections using the fact that if $x_i^n(t) = 0$, then some $x_j^n(t)$ must be greater than $1/K$. Select the appropriate subset of the coordinates on each section.

Consider the open network that is obtained when queue 2 is deleted. For this open network, the reflection directions are $d_1 = (1 - q_{11}, -q_{13})/|q_1|$, $d_3 = (-q_{31}, 1 - q_{33})/|q_3|$, where the $|q_i|$ are the normalizing factors that ensure unit length. Since Q is irreducible, the spectral radius of the reduced matrix $\begin{bmatrix} q_{11} & q_{13} \\ q_{31} & q_{33} \end{bmatrix}$ is less than unity. Thus (see comment below (A3.5.4)) there are $a_1 > 0$ and $a_3 > 0$ such that

$$a_1 \langle n_1, d_1 \rangle > a_3 \left| \langle n_1, d_3 \rangle \right|, \tag{2.3a}$$

$$a_3 \langle n_3, d_3 \rangle > a_1 \left| \langle n_3, d_1 \rangle \right|. \tag{2.3b}$$

This is equivalent to

$$\frac{a_1}{|q_1|}(1 - q_{11}) > \frac{a_3}{|q_3|} q_{31}, \tag{2.4a}$$

$$\frac{a_3}{|q_3|}(1 - q_{33}) > \frac{a_1}{|q_1|} q_{13}. \tag{2.4b}$$

Now return to the closed network, whose two-dimensional representation is illustrated in Figure 2.1. On the $x_1 = 0$ axis, the reflection direction is $\bar{d}_1 = (1 - q_{11}, -q_{12})/|\bar{q}_1|$, where the $|\bar{q}_i|$ are normalizing factors. On the diagonal, the normal unit vector is $\bar{n}_3 = (-1, -1)/\sqrt{2}$, and the reflection direction is $\bar{d}_3 = (-q_{31}, -q_{32})/|\bar{q}_3|$. To verify (A3.5.4), we need to find $\bar{a}_i > 0$ such that

$$\bar{a}_1 \langle n_1, \bar{d}_1 \rangle > \bar{a}_3 \left| \langle n_1, \bar{d}_3 \rangle \right|,$$

$$\bar{a}_3 \langle \bar{n}_3, \bar{d}_3 \rangle > \bar{a}_1 \left| \langle \bar{n}_3, \bar{d}_1 \rangle \right|.$$

Equivalently, find $\bar{a}_i$ such that

$$\frac{\bar{a}_1}{|\bar{q}_1|}(1 - q_{11}) > \frac{\bar{a}_3}{|\bar{q}_3|} q_{31}, \tag{2.5a}$$

$$\frac{\bar{a}_3}{|\bar{q}_3|}(q_{31} + q_{32}) > \frac{\bar{a}_1}{|\bar{q}_1|}(1 - q_{11} - q_{12}). \tag{2.5b}$$

But defining $\bar{a}_i$ such that $a_1/|q_1| = \bar{a}_1/|\bar{q}_1|$ and $a_3/|q_3| = \bar{a}_3/|\bar{q}_3|$ yields (2.5).

6.2.2 *Nonbottleneck Processors*

Up to this point it has been supposed that all processors are working in heavy traffic. If some processor is not, then it can effectively be eliminated from the network, as will now be seen.

Assume the conditions of Theorem 1.1, but suppose that some processor P_i does not have heavy traffic in the sense that the right-hand side of (1.2) is minus infinity for that i. Then $x_i^n(\delta + \cdot)$ converges weakly to the "zero" process for any $\delta > 0$. Suppose that $q_{ii}^n = 0$. Otherwise, eliminate the direct feedback by adjusting the service-time distribution at P_i to account for the feedback and replace $q_{ij}^n, j \neq i$, by $q_{ij}^n/(1-q_{ii}^n)$ for the routing probabilities. Then for any $\delta > 0$,

$$D_i^n(\delta + \cdot) - \left[A_i^n(\delta + \cdot) + \sum_{j \neq i} D_{ji}^n(\delta + \cdot)\right]$$

converges weakly to the "zero" process. This implies that asymptotically, the output process from P_i is equivalent to its input process. If $x_i^n(0)$ converges weakly to zero, then we can set $\delta = 0$. The processor P_i can be eliminated from the network, and its inputs simply routed to the other processors according to the routing rule, without changing the weak-sense limit. The service times at P_i do not contribute to the limit.

The redistribution can be done as follows, without changing the weak-sense limits. For $j \neq i$, $q_{ij}^n x_i^n(0)$ is added to $x_j^n(0)$. For $j, k \neq i$, $D_{jk}^n(\cdot)$ is computed by using the routing probabilities $q_{kj}^n + q_{ki}^n q_{ij}^n$. If $q_{ij}^n = 1$ for some j, then add $A_i^n(\cdot)$ to $A_j^n(\cdot)$. Otherwise, $A_i^n(\cdot)$ must be split among the $A_j^n(\cdot), j \neq i$, according to the routing probabilities. To simplify matters, suppose that for each n the $\Delta_{i,l}^{a,n}$ are mutually independent and identically distributed with mean value $\bar{\Delta}_i^{a,n}$ and that $\{|\Delta_{i,l}^{a,n}|^2; l, n\}$ is uniformly integrable. Then distribute the exogenous arrivals to P_i to the $P_j, j \neq i$, at random with probabilities q_{ij}^n.

The above redistribution was simple due to the mutual independence of the routing variables $\{I_{i,l}^n, l < \infty, i \leq K\}$. The problem of redistribution is more difficult under (A1.6), since there the order of arrivals to P_i is taken into account, so one cannot arbitrarily redistribute its inputs independently.

6.2.3 Batch Arrivals, Multiple Servers, Multicast and Fork-Join Queues

Batch Arrivals. Suppose that the exogenous arrivals occur in batches, but (for simplicity only) suppose that only one job is served at a time. Adapt the scheme of Theorem 5.4.1. The following are analogues of (A5.4.1)–(A5.4.2). Let $\Delta_{b,i,l}^{a,n}$ denote the intervals between arrivals of exogenous batches, and $\beta_{i,l}^{a,n}$ the number of jobs contained in the lth exogenous batch arrival to P_i. For some centering constants $\bar{\beta}_i^{a,n}, \bar{\Delta}_{b,i}^{a,n}$, define

$$\begin{aligned} w_{b,i}^n(t) &= \frac{1}{\sqrt{n}} \sum_{l=1}^{nt} \left[\beta_{i,l}^{a,n} - \bar{\beta}_i^{a,n}\right], \\ w_{b,i}^{a,n} &= \frac{1}{\sqrt{n}} \sum_{l=1}^{nt} \left[1 - \frac{\Delta_{b,i,l}^{a,n}}{\bar{\Delta}_{b,i}^{a,n}}\right]. \end{aligned}$$

A2.3. *The set of batch sizes is independent of all of the other driving random variables. There are* $\bar{\beta}_i^a$ *such that* $\bar{\beta}_i^{a,n} \to \bar{\beta}_i^a$. *The* $w_{b,i}^n(\cdot)$, $i \le K$, *converge weakly to a set of mutually independent Wiener processes* $w_{b,i}(\cdot)$, $i \le K$, *with variances* $\sigma_{b,i}^2$.

A2.4.

$$\text{There are real } b_i \text{ such that } \sqrt{n}\left[\frac{\bar{\beta}_i^{a,n}}{\bar{\Delta}_{b,i}^{a,n}} + \sum_j \frac{q_{ji}^n}{\bar{\Delta}_j^{d,n}} - \frac{1}{\bar{\Delta}_i^{d,n}}\right] = b_i^n \to b_i.$$

For the model with many servers at each processor, there is an obvious analogue of (A5.4.3)–(A5.4.4), with or without batch arrivals.

Theorem 2.2. *Assume* (A1.0)–(A1.2), *where* $w_{b,i}^{a,n}(\cdot)$ *replaces* $w_i^{a,n}(\cdot)$, *and* (A2.3), (A2.4). *Then the conclusions of Theorem* 1.1 *hold with the* b_i *given by* (A2.4) *and* $w_i^a(\bar{\lambda}_i^a t)$ *replaced by*

$$\bar{\beta}_i^a w_{b,i}^a(\bar{\lambda}_{b,i}^a t) + w_{b,i}(\bar{\lambda}_{b,i}^a t). \tag{2.6}$$

Remark on the proof. Let $S_{b,i}^{a,n}(t)$ denote $1/n$ times the number of exogenous batches that have arrived by real time nt. The proof is that of Theorem 1.1, with the ideas of Theorem 5.4.1 used to account for the batches. We note only that (1.19) is replaced by

$$\begin{aligned} A_i^n(t) &= \frac{1}{\sqrt{n}} \sum_{l=1}^{nS_{b,i}^{a,n}(t)} \beta_{i,l}^{a,n} = \frac{1}{\sqrt{n}} \sum_{l=1}^{nS_{b,i}^{a,n}(t)} \left[\beta_{i,l}^{a,n} - \bar{\beta}_i^{a,n}\right] \\ &+ \frac{1}{\sqrt{n}} \bar{\beta}_i^{a,n} \sum_{l=1}^{nS_{b,i}^{a,n}(t)} \left[1 - \frac{\Delta_{b,i,l}^{a,n}}{\bar{\Delta}_{b,i}^{a,n}}\right] + \frac{1}{\sqrt{n}} \frac{\bar{\beta}_i^{a,n}}{\bar{\Delta}_i^{a,n}} \sum_{l=1}^{nS_{b,i}^{a,n}(t)} \Delta_{b,i,l}^{a,n}. \end{aligned} \tag{2.7}$$

Multiple servers at each P_i. Return to the setup of Theorem 1.1, but suppose that processor P_i has $\bar{s}_i$ servers (but still only one queue), each of which can do the work on any of the queued jobs. The servers are not necessarily identical. When a server is free, it takes the next job in the queue, as in Section 5.4.2. The result and proof are essentially the same as in Theorem 1.1. The main difference is the heavy traffic condition. Let $\Delta_{i,l}^{d,k,n}$ denote the lth service interval for server k at P_i, and let $S_i^{d,k,n}(t)$ denote $1/n$ times the number of jobs that have completed service there by real time nt. Analogously to (5.4.4), for centering constants $\bar{\Delta}_i^{d,k,n}$, define

$$w_i^{d,k,n}(t) = \frac{1}{\sqrt{n}} \sum_{l=1}^{nS_i^{d,k,n}(t)} \left[1 - \frac{\Delta_{i,l}^{d,k,n}}{\bar{\Delta}_i^{d,k,n}}\right].$$

If there are an infinite number of servers at P_i, then all customers at P_i are being served simultaneously, and we say that there is "self-service." This is treated in Chapter 8. One can have self-service at some nodes and only one or some finite number of servers at others. The reader should be able to write the equations for such combinations from the results in this section and in Chapter 8. The following conditions can be used.

A2.5. *There are constants* $\bar{\Delta}_i^{d,k} = 1/\bar{\lambda}_i^{d,k}$, $\bar{\Delta}_i^a$, *and* $\sigma_{a,i}^2, \sigma_{d,k,i}^2$, $k \leq \bar{s}_i$, $i \leq K$, *such that* $\bar{\Delta}_i^{d,k,n} \to \bar{\Delta}_i^{d,k}$ *and* $\bar{\Delta}_i^{a,n} \to \bar{\Delta}_i^a$. *The* $w_i^{a,n}(\cdot), w_i^{d,k,n}(\cdot); k \leq \bar{s}_i$, $i \leq K$, *converge to mutually independent Wiener processes* $w_i^a(\cdot)$, $w_i^{d,k}(\cdot)$; $k \leq \bar{s}_i$, $i \leq K$, *with variance parameters* $\sigma_{a,i}^2, \sigma_{d,k,i}^2$, *respectively.*

A2.6. *There are real* b_i *such that*

$$\sqrt{n}\left[\frac{1}{\bar{\Delta}_i^{a,n}} + \sum_j q_{ji}^n \sum_{k=1}^{\bar{s}_j} \frac{1}{\bar{\Delta}_j^{d,k,n}} - \sum_{k=1}^{\bar{s}_i} \frac{1}{\bar{\Delta}_i^{d,k,n}}\right] = b_i^n \to b_i.$$

Replace (A1.1) and (A1.3) by (A2.5) and (A2.6). Then Theorem 1.1 holds with $w_i^d(\bar{\lambda}_i^d t)$ replaced by

$$\sum_k w_i^{d,k}(\bar{\lambda}_i^{d,k} t).$$

Services in batches. Return to the setup of Theorem 1.1, but suppose that jobs are served in groups. The only new issue concerns the routing policy. They might be all routed together or individually. If they are routed together, we need to decide what to do if the receiving buffer cannot accommodate the entire group. The final results are simple variations of what has been done. The form depends on the specific assumptions, and is left to the reader.

Multicast. A file in a communications system might be transmitted to several destinations. Similarly, a job exiting a computer might be sent simultaneously to more than one destination for further processing. This situation is referred to as *multicast.* The development is exactly as for Theorem 1.1, except for the form of Q. Consider a simple three-processor network, where the output of P_1 is sent to both P_2 and P_3 and there is no feedback (although this can be allowed). For this example

$$Q = \begin{bmatrix} 0 & 1 & 1 \\ 0 & 0 & 0 \\ 0 & 0 & 0 \end{bmatrix},$$

which is not a probability transition matrix, although the spectral radius is less than unity. Under its other conditions, Theorem 1.1 holds, and the Wiener process has components

$$\begin{aligned} w_1(t) &= w_1^a(\bar{\lambda}_1^a t) - w_1^d(\bar{\lambda}_1^a t), \\ w_2(t) &= w_2^a(\bar{\lambda}_2^a t) + w_1^d(\bar{\lambda}_1^a t) - w_2^d(\bar{\lambda}_2^d t) + w_{12}^r(\bar{\lambda}_1^a t), \\ w_3(t) &= w_3^a(\bar{\lambda}_3^a t) + w_1^d(\bar{\lambda}_1^a t) - w_3^d(\bar{\lambda}_3^d t) + w_{13}^r(\bar{\lambda}_1^a t). \end{aligned}$$

The fork-join queue. The same job might require processing by more than one processor at the same or at arbitrary times. In the simplest model, which is used to model parallel computer processing, each job is "split" (or "forked") on arrival and goes to K queues simultaneously. The processors for each queue act independently. A job is said to be completed only after all of the K parts have been completed. There is a large literature on this and related applications. See [232] for additional references. In certain symmetric cases, one can approximate moments of the stationary distributions [75, 232, 241]. In the general problem, the outputs of any processor might further split and then recombine in various ways. Apart from some simple cases, where each of the K processors has the same input stream, the problem is difficult, and heavy traffic analysis can simplify it greatly. An elegant and thorough analysis of a general case is in [203].

The simplest "parallel processing" model described above can be treated by the methods of Theorem 1.1. The analysis would yield the limit of the K scaled-queue-length processes. Since a job is not completed until all parts are completed, one would like an estimate of the total waiting time. By Theorem 4.1, the (scaled) waiting time of a customer arriving when the queue lengths are $x_i^n(t)$, $i \leq K$, will be (asymptotically)

$$\max\left\{\bar{\Delta}_i^d x_i^n(t),\, i \leq K\right\}.$$

The excess scaled waiting time between the first and last completion is (asymptotically)

$$\max\left\{\bar{\Delta}_i^d x_i^n(t),\, i \leq K\right\} - \min\left\{\bar{\Delta}_i^d x_i^n(t),\, i \leq K\right\}. \tag{2.8}$$

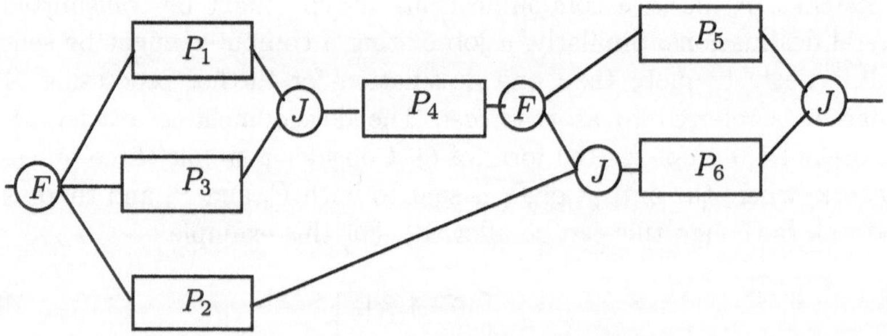

Figure 2.2. A fork-join system.

A more complicated case is illustrated in Figure 2.2. In the figure, F signifies a fork, and J a join. A job cannot proceed out of a join until all of the relevant offspring of the ancestor of that job have arrived there. An alternative view of the general fork–join queue is in terms of a critical path/project evaluation type of procedure. With this interpretation, there is a sequence of jobs into the system. Each job consists of a number of tasks, and each task is identified with a processor. Any task P_i has predecessor tasks that must be completed before P_i can be done. For example, in the Figure, tasks P_1 and P_3 must be completed before P_4 can be started. The general problem is difficult from a notational point of view, since one must keep track of the queue contents as well as the numbers of the outputs of the processors that are waiting for a join to be completed. The reader is referred to [203] for the general result.

6.3 Blocking

Up to this point, it has been supposed that if the output of some processor is routed to another whose (finite) buffer is full, then that output is lost. This is a frequent situation in communications systems. In manufacturing problems, it often happens that such an output will be held at the source processor until space becomes available for it, but holding it at the source processor might prevent that processor from doing further work until the completed job leaves. In the following example, idle time and the consequent effect on the reflection process can occur even if there is work to be done, if it is not possible to do it owing to the physical constraint.

Consider the model of Figure 3.1. There are two queues P_i, $i = 1, 2$, in tandem, without feedback. There are exogenous inputs only to P_1, the output of which goes to P_2, and the output of P_2 exits the system. Processor P_1 has an infinite buffer. The processor P_2 has a finite buffer of size $B_2\sqrt{n}$ where $B_2 > 0$. If it is full on completion of a service at P_1, then P_1 is *blocked*. It holds the completed item until space is available in the buffer of P_2, and cannot process further during this "blocked time." Heavy traffic work on such systems goes back to [252].

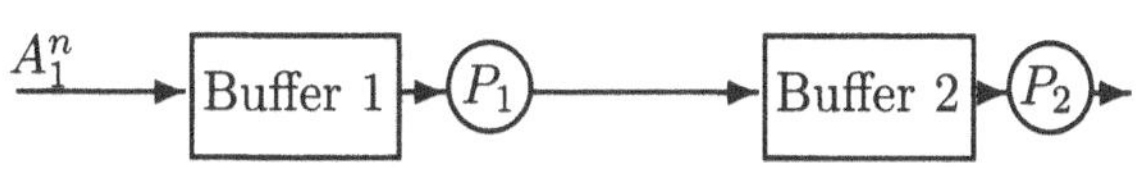

Figure 3.1. A tandem system with blocking. Buffer of P_2 is $\sqrt{n}B_2$.

We will work with the problem of Figure 3.1, and the following conditions will replace those of Theorem 1.1.

A3.1. *The members of each set of intervals* $\{\Delta_{i,l}^{a,n}\}$, $\{\Delta_{i,l}^{d,n}\}$ *are mutually independent and identically distributed, and the sets are mutually indepen-*

dent. Also, for $\alpha = a, d$,

$$\bar{\Delta}_i^{\alpha,n} = E\Delta_{i,l}^{\alpha,n} \to \bar{\Delta}_i^{\alpha}, \quad \lim_{n\to\infty, l\to\infty} E\left[1 - \frac{\Delta_{i,l}^{\alpha,n}}{\bar{\Delta}_i^{\alpha,n}}\right]^2 = \sigma_{\alpha,i}^2.$$

A3.2. $\left\{|\Delta_{i,l}^{a,n}|^2, |\Delta_{i,l}^{d,n}|^2; \, l, n, \, i = 1, 2\right\}$ *is uniformly integrable.*

A3.3. *For real* b_i,

$$\sqrt{n}\left[\frac{1}{\bar{\Delta}_1^{a,n}} - \frac{1}{\bar{\Delta}_1^{d,n}}\right] = b_1^n \to b_1,$$

$$\sqrt{n}\left[\frac{1}{\bar{\Delta}_1^{d,n}} - \frac{1}{\bar{\Delta}_2^{d,n}}\right] = b_2^n \to b_2.$$

Theorem 3.1. *Assume* (A1.0) *and* (A3.1)–(A3.3). *Let* $x^n(0) \Rightarrow x(0)$. *Then the conclusions of Theorem* 1.1 *hold, with limit process*

$$\begin{aligned}
&x_1(t) = x_1(0) + b_1 t + \left[w_1^a(\bar{\lambda}_1^a t) - w_1^d(\bar{\lambda}_1^a t)\right] + y_1(t) + U_1^b(t), \\
&x_2(t) = x_2(0) + b_2 t + \left[w_1^d(\bar{\lambda}_1^a t) - w_2^d(\bar{\lambda}_1^a t)\right] + y_2(t) - y_1(t) - U_1^b(t), \\
&x_1(t) \geq 0, \; x_2(t) \in [0, B_2].
\end{aligned} \tag{3.1}$$

The processes $y_i(\cdot)$, $i = 1, 2$, *and* $U_1^b(\cdot)$ *satisfy* $y_i(0) = U_1^b(0) = 0$ *and are continuous and nondecreasing. The* $y_i(\cdot)$ *can increase only at* t *where* $x_i(t) = 0$, *and* $U_1^b(\cdot)$ *can increase only at* t *where* $x_2(t) = B_2$.

Remark. The reflection directions are illustrated in Figure 3.2. The reflection at the upper boundary arises from the idle time at P_1, which is imposed by the blocking.

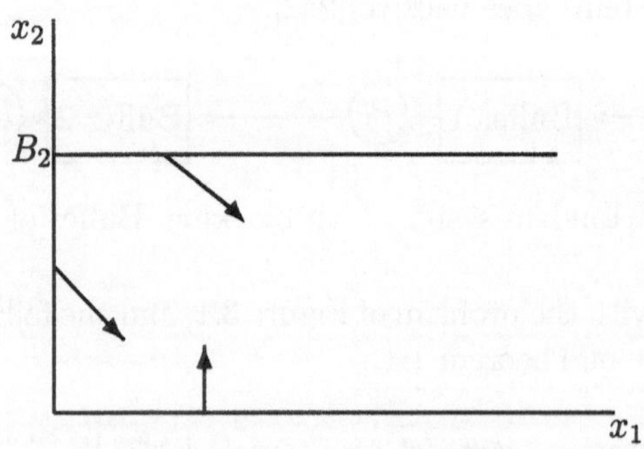

Figure 3.2: Blocking. Reflection directions for the problem of Figure 3.1.

Comments on the proof. Unless otherwise noted, we use the terminology of Theorem 1.1. If there is no blocking at scaled time t, then

$$x_1^n(t) = x_1^n(0) + A_1^n(t) - D_1^n(t),$$

$$x_2^n(t) = x_2^n(0) + D_1^n(t) - D_2^n(t).$$

If there is blocking at scaled time t, then add $1/\sqrt{n}$ to the first equation and subtract it from the second. Without loss of generality, we ignore this latter $1/\sqrt{n}$ quantity henceforth. The $A_1^n(\cdot)$ and $D_2^n(\cdot)$ are treated as in Theorem 1.1. Write, as usual,

$$D_1^n(t) = \frac{1}{\sqrt{n}} \sum_{l=1}^{nS_1^{d,n}(t)} \left[1 - \frac{\Delta_{1,l}^{d,n}}{\bar{\Delta}_1^{d,n}}\right] + \frac{1}{\sqrt{n}} \sum_{l=1}^{nS_1^{d,n}(t)} \frac{\Delta_{1,l}^{d,n}}{\bar{\Delta}_1^{d,n}}.$$

The second term on the right can be written as (modulo a residual-time error term)

$$\frac{1}{\sqrt{n}\bar{\Delta}_1^{d,n}} \left[nt - \text{idle time}\right].$$

The idle time is broken into two parts. The first part is due to a lack of work at P_1 (i.e., when $x_1^n(t) = 0$) and is $y_1^n(t)$). The second part is due to the forced idling due to the blocking. Let $\bar{\Delta}_1^{d,n} U_1^{b,n}(t)$ denote $1/\sqrt{n}$ times the idle time forced by blocking by real time nt. From this point on, the method of Theorem 1.1 can be used, with tightness of the reflection terms being given by Theorem 3.6.1.

If P_2 has its own exogenous arrival stream, then one subtracts $U_2(t)$ from the second line of (3.1), where $U_2(t)$ is the weak-sense limit of $1/\sqrt{n}$ times the number of exogenous arrivals to P_2 that were denied admission due to a full buffer. The reflection direction when $x_2 = B_2$ is harder to compute, unless the interarrival times to P_2 and all service times are exponentially distributed.

Comment on the blocking intervals. Let $\Delta_l^{b,n}$ denote the real time duration of the lth blocking interval. Consider the set of blocking intervals for which P_1 has work to do when those blocks end. When such a block ends, both processors start processing their next jobs at the same time. Let the next service interval for P_1 be $\Delta_{1,l}^{d,n}$, and that for P_2, $\Delta_{2,k(l)}^{d,n}$. There will be a block at the end of the next service completion at P_1, if $\Delta_{1,l}^{d,n} < \Delta_{2,k(l)}^{d,n}$. Then the next blocking time will be just the residual time for the service at P_2, namely, $\left[\Delta_{2,k(l)}^{d,n} - \Delta_{1,l}^{d,n}\right]^+$, and it will be independent of all past blocking times. The characterization of the next blocking interval is more complicated if P_1 does not have work to do when a block ends, but the blocking intervals are still mutually independent.

Remark on a more general network. The scheme for a network of the type of Theorem 1.1 is handled in a similar way. For simplicity, we confine ourselves to a network without feedback and with deterministic routing. Assume the weak convergence of $x^n(0)$, (A1.0), (A1.2), (A1.3), (A3.1), and (A3.2). Let the buffer for P_i be $\sqrt{n}B_i$, $0 < B_i \le \infty$. Suppose that for each i, there is a unique $k = k(i) \neq i$ such that $q^n_{ik(i)} > 0$ and let $q_{ik(i)} > 0$. Suppose that any processor that can receive inputs from another processor does not have an exogenous arrival stream. Proceed as follows. Let $T_i^{d,n}(\cdot)$ denote the scaled idle-time process at P_i that is not due to a block, and $\bar{\Delta}_i^{d,n} U_i^{b,n}(\cdot)$ that which is due to a block. Then (1.22) is replaced by

$$\frac{1}{\bar{\Delta}_i^{d,n}}\left[\sqrt{n}t - T_i^{d,n}(t) - \bar{\Delta}_i^{d,n} U_i^{b,n}(t) - \frac{1}{\sqrt{n}}\tilde{\Delta}_i^{d,n}(t)\right].$$

The limit equation is

$$x(t) = x(0) + bt + w(t) + [I - Q']\, y(t) + [I - Q']\, U^b(t) - U(t),$$

where $U_i^b(\cdot)$ can increase only at t where $x_{k(i)}(t) = B_{k(i)}$. If $q_{ij} > 0$ for more than one value of j, then the problem is more subtle, since $U_i^b(\cdot)$ can increase at t where $x_j(t) = B_j$ for any such j. The $U_i^n(\cdot)$ will still be asymptotically continuous.

The tandem queue with $B_2 = 0$. Return to the tandem queue model in Figure 3.1, but now suppose that P_2 has no buffer, so that if it is serving any customer when P_1 completes service, then P_1 will be blocked. For simplicity, let $x_2^n(0) = 0$, although the result holds if $x_2^n(0)$ converges weakly to zero. The problem is one-dimensional, since only P_1 has a queue. We say that P_1 is in a busy period at t if $x_1^n(t) > 0$. Except for the first service in a busy period, the effective service times for P_1 are $\Delta_{1,l}^{e,n} = \max\{\Delta_{1,l}^{d,n}, \Delta_{2,l-1}^{d,n}\}$. For some centering constant $\bar{\Delta}_1^{e,n}$, define

$$w_i^{e,n}(t) = \frac{1}{\sqrt{n}} \sum_{l=1}^{nt} \left[1 - \frac{\Delta_{1,l}^{e,n}}{\bar{\Delta}_1^{e,n}}\right].$$

In Theorem 5.1.1, replace $\Delta_l^{a,n}$ by $\Delta_{1,l}^{a,n}$ and $\Delta_l^{d,n}$ by $\Delta_{1,l}^{e,n}$, and similarly for their centering constants. If the conditions of that theorem hold with these replacements, then its conclusions also hold. The only complication is due to the fact that the service times at P_1 are not $\Delta_{1,l}^{e,n}$ for the first services in a busy period, but this involves only a revision (a reduction) in the effective service time for some of the services that begin when $x_1^n(t) = 1/\sqrt{n}$ and can be incorporated into the "idle-time" term. Theorems 5.2.3 or 3.6.1 can still be used to prove the asymptotic continuity of this revised idle-time term.

6.4 The Workload Formulation and Priorities

The next subsection is concerned with the single-class problem of Theorem 1.1, but from the point of view of workload in each queue and Theorem 5.3.1. It obtains the asymptotic linear relationship between the scaled size and work in each queue. There is a much more important definition of workload, which is critical for dealing with multiclass networks [87, 92, 100, 173, 206, 248] and discussed in subsequent subsections and in Chapter 12. It is a very effective method for dealing with multiple job classes in a network, and can handle applications that are considerably more complex than the single-job-class model of Theorem 1.1.

6.4.1 Queued Workload

Define the queued workload $WQ_i^n(t)$ to be $1/\sqrt{n}$ times the amount of real time that P_i needs to work to complete all of the jobs that are in its (and only its) queue at real time nt. The next theorem is a simple network version of Theorem 5.3.3.

Theorem 4.1. *Assume the system model that was used in Theorem* 1.1 *but with infinite buffers. Suppose that* $w_i^{d,n}(\cdot)$, $i \leq K$, *converge weakly to continuous processes and that* $\bar{\Delta}_i^{d,n} \to \bar{\Delta}_i^d$. *Let* $x^n(\cdot)$ *be bounded in probability in that for each* $T > 0$,

$$\lim_N \limsup_n P\left\{\sup_{t\leq T} |x^n(s)| \geq N\right\} = 0. \tag{4.1}$$

Then the differences

$$WQ_i^n(\cdot) - \bar{\Delta}_i^{d,n} x_i^n(\cdot),\ i \leq K, \tag{4.2}$$

converge weakly to the "zero" process. Now drop the assumption (4.2), *but suppose that, asymptotically, with an arbitrarily high probability and on any bounded time interval, the* $S_i^{a,n}(\cdot)$ *are bounded by a function with a linear growth plus a constant and that* $x^n(0)/\sqrt{n}$ *converges weakly to zero. Then* (4.2) *holds if it holds for* $WQ_i^n(\cdot)$, $i \leq K$.

Proof. Let $I_i^{d,n}(t)$ denote the index of the last customer to complete service in queue i at or before real time nt. The processes $x_i^n(\cdot)$ and $WQ_i^n(\cdot)$ satisfy

$$\frac{1}{\sqrt{n}} \sum_{l=I_i^{d,n}(t)+2}^{I_i^{d,n}(t)+\sqrt{n}x_i^n(t)} \Delta_{i,l}^{d,n} \leq WQ_i^n(t) \leq \frac{1}{\sqrt{n}} \sum_{l=I_i^{d,n}(t)+1}^{I_i^{d,n}(t)+\sqrt{n}x_i^n(t)} \Delta_{i,l}^{d,n}. \tag{4.3}$$

By the hypotheses, the left and right-hand terms in (4.3) differ by a residual-time error term. Now use the method of Theorem 5.3.3.

6.4.2 A Multiclass Feedforward System.

The result of Theorem 4.1 is interesting in that it shows that if there is only a single class of jobs, then there is a law of large numbers effect that allows us to (asymptotically) compute the work in any queue from the number queued, and conversely. There is a much more useful definition of workload, due to Harrison [92], which can greatly simplify the analysis when there are many job classes. Consider a *feedforward* network where each P_i can have a finite number of exogenous input streams $A_{ik}^n, k = 1, \ldots$, each with its own service requirements. A completed job of any given class at P_i will be routed randomly, but according to its class, and it will change class either in some random or deterministic way. The kth class at P_i is denoted by (ik). Let $x_{ik}^n(\cdot)$ denote $1/\sqrt{n}$ times the number of class (ik) at P_i at real time nt. The classes are queued separately, and there is a well-defined priority hierarchy at each P_i. Let $q_{ik,jp}^n$ denote the probability that a completed class (ik) job becomes a class (jp) job; i.e., that it is routed to the pth class at P_j. Some of the basic ideas were introduced for a two-class feedforward system in [120] and for a multiple class feedforward system in [206].

The "(processor, queue)" notation in the indices is used so that we can keep a sense of the structure of the physical system in the analysis. It becomes cumbersome for the general feedforward problem and for the scheduling problems of Chapter 12, where we will use the "job class" and "activity" terminology of [92, 100, 96], which is standard for the multiclass network problem at this time.

Suppose that we stop the arrival of the exogenous inputs at some scaled time t. The various queued jobs will be processed and then go on to other processors before eventually leaving the system. Eventually the system will be emptied. Define the workload $WL_i^n(t)$ to be the total (always scaled) work that P_i must do until the time that the system empties. Thus, $WL_i^n(t)$ depends on whatever future service intervals are needed to complete all of the jobs that are in the system at real time nt. Suppose that a processor is working on a low-priority job when a high-priority job arrives. Then we suppose that (A4.0) holds. The limit *form* of the workload equation will not depend on the priorities or the service disciplines, provided that (A4.0) holds.

A priority system. To avoid excess notation, the ideas will be illustrated for the particular three-processor feedforward priority system of Figure 4.1.

Notation. The exogenous input streams to P_1 are A_{11}^n and A_{12}^n, with A_{11}^n having high priority. The exogenous arrival stream to P_2 is called A_{21}^n, and it has high priority there. The exogenous arrival stream to P_3 is called A_{31}^n and it has low priority there. Upon completing service at P_1, a class $(1,2)$ job is routed at random to either P_2 or P_3. Those going to P_2 merge with the served $(1,1)$ class jobs waiting there and all become class $(2,2)$.

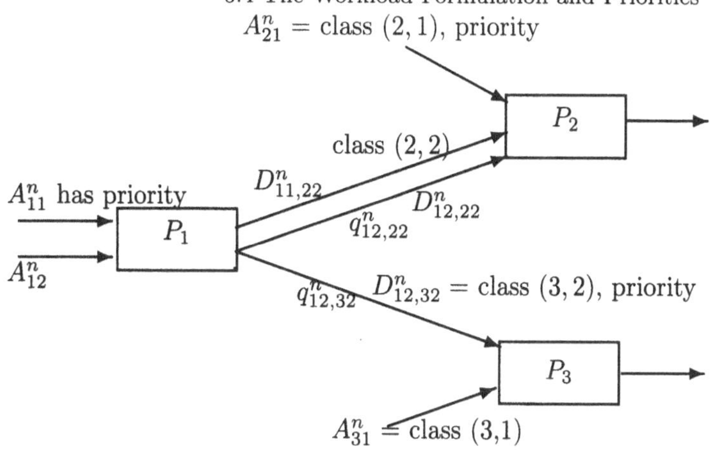

Figure 4.1. A feedforward multiclass network with priorities.

The interarrival intervals for the exogenous arrival streams are denoted by $\Delta^{a,n}_{ik,l}$, where (ik) is the class, with $k = 1, 2$, for $i = 1$ and $k = 1$ for $i = 2, 3$. Keeping track of the service times is notationally complicated since classes can merge, as in Figure 4.1, where all jobs sent from P_1 to P_2 are class $(2, 2)$, irrespective of their original exogenous source. If the lth member of exogenous arrival class (ik) is ever served at P_j as a class (jp), then we write $\Delta^{d,n}_{ik,jp,l}$ for its service time.

Let $I^n_{ik,jp,l}$ denote the indicator function of the event that the lth member of exogenous arrival class (ik) will become a class (jp) job at some time. For centering constants $\bar{\Delta}^{a,n}_{ik}$ and $\bar{\Delta}^{d,n}_{ik}$ define, for the ik, jp of interest in this problem,

$$\begin{aligned} w^{a,n}_{ik}(t) &= \frac{1}{\sqrt{n}} \sum_{l=1}^{nt} \left[1 - \frac{\Delta^{a,n}_{ik,l}}{\bar{\Delta}^{a,n}_{ik}} \right], \\ w^{d,n}_{ik,jp}(t) &= \frac{1}{\sqrt{n}} \sum_{l=1}^{nt} I^n_{ik,jp,l} \left[1 - \frac{\Delta^{d,n}_{ik,jp,l}}{\bar{\Delta}^{d,n}_{jp}} \right], \\ w^{r,n}_{ik,jp}(t) &= \frac{1}{\sqrt{n}} \sum_{l=1}^{nt} \left[I^n_{ik,jp,l} - q^n_{ik,jp} \right]. \end{aligned} \tag{4.4}$$

Write $\Delta^{d,n}_{ik,ik,l}$ and $w^{d,n}_{ik,ik}(\cdot)$ simply as $\Delta^{d,n}_{ik,l}$ and $w^{d,n}_{ik}(\cdot)$, respectively. Define the vector $w^{r,n}_{ik}(\cdot) = \{w^{r,n}_{ik,jp}(\cdot); jp\}$. In the present case, the only one of interest is $w^{r,n}_{12}(\cdot) = (w^{r,n}_{12,22}(\cdot), w^{r,n}_{12,32}(\cdot))$. We will use the following assumptions.

A4.0. *Suppose that a processor is working on a low-priority job when a high priority job arrives. Then either the low-priority job is allowed to be completed or else it is interrupted, but when work on it resumes later, only the "residual" work needs to be done.*

A4.1. *$WL^n(0)$ converges weakly to $WL(0)$, and the initial work for the high priority jobs at each processor converges weakly to zero.*

A4.2. *The centering constants converge: $\bar{\Delta}_{ik}^{\alpha,n} \to \bar{\Delta}_{ik}^{\alpha}$, $\alpha = a, d$, $i \le K$. The set of processes $w_{ik}^{a,n}(\cdot)$ and $w_{ik,jp}^{d,n}(\cdot)$, for all relevant ik, jp, converge weakly to mutually independent Wiener processes with variances $\sigma_{a,ik}^2, \sigma_{d,ik,jp}^2$, respectively. The weak-sense limit is independent of the weak-sense limit of $WL^n(0)$.*

A4.3. *The routings are mutually independent and independent of the other driving random variables, and there are $q_{ik,jp}$ such that $P\{I_{ik,jp,l}^n = 1\} = q_{ik,jp}^n \to q_{ik,jp}$.*

A4.4. *Define*

$$b_1^n = \sqrt{n}\left[\frac{\bar{\Delta}_{11}^{d,n}}{\bar{\Delta}_{11}^{a,n}} + \frac{\bar{\Delta}_{12}^{d,n}}{\bar{\Delta}_{12}^{a,n}} - 1\right],$$

$$b_2^n = \sqrt{n}\left[\frac{\bar{\Delta}_{21}^{d,n}}{\bar{\Delta}_{21}^{a,n}} + \frac{\bar{\Delta}_{22}^{d,n}}{\bar{\Delta}_{11}^{a,n}} + \frac{q_{12,22}^n \bar{\Delta}_{22}^{d,n}}{\bar{\Delta}_{12}^{a,n}} - 1\right],$$

$$b_3^n = \sqrt{n}\left[\frac{\bar{\Delta}_{31}^{d,n}}{\bar{\Delta}_{31}^{a,n}} + \frac{q_{12,32}^n \bar{\Delta}_{32}^{d,n}}{\bar{\Delta}_{12}^{a,n}} - 1\right].$$

Then there are real b_i such that $\lim_n b_i^n = b_i$.

Remark on (A4.4). An alternative to (A4.4) could be based on an assumption of mutual independence of the set of all intervals and routing indicators. But in the proof the various processes appear in forms of (4.4) only, and so it is simpler to put conditions on them directly.

Remark on the workload results. Theorems 4.2 and 4.3 illustrate some of the simplicity allowed by the workload approach. The workload approach avoids the problems of dealing explicitly with the departure processes from any processor as inputs to any other, which is what gave rise to the "coupled" form of the reflection term in Theorem 1.1, for the single-class system. The heavy traffic conditions are also simple. The various fractions in the ith line in (A4.4) are simply the mean traffic intensities for the various exogenous inputs that will affect the workload at P_i: Each is written as the product of the mean exogenous arrival rate, the mean work per job for that class when served at P_i, and the fraction going to P_i. The $y_i(\cdot)$ in (4.5) are *idle-time* terms and *not necessarily* reflection terms. The process $y_i(\cdot)$ can increase at t where P_i has no immediate work to do. For example, $y_3(\cdot)$ can increase at any t if $x_{31}(t) = x_{32}(t) = 0$, even if $x_{12}(t) > 0$, hence when $WL_3(t) > 0$. However, since the queues for the lowest-priority classes at the P_i are the only ones that can be positive in the limit, Theorem 4.3 shows that we can get the evolution equations for the lowest-priority classes from (4.5) and (4.13), and that $y_i(\cdot)$ can only increase at t where the queue for the lowest-priority class at P_i is zero. *Thus, even if the $y_i(\cdot)$ increase when $WL_i(t) > 0$, the workload formulation still provides a convenient way of getting the limit equations for the individual low-priority queues.*

The form of the $w_i(\cdot)$, as given in Theorem 4.2, also follows a simple rule. Each exogenous arrival process that affects the workload at P_i contributes a term, and all random routings contribute a term. Consider $w_2(\cdot)$, for example. The component

$$\bar{\Delta}_{22}^d \left[q_{12,22} w_{12}^a(\bar{\lambda}_{12}^a t) - w_{12,22}^d(\bar{\lambda}_{12}^a t)\right]$$

of $w_2(t)$ arising from exogenous input class $(1,2)$ is typical. It is the product of the mean work per job $(\bar{\Delta}_{22}^d)$ for class $(2,2)$ and $q_{12,22} w_{12}^a(\bar{\lambda}_{12}^a t) - w_{12,22}^d(\bar{\lambda}_{12}^a t)$. The first term in the latter quantity is due to the randomness in the arrivals for class $(1,2)$ times the mean fraction of the class that goes to P_2. The second term is the randomness in the service intervals at P_2 of the class (1,2) customers that go to P_2. The component (of $w_2(\cdot)$) $\bar{\Delta}_{22}^d w_{12,22}^r(\bar{\lambda}_{12}^a t)$ is due to the randomness in the routing.

Theorem 4.3 is an extension to networks of Theorem 5.3.3 and Theorem 5.6.2 concerning priorities. It simply says that, asymptotically, the workload at P_i is the sum over j of the low-priority queue size at P_j times the fraction of the queue that will eventually reach P_i times the mean work per job at P_i for those that reach P_i. Define $y_i^n(t) = T_i^{d,n}(t)$.

Theorem 4.2. *Assume* (A4.0)–(A4.4) *and suppose that the buffers are infinite. Then*

$$\left\{WL_i^n(\cdot), S_{ik}^{a,n}(\cdot), w_{ik}^{\alpha,n}(\cdot),\ \alpha = a, d, w_{ik,jp}^{d,n}(\cdot), w_{ik,jp}^{r,n}(\cdot), y_i^n(\cdot),\ \textit{all } ik, jp, i\right\}$$

is tight and converges weakly to continuous processes

$$\left\{WL_i(\cdot), S_{ik}^a(\cdot), w_{ik}^\alpha(\cdot),\ \alpha = a, d, w_{ik,jp}^d(\cdot), w_{ik,jp}^r(\cdot), y_i(\cdot),\ \textit{all } ik, jp, i\right\}.$$

The limit w-processes $w_{ik}^a(\cdot), w_{ik,jp}^d(\cdot)$ for all ik, jp, and $w_{12}^r(\cdot) = (w_{12,22}^r(\cdot), w_{12,32}^r(\cdot))$ are mutually independent Wiener processes, the first two sets with the variances in (A4.2), *and the covariance of $w_{12}^r(\cdot)$ is*

$$\Sigma_{r,12} = \begin{bmatrix} q_{12,22}(1-q_{12,22}) & -q_{12,22}q_{12,32} \\ -q_{12,22}q_{12,32} & q_{12,32}(1-q_{12,32}) \end{bmatrix}.$$

The $y_i(\cdot)$ are continuous and nondecreasing, and they satisfy $y_i(0) = 0$: The limit process $S_{ik}^a(\cdot)$ has values $\bar{\lambda}_{ik}^a t$. Also,

$$WL_i(t) = WL_i(0) + b_i t + w_i(t) + y_i(t), \tag{4.5}$$

where

$$w_1(t) = \bar{\Delta}^d_{11}\left[w^a_{11}(\bar{\lambda}^a_{11}t) - w^d_{11}(\bar{\lambda}^a_{11}t)\right] + \bar{\Delta}^d_{12}\left[w^a_{12}(\bar{\lambda}^a_{12}t) - w^d_{12}(\bar{\lambda}^a_{12}t)\right],$$

$$\begin{aligned} w_2(t) = \bar{\Delta}^d_{21}&\left[w^a_{21}(\bar{\lambda}^a_{21}t) - w^d_{21}(\bar{\lambda}^a_{21}t)\right] + \bar{\Delta}^d_{22}\left[w^a_{11}(\bar{\lambda}^a_{11}t) - w^d_{11,22}(\bar{\lambda}^a_{11}t)\right] \\ &+ \bar{\Delta}^d_{22}\left[q_{12,22}w^a_{12}(\bar{\lambda}^a_{12}t) - w^d_{12,22}(\bar{\lambda}^a_{12}t)\right] + \bar{\Delta}^d_{22}w^r_{12,22}(\bar{\lambda}^a_{12}t), \end{aligned}$$

$$\begin{aligned} w_3(t) = \bar{\Delta}^d_{31}&\left[w^a_{31}(\bar{\lambda}^a_{31}t) - w^d_{31}(\bar{\lambda}^a_{31}t)\right] \\ &+ \bar{\Delta}^d_{32}\left[q_{12,32}w^a_{12}(\bar{\lambda}^a_{12}t) - w^d_{12,32}(\bar{\lambda}^a_{12}t)\right] + \bar{\Delta}^d_{32}w^r_{12,32}(\bar{\lambda}^a_{12}t). \end{aligned}$$

Proof. We will use the following standard expansions:

$$\begin{aligned} \Delta^{d,n}_{ik,jp,l}I^n_{ik,jp,l} = -\bar{\Delta}^{d,n}_{jp}I^n_{ik,jp,l}&\left[1 - \frac{\Delta^{d,n}_{ik,jp,l}}{\bar{\Delta}^{d,n}_{jp}}\right] + \bar{\Delta}^{d,n}_{jp}\left[I^n_{ik,jp,l} - q^n_{ik,jp}\right] \\ &+ \bar{\Delta}^{d,n}_{jp}q^n_{ik,jp}, \end{aligned} \tag{4.6}$$

and

$$\frac{1}{\sqrt{n}}\sum_{l=1}^{nS^{a,n}_{ik}(t)} 1 = \frac{1}{\sqrt{n}}\sum_{l=1}^{nS^{a,n}_{ik}(t)}\left[1 - \frac{\Delta^{a,n}_{ik,l}}{\bar{\Delta}^{a,n}_{ik}}\right] + \frac{1}{\sqrt{n}\bar{\Delta}^{a,n}_{ik}}\sum_{l=1}^{nS^{a,n}_{ik}(t)}\Delta^{a,n}_{ik,l}. \tag{4.7}$$

The last term on the right of (4.7) is $[\sqrt{n}/\bar{\Delta}^{a,n}_{ik}]t$, modulo a residual-time error term.

We can now write (all expressions below are modulo residual-time error terms),

$$\begin{aligned} \frac{1}{\sqrt{n}}\sum_{l=1}^{nS^{a,n}_{ik}(t)}\Delta^{d,n}_{ik,jp,l}I^n_{ik,jp,l} &= -\bar{\Delta}^{d,n}_{jp}w^{d,n}_{ik,jp}(S^{a,n}_{ik}(t)) + \bar{\Delta}^{d,n}_{jp}w^{r,n}_{ik,jp}(S^{a,n}_{ik}(t)) \\ &\quad + q^n_{ik,jp}\bar{\Delta}^{d,n}_{jp}w^{a,n}_{ik}(S^{a,n}_{ik}(t)) + \frac{\sqrt{n}q^n_{ik,jp}\bar{\Delta}^{d,n}_{jp}}{\bar{\Delta}^{a,n}_{ik}}t. \end{aligned} \tag{4.8}$$

The workloads at scaled time t are

$$WL^n_1(t) = WL^n_1(0) + \frac{1}{\sqrt{n}}\sum_{l=1}^{nS^{a,n}_{11}(t)}\Delta^{d,n}_{11,l} + \frac{1}{\sqrt{n}}\sum_{l=1}^{nS^{a,n}_{12}(t)}\Delta^{d,n}_{12,l} - \sqrt{n}t + y^n_1(t), \tag{4.9a}$$

$$\begin{aligned} WL^n_2(t) &= WL^n_2(0) + \frac{1}{\sqrt{n}}\sum_{l=1}^{nS^{a,n}_{21}(t)}\Delta^{d,n}_{21,l} + \frac{1}{\sqrt{n}}\sum_{l=1}^{nS^{a,n}_{11}(t)}\Delta^{d,n}_{11,22,l} \\ &\quad + \frac{1}{\sqrt{n}}\sum_{l=1}^{nS^{a,n}_{12}(t)}\Delta^{d,n}_{12,22,l}I^n_{12,22,l} - \sqrt{n}t + y^n_2(t), \end{aligned} \tag{4.9b}$$

$$WL_3^n(t) = WL_3^n(0) + \frac{1}{\sqrt{n}} \sum_{l=1}^{nS_{31}^{a,n}(t)} \Delta_{31,l}^{d,n} + \frac{1}{\sqrt{n}} \sum_{l=1}^{nS_{12}^{a,n}(t)} \Delta_{12,32,l}^{d,n} I_{12,32,l}^n - \sqrt{n}t + y_3^n(t). \tag{4.9c}$$

Now we can write

$$\begin{aligned} WL_1^n(t) = WL_1^n(0) &+ \bar{\Delta}_{11}^{d,n}\left[w_{11}^{a,n}(S_{11}^{a,n}(t)) - w_{11}^{d,n}(S_{11}^{a,n}(t))\right] \\ &+ \bar{\Delta}_{12}^{d,n}\left[w_{12}^{a,n}(S_{12}^{a,n}(t)) - w_{12}^{d,n}(S_{12}^{a,n}(t))\right] + b_1^n t + y_1^n(t), \end{aligned} \tag{4.10a}$$

$$\begin{aligned} WL_2^n(t) = WL_2^n(0) &+ \bar{\Delta}_{21}^{d,n}\left[w_{21}^{a,n}(S_{21}^{a,n}(t)) - w_{21}^{d,n}(S_{21}^{a,n}(t))\right] \\ &+ \bar{\Delta}_{22}^{d,n}\left[w_{11}^{a,n}(S_{11}^{a,n}(t)) - w_{11,22}^{d,n}(S_{11}^{a,n}(t))\right] \\ &+ \bar{\Delta}_{22}^{d,n}\left[q_{12,22}^n w_{12}^{a,n}(S_{12}^{a,n}(t)) - w_{12,22}^{d,n}(S_{12}^{a,n}(t))\right] \\ &+ \bar{\Delta}_{22}^{d,n} w_{12,22}^{r,n}(S_{12}^{a,n}(t)) + b_2^n t + y_2^n(t), \end{aligned} \tag{4.10b}$$

$$\begin{aligned} WL_3^n(t) = WL_3^n(0) &+ \bar{\Delta}_{31}^{d,n}\left[w_{31}^{a,n}(S_{31}^{a,n}(t)) - w_{31}^{d,n}(S_{31}^{a,n}(t))\right] \\ &+ \bar{\Delta}_{32}^{d,n}\left[q_{12,32}^n w_{12}^{a,n}(S_{12}^{a,n}(t)) - w_{12,32}^{d,n}(S_{12}^{a,n}(t))\right] \\ &+ \bar{\Delta}_{32}^{d,n} w_{12,32}^{r,n}(S_{12}^{a,n}(t)) + b_3^n t + y_3^n(t). \end{aligned} \tag{4.10c}$$

From this point on, the proof is completed as in Theorem 1.1, except that one needs to prove the tightness and asymptotic continuity of the $y_i^n(\cdot)$. The asymptotic continuity of $y_1^n(\cdot)$ holds, since it is also a reflection term; it can increase only where $WL_1^n(t) = 0$. The main problem in showing asymptotic continuity of $(y_2^n(\cdot), y_3^n(\cdot))$ is showing that there cannot be an asymptotic discontinuity in $(WL_2^n(\cdot), WL_3^n(\cdot))$ (which would have to be caused by idle time in the respective P_i) when $WL_1^n(t) \geq \epsilon$ for an arbitrary $\epsilon > 0$, and the details are left to the reader. ■

Suppose that a low-priority job is stopped on arrival of a high-priority job, but contrary to (A4.0), when work resumes on it later, there is a penalty, so that the work to be done is greater than the residual work for that job. Then (4.9) will no longer hold, even modulo a residual-time error term. One must account for the effects of the penalties.

Theorem 4.3. *Assume the conditions of Theorem 4.2. Then the $x_{ik}^n(\cdot)$ for the high-priority classes (ik) converge weakly to the zero process. The differences*

$$\begin{aligned} &WL_1^n(\cdot) - \bar{\Delta}_{12}^d x_{12}^n(\cdot), \\ &WL_2^n(\cdot) - q_{12,22}\bar{\Delta}_{22}^d x_{12}^n(\cdot) - \bar{\Delta}_{22}^d x_{22}^n(\cdot), \\ &WL_3^n(\cdot) - q_{12,32}\bar{\Delta}_{32}^d x_{12}^n(\cdot) - \bar{\Delta}_{31}^d x_{31}^n(\cdot), \end{aligned} \tag{4.11}$$

converge to the "zero" process.

Remark on the proof. The fact that the $x_{ik}^n(\cdot)$ converge weakly to zero for the high-priority classes follows the proof of a similar result in Theorem 5.6.2. Then use the method of Theorem 5.3.3 to show that (4.13) converges weakly to the "zero" process.

Remark on feedback systems. Theorems 4.2 and 4.3 can be extended to general feedforward systems, but systems with feedback are substantially more subtle. Some of the issues concerning stability and scheduling are discussed in [28, 31, 54, 144, 145, 217].

6.5 Fluid Scaling and Limits

6.5.1 A Single-Class Network

So far, we have used a scaling that is like that which is used in the central limit theorem, and the excess system capacity is given by the various heavy traffic conditions. This led to a reflected diffusion model for the weak-sense limit. On the other hand, in dealing with the behavior during vacation intervals in Subsections 5.6.4 and 7.3.2, a scaling like that of the law of large numbers was used to expose the structure of the local behavior. Such "mean flow" or "law of large numbers" approximations are also very important for systems such as those of Section 1. The basic ideas are actually simpler than what has already been done.

We keep the network structure of Theorem 1.1, but use a different scaling and replace (A1.1) and (A1.3) by the following. Let $\bar{x}_i^n(t)$ denote $1/n$ times the queue size at P_i at real time nt. Unless otherwise noted, we use the terminology of Theorem 1.1. Define $\bar{y}^n(\cdot) = y^n(\cdot)/\sqrt{n}$ and $\bar{U}^n(\cdot) = U^n(\cdot)/\sqrt{n}$.

A5.1. *There are $\bar{\Delta}_i^{\alpha,n} \to \bar{\Delta}_i^{\alpha}$, $\alpha = a, d$, $i \le K$, such that the processes defined by*

$$\frac{w_i^{\alpha,n}(t)}{\sqrt{n}} = \frac{1}{n\bar{\Delta}_i^{\alpha,n}} \sum_{l=1}^{nt} \left[\bar{\Delta}_i^{\alpha,n} - \Delta_{i.l}^{\alpha,n}\right]$$

converge weakly to the "zero" process.

A5.2. *There are real $\bar{b}_i$ such that*

$$\left[\frac{1}{\bar{\Delta}_i^{a,n}} + \sum_j \frac{q_{ji}^n}{\bar{\Delta}_j^{d,n}} - \frac{1}{\bar{\Delta}_i^{d,n}}\right] = \bar{b}_i^n \to \bar{b}_i.$$

In this section define the *residual-time error terms $\bar{\epsilon}_i^n(t)$, or $\bar{\epsilon}^n(t)$,* to be $1/n$ times the time that any interarrival or service interval covering nt has been or will be active plus anything that converges weakly to the "zero" process.

Theorem 5.1. *Assume* (A1.2), (A5.1), *and* (A5.2). *Let* $\bar{x}^n(0)$ *converge weakly to* $\bar{x}(0)$. *Then* $\bar{x}^n(\cdot)$ *is tight and converges weakly to the solution of*

$$\bar{x}(t) = \bar{x}(0) + \bar{b}t + [I - Q'] \bar{y}(t) - \bar{U}(t), \quad 0 \le x_i(t) \le B_i, \tag{5.1}$$

where $y(0) = U(0) = 0$, *and* $\bar{y}_i(\cdot)$ (*respectively,* $\bar{U}_i(\cdot)$) *is nonnegative, continuous, nondecreasing, and can increase only at* t *where* $\bar{x}_i(t) = 0$ (*respectively, where* $x_i(t) = B_i$).

Proof. The proof is parallel to, but simpler than, that of Theorem 1.1. Define $S_{ij}^{d,n}(t) = D_{ij}^{d,n}(t)/\sqrt{n}$. Then the replacement of (1.13) is

$$\bar{x}_i^n(t) = \bar{x}_i^n(0) + S_i^{a,n}(t) + \sum_j S_{ji}^{d,n}(t) - S_i^{d,n}(t) - \bar{U}_i^n(t). \tag{5.2}$$

Expressions (1.8) and (1.9) are unchanged. The proof of Theorem 5.1.1 implies that $S_i^{a,n}(\cdot)$ converges weakly to the process with values $\bar{\lambda}_i^a t$. The proof of Theorem 1.1 implies that $S_i^{d,n}(\cdot)$ and $S_{ij}^{d,n}(\cdot)$ are asymptotically continuous, that any weak-sense limit is Lipschitz continuous with Lipschitz constants no greater than $\bar{\lambda}_i^d$, and that the weak-sense limit of $S_{ij}^{d,n}(\cdot)$ is q_{ij} times that of $S_i^{d,n}(\cdot)$.

Write

$$S_i^{d,n}(t) = \frac{1}{n} \sum_{l=1}^{nS_i^{d,n}(t)} \left[1 - \frac{\Delta_{i,l}^{d,n}}{\bar{\Delta}_i^{d,n}}\right] + \frac{1}{n} \sum_{l=1}^{nS_i^{d,n}(t)} \frac{\Delta_{i,l}^{d,n}}{\bar{\Delta}_i^{d,n}}.$$

The first term is $w_i^{d,n}(S_i^{d,n}(t))/\sqrt{n}$. By (A5.1), the tightness of the initial conditions and of $S_i^{a,n}(\cdot)$ implies that this first term converges weakly to the "zero" process. The second term is

$$\frac{1}{n\bar{\Delta}_i^{d,n}} \left[\text{time devoted to completed services at } P_i \text{ by real time } nt\right]. \tag{5.3}$$

Modulo a residual time error term, (5.3) equals $t/\bar{\Delta}_i^{d,n} - \bar{y}_i^n(t)$. Repeating this analysis for $S_i^{d,n}(\cdot)$ and $S_{ji}^{d,n}(\cdot)$ and using the routing assumption (A1.2) yields the following replacement for (1.25):

$$\bar{x}_i^n(t) = \bar{x}_i^n(0) + \left[\frac{1}{\bar{\Delta}_i^{a,n}} + \sum_j \frac{q_{ji}^n}{\bar{\Delta}_j^{d,n}} - \frac{1}{\bar{\Delta}_i^{d,n}}\right] t$$
$$+ \frac{1}{\sqrt{n}}\left[w_i^{a,n}(S_i^{a,n}(t)) - w_i^{d,n}(S_i^{d,n}(t))\right]$$
$$+ \frac{1}{\sqrt{n}}\left[\sum_j q_{ji}^n w_j^{d,n}(S_j^{d,n}(t)) + \sum_j w_{ji}^{r,n}(S_j^{d,n}(t))\right] + \bar{z}_i^n(t) + \bar{\epsilon}_i^n(t), \tag{5.4}$$

$$\bar{z}^n(t) = [I - Q_n'] \bar{y}^n(t) - \bar{U}^n(t). \tag{5.5}$$

The process $\bar{\epsilon}^n(\cdot)$ and all of the $w/\sqrt{n}$-processes converge weakly to the "zero" process, $\bar{y}_i^n(\cdot)$ can increase only when $\bar{x}_i^n(t)$ is arbitrarily close to zero, and $\bar{U}_i^n(\cdot)$ can increase only when $x_i^n(t)$ is arbitrarily close to B_i. Tightness and asymptotic continuity of $\bar{y}^n(\cdot), \bar{U}_i^n(\cdot)$ follow from either Theorem 3.6.1 or 3.6.2. By the properties of $\bar{y}_i^n(\cdot)$, for any weak-sense limit $(\bar{x}(\cdot), \bar{y}(\cdot), \bar{U}(\cdot), \bar{y}_i(\cdot))$ can increase only at t where $\bar{x}_i(t) = 0$, and $\bar{U}_i(\cdot)$ can increase only at t where $\bar{x}_i(t) = B_i$. ■

The general Skorohod-type problem and fluid limits. The reflection term often arises from factors other than idle time, when the network is not of the form used in Theorem 1.1, and one is obliged to use the Skorohod problem formulation. The idea for the diffusion limit case was discussed in Subsection 1.4, and the method for the fluid limit problem is analogous. Suppose that for some process $\bar{h}^n(\cdot)$, the system can be represented in the form

$$\bar{x}^n(t) = \bar{x}^n(0) + \bar{h}^n(t) + \int_0^t b(x^n(s))ds + \bar{z}^n(t). \tag{5.6}$$

A5.3. *$\bar{h}^n(\cdot)$ is asymptotically continuous and converges weakly to the "zero" process if $\{\bar{z}^n(\cdot)\}$ is tight.*

Then the following theorem holds.

Theorem 5.2. *Assume the form* (5.6), *assumptions* (A5.3), (A1.9), *that* (A1.10) *holds for $\bar{z}^n(\cdot)$, and* (A3.5.1)–(A3.5.3). *Let $\bar{x}^n(0)$ converge weakly to $\bar{x}(0)$. Then $\bar{x}^n(\cdot)$ and $\bar{z}^n(\cdot)$ are asymptotically continuous, and any weak-sense limit satisfies the Skorohod problem*

$$\bar{x}(t) = \bar{x}(0) + \int_0^t b(\bar{x}(s))ds + \bar{z}(t), \quad \bar{x}(t) \in G, \tag{5.7}$$

of Subsection 3.5.1, *where $\bar{z}(\cdot)$ is the reflection term.*

6.6 An Alternative Scaling: Fast Arrivals and Services

Recall the comments of Subsection 5.7 concerning the scaling when the arrivals and services are "fast." The same results hold for the network case, and are quite important in applications to communications theory.

Denote the actual exogenous interarrival and service times by $\hat{\Delta}^{a,n}_{i,l}$ and $\hat{\Delta}^{d,n}_{i,l}$, respectively. Define them in terms of random variables $\Delta^{a,n}_{i,l}$ and $\Delta^{d,n}_{i,l}$ by

$$\hat{\Delta}^{a,n}_{i,l} = \frac{1}{n}\Delta^{a,n}_{i,l}, \quad \hat{\Delta}^{d,n}_{i,l} = \frac{1}{n}\Delta^{d,n}_{i,l}. \tag{6.1}$$

If there are vacations, then rescale the intervacation intervals and the vacation durations in the same way. Redefine $x^n_i(t)$ to be $1/\sqrt{n}$ times the number of customers (jobs) in the queue at P_i at real time t. Also, redefine $S^{a,n}_i(t)$ and $S^{d,n}_i(t)$ to be $1/n$ times the number of exogenous arrivals and the number of departures at P_i, respectively, by real time t. For the queued workload, we need to rescale as in Section 5.7. Thus, let $WL^n_i(t)$ denote $\sqrt{n}$ times the time required for the system to complete all of the work that is queued at real time t. Then the results of the previous sections hold.

7
Uncontrolled Networks, Continued

Chapter Outline

This chapter continues the themes of Chapter 6, concentrating on several canonical classes of problems. Section 1 deals with some models that arise in scheduling and inventory management in manufacturing. The reflection process arises from inventory shortages and not from processor idle time, and new issues arise concerning its computation in the more sophisticated examples. When there are several sources of demands on the various inventories and each source has a different ratio of inventory needs, the simple way of dealing with the reflection term of Theorem 6.1.1 is no longer adequate. Under appropriate modifications of the conditions, one can compute the reflection directions. But the general problem is still open and the discussion illustrates how apparently small changes in the model can have significant effects on the weak-sense limit. Similar issues arise in communications systems, where a single buffer resource is shared by users with different needs. In the shared -buffer problem of Section 2, the state space is a triangle, and complications in the computation of the reflection direction can also arise.

Section 3 generalizes the results in Section 5.6 on processor interruptions and vacations to networks. The problem of vacations is more subtle in the network case (particularly when they are "long") due to the effect of a vacation of one processor on the input processes of others, and so on. For the case of long vacations, it turns out that the asymptotic discontinuities of the scaled-queue-length processes can be obtained via the solution of a natural Skorohod problem.

So far, if a job visits several processors or revisits the same processor before leaving the system, there was essentially no correlation allowed between the work requirements at the various places. Introducing correlation complicates the notation, but follows a systematic procedure, and is dealt with in Section 4. In many applications to communications and computer systems the processor divides its time essentially continuously between jobs from several different users. This so-called processor sharing raises some new issues, since the reflection directions do not satisfy what has been a crucial condition, namely (A3.5.3). Nevertheless, an analysis similar to that for the network of Section 6.1 can be done, and an outline is given in Section 5.

7.1 A Manufacturing System

In the problem of Theorem 6.1.1, the reflection term arises due to the effects of idle time owing to the particular queueing structure that was used. In many manufacturing scheduling and inventory problems, the idle time arises due to inventory shortage, although a parallel analysis can be done. A basic form of such a problem will be developed in this section. It will be seen that incompletely resolved complications arise in the characterization of the reflection direction for the more complicated cases. Such complications arise in communications problems as well where different types of users require different amounts of buffer space. See also the next section. The chosen examples are canonical in that each is illustrative of a unique class of problems. For expository simplicity only, the examples concern products that are composed of two part types.

7.1.1 The Basic Model

Consider the following production system, which is illustrated in Figure 1.1. Requests for a final product occur at intervals $\{\Delta_l^{d,n}, l < \infty\}$. Each final product requires one of each of two part types. Each of the two part types is produced by a dedicated processor P_i, $i = 1, 2$. The times required for production of the items at each P_i might be random and are denoted by $\{\Delta_{i,l}^{a,n}, l < \infty\}$. The part types are manufactured continuously and stored in separate buffers until needed. When a request arrives, if there is enough of each part type to assemble it (that is, at least one of each is available), then it is immediately assembled and sent out. Until further notice, it is assumed that requests that cannot be met are lost. In this section we redefine $y^n(t)$ to be $1/\sqrt{n}$ times the number of requests that are lost by real time nt. The buffers are finite, of sizes $\sqrt{n}B_i$, $B_i < \infty$, $i = 1, 2$. The limit results imply that a buffer of this size is correct for good behavior.

The definitions and conditions of Theorems 5.1.1 and 6.1.1 will be used where convenient, with the obvious adjustments; for example, for $i = 1, 2$, use

$$\begin{aligned} D^n(t) &= \frac{1}{\sqrt{n}}[\text{total number of requests by real time } nt], \\ A_i^n(t) &= \frac{1}{\sqrt{n}}[\text{number of part type } i \text{ produced by real time } nt], \\ x_i^n(t) &= \frac{1}{\sqrt{n}}[\text{content of buffer } i \text{ at real time } nt], \\ U_i^n(t) &= \frac{1}{\sqrt{n}}[\text{number of part type } i \text{ lost due to buffer overflow} \\ &\qquad \text{by real time } nt], \end{aligned}$$

$$w_i^{a,n}(t) = \frac{1}{\sqrt{n}} \sum_{l=1}^{nt} \left[1 - \frac{\Delta_{i,l}^{a,n}}{\bar{\Delta}_i^{a,n}} \right], \quad w^{d,n}(t) = \frac{1}{\sqrt{n}} \sum_{l=1}^{nt} \left[1 - \frac{\Delta_l^{d,n}}{\bar{\Delta}^{d,n}} \right].$$

The production process is uncontrolled. The optimal control (of the production process) is dealt with in [149] using methods of Chapter 10, where it is seen that the optimal policy is to idle P_i if the state of both queues is "outside" of a switching curve. The state equation is

$$x_i^n(t) = x_i^n(0) + A_i^n(t) - D^n(t) + y^n(t) - U_i^n(t), \; i = 1, 2. \tag{1.1}$$

The (scaled by $1/\sqrt{n}$) demand that is satisfied by real time nt is $D^n(t) - y^n(t)$.

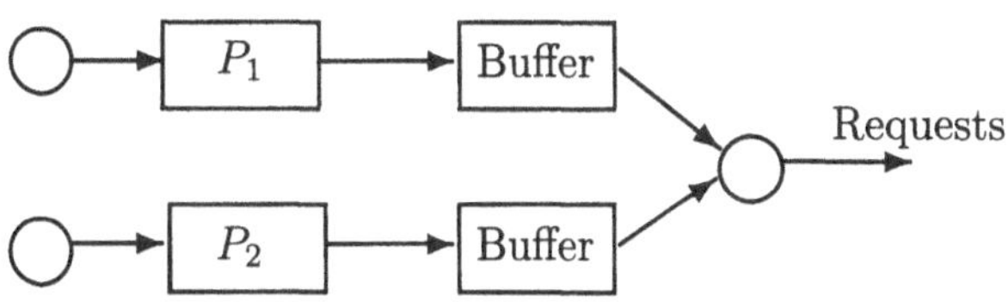

Figure 1.1. A manufacturing system: Two part types.

The heavy traffic condition is the following.

A1.1. *There are real* b_i *such that* $\sqrt{n}\left[\frac{1}{\bar{\Delta}_i^{a,n}} - \frac{1}{\bar{\Delta}^{d,n}} \right] = b_i^n \to b_i$.

The following theorem is an analogue of Theorem 6.1.1.

Theorem 1.1. *Assume* (A1.1), (A6.1.0), *and* (A6.1.1), *adjusted for this problem. Let* $x^n(0)$ *converge weakly to* $x(0)$. *Then*

$$\{x_i^n(\cdot), y^n(\cdot), U_i^n(\cdot), w^{d,n}(\cdot), w_i^{a,n}(\cdot), \; i = 1, 2\}$$

is tight, and any weak-sense limit satisfies

$$x_i(t) = x_i(0) + w_i^a(\bar{\lambda}_i^a t) - w^d(\bar{\lambda}^d t) + b_i t + y(t) - U_i(t),\ i = 1, 2, \quad (1.2)$$

where $y(\cdot)$ and $U_i(\cdot)$ are the reflection terms: $U_i(\cdot)$ can increase only at t such that $x_i(t) = B_i$, and $y(\cdot)$ can increase only at t such that $x_i(t) = 0$ for some i. The $w^d(\cdot)$ and $w_i^a(\cdot)$, $i = 1, 2$, are mutually independent Wiener processes with variances σ_d^2 and $\sigma_{a,i}^2$, $i = 1, 2$, respectively, and are independent of $x(0)$.

Proof. The proof is nearly identical to that of Theorem 6.1.1, with only a slight difference in the treatment of the reflection term $y^n(\cdot)$, since its origin is different. In Section 6.1 it arose as a representation of idle time via the expression (6.1.22). Here, there is no idle time, and $y^n(\cdot)$ represents unsatisfied requests: It appears directly in the formula (1.1). It can increase only when at least one of the buffers is empty at the time of arrival of a request.

The processes $S^{d,n}(\cdot)$ and $S_i^{a,n}(\cdot)$ are all treated exactly as $S^{a,n}(\cdot)$ was treated in Theorem 5.1.1, since the reflection process does not influence their values. They converge weakly to processes with the values $\bar{\lambda}^d t$ and $\bar{\lambda}_i^a t$, respectively. Then the procedure leading to (6.1.25), when applied to (1.1), yields

$$\begin{aligned} x_i^n(t) &= x_i^n(0) + w_i^{a,n}(S_i^{a,n}(t)) - w^{d,n}(S^{d,n}(t)) + \sqrt{n}\left[\frac{1}{\bar{\Delta}_i^{a,n}} - \frac{1}{\bar{\Delta}^{d,n}}\right] t \\ &\quad + y^n(t) - U_i^n(t) + \epsilon_i^n(t), \end{aligned} \tag{1.3}$$

where $\epsilon_i^n(t)$ is a residual-time error term, and it converges weakly to the "zero" process. The upper reflection term $U_i^n(\cdot)$ can increase only at t where $x_i^n(t) = B_i$. The processes $w_i^{a,n}(S_i^{a,n}(\cdot))$ and $w^{d,n}(S^{d,n}(\cdot))$ are asymptotically continuous, as in Theorem 5.1.1. Now, either Theorem 3.6.1 or 3.6.2 implies that $\{y^n(\cdot), U^n(\cdot)\}$ is tight and that the weak-sense limit of any weakly convergent subsequence is continuous with probability one. From this point on, the proof is identical to that of Theorem 6.1.1. The form of the limits of the reflection terms $y^n(\cdot)$ is obvious in this case, since $y^n(\cdot)$ can increase only at t where some $x_i^n(t) = 0$. ■

7.1.2 Multiple Types of Final Products I

Up to now, there was a single final product, which required one of each of the part types. If there are many final products, each requiring a different mixture of the part types, the main change in the development concerns the reflection direction when some $x_i(t) = 0$. In the last subsection the reflection direction was that in Figure 3.6.2, and was the same no matter which inventory was zero. This changes when there is more than one final product. We now consider the case where each final product requires either zero or one unit of each part type. As before, we work with two part types for notational simplicity only. In the next subsection the number of units of a part type that are required for the assembly of a final product is allowed to be arbitrary. Then additional issues arise in the characterization of the reflection directions. Other than the modifications introduced below, for simplicity we retain the conditions used in Theorem 1.1.

In the present example, we suppose that there are two final products. Let $D_j^n(\cdot)$, $j=1,2$, denote $1/\sqrt{n}$ times the demand for final product type j by real time nt, and let the interdemand intervals be $\Delta_{i,l}^{d,n}$. Let $y_j^n(\cdot)$ represent the (scaled by $1/\sqrt{n}$) process of unsatisfied requests for final product type j. See Figure 1.2 for an illustration of a case where the first final product requires no parts of part type 1 and one of part type 2, and the second final product requires one of each of the part types.

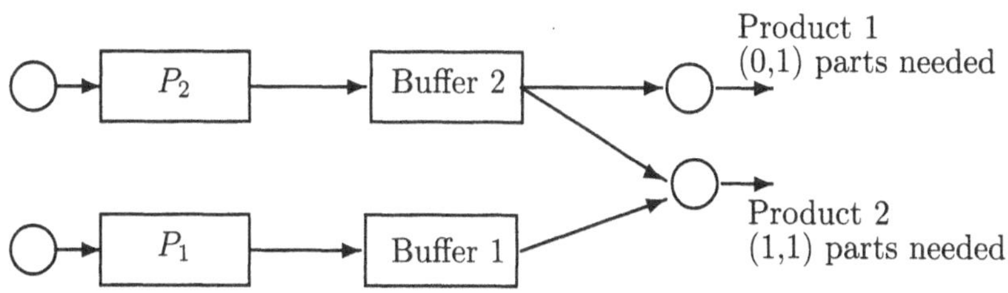

Figure 1.2. A system with two part types and two final products.

In general, suppose that final product type j requires r_{ji} of part type i, and set $r_j = (r_{j1}, r_{j2})$. *In this section the r_{ji} equal zero or one.* Define the "reflection" vector $z^n(\cdot) = (z_i^n(\cdot), i=1,2)$ by

$$z^n(t) = \sum_j r_j y_j^n(t) - U^n(t). \tag{1.4}$$

Equation (1.1) is replaced by $(i=1,2)$

$$x_i^n(t) = x_i^n(0) + A_i^n(t) - \sum_j r_{ji} D_j^n(t) + z_i^n(t). \tag{1.5}$$

We will use the following conditions:

A1.2. *For each n, the interrequest times $\Delta_{i,l}^{d,n}$ are independent of the production intervals. They are mutually independent and exponentially distributed with mean intervals (for final product i) $E\Delta_{i,l}^{d,n} = \bar{\Delta}_i^{d,n} = 1/\bar{\lambda}_i^{d,n}$. Also, there are $\bar{\Delta}_i^d$ such that $\bar{\Delta}_i^{d,n} \to \bar{\Delta}_i^d = 1/\bar{\lambda}_i^d$.*

The reasons for using the exponential distribution are discussed below. Owing to the exponential distribution, $E\left[1 - \Delta_{i,l}^{d,n}/\bar{\Delta}_i^{d,n}\right]^2 = 1$. The heavy traffic condition is the following.

A1.3. *There are real b_i such that* $\sqrt{n}\left[\frac{1}{\bar{\Delta}_i^{a,n}} - \sum_j \frac{r_{ji}}{\bar{\Delta}_j^{d,n}}\right] = b_i^n \to b_i.$

Comment on the special case of Figure 1.2. Let us examine the model of Figure 1.2 in more detail. The possible reflections are illustrated in Figure 1.3. Suppose that buffer 2 is empty at time t (that is, $x_2^n(t) = 0$), but $x_1^n(t) > 0$, and a request for final product type 1 arrives. In particular, let the state be point (a) in Figure 1.3. Final products of type 1 require one part of type 2 and no parts of type 1. Satisfying this demand would put the state at point (b), which is infeasible; hence there will be a "vertical" reflection from (b) to (a), as indicated by the line D_1 in the figure. Suppose that a demand for final product type 2 arrives instead. Since this final product requires one of each part type, the reflection will be from (c) to (a), as indicated by the line D_2 in the figure. The probability of a simultaneous request for both final products is zero.

In general, the actual reflection direction for the limit process at a given point on the boundary $\{x : x_2 = 0\}$ is a "mean value" of the two directions D_1 and D_2 drawn in the figure, each being weighted according to the "frequency" of that request type at that boundary point. For general request processes, this weighing appears to be quite difficult to compute. But it is trivial to compute when the interrequest times are exponentially distributed. Due to the "no memory" property of the exponential distribution, the mean direction is just that given by the relative rates of the two request types. It turns out (see Theorem 1.2 and the remark after the theorem) that the mean direction is what appears as the direction of reflection in the limit.

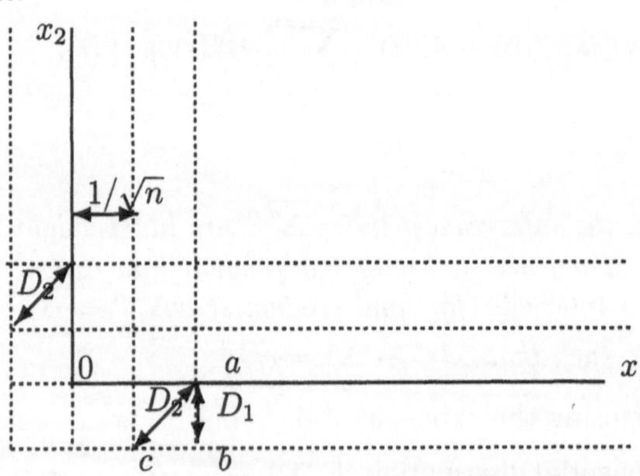

Figure 1.3. The reflection process for the case of Figure 1.2.

Theorem 1.2. *Assume* (A1.2), (A1.3), (A6.1.0), *and* (A6.1.1) *(for the* $w_i^{a,n}(\cdot)$ *processes). Let* $x^n(0)$ *converge weakly to* $x(0)$. *Then*

$$\left\{x_i^n(\cdot), z_i^n(\cdot), U_i^n(\cdot), w_i^{d,n}(\cdot), w_i^{a,n}(\cdot)\right\}, \; i = 1, 2,$$

is tight and the weak-sense limits satisfy, for $i = 1, 2$,

$$\begin{aligned} x_i(t) &= x_i(0) + w_i^a(\bar{\lambda}_i^a t) - \textstyle\sum_j r_{ji} w_j^d(\bar{\lambda}_j^d t) + b_i t + z_i(t), \\ z(t) &= \textstyle\sum_j r_j y_j(t) - U(t), \end{aligned} \tag{1.6}$$

where $z(\cdot)$ *is the reflection term, and* $w_i^\alpha(\cdot)$, $\alpha = a, d$, $i = 1, 2$, *are mutually independent Wiener processes,* $w_i^d(\cdot)$ *with unit variance and* $w_i^a(\cdot)$ *with variance* $\sigma_{a,i}^2$. *The process* $y_1(\cdot)$ *(respectively,* $y_2(\cdot)$*) can increase only at* t *where* $x_2(t) = 0$ *(respectively,where* $x_1(t)$ *or* $x_2(t)$ *are zero). On the boundary where* $x_i = 0$, *the reflection direction is defined by the vector*

$$\sum_{j: r_{ji} > 0} \bar{\lambda}_j^d r_j. \tag{1.7}$$

Proof. By using (1.5) and following the usual expansion procedure leading to (6.1.25), modulo a residual-time error term we get

$$\begin{aligned} x_i^n(t) = x_i^n(0) &+ w_i^{a,n}(S_i^{a,n}(t)) - \sum_j r_{ji} w_j^{d,n}(S_j^{d,n}(t)) \\ &+ \sqrt{n}\left[\frac{1}{\bar{\Delta}_i^{a,n}} - \sum_j \frac{r_{ji}}{\bar{\Delta}_j^{d,n}}\right] t + z_i^n(t). \end{aligned} \tag{1.8}$$

This can be written in the form $x^n(t) = x^n(0) + \tilde{h}^n(t) + z^n(t)$, where by Theorem 6.1.1, $\tilde{h}^n(\cdot)$ is asymptotically continuous in the sense of (3.6.4). We need to find the asymptotic properties of $z_i^n(\cdot)$.

Let $I_i^n(t)$ be the indicator function of the event that buffer i is empty at real time t and let $T_i^{d,n}(t)$ denote $1/\sqrt{n}$ times the total time that queue i is empty by real time nt. Define $\bar{D}_j^n(t) = \sqrt{n} D_j^n(t)$, the unscaled number of requests for final product type j by real time t. We now write $z^n(\cdot)$ in a more convenient way. By the assumption concerning the exponential distribution (A1.2), the probability is zero that there will be a simultaneous arrival of a request and the completion of the production of a part type. The contribution to the reflection vector $z^n(t)$ due to the times that $x_i^n(s) = 0$, $s \le t$, can be written as

$$\sum_{j: r_{ji} > 0} r_j \frac{1}{\sqrt{n}} \int_0^{nt} I_i^n(s-) d\bar{D}_j^n(s). \tag{1.9}$$

The compensator of the process (1.9) is (see Subsection 2.3.2 for the definition) the ith summand in the definition

$$\bar{z}^n(t) = \sum_i \sum_{j:r_{ji}>0} r_j \frac{1}{\sqrt{n}} \int_0^{nt} I_i^n(s)\bar{\lambda}_j^{d,n} ds = \sum_i \left[\sum_{j:r_{ji}>0} r_j \bar{\lambda}_j^{d,n} T_i^{d,n}(t) \right]. \tag{1.10}$$

Using the decomposition of the integral of a jump process of Section 2.3.2, we can write the sum of (1.9) over i as

$$\bar{z}^n(t) + \sum_i \sum_{j:r_{ji}>0} r_j M_{ij}^{d,n}(t), \tag{1.11}$$

where $M_{ij}^{d,n}(\cdot)$ is a martingale with Doob–Meyer process (see Section 2.4)

$$\langle M_{ij}^{d,n} \rangle(t) = \frac{1}{n} \int_0^{nt} I_i^n(s)\bar{\lambda}_j^{d,n} ds. \tag{1.12}$$

The form of (1.12) and Theorem 2.8.3 imply the tightness of $\{M_{ij}^{d,n}(\cdot)\}$. Since the discontinuities of the $M_{ij}^{d,n}(\cdot)$ are $1/\sqrt{n}$, any weak-sense limit has continuous paths with probability one. Then by the above decomposition and the tightness and asymptotic continuity of the $M_{ij}^{d,n}(\cdot)$, we can write

$$x^n(t) = x^n(0) + h^n(t) + \bar{z}^n(t) - U^n(t), \tag{1.13}$$

were $h^n(\cdot)$ is asymptotically continuous. The ith summand in the definition (1.10) of $\bar{z}^n(\cdot)$ can grow only at t where $x_i^n(t) = 0$. The time that all $x_i^n(t)$ equal zero simultaneously are counted twice in (1.13), but this is not important for the characterization of the tightness and the reflection terms $\bar{z}^n(\cdot)$.

Now, the form (1.13) and Theorem 3.6.1 imply the asymptotic continuity of $\bar{z}^n(\cdot)$ and $U^n(\cdot)$, hence of $T_i^{d,n}(\cdot)$, $i = 1, 2$. This, in turn, implies that the processes $M_{ij}^{d,n}(\cdot)$ converge weakly to the "zero" process as $n \to \infty$. To see this last assertion, note that (2.1.2) implies that, for any $\epsilon > 0$,

$$P\left\{ \sup_{s \le t} \left| M_{ij}^{d,n}(s) \right| \ge \epsilon \right\} \le \frac{E[\text{r.h.s. of (1.12)}]}{\epsilon^2}.$$

By the asymptotic continuity of the $T_i^{d,n}(\cdot)$, $i = 1, 2$, we can suppose (using a stopping-time argument if necessary) that the right side goes to zero as $n \to \infty$. Thus, the difference $z^n(\cdot) - (\bar{z}^n(\cdot) - U^n(\cdot))$ converges weakly to the "zero" process. All the other details are as in Theorem 6.1.1. ■

An intuitive explanation of the form (1.7). Since $U_i^n(\cdot)$ plays no role in this discussion, drop it. The r_j are vectors: r_j is the vector (# part type

1, # part type 2) needed for final product j. The reflection vector is the sum of the "corrections" when a buffer is empty and there is a demand for a product that needs the content of that buffer. Thus, $z^n(t)$ is the sum of two components, the corrections when buffer 1 is empty plus the corrections when buffer 2 is empty. The expression (1.7) is the (average rate of) corrections when buffer i is empty. For that term, we need only work with the final products that require part type i. That explains the $\{j : r_{ji} > 0\}$ in the lower index of summation. Any component of the vector $z^n(t)$ might increase on any boundary, depending on the requirements of the requests that require that part type.

7.1.3 *Multiple Types of Final Products: II*

We continue the development of the last subsection, but now allow the number of units r_{ji} of part type i that is needed by final product j to be an arbitrary integer. Under the conditions used in Theorem 1.2, there is no new problem in proving tightness or asymptotic continuity of the reflection terms or the characterization of the weak-sense limit Wiener processes. But there are new and not yet resolved difficulties in determining the reflection *directions.*

Consider the problem where final product 1 requires $(0, 1)$ units of the two part types, and final product type 2 requires $(1, 2)$ units of the two part types. Refer to Figure 1.4 for an illustration of the new issue. Suppose that we are at point (a) in the figure, and a request for final product type 2 arrives. At point (a), there are two items of part type 1 and one of part type 2 in inventory. Satisfying this request would push the state to point (b), which is infeasible, since $x_2 < 0$ there. Thus there is a reflection from (b) back to (a), as indicated in the figure, but this reflection occurs at a point (namely, (a)) that is *not on the boundary.* Now suppose that we are at point (c) on the lower boundary, and a request for final product type 2 arrives. Satisfying this request would push the state to point (e), which is infeasible, and hence there is a reflection back to (c). If a request for final product type 1 arrives while at point (c), the state would move to point (d), and then be reflected back to (c). In the limit, the direction of reflection at the lower boundary depends on the relative frequencies of requests of final product type 2 arriving when $x_2^n(t) = 1/\sqrt{n}$, and of the requests of both types 1 and 2 arriving when $x_2^n(t) = 0$.

Suppose that the times between the requests are exponentially distributed as in (A1.2). We can then compute the relative rates of arrival of the two types of requests when $x_2^n(t) = 0$, and the rate of arrival of type 2 requests when $x_2^n(t) = 1/\sqrt{n}$. But we cannot readily compute the amount of time that $x_2^n(t) = 0$ relative to the amount of time that $x_2^n(t) = 1/\sqrt{n}$. Because of this, it is hard to compute the mean direction of reflection.

The basic problem is that only one request type can cause a reflection when the state is (a), but both request types can cause a reflection when the state is (c). Such phenomena occur in examples where a set of resources is shared by different classes of users, and each class requires a different amount of each resource. The problem is not one of proving tightness, but of characterizing the reflection direction of any weak-sense limit process.

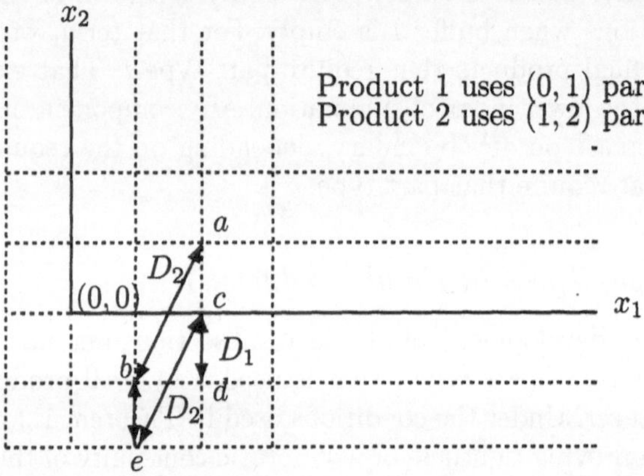

Figure 1.4. Reflections when some final product needs two of part type 2.

One way of resolving the problem of determining the reflection direction for the limit process is to modify the problem and allow a partial backlogging, as follows. If a request arrives and *at least one of each part type that is needed is available*, then let the request be backlogged. The *available* parts that are needed will be held in reserve (subtracted from the buffer), and the unavailable parts backlogged, so that some states might become slightly negative. If there is no availability at all of any part type that is needed for that request, then reject that request. With this approach, the request for final product type 2 arriving at interior point (a) will take the state to point (b). There will be no reflection. There can be a reflection only when a request arrives and there are no units available of any part type that is needed for that request. With this structure, at any point at which the arrival of one type of request causes a reflection due to the unavailability of some part type, so will an arrival of any other type of request that uses that part type. Then the problem can be treated exactly as was done in Theorem 1.2, with the same results, provided that the correct values are used for the r_j vectors. One can backlog more requests with the same result, provided that any point that yields a reflection for a request for any one final product type, owing to an inadequate supply of some part, will yield it for all other request types that need that part as well.

Although a possibly useful policy, the suggested "backlog" modification is arbitrary. But it illustrates the sensitivity of the results to even reasonable variations of the policy when a demand cannot be filled immediately. If infinite backlogging is allowed, then, of course, there is no reflection term.

7.2 Shared Buffers

We next illustrate a different way that the reflection term can originate as well as a nonhyperrectangular state space. The situation is somewhat similar to the manufacturing system problem of Subsection 1.3, where more than one of some part type was needed for the assembly of some final product. Suppose that there are two input streams, A_i^n, $i = 1, 2$, and each A_i^n is to be served by its own processor P_i. However, the two processors share a buffer of size $\sqrt{n}B$ where $B > 0$. Buffer overflows are rejected, no matter what the source. Figure 2.1 depicts the state space.

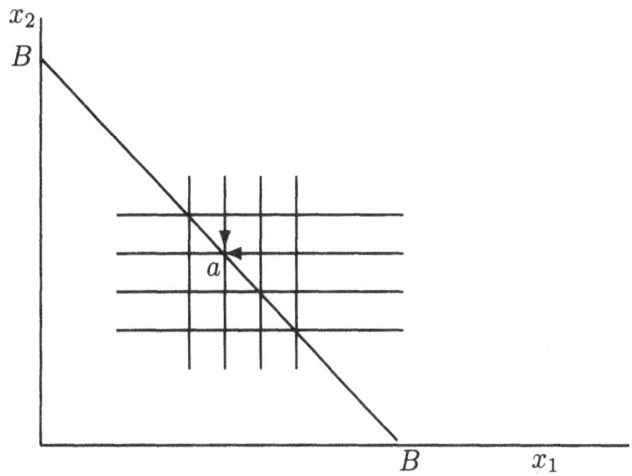

Figure 2.1. Shared buffer.

The main problem is the determination of the reflection direction on the diagonal part of the state space. The state equations are

$$x_i^n(t) = x_i^n(0) + A_i^n(t) - D_i^n(t) - U_i^n(t), \tag{2.1}$$

where $U_i^n(t)$ denotes $(1/\sqrt{n})\times$[number of customers lost from A_i^n due to a full buffer on arrival]. For general A_i^n-processes, the characterization of the reflection direction on the diagonal boundary for the weak-sense limit process is still an open problem. It is difficult, in general, to compute the relative number of rejections of the two input streams, on any interval on the diagonal boundary. However, the problem is simple when the A_i^n-streams are Poisson and independent. Let $y_i^n(t)$ be defined as in Theorem 6.1.1, namely $[1/\sqrt{n}\bar{\Delta}_i^{d,n}]$ times [the idle time of P_i by real time nt].

Theorem 2.1. *For the shared-buffer problem, assume* (A6.1.0) *and* (A6.1.1) (*except for the convergence of the* $w_i^{a,n}(\cdot)$). *Suppose that the* (*unscaled*) *arrival processes are mutually independent, independent of the other driving variables, and Poisson with rates* $\bar{\lambda}_i^{a,n} = 1/\bar{\Delta}_i^{a,n} \to \bar{\lambda}_i^a$. *Then* $w_i^{a,n}(\cdot)$ *converges weakly to* $w_i^a(\cdot)$, *with unit variance and the limit Wiener processes* *are mutually independent. Let $x^n(0)$ converge weakly to $x(0)$ and assume that there are b_i, $i = 1, 2$, such that*

$$\sqrt{n}\left[\frac{1}{\bar{\Delta}_i^{a,n}} - \frac{1}{\bar{\Delta}_i^{d,n}}\right] = b_i^n \to b_i.$$

Then $\{x^n(\cdot), U^n(\cdot), y^n(\cdot), w_i^{a,n}(\cdot), w_i^{d,n}(\cdot), i = 1, 2\}$ is tight, the weak-sense limit of any weakly convergent subsequence is continuous and satisfies

$$x(t) = x(0) + bt + w(t) + y(t) - U(t), \tag{2.2}$$

where $w_i(t) = w_i^a(\bar{\lambda}_i^a t) - w_i^d(\bar{\lambda}_i^a t)$, and $y(\cdot)$ is the reflection term for the lower boundaries in that it is nonnegative, nondecreasing, with $y(0) = 0$, and $y_i(\cdot)$ can increase only at t such that $x_i(t) = 0$. Also, $U(0) = 0$, $U(\cdot)$ is nonnegative, continuous, and nondecreasing, and its components can increase only at t where $x(t)$ is on the diagonal. The process $U(\cdot)$ has the representation

$$U_i(t) = \bar{\lambda}_i^a v(t), \tag{2.3}$$

where the real-valued process $v(\cdot)$ is nonnegative, nondecreasing with $v(0) = 0$, and it can increase only at t where $x(t)$ is on the diagonal. The solution to (2.2) is unique in the strong-sense, and the original sequence converges weakly.

Proof. The only difference between the situation here and that in Theorem 5.1.1 or Theorem 6.1.1 is the behavior when the buffer is full. Following the development and notation of Theorem 5.1.1, (2.1) can be written as

$$x_i^n(t) = x_i^n(0) + b_i^n t + w_i^{a,n}(S_i^{a,n}(t)) - w_i^{d,n}(S_i^{d,n}(t)) + z_i^n(t) + \epsilon_i^n(t), \tag{2.4}$$

where $z_i^n(t) = y_i^n(t) - U_i^n(t)$ and $\epsilon_i^n(\cdot)$ is a residual-time error term that converges weakly to the "zero" process. The differences $y_i^n(\cdot) - U_i^n(\cdot)$ can readily be shown to be asymptotically continuous.

Following the logic used in Theorem 1.2, define $\bar{A}_i^n(\cdot) = \sqrt{n}A_i^n(\cdot)$, the unscaled number of arrivals to P_i by real time nt, and let $I^n(t)$ denote the indicator function of the event that the buffer is full at real time nt. Then as in Theorem 2.1 we can write

$$U_i^n(t) = \frac{1}{\sqrt{n}}\int_0^{nt} I^n(s-)d\bar{A}_i^n(s). \tag{2.5}$$

Define

$$v^n(t) = \frac{1}{\sqrt{n}}\int_0^{nt} I^n(s)ds. \tag{2.6}$$

The decomposition used in Theorem 2.1 yields

$$U_i^n(t) = \bar{\lambda}_i^{a,n} v^n(t) + M_i^n(t), \tag{2.7}$$

where $M_i^n(\cdot)$ is a martingale with discontinuities of order $O(1/\sqrt{n})$, and Doob–Meyer process $\bar{\lambda}_i^{a,n} v^n(t)/\sqrt{n}$. As in Theorem 2.1, we can conclude that $M_i^n(\cdot)$ converges weakly to the "zero" process; hence so does the difference $\bar{\lambda}_i^{a,n} v^n(\cdot) - U_i^n(\cdot)$.

We can conclude that $(x^n(\cdot), y^n(\cdot), U^n(\cdot))$ has asymptotically continuous paths. Also, if a weak-sense limit of some weakly convergent subsequence of $(x^n(\cdot), y^n(\cdot), U^n(\cdot))$ is denoted by $(x(\cdot), y(\cdot), U(\cdot))$, then $y(\cdot)$ must be the asserted reflection term on the lower boundaries, and $U(\cdot)$ can increase only when the state is on the diagonal. The rest of the details are left to the reader. ■

Different buffer requirements. Let us complicate the situation a little, by changing the buffer requirements of the two input streams. Suppose that each arrival from input stream 1 requires two units of buffer space, and those of input stream 2 just one unit. Continue to suppose that the input streams are Poisson and consider the transitions depicted in Figure 2.2.

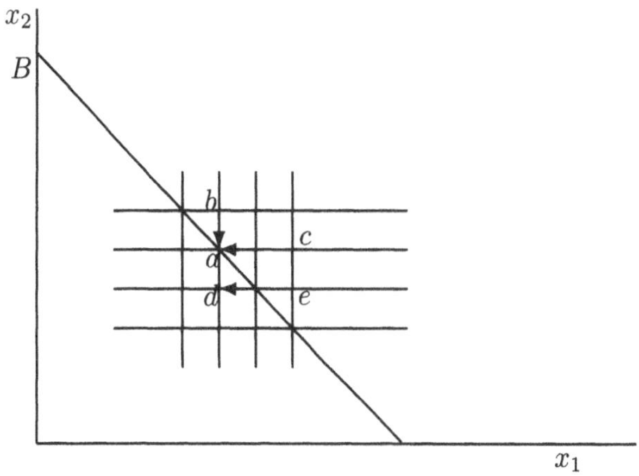

Figure 2.2. Shared buffer, different space requirements.

An arrival from input stream A_1^n occurs at point (d) and wants to take the system to point (e). Since (e) is out of bounds, the arrival is rejected, and the state stays at (d). This corresponds to a reflection from (e) to (d). Similarly, an arrival from A_1^n when the state is at (a) causes a reflection from (c) to (a). An arrival from A_2^n at point (a) similarly causes a reflection from (b) to (a).

The proof of the asymptotic continuity of the reflection terms $U^n(\cdot)$ as done in Theorem 2.1 can be used here as well. But the characterization of the mean reflection direction on the diagonal part of the boundary is still an open problem. Analogously to the problem in Subsection 1.3, the computation involves the relative times during which the rejections can occur for the two input streams, and this entails knowing the relative time that the state spends at points that are located one unit inside the diagonal (that is, at points such as d) and at points on the diagonal. This problem occurs whenever there are different user classes with different buffer requirements and that share a common buffer. For example, the buffer might be a communications channel of bandwidth $\sqrt{n}B$, the different user classes require different bandwidths, and an arrival whose requirements cannot be met immediately is rejected.

The main problem is that the points at which the two input streams can face rejection are not the same, unlike the situation dealt with in Theorem 2.1, but analogous to the model in Subsection 1.3. The difficulty can be eliminated by a slight modification of the model, analogous to what was discussed in Subsection 1.3. Suppose that an arrival from input stream 1 at points such as (d) (where there is some but not enough storage) is held in temporary storage until space becomes available. Then members of the two input streams are rejected at precisely the same points. With this modification, $v^n(t)$ is redefined to be the scaled time spent on or above the diagonal by real time nt. From the point of view of the $U^n(\cdot)$ process, the arrivals and services for input stream A_1^n occur in batches of size 2. (The size of the batches can be any fixed integer, or even a sequence of random variables.) For this example, the weak-sense limit reflection term on the diagonal boundary has components $U_i(t) = \bar{b}_i^a \bar{\lambda}_i^a v(t)$, $i = 1, 2$, where $\bar{b}_i^a$ is the mean batch size for input stream i, and $v(\cdot)$ is the weak-sense limit of $v^n(\cdot)$.

This example shows, once again, that apparently small changes in the model can have serious implications for the reflection direction and the relative rejection rates.

7.3 Process Interruptions and Vacations

This section treats the analogue of the processor interruption problem of Section 5.6 for the network of Section 6.1. The heavy traffic approach to processor interruption was first treated in [38, 123]. For each processor, time is divided into alternating intervals where the processor is operating and where it is not. As noted in Section 5.6, such periods of interruption are called *vacations*. The vacations need not correspond to breakdowns. They might be times at which some priority job is being done. The preempt-resume discipline will be used, where the vacation can interrupt a job in process, and when work on that job resumes at the end of the vacation, only the residual work needs to be done. The types of vacations will be categorized analogously to what was done in Section 5.6. Any combination of the scenarios is possible. Some processors might have one type of interruption and others another type, or each processor might have more than one type. The results for the combinations should be clear from the given results.

7.3.1 *Short Vacations*

Let us first consider the problem of the network version of the problem in Subsections 5.6.1 and 5.6.3, where the order of the durations of the vacations is $O(1)$.

Short and frequent vacations. Let $\Delta_{i,l}^{v,n}$ denote the duration (real time) of the lth vacation at P_i, and $\Delta_{i,l}^{s,n}$ the interval between the end of the $(l-1)$st vacation and the beginning of the lth. Suppose, for specificity, that the processor is operating at time 0. For some centering constants $\bar{\Delta}_i^{s,n}, \bar{\Delta}_i^{v,n}$, define, analogously to (5.6.1) and (5.6.2),

$$w_i^{v,n}(t) = \frac{1}{\sqrt{n}} \sum_{l=1}^{nt} \left(\Delta_{i,l}^{v,n} - \bar{\Delta}_i^{v,n} \right), \tag{3.1}$$

$$w_i^{s,n}(t) = \frac{1}{\sqrt{n}} \sum_{l=1}^{nt} \left[1 - \frac{\Delta_{i,l}^{v,n} + \Delta_{i,l}^{s,n}}{\bar{\Delta}_i^{v,n} + \bar{\Delta}_i^{s,n}} \right]. \tag{3.2}$$

We will require the following assumptions.

A3.1. *The vacation and intervacation intervals* $\{\Delta_{i,l}^{v,n}, \Delta_{i,l}^{s,n}\}$ *are independent of the other driving variables and independent in* i. *Also,* $\bar{\Delta}_i^{\alpha,n} \to \bar{\Delta}_i^{\alpha} > 0$, $\alpha = v, s$, *and, for each* i, $(w_i^{v,n}(\cdot), w_i^{s,n}(\cdot))$ *converges weakly to a Wiener process* $(w_i^v(\cdot), w_i^s(\cdot))$ *with covariance matrix* $\Sigma_{v,i}$.

A3.2. The heavy traffic condition. *There are real* b_i *such that*

$$\sqrt{n} \left[\frac{1}{\bar{\Delta}_i^{a,n}} + \sum_j \frac{q_{ji}^n}{\bar{\Delta}_j^{d,n}} \frac{\bar{\Delta}_j^{s,n}}{\bar{\Delta}_j^{v,n} + \bar{\Delta}_j^{s,n}} - \frac{1}{\bar{\Delta}_i^{d,n}} \frac{\bar{\Delta}_i^{s,n}}{\bar{\Delta}_i^{v,n} + \bar{\Delta}_i^{s,n}} \right] = b_i^n \to b_i, \tag{3.3}$$

where the routing probability q_{ij}^n *is defined in* (A6.1.2).

Remark. Condition (A3.1) asserts the mutual independence of the vacation and intervacation intervals between the different P_i. With some extra complication in the notation, we can allow the intervals to be correlated in i. For example, the interruptions could start at several processors at the same times, but be of different lengths. For more general correlations, it is just a question of dealing with the correlations among the pairs of the first and second terms of (3.8) for the different i, and we will not deal with it further. In (3.3) the service rates are multiplied by the fraction of time that the server is not on vacation.

Define

$$\bar{\lambda}_i^{vac} = \frac{1}{\bar{\Delta}_i^v + \bar{\Delta}_i^s}, \quad \bar{\lambda}_i^{dv} = \frac{\bar{\lambda}_i^d \bar{\Delta}_i^s}{\bar{\Delta}_i^v + \bar{\Delta}_i^s}. \tag{3.4}$$

The first term is the asymptotic mean rate of vacations, and the second is the asymptotic effective mean rate of service.

Theorem 3.1. *Assume the model of Theorem* 6.1.1, *but with vacations. Let* $x^n(0)$ *converge weakly to* $x(0)$. *Assume* (A6.1.0)–(A6.1.2), (A3.1) *and* (A3.2). *Then the conclusions of Theorem* 6.1.1 *hold, but with the vector* b *defined by* (3.3), *and the* $w(\cdot)$ *in* (6.1.18) *defined by*

$$w(t) = \bar{w}^a(t) - [I - Q']\,\bar{w}^d(t) + \bar{w}^r(t) + [I - Q']\,\bar{w}^v(t), \tag{3.5}$$

where the i*th component of* $\bar{w}^v(t)$ *is*

$$\bar{\lambda}_i^d \left[w_i^v(\bar{\lambda}_i^{vac} t) + \bar{\Delta}_i^v w_i^s(\bar{\lambda}_i^{vac} t) \right],$$

and the i*th component of* $\bar{w}^d(t)$ *is now* $w_i^d(\bar{\lambda}_i^{dv} t)$. *Let* $\mathcal{F}_t$ *denote the filtration engendered by*

$$\left(x(u), y(u), U(u), w_i^a(\bar{\lambda}_i^a u), w_i^d(\bar{\lambda}_i^{dv} u), w_i^v(\bar{\lambda}_i^{vac} u), w_i^s(\bar{\lambda}_i^{vac} u);\ i \le K,\ u \le t \right). \tag{3.6}$$

Then the w*-processes in* (3.6) *are all* $\mathcal{F}_t$*-Wiener processes. They are independent in* i. *For each* i, *the* $w_i^\alpha(\cdot)$, $\alpha = a, d, r$, *are as in Theorem* 6.1.1 *and are independent of* $w_i^\alpha(\cdot)$, $\alpha = s, v$, *that are defined in* (A3.1).

Proof. The proof is essentially that of Theorem 6.1.1 with modifications of the type in Theorem 5.6.1 used to account for the vacations. Equations (6.1.19)–(6.1.21) still hold. Let $S_i^{s,n}(t)$ denote $1/n$ times the number of completed vacations by real time nt, and redefine $T_i^{d,n}(t)$ to be $1/\sqrt{n}$ times the idle time at P_i by real time nt that is not during a vacation interval. For the present problem, the right-hand term of (6.1.20) can be written as

$$\frac{1}{\bar{\Delta}_i^{d,n}} \left[\sqrt{n}t - T_i^{d,n}(t) - \frac{1}{\sqrt{n}} \tilde{\Delta}_i^{d,n}(t) - \frac{1}{\sqrt{n}} \sum_{l=1}^{nS_i^{s,n}(t)} \Delta_{i,l}^{v,n} - \frac{1}{\sqrt{n}} \tilde{\Delta}_i^{v,n}(t) \right], \tag{3.7}$$

where $\tilde{\Delta}_i^{v,n}(t)$ is the time that a vacation (if any) that is active at P_i at real time nt has been in progress.

Rewrite the fourth term inside the brackets in (3.7) as

$$-\frac{1}{\sqrt{n}} \sum_{l=1}^{nS_i^{s,n}(t)} \left(\Delta_{i,l}^{v,n} - \bar{\Delta}_i^{v,n} \right) - \frac{1}{\sqrt{n}} \bar{\Delta}_i^{v,n} \sum_{l=1}^{nS_i^{s,n}(t)} \left[1 - \frac{\Delta_{i,l}^{v,n} + \Delta_{i,l}^{s,n}}{\bar{\Delta}_i^{v,n} + \bar{\Delta}_i^{s,n}} \right]$$
$$- \frac{1}{\sqrt{n}} \frac{\bar{\Delta}_i^{v,n}}{\bar{\Delta}_i^{v,n} + \bar{\Delta}_i^{s,n}} \sum_{l=1}^{nS_i^{s,n}(t)} \left(\Delta_{i,l}^{v,n} + \Delta_{i,l}^{s,n} \right). \tag{3.8}$$

The last term on the right of (3.8) is, modulo a residual-time error term,

$$-\frac{\sqrt{n}\bar{\Delta}_i^{v,n}t}{\bar{\Delta}_i^{v,n}+\bar{\Delta}_i^{s,n}}.$$

Thus, the "pre-Wiener process" term arising from the contribution of the vacations to (3.7) is

$$-\frac{1}{\bar{\Delta}_i^{d,n}}\left[w_i^{v,n}(S_i^{s,n}(t))+\bar{\Delta}_i^{v,n}w_i^{s,n}(S_i^{s,n}(t))\right]. \tag{3.9}$$

The proof that the process $S_i^{d,n}(\cdot)$ converges weakly to the process with values $\bar{\lambda}_i^{dv}t$, and the rest of the details are completed as in Theorem 6.1.1 in the same way that the proof of Theorem 5.6.1 was completed by appealing to the method of Theorem 5.1.1. ■

Short but infrequent vacations. Now, consider the network analogue of the problem in Subsection 5.6.3, where the order of the rate of vacations is much less than the order of the rate of exogenous arrivals, and the vacations are $O(1)$ long. Under conditions analogous to those used in Subsection 5.6.3, the asymptotic effect is just a change in the value of b.

Suppose that the number of vacations on the real time interval $[0, nt]$ is $O(\sqrt{n})$ in the sense that there are numbers $\bar{\lambda}_i^s$ and $\bar{\Delta}_i^v$ such that

$$\frac{1}{\sqrt{n}}\sum_{l=1}^{\sqrt{n}t}\Delta_{i,l}^{v,n} \tag{3.10}$$

converges weakly to the process with values $\bar{\Delta}_i^v t$ and $\sqrt{n}S_i^{s,n}(\cdot)$ converges weakly to the process with values $\bar{\lambda}_i^s t$. Then under the additional conditions of Theorem 6.1.1, its conclusions hold, but with the b defined in (A6.1.3) replaced by

$$b + [I - Q']\, b^v, \tag{3.11}$$

where the ith component of b^v is

$$b_i^v = \frac{\bar{\Delta}_i^v\bar{\lambda}_i^s}{\bar{\Delta}_i^d}. \tag{3.12}$$

7.3.2 Long But Infrequent Interruptions

Introductory comments and assumptions. The network analogue of the problem in Subsection 5.6.4 is harder owing to the interactions among the processors. The basic problem can be seen from the following example. Suppose that the vacation duration is $O(\sqrt{n})$, and the intervacation interval is $O(n)$ in real time. Let a vacation at P_i start at real time zero, and suppose that there are no other vacations and that $x^n(0) = x(0)$. Suppose that $q_{ij}^n = q_{ij}$, where $q_{ii} = 0$, and define the set $S_i = \{j : q_{ij} > 0\}$. The two *immediate* effects of this vacation at P_i are the growth in $x_i^n(\cdot)$ and the decrease in the $x_j^n(\cdot)$, $j \in S_i$. By the heavy traffic assumption, for large n these $x_j^n(\cdot)$ will eventually be reduced to zero if the vacation is long enough. Define $S_i^1 = \{k : q_{jk} > 0, j \in S_i\}$. The next effect, if the vacation lasts long enough, is the decrease in $x_k^n(\cdot)$, $k \in S_i^1$, and so forth. In the limit and in the scaled time, these changes become jumps and seem to occur "simultaneously" as "impulses," even though they only partially overlap or are sequential in *real* time. Clearly, the precise asymptotic dynamics depend on the connectivity and the "order" of the impulses. This requires a detailed "local" analysis of the behavior during the vacation. The situation is more complex, but follows the same general logic, if vacations of several processors can overlap. For the sake of simplicity of development, we will treat the case where the vacations of different processors do not (asymptotically) overlap. An interesting example of a problem where the vacations are caused by a control shutting down a processor is in [174], which leads to a control problem with "multiple simultaneous impulses" in the limit.

We will use the assumptions of Theorem 6.1.1, with the following analogues of (A5.6.6) and (A5.6.7).

A3.3. *For each n, i, the intervals between the end of a vacation and the start of the next one for P_i are denoted by $n\tau_{i,l}^{s,n}$, $l = 1, \ldots$. The $\tau_{i,l}^{s,n}$ are mutually independent, exponentially distributed, independent of all the other driving random variables and have rate $\bar{\lambda}_i^{s,n}$, where $\bar{\lambda}_i^{s,n}$ converges to $\bar{\lambda}_i^s > 0$ as $n \to \infty$. The intervals for the different P_i are mutually independent. By convention, we define $\tau_{i,0}^{s,n} = 0$.*

A3.4. *For each n, i, there are mutually independent and identically distributed random variables $\tau_{i,l}^{v,n}$, $l = 1, \ldots$, such that the duration of the lth vacation interval for P_i is $\sqrt{n}\tau_{i,l}^{v,n}$. Also, $\tau_{i,l}^{v,n}$ converges weakly to a random variable τ_i^v as $n \to \infty$. For each i, the $\tau_{i,l}^{v,n}$, $l = 1, \ldots$, are independent of all other driving random variables. The intervals for the different P_i are mutually independent. No processor is on vacation at time zero.*

To motivate the general result of Theorem 3.2, consider the case of the problem of Figure 3.1, when there is a vacation of P_1 of (real time) length $\sqrt{n}\tau_{1,1}^{v,n}$ starting at time zero, with no other vacations. The buffers are assumed to be infinite. By the logic used in Theorem 5.6.4, the initial (asymptotic) effect of an interruption at P_1 is an increase in $x_1^n(\cdot)$ at a rate $\sqrt{n}/\bar{\Delta}_1^{a,n} + \sqrt{n}q_{21}^n/\bar{\Delta}_2^{d,n}$, and a decrease in $x_2^n(\cdot)$ at a rate $\sqrt{n}/\bar{\Delta}_2^{d,n} - \sqrt{n}/\bar{\Delta}_2^{a,n}$. This continues until $x_2^n(\cdot)$ first becomes zero, after which the asymptotic rate of increase of $x_1^n(\cdot)$ slows down to $\sqrt{n}/\bar{\Delta}_1^{a,n} + \sqrt{n}q_{21}^n/\bar{\Delta}_2^{a,n}$, and $x_2^n(\cdot)$ remains arbitrarily close to zero. The rate slows down, since the only customers that can be fed back to P_1 from P_2 are those from the exogenous arrival stream A_2^n. This rough discussion is illustrated in Figure 3.2 and will be formalized in the theorem.

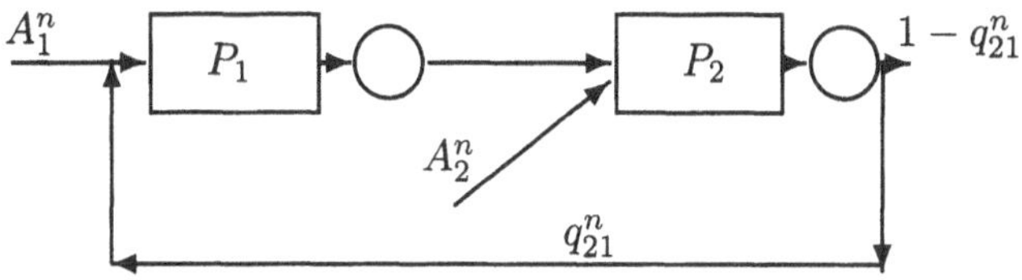

Figure 3.1. A feedback system with a vacation at P_1.

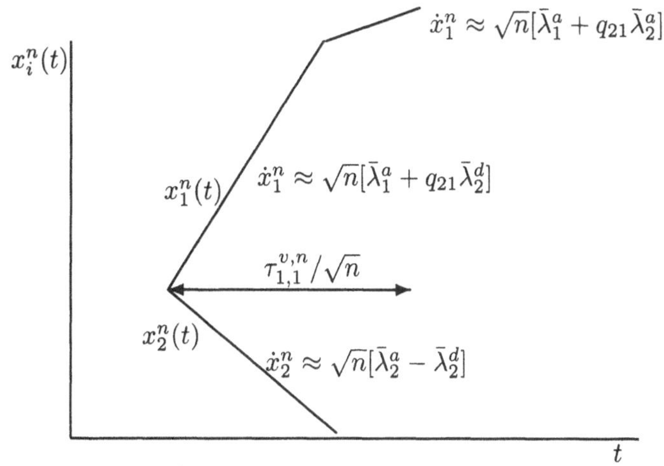

Figure 3.2. The limits of the $x_i^n(\cdot)$-paths during a vacation, in scaled time.

For a more careful development, one follows the method of Theorem 5.6.4, where the vacation intervals are rescaled so that they are $O(1)$ and the analysis is in the *local-fluid time scale*. For $k > 0$, define

$$\nu_{i,k}^n = \sum_{l=1}^{k} \left[\tau_{i,l}^{s,n} + \tau_{i,l-1}^{v,n}/\sqrt{n} \right], \tag{3.13}$$

which is $1/n$ times the real time of the start of the kth vacation at P_i. Let $\bar{\nu}_{i,k}^n$ denote $1/n$ times the end of the kth vacation at P_i in real time. By (A3.3), the vacations at different P_i can be assumed to not overlap for large n. Let ν_k^n (respectively, $\bar{\nu}_k^n$) denote $1/n$ times the real time of the start (respectively, end) of the kth vacation, no matter where. Define $\bar{\nu}_0^n = 0$. For $k \geq 1$, define the kth *section* of $x^n(\cdot)$ as $x^n((\bar{\nu}_{k-1}^n + t) \wedge \nu_k^n)$. For each $p \leq K$, define $\bar{b}_i^p$ by setting $\bar{\lambda}_p^d = 0$ in

$$\bar{b}_i^p = \bar{\lambda}_i^a + \sum_j q_{ji}\bar{\lambda}_j^d - \bar{\lambda}_i^d.$$

Define $\bar{b}^p = (\bar{b}_i^p, i \le K)$. In detail,

$$\bar{b}_i^p = \bar{\lambda}_i^a + \sum_{j \neq p} q_{ji}\bar{\lambda}_j^d - \bar{\lambda}_i^d, \quad i \neq p, \quad \bar{b}_p^p = \bar{\lambda}_p^a + \sum_{j \neq p} q_{jp}\bar{\lambda}_j^d. \tag{3.14}$$

In proving the weak convergence in the next theorem, it is simplest to work with the sections of the process in sequence. Start with the first section and first vacation. The proof of the weak convergence of the first section is given by Theorem 6.1.1, using the fact that the first vacation occurs at a time that is exponentially distributed and tight in n. Thus, the set $\{x^n(\nu_1^n)\}$ is tight. In fact, it converges weakly to a value to be called $x(\nu_1-)$. As seen in the theorem, the difference between the value of the state at the start and at the end of the first vacation converges weakly, and is given by the solution to the Skorohod problem (3.16). In scaled time, this vacation takes place on an interval that collapses to a point as $n \to \infty$. Consequently, the set of initial conditions at the start of the second section is tight, indeed converges weakly. Now, continue the procedure, and construct the "limit" process $x(\cdot)$ from the limits of the intervacation sections and the jumps due to the vacations. This is the procedure of the theorem.

Theorem 3.2. *Assume the system setup of Theorem* 6.1.1, *but with vacations. Assume* (A6.1.0)–(A6.1.3), (A3.3), (A3.4), *and suppose that* $x^n(0)$ *converges weakly to* $x(0)$. *The set*[1]

$$\{\nu_k^n - \bar{\nu}_{k-1}^n, \tau_{i,k}^{v,n}, \tau_{i,k}^{s,n}, \nu_{i,k}^n - \bar{\nu}_{i,k-1}^n, \ i \le K,$$
$$\Psi^n((\bar{\nu}_{k-1}^n + \cdot) \wedge \nu_k^n) - \Psi^n(\bar{\nu}_{k-1}^n); k \ge 1\}$$

converges weakly. Denote the weak-sense limit by

$$\{\nu_k - \bar{\nu}_{k-1}, \tau_{i,k}^v, \tau_{i,k}^s, \nu_{i,k} - \bar{\nu}_{i,k-1}, \ i \le K,$$
$$\Psi((\bar{\nu}_{k-1} + \cdot) \wedge \nu_k) - \Psi(\bar{\nu}_{k-1}); \ k \ge 1\}.$$

Defining $x(\cdot)$ *by putting together the sections and jumps in sequence yields*

$$x(t) = x(0) + w(t) + bt + J(t) + z(t), \tag{3.15}$$

where $w(\cdot)$ *and* $z(\cdot) = [I - Q']y(\cdot) - U(\cdot)$ *are as in Theorem* 6.1.1 and $J(\cdot) = \sum_i J_i(\cdot)$, *where*

$$J_i(t) = \sum_{k:\nu_{i,k} \le t} \delta x(\nu_{i,k}), \quad \nu_{i,k} = \sum_{l \le k} \tau_{i,l}^s.$$

For each i, k, $\delta x(\nu_{i,k})$ *is the solution at time* $\tau_{i,k}^v$ *to the Skorohod problem*

$$\delta x(t) = x(\nu_{i,k}-) + \bar{b}^i t + [I - Q']\,\delta y(t) - \delta U(t), \tag{3.16}$$

[1] See Theorem 6.1.1 for the definition of $\Psi^n(\cdot)$.

where $[I - Q']\delta y(\cdot)$ *and* $\delta U(\cdot)$ *are the reflection terms. The* $\{\tau^s_{i,k}; i, k\}$ *are mutually independent, independent of the* $\{\tau^v_{i,k}; i, k\}$, *and exponentially distributed with rate* $\bar{\lambda}^s_i$ *for* $\tau^s_{i,k}$. *The* $\{\tau^v_{i,k}; i, k\}$ *are mutually independent. The set of these times are independent of the weak-sense limit Wiener processes and of* $x(0)$. *The solution to* (3.15) *is nonanticipative in the sense that for each* t *the increments of the Wiener processes and jump times and values after* t *are independent of these processes and the values of* $x(\cdot), y(\cdot), U(\cdot)$ *before or at* t.

Proof. Asymptotically, the vacations of the different P_i do not overlap. To analyze the change in $x^n(\cdot)$ during the kth vacation of P_i, we can ignore overlap and work with the local-fluid time scale that is implied by the form $x^n(\nu^n_{i,k} + (t \wedge \tau^{v,n}_{i,k})/\sqrt{n})$.

The hypotheses (A6.1.0), (A3.3), and (A3.4) imply the conclusions concerning the times $\nu_{i,k}, \bar{\nu}_{i,k}, \nu_k, \bar{\nu}_k, \tau^s_{i,k}, \tau^v_{i,k}$. The result for the first section $\Psi^n(t \wedge \nu^n_1)$ is obtained as in Theorem 6.1.1. It will be shown in the next two paragraphs that this together with (A3.3) and (A3.4) imply the tightness of the set $\{x^n(\bar{\nu}^n_1)\}$. With this given, use the method of Theorem 6.1.1 and (A3.3), (A3.4) sequentially to get the tightness and asymptotic continuity of the sections $\Psi^n((\bar{\nu}^n_{k-1} + t) \wedge \nu^n_k)$ and the tightness of the set of their initial and terminal values for each $k < \infty$.

Let us compute the increment in $x^n(\cdot)$ during the first vacation. Without loss of generality, we can suppose that the first vacation is that of some given P_p. Thus, $x^n(\tau^{s,n}_{p,1})$ converges weakly to a limit, which we call $x(\tau^s_{p,1}-)$. For any function $f(\cdot)$ on $[0, \infty)$ define $\delta f^n(t) = f(\tau^{s,n}_{p,1} + t/\sqrt{n}) - f(\tau^{s,n}_{p,1})$ and $\bar{x}^n_i(t) = x^n(\tau^{s,n}_{p,1} + t/\sqrt{n})$. Then (6.1.25) yields, for $t \le \tau^{v,n}_{p,1}$, $i \le K$, and where $\bar{\Delta}^{d,n}_p$ is set to infinity,

$$\begin{aligned} \bar{x}^n_i(t) = x^n(\tau^{s,n}_{p,1}) + \left[\frac{1}{\bar{\Delta}^{a,n}_i} + \sum_j \frac{q^n_{ji}}{\bar{\Delta}^{d,n}_j} - \frac{1}{\bar{\Delta}^{d,n}_i} \right] t + \delta w^{a,n}_i(S^{a,n}_i(t)) \\ - \delta w^{d,n}_i(S^{d,n}_i(t)) + \sum_j q^n_{ji} \delta w^{d,n}_j(S^{d,n}_j(t)) + \sum_j \delta w^{r,n}_{ji}(S^{d,n}_j(t)) \\ + \delta z^n_i(t) + \epsilon^n_i(t), \end{aligned} \tag{3.17}$$

where $\epsilon^n_i(\cdot)$ is a residual-time error term and

$$\delta z^n_i(t) = \delta y^n_i(t) - \sum_j q^n_{ji} \delta y^n_j(t) - \delta U^n_i(t).$$

In vector form, this is

$$\delta z^n(t) = [I - Q'_n] \, \delta y^n(t) - \delta U^n(t). \tag{3.18}$$

The processes $\delta y^n_i(\cdot)$ and $\delta U^n_i(\cdot)$ are nonnegative and nondecreasing, with zero initial condition. The $\delta y^n_i(\cdot)$ (respectively, $\delta U^n_i(\cdot)$) can increase only at t where $\bar{x}^n_i(t) = 0$ (respectively, where $\bar{x}^n_i(t) = B_i$).

The weak convergence of $\tau_{p,1}^{s,n}$ and of the first section of $x^n(\cdot)$ together with (A6.1.1)–(A6.1.3) imply that the $\delta w_i^{d,n}(S_i^{d,n}(\cdot))$ and the other δw-processes in (3.17) converge weakly to the "zero" process. Then (3.17) and the proof of Theorem 6.1.1 imply that the sequence $\bar{x}^n(\cdot \wedge \tau_{p,1}^{v,n})$ is tight and the weak-sense limits satisfy, for $i \leq K$,

$$\bar{x}_i(t \wedge \tau_{p,1}^v) = x_i(\tau_{p,1}^s -) + (t \wedge \tau_{p,1}^v)\bar{b}_i^p + \delta z_i(t \wedge \tau_{p,1}^v), \tag{3.19}$$

where

$$\delta z(t) = [I - Q']\, \delta y(t) - \delta U(t). \tag{3.20}$$

The $\delta y_i(\cdot)$ is nondecreasing, continuous, can increase only at t where $\bar{x}_i(t) = 0$, and $\delta y(0) = 0$, and analogously for $\delta U(\cdot)$. We have characterized the first jump as stated in the theorem. Now, proceed section by section to get the general result.

The nonanticipativity assertion follows from the mutual independence in the limit of the initial condition, the Wiener processes, and the interjump intervals and jump durations. ■

Remark on the motivational example and (3.16). For the case of Figure 3.1, where the buffers were assumed to be infinite, $\bar{b}_1^1 = \bar{\lambda}_1^a + q_{21}\bar{\lambda}_2^d$ and $\bar{b}_2^1 = \bar{\lambda}_2^a - \bar{\lambda}_2^d$. Thus, until $\bar{x}_2(t) = 0$ there is no reflection, and $\dot{\bar{x}}_i(t) = \bar{b}_i^1$, $i = 1, 2$. Once $\bar{x}_2(t)$ reaches zero, the reflection kicks in. The value of $\bar{x}_2(t)$ remains at zero, and $\delta y_2(\cdot)$ increases at a rate of $-\bar{b}_2^1$. By the heavy traffic condition (A6.1.3), $\bar{b}_2^1 = -\bar{\lambda}_1^d$ and $\bar{\lambda}_2^a + \bar{\lambda}_1^d = \bar{\lambda}_2^d$. Thus, $\bar{x}_1(\cdot)$ now increases at the rate $\bar{b}_1^1 + q_{21}\bar{b}_2^1 = \bar{\lambda}_1^a + q_{21}\bar{\lambda}_2^a$, as shown in Figure 3.2. Since there is an infinite buffer for P_1, this rate of increase continues until the end of the vacation interval.

7.4 Correlated Service Times

In Theorem 6.1.1 the relationships between the service times of the sequence of processings of the same job in either different or in the same processor were not specified explicitly, but were buried in the general assumption (A6.1.1). If the successive processing times are correlated in some explicit way, then (A6.1.1) needs to be replaced by some condition that describes the form of the correlation. Several such cases will be discussed in this section. The discussion is confined to special cases, since the notation can get complicated. But these cases will illustrate the type of results to be expected. The buffers will be assumed to be infinite, but buffers of size $B_i\sqrt{n}$ can be added with the obvious changes.

7.4.1 A Simple Feedforward System

Let us first consider the tandem system illustrated in Figure 4.1 with A_2^n deleted and $q_{12}^n = 1$. Since the buffer of each processor might have a non zero initial condition, the index of the service time of any individual customer will, in general, be different in the two processors. Let the lth customer served by P_1 be the k_l^nth customer served by P_2. We will use the following assumptions.

A4.1. *The* $w_i^{a,n}(\cdot)$, $i = 1, 2$, *satisfy* (A6.1.1), *and are independent of the service times.*

A4.2. *The* $(\Delta_{1,l}^{d,n}, \Delta_{2,k_l^n}^{d,n})$ *are independent in* l, *and identically (in* l*) distributed, with* $E\Delta_{i,l}^{d,n} = \bar{\Delta}_i^{d,n}$ *and the variance of* $\Delta_{i,l}^{d,n}/\bar{\Delta}_i^{d,n}$ *converges to* $\sigma_{d,i}^2$ *as* $n \to \infty$. *Also,*

$$\frac{E\Delta_{1,l}^{d,n}\Delta_{2,k_l^n}^{d,n}}{\bar{\Delta}_1^{d,n}\bar{\Delta}_2^{d,n}} - 1 = c^n \to c.$$

The set $\left\{|\Delta_{i,l}^{d,n}|^2; n, i, l\right\}$ *is uniformly integrable.*

By (6.1.25), the state equation can be written as

$$\begin{aligned} x_1^n(t) &= x_1^n(0) + b_1^n t + w_1^{a,n}(S_1^{a,n}(t)) - w_1^{d,n}(S_1^{d,n}(t)) + y_1^n(t) + \epsilon_1^n(t), \\ x_2^n(t) &= x_2^n(0) + b_2^n t + w_1^{d,n}(S_1^{d,n}(t)) - w_2^{d,n}(S_2^{d,n}(t)) \\ &\quad + y_2^n(t) - y_1^n(t) + \epsilon_2^n(t), \end{aligned} \tag{4.1}$$

where the $\epsilon_i^n(\cdot)$ are residual-time error terms.

Theorem 4.1. *For the system of Figure* 4.1 *with* A_2^n *deleted, assume* (A6.1.0), (A4.1), (A4.2) *and* (A6.1.3). *Let* $q_{12}^n = 1$ *with the other* q_{ij}^n *being zero. Let* $x^n(0)$ *converge weakly to* $x(0)$. *Then the weak-sense limit of the processes in* (4.1) *satisfies*

$$\begin{aligned} x_1(t) &= x_1(0) + b_1 t + w_1^a(\bar{\lambda}_1^a t) - w_1^d(\bar{\lambda}_1^a t) + y_1(t), \\ x_2(t) &= x_2(0) + b_2 t + w_1^d(\bar{\lambda}_1^a t) - w_2^d(\bar{\lambda}_1^a t) + y_2(t) - y_1(t). \end{aligned} \tag{4.2}$$

Let $\mathcal{F}_t$ *denote the minimal* σ*-algebra that measures* $\{x(0), w_i^\alpha(\bar{\lambda}_i^\alpha s), y_i(s), s \le t;\ i = 1, 2,\ \alpha = a, d\}$. *Then the* $w_i^\alpha(\bar{\lambda}_i^\alpha \cdot)$ *are* $\mathcal{F}_t$*-Wiener processes. The* $(w_i^a(\cdot), 1 = 1, 2)$ *are independent of the other Wiener processes and have variances* $\sigma_{a,i}^2$. *Each* $w_i^d(\cdot)$ *has variance* $\sigma_{d,i}^2$, *and* $Ew_1^d(1)w_2^d(1) = c$.

Proof. By Theorem 2.8.8, $(w_i^{d,n}(\cdot), i = 1, 2)$ converges weakly to a limit $(w_i^d(\cdot), i = 1, 2)$, the components of which are Wiener processes with variances $\sigma_{d,i}^2$, $i = 1, 2$, and each $w_i^d(\cdot)$ is independent of the $w_i^a(\cdot)$, $i = 1, 2$.

Then owing to the continuity of these limit processes, the method of Theorem 6.1.1 can be used almost without change to get the desired result. By the proof of Theorem 6.1.1, $S_1^{a,n}(\cdot), S_1^{d,n}(\cdot)$, and $S_2^{d,n}(\cdot)$ converge weakly to the process with values $\bar{\lambda}_1^a t$. This implies that we can suppose that $k_l^n = l$ for purposes of computing the relations between the $w_i^d(\cdot)$, $i = 1, 2$. Then Theorem 2.8.8 can be used to show that the pair $(w_1^d(\cdot), w_2^d(\cdot))$ is a Wiener process with $Ew_1^d(1)w_2^d(1) = c$. ∎

7.4.2 Random Routing in a Feedforward System

The correlated service time problem becomes more interesting if there is random routing. In order to illustrate the main point without excessive notation, we next consider the case illustrated in Figure 4.1.

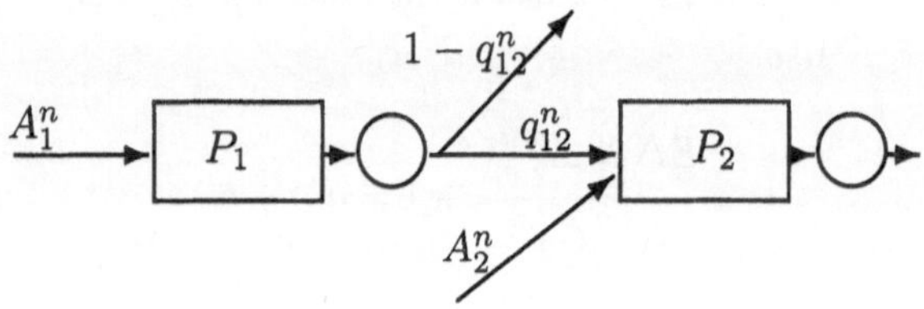

Figure 4.1. A tandem network with random routing.

A4.3. (A4.2) *holds with the following addition. The service intervals for the A_2^n stream are mutually independent, have the same first and second moments as those for the jobs coming from P_1, and are independent of all of the other service intervals.*

Theorem 4.2. *Assume the system of Fig.* 4.1, (A6.1.0), (A6.1.2), (A6.1.3), (A4.1) *and* (A4.3). *Let $x^n(0)$ converge weakly to $x(0)$. Then the conclusions of Theorem* 6.1.1 *hold, except that $w_1^d(\cdot)$ and $w_2^d(\cdot)$ are correlated with*

$$Ew_1^d(1)w_2^d(1) = cq_{12}.$$

Proof. Equation (6.1.25) holds with the "pre–Wiener" processes for $x_1^n(\cdot)$ and $x_2^n(\cdot)$ being, respectively,

$$\begin{gathered} w_1^{a,n}(S_1^{a,n}(t)) - w_1^{d,n}(S_1^{d,n}(t)), \\ w_2^{a,n}(S_2^{a,n}(t)) + q_{12}^n w_1^{d,n}(S_1^{d,n}(t)) + w_{12}^{r,n}(S_1^{d,n}(t)) - w_2^{d,n}(S_2^{d,n}(t)). \end{gathered} \tag{4.3}$$

The procedure is virtually the same as that in Theorem 4.1. The individual w-processes converge weakly to weak-sense limits, each of which is a Wiener process with the correct individual variance. The $S_i^{\alpha,n}(\cdot)$, $\alpha = a, d$, $i = 1, 2$, converge weakly to processes with values $\bar{\lambda}_i^\alpha t$, respectively. The $x^n(\cdot), y^n(\cdot)$

have continuous weak-sense limits $x(\cdot), y(\cdot)$, with $z(\cdot) = [I - Q']y(\cdot)$ being the reflection term. As in Theorem 4.1, the only issue is showing that the pair $(w_1^d(\cdot), w_2^d(\cdot))$ is a Wiener process with the correct covariance. The weak convergence of the $S_i^{\alpha,n}(\cdot)$ allows us to suppose, for the purposes of characterizing the weak-sense limit Wiener process, that the initial condition $x^n(0)$ is zero.

Let $\Delta_{12,l}^{d,n}$ and $\Delta_{a2,l}^{d,n}$ denote the service time at P_2 for the lth output from P_1 if it goes to P_2, and of the lth member of the A_2^n stream, respectively. Let $I_{12,l}^{d,n}$ denote the indicator function of the event that the lth output from P_1 goes to P_2. Define

$$w_{12}^{d,n}(t) = \frac{1}{\sqrt{n}} \sum_{l=1}^{nt} \left[1 - \frac{\Delta_{12,l}^{d,n}}{\bar{\Delta}_2^{d,n}} \right] I_{12,l}^{d,n}, \quad w_{a2}^{d,n}(t) = \frac{1}{\sqrt{n}} \sum_{l=1}^{nt} \left[1 - \frac{\Delta_{a2,l}^{d,n}}{\bar{\Delta}_2^{d,n}} \right],$$

$$\begin{aligned} M_1^{d,n}(t) &= \frac{q_{12}^n}{\sqrt{n}} \sum_{l=1}^{nt} \left[1 - \frac{\Delta_{12,l}^{d,n}}{\bar{\Delta}_2^{d,n}} \right], \\ M_2^{d,n}(t) &= \frac{1}{\sqrt{n}} \sum_{l=1}^{nt} \left[1 - \frac{\Delta_{12,l}^{d,n}}{\bar{\Delta}_2^{d,n}} \right] \left[I_{12,l}^{d,n} - q_{12}^n \right]. \end{aligned}$$

We claim that the $w_2^{d,n}(S_2^{d,n}(t))$ in (4.3) can be replaced by the right hand side of

$$w_2^{d,n}(S_2^{d,n}(t)) \approx w_{12}^{d,n}(S_1^{d,n}(t)) + w_{a2}^{d,n}(S_2^{a,n}(t)), \tag{4.4}$$

in the sense that the weak-sense limit of the difference is the "zero" process. The left side of (4.4) would equal the right side if the $S_2^{d,n}(t)$ were replaced by $1/n$ times the number of inputs to P_2 by real time nt. But the difference between this quantity and $S_2^{d,n}(t)$ is bounded by $x_2^n(t)/\sqrt{n}$, which converges weakly to the zero process. The processes defined by the two terms on the right of (4.4) are mutually independent.

Next, write

$$w_{12}^{d,n}(S_1^{d,n}(t)) = M_1^{d,n}(S_1^{d,n}(t)) + M_2^{d,n}(S_1^{d,n}(t)). \tag{4.5}$$

For the purposes of characterizing the weak-sense-limit Wiener processes, we can set $S_1^{d,n}(t) = \bar{\lambda}_i^a t$, with no loss of generality, since the weak-sense limits of $w_1^{d,n}(S_1^{d,n}(\cdot))$ and of the processes in (4.5) will not change. Then Theorem 2.8.8 implies that $(M_1^{d,n}(\cdot), M_2^{d,n}(\cdot), w_1^{d,n}(\cdot))$ converges weakly to a Wiener process $(M_1^d(\cdot), M_2^d(\cdot), w_1^d(\cdot))$ and that $M_2^d(\cdot)$ is independent of $(M_1^d(\cdot), w_1^d(\cdot))$. Thus, to get the relation between $w_1^d(\cdot)$ and $w_2^d(\cdot)$, we need only get the relations between $w_1^d(\cdot)$ and $M_1^d(\cdot)$. The correlation is $q_{12}c$. Hence $Ew_1^d(1)w_2^d(1) = q_{12}c$. The independence and nonanticipativity assertions follow as in Theorem 6.1.1. ■

Summary of the general procedure for a feedforward network. The proof of tightness and the continuity of the limit processes or the characterization of the reflection directions is not affected by the correlations in the service times. Also, each of the individual "service noise" processes in question converges weakly to a Wiener process. The only question concerns the relations among the Wiener processes associated with the service processes. For the purpose of characterizing the weak-sense Wiener processes and computing their covariances, we can ignore the initial condition, and assume as well that all arrivals have been processed. Then with these approximations, decompose the departure process for the queue in question into the sum of those associated with the exogenous arrivals (which is independent of the rest) and those associated with the arrivals from the other processors. Further decompose the latter process into orthogonal parts such as our $M_i^{d,n}(\cdot)$, $i = 1, 2$, where the second is centered about the routing variables and does not affect the computation of the correlation. Then the correlation can be read off.

7.4.3 A Feedback System with Correlated Service Times

Theorems 4.1 and 4.2 concerned a tandem system where the service times of a job in the various processors were correlated, but independent from job to job. For notational simplicity, we dealt with a special case. A similar procedure can be carried out for systems of the type of Theorem 6.1.1. Since the notation can become unpleasant, we will illustrate the procedure for the case of Figure 4.2. In Theorem 4.2, a job leaving P_1 could either exit the system, or else go to P_2, and then exit the system. With feedback, the same job can circulate several times before exiting. Thus, if we allow the times required for different services for the same job to be correlated, we need to define the correlation for all the possible services for returns to the same processor as well among the two processors. To simplify the notation, we suppose that the times needed for services of the same job at different processors are mutually independent, but the times required for the repeated service of any job at the same processor are dependent. The times required for different jobs are mutually independent, and the means and variances of the service times depend only on the serving processor.

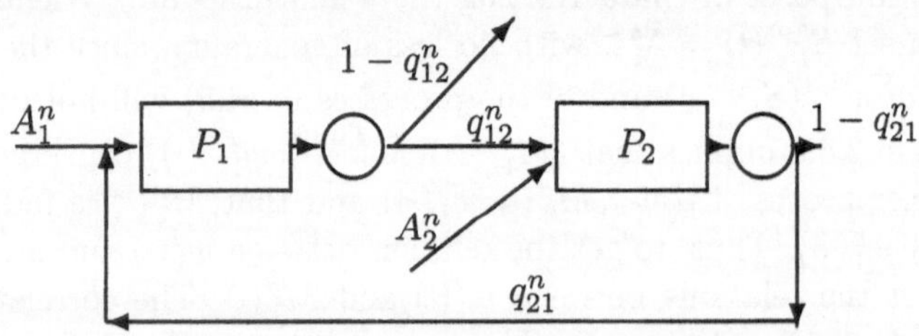

Figure 4.2. A feedback system with correlated service times.

Let $I^n_{k,ji,l}$ denote the indicator function of the event that the lth member of A^n_j is served at P_i at its kth routing. By convention, the zeroth routing is the initial service at P_j. Let $\Delta^{d,n}_{k,ji,l}$ denote the service time of the kth routing of the lth member of A^n_j if it is served at P_i at that time. Let $I^n_{k,ji,l}(t)$ denote the indicator function of the event that the kth routing of the lth member of A^n_j is served at P_i by real time nt. Define

$$\xi^{d,n}_{k,ji,l} = \left[1 - \frac{\Delta^{d,n}_{k,ji,l}}{\bar{\Delta}^{d,n}_i}\right], \quad \xi^{d,n}_{i,l} = \left[1 - \frac{\Delta^{d,n}_{i,l}}{\bar{\Delta}^{d,n}_i}\right].$$

Condition (A4.3) is to be replaced by (A4.4), where the correlation of the service times of a job at each P_i does not depend on whether the job originated from A^n_1 or A^n_2. Such dependence can be allowed, but the variance of the weak-sense limits cannot be represented as simply as for the case of Theorem 4.3. We suppose that $x^n(0) = 0$. To include an arbitrary initial condition simply requires that we specify the service times for these initial jobs, but we wish to avoid the extra notation.

A4.4. *For each n, the sets of potential service times $\{\Delta^{d,n}_{k,ji,l}, k < \infty\}$ are independent in l, j, i, and identically (in j, l) distributed for each i. There are constants $\bar{\Delta}^{d,n}_i$ such that $E\Delta^{d,n}_{k,ji,l} = \bar{\Delta}^{d,n}_i$. There are symmetric functions $c^n_j(\cdot)$ and $c_i(\cdot)$ defined on $\{0, \pm 1, \pm 2, \ldots\}$, with $c_i(0)$ also written as $\sigma^2_{d,i}$, such that*

$$E\xi^{d,n}_{k,ji,l}\xi^{d,n}_{v,ji,l} = c^n_i(v-k),$$

where $c^n_i(v) \to c(v)$ for each v, and

$$\lim_m \sup_n \sum_{v \geq m} |c^n_i(v)| = 0, \; i = 1, 2.$$

Also,

$$\left\{ \left|\Delta^{d,n}_{k,ji,l}\right|^2 ; n, j, i, k, l \right\} \text{ is uniformly integrable.}$$

Theorem 4.3. *Assume* (A6.1.0), (A6.1.2), (A6.1.3), (A4.1), *and* (A4.4) for the system of Figure 4.2. *Let $x^n(0) = 0$. Then the conclusions of Theorem* 6.1.1 *hold, except that $w^d_i(\cdot)$ has variance*

$$\sum_{u=-\infty}^{\infty} q^{(|u|)}_{ii} c_i(u),$$

where q^u_{ii} is the (i,i)th element of the matrix Q^u, $u \geq 0$, and $q^0_{ii} = 1$.

Proof. By hypothesis, $\{w_i^{a,n}(\cdot)\}$ is tight for each i, and the weak-sense limits are mutually independent Wiener processes $w_i^a(\cdot)$. By Theorem 6.1.1, $S_i^{a,n}(\cdot)$ converges weakly to the process with values $\bar{\lambda}_i^a t$. Also, $\mathcal{T}_i^{d,n}(\cdot)$ can be shown to converge to the process with values $\bar{\Delta}_i^d t$. Hence, as in Theorem 6.1.1, $S_i^{d,n}(\cdot)$ is asymptotically continuous. The main problem concerns the tightness and asymptotic continuity of the $w_i^{d,n}(\cdot)$, $i = 1, 2$, and the characterization of their weak-sense limits. Such limits will be independent in i and independent of the weak-sense limits of the $w_i^{a,n}(\cdot)$, $i = 1, 2$. Once the asymptotic continuity of $w_i^{d,n}(\cdot)$, $i = 1, 2$, is shown, the proof of convergence of the $S_i^{d,n}(\cdot)$ and the rest of the details are as in Theorem 6.1.1, except for the characterization of the variance of the $w_i^d(\cdot)$.

Let us decompose $\bar{w}_i^{d,n}(t) = w_i^{d,n}(S_i^{d,n}(t))$ as follows:

$$\bar{w}_i^{d,n}(t) = \frac{1}{\sqrt{n}} \sum_{l=1}^{nS_i^{d,n}(t)} \xi_{i,l}^{d,n} = \sum_j \sum_{k=0}^{\infty} \left[\frac{1}{\sqrt{n}} \sum_{l=1}^{\infty} \xi_{k,ji,l}^{d,n} I_{k,ji,l}^n(t) \right]. \tag{4.6}$$

In obtaining the asymptotic properties of the $w_i^{d,n}(\cdot)$ we need only work on $[0, T]$ for an arbitrary $T > 0$. In view of the weak convergence of the $S_i^{a,n}(\cdot)$, we can suppose (without loss of generality) that $S_i^{a,n}(t) \le K_0$ on this interval for some suitably large K_0 not depending on n. Then consider the bracketed term in (4.6). The supremum of its square is bounded by

$$\frac{1}{n} \sup_{m \le nK_0} \left| \sum_{l=1}^{m} \xi_{k,ji,l}^{d,n} I_{k,ji,l}^n \right|^2 . \tag{4.7}$$

The summands in (4.7) are mutually independent. By the martingale property of the process defined by the sum, the probability that (4.7) is greater than any $\delta_k > 0$ is bounded by

$$O\left(\frac{1}{n}\right) O(1) \sum_{l=1}^{nK_0} EI_{k,ji,l}^n / \delta_k^2, \tag{4.8}$$

where the $O(1)$ is a bound on the variance of the $\xi_{k,ji,l}^{d,n}$. By the spectral radius condition on $\{q_{ij}; i, j\}$, there are $K_1 < \infty$ and $\mu \in [0, 1)$ (neither depending on n) such that for large n

$$EI_{k,ji,l}^n \le \max_j P\left\{\text{there are} \ge k \text{ routings} \middle| \text{customer is from } A_j^n\right\} \le K_1 \mu^k. \tag{4.9}$$

Thus, there is K_2 (not depending on n) such that (4.8) is bounded by $K_2 \mu^k / \delta_k^2$. There are δ_k such that $\sum_k \delta_k < \infty$ and $\sum_k \mu^k / \delta_k^2 < \infty$. This and the Borel–Cantelli theorem imply that the process defined by the tail $\sum_{k_0}^{\infty}$ of the sum $\sum_k$ in (4.6) converges weakly to the "zero" process (uniformly in n) as $k_0 \to \infty$. This implies that in the proof of tightness of $\{w_i^{d,n}(\cdot)\}$, and in the characterization of its weak-sense limit, we can neglect more than a fixed, but arbitrary, number of returns of any customer to each P_i. In particular, $\bar{w}_i^{d,n}(\cdot)$ is asymptotically continuous if each of the processes defined by the bracketed term in (4.6) is asymptotically continuous.

The asymptotic continuity of $S_i^{d,n}(\cdot)$ and the mutual independence, zero mean, and bounded variances of the $\xi_{k,ji,l}^{d,n}$ imply the tightness of the process defined by the bracketed term in (4.6) for each k. Now the uniform integrability of the squares of the service intervals implies that the weak-sense limits of $\bar{w}_i^{d,n}(\cdot)$ are continuous with probability one. Thus, as in Theorem 6.1.1, the fact that $(w_i^{\alpha,n}(S_i^{\alpha,n}(\cdot)), w_i^{r,n}(S_i^{d,n}(\cdot)),\ \alpha = a, d,\ i = 1, 2)$ have continuous weak-sense limits together with the heavy traffic condition (A6.1.3) implies that $(x^n(\cdot), y^n(\cdot))$ has continuous weak-sense limits. Hence, as in Theorems 5.1.1 and 6.1.1, $S_i^{d,n}(\cdot)$ converges weakly to the process with values $\bar{\lambda}_i^d t$, $i = 1, 2$. The difference between the bracketed term in (4.6) and

$$M_{k,ji}^{d,n}(t) = \frac{1}{\sqrt{n}} \sum_{l=1}^{nS_j^{a,n}(t)} \xi_{k,ji,l}^{d,n} I_{k,ji,l}^n \tag{4.10}$$

converges to the "zero" process, since $1/n$ times the number from A_j^n that have arrived but not departed the system by real time nt converges weakly to the "zero" process. Define

$$w_{k,ji}^{d,n}(t) = \frac{1}{\sqrt{n}} \sum_{l=1}^{nt} \xi_{k,ji,l}^{d,n} I_{k,ji,l}^n. \tag{4.11}$$

The difference between the processes defined by (4.10) and $w_{k,ji}^{d,n}(\bar{\lambda}_j^a \cdot)$ also converges weakly to the "zero" process.

Let $q_{ji}^{(k)}$ denote the (j, i)th element of Q^k. By Theorem 2.8.8, for each i, j and $k \geq 0$, $w_{k,ji}^{d,n}(\bar{\lambda}_j^a \cdot)$ converges weakly to a Wiener process $w_{k,ji}^d(\bar{\lambda}_j^a \cdot)$ with variance $\sigma_{d,i}^2 \bar{\lambda}_j^a q_{ji}^{(k)}$, and the limits are independent in i, j. Consider the set of lth summands of the $(w_{k,ji}^{d,n}(\cdot), k < \infty)$, namely, $\left(\xi_{k,ji,l}^{d,n} I_{k,ji,l}^n, k < \infty\right)$. These sets are independent in l, j, i. Then Theorem 2.8.8 can be applied to show that for any $\bar{k} < \infty$, the weak-sense limit of the vector-valued process $\left(w_{k,ji}^{d,n}(\cdot), k < \bar{k}\right)$ is a Wiener process. We need only compute the covariance of the weak-sense limit of

$$\sum_j \sum_{k=0}^{\bar{k}} w_{k,ji}^{d,n}(\bar{\lambda}_j^a \cdot) \tag{4.12}$$

and take limits as $\bar{k} \to \infty$. This can be done via Theorem 2.8.8. Conditions (A6.1.2) and (A4.4) will be used. Owing to the independence in j, we can work with each value of j separately.

By Theorem 2.8.8, the covariance between $w^d_{k,ji}(\bar{\lambda}^a_j)$ and $w^d_{v,ji}(\bar{\lambda}^a_j)$ is the limit (as $n \to \infty$ and for any l) of

$$\bar{\lambda}^a_j P\left\{\text{job } l \text{ from } A^n_j \text{ is at } P_i \text{ at routings } k \text{ and } v\right\} c^n_i(v-k).$$

For $v \geq k$, the limit of the above term is $\bar{\lambda}^a_j q^{(k)}_{ji} q^{(v-k)}_{ii} c_i(v-k)$. Finally, we can sum all the covariances to get that the variance of $\bar{w}^d_i(1)$ is the limit of

$$\sum_j \sum_{k=0}^{\infty} \sum_{v=0}^{\infty} \bar{\lambda}^a_j P\left\{\text{job } l \text{ from } A^n_j \text{ is at } P_i \text{ at routings } k \text{ and } v\right\} c^n_i(v-k).$$

The limit is

$$\sum_j \bar{\lambda}^a_j \sum_{k=0}^{\infty} \sum_{u=-\infty}^{\infty} q^{(k)}_{ji} q^{(|u|)}_{ii} c_i(u). \tag{4.13}$$

Define the vectors $\bar{\lambda}^\alpha = (\bar{\lambda}^\alpha_i,\ i = 1,2)$, $\alpha = a, d$. The sum

$$\sum_j \sum_{k=0}^{\infty} q^{(k)}_{ji} \bar{\lambda}^a_j$$

is the ith entry of the vector $\left[I - Q'\right]^{-1} \bar{\lambda}^a$. By the heavy traffic condition (A6.1.3), $\bar{\lambda}^a = \left[I - Q'\right] \bar{\lambda}^d$. Hence, the variance of $\bar{w}^d_1(1)$ is

$$\bar{\lambda}^d_i \sum_{u=-\infty}^{\infty} q^{(|u|)}_{ii} c_i(u).$$

The $\bar{w}^d_i(\cdot)$ are independent in i, since the service times at the different processors are assumed to be independent of each other. ■

Note. If $c_i(u)$ is replaced by $c_{ji}(u)$, then the calculation can be done in a similar way, but the results cannot be represented in such a simple way.

7.5 Processor Sharing

In many applications to communications and computer systems, a single processor divides its time essentially continuously between jobs coming from several users. For example, the data might be divided into "cells" of small size. If the server can switch from source to source without a switching-time penalty, then one can suppose that time is infinitely divisible. Processor sharing, also called rate-based flow control, is motivated by "fairness," where the proportion of the capacity that is continuously allocated to any one user depends on the needs of that user, and is roughly proportional to the mean rate of creation of work by that user. Such processor sharing is an alternative to statistical mutltiplexing, where all work is queued in a single buffer in order of arrival and is handled on a FCFS basis. The subject of processor sharing is of great current interest; see, for example, [66, 65, 73, 135, 200]. See [205] for a good overall discussion of the advantages of processor sharing, and [240] for a discussion of practical realization. Fluid models are analyzed in [29, 36].

The system of concern has a single processor and $K > 1$ input streams, each queued separately. In the ideal model, the processor divides its time between the nonempty queues roughly proportionally to the rates of arrival of work to those queues. In our model, it is assumed that time is infinitely divisible and that the processor can exploit this to work on any number of queues "simultaneously." Although unrealistic in this form, this model captures the essential flavor of more realistic cases. For such applications, the arrivals and services are fast, and the scaling of Section 5.7 is the most appropriate one. Let $E^n(t)$ denote the set of queues that are empty at t. Let there be positive numbers β_i^n such that the "instantaneous" proportion of time at t that is allocated to queue i when it is nonempty is $\beta_i^n / \sum_{j \notin E^n(t)} \beta_j^n$. Normalize such that $\sum_i \beta_i^n = 1$. Redefine $S_i^{a,n}(t)$ to be $1/n$ times the number of arriving jobs at queue i on the real time interval $[0, t]$. The server works at a unit rate.

Following the scaling used in Section 5.7, let $\Delta_{i,l}^{a,n}/n$ (respectively, $\Delta_{i,l}^{d,n}/n$) denote the interarrival (respectively, service) intervals for the ith input stream. The centering constants are denoted by $\bar{\Delta}_i^{\alpha,n}$, $\alpha = a, d$, Recall that in this scaling $WL_i^n(t)$ is defined to be $\sqrt{n}$ times the time required to complete all of the work in queue i at real time t. The following assumptions will be used.

A5.1. *The processes $w_i^{\alpha,n}(\cdot)$, $\alpha = a, d$, $i \le K$, converge weakly to mutually independent Wiener processes with variances $\sigma_{\alpha,i}^2$ and there are $\bar{\Delta}_i^\alpha$ such that $\bar{\Delta}_i^{\alpha,n} \to \bar{\Delta}_i^\alpha$ as $n \to \infty$. The weak-sense limits are independent of any weak-sense limit of $WL_i^n(0)$.*

A5.2. *There are β_i, b_i, $i \le K$, such that $\beta_i^n \to \beta_i$ and*

$$\sqrt{n}\left[\frac{\bar{\Delta}_i^{d,n}}{\bar{\Delta}_i^{a,n}} - \beta_i^n\right] = b_i^n \to b_i.$$

The scaled system workload equations are, for $i \le K$,

$$WL_i^n(t) = WL_i^n(0) + \frac{1}{\sqrt{n}} \sum_{l=1}^{nS_i^{a,n}(t)} \Delta_{i,l}^{d,n} - \sqrt{n}\beta_i^n t + \text{ correction term.} \quad (5.1)$$

In (5.1), β_i^n is the fraction of the processor time that is allocated to queue i when all queues are nonempty. The correction term corrects for the fact that the distribution of time will vary depending on which (if any) queues are empty. *Redefine* $y_i^n(t)$ to be $\sqrt{n}\beta_i^n$ times the cumulative time on $[0,t]$ that queue i is empty. Suppose for the moment that one and only one queue, called queue j, is empty at time t. Then since the instantaneous proportion of time devoted to queue $i \neq j$ is now $\beta_i^n/(1-\beta_j^n)$, the derivative of the correction term for $WL_i^n(\cdot)$ at time t is

$$\sqrt{n}\beta_i^n\left[1-\frac{1}{1-\beta_j^n}\right] = -\frac{\beta_i^n}{1-\beta_j^n}\dot{y}_j^n(t),\ i \neq j. \tag{5.2}$$

For $WL_j^n(\cdot)$, it is $\sqrt{n}\beta_j^n = \dot{y}_j^n(t)$.

Define the matrix Q_n, whose jth row is

$$\frac{1}{1-\beta_j^n}\left(\beta_1^n,\ldots,0,\ldots,\beta_K^n\right),$$

where the zero is in the jth position. Then under the assumption that there is at most one empty queue at a time, the correction term can be written in vector form as

$$z^n(t) = [I - Q_n']\, y^n(t). \tag{5.3}$$

Define $Q = \lim_n Q_n$ and let d_i denote the ith column of $[I - Q']$. Since $\sum \beta_i^n = 1$, the spectral radius of Q_n is unity. The criterion (A3.5.3) holds on the edges but not at the origin. See Figure 5.1 for a two dimensional illustration.

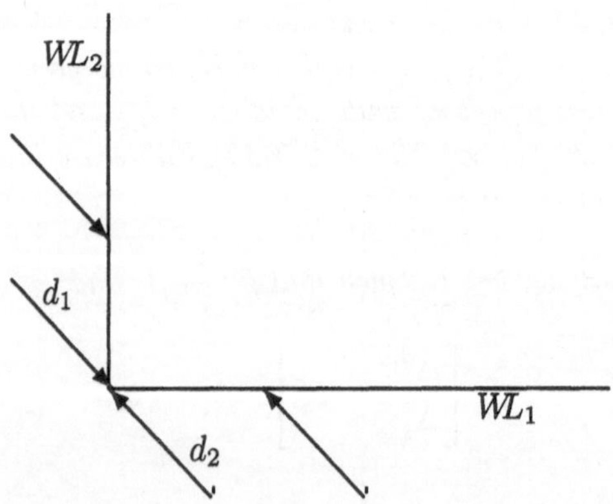

Figure 5.1. Reflection directions for a processor sharing system.

Now suppose that two distinct queues j and k are empty at t. Then the derivatives of the correction terms are

$$\sqrt{n}\beta_i^n\left[1-\frac{1}{\sum_{l\neq j,k}\beta_l^n}\right],\ i \neq j,k,$$

and for queues j and k they are $\sqrt{n}\beta_j^n$ and $\sqrt{n}\beta_k^n$, respectively. It turns out that this vector of derivatives is a nonnegative linear combination of the jth and kth columns of $[I - Q_n']$. The analogous result holds when there are any number of empty queues at time t. Thus, there are nondecreasing and continuous processes $y_i^n(\cdot)$ where $y^n(0) = 0$ and $y_i^n(\cdot)$ can increase only at t where queue i is empty and is such that

$$WL_i^n(t) = WL_i^n(0) + \frac{1}{\sqrt{n}} \sum_{l=1}^{nS_i^{a,n}(t)} \Delta_{i,l}^{d,n} - \sqrt{n}\beta_i^n t + z_i^n(t), \tag{5.4}$$

where

$$z_i^n(t) = y_i^n(t) - \sum_{j \neq i} \frac{\beta_i^n}{1 - \beta_j^n} y_j^n(t). \tag{5.5}$$

The next step is to expand the sum term in (5.4) in the usual way as

$$-\bar{\Delta}_i^{d,n} w_i^{d,n}(S_i^{a,n}(t)) + \frac{\bar{\Delta}_i^{d,n}}{\sqrt{n}} \sum_{l=1}^{nS_i^{a,n}(t)} 1,$$

and the second term above equals

$$\bar{\Delta}_i^{d,n} w_i^{a,n}(S_i^{a,n}(t)) + \frac{\bar{\Delta}_i^{d,n}}{\sqrt{n}\bar{\Delta}_i^{a,n}} \sum_{l=1}^{nS_i^{a,n}(t)} \Delta_i^{a,n}.$$

Thus, modulo a residual-time error term,

$$WL_i^n(t) = WL_i^n(0) + \bar{\Delta}_i^{d,n} \left[w_i^{a,n}(S_i^{a,n}(t)) - w_i^{d,n}(S_i^{a,n}(t)) \right] + b_i^n t + z_i^n(t). \tag{5.6}$$

The processes $z_i(\cdot)$ in (5.7) are not of bounded variation, as might be guessed, since the derivative of the correction term is infinite at the origin. However, the $z^n(\cdot)$ processes are still well behaved. The following lemma from [67] will be needed. The reflection direction vectors on the face where $WL_i = 0$ in [66] are the same as ours, but divided by β_i^n.

Lemma 5.1. [67, Theorem 3.7] *Consider the generalized Skorohod problem of Section* 3.5,

$$x(t) = \psi(t) + z(t), \quad x_i \geq 0, x \in I\!R^K,$$

with reflection directions d_i (defined below (5.3)) on the face $\{x : x_i = 0\}$ and where $\psi(\cdot) \in D(I\!R^K; 0, \infty)$. There is a real K_1 such that for any two solutions $(x^i(\cdot), \psi^i(\cdot), z^i(\cdot))$ and all t,

$$\sup_{s \leq t} \left| z^1(s) - z^2(s) \right| + \sup_{s \leq t} \left| x^1(s) - x^2(s) \right| \leq K_1 \sup_{s \leq t} \left| \psi^1(s) - \psi^2(s) \right|.$$

Remark. Consider the reflected SDE (3.5.4), where $b(\cdot)$ and $\sigma(\cdot)$ satisfy a uniform Lipschitz condition. Then Lemma 5.1 implies that there is a strong-sense unique solution for each initial condition.

Theorem 5.1. *Assume* (A5.1), (A5.2), *and let* $WL^n(0) \to WL(0)$. *Then* $\{WL^n(\cdot), z^n(\cdot), w_i^{\alpha,n}(\cdot), \alpha = a, d, i \leq K\}$ *is tight, and the weak-sense limit satisfies the generalized Skorohod problem of Subsection* 3.5.1, *with reflection directions* d_i *and satisfying*

$$WL_i(t) = WL_i(0) + b_i t + \bar{\Delta}_i^d \left[w_i^a(\bar{\lambda}_i^a t) - w_i^d(\bar{\lambda}_i^a t)\right] + z_i(t),\ i \leq K. \quad (5.7)$$

Equation (5.7) *has a unique strong-sense solution for each initial condition* $WL(0)$, *and* $(WL(\cdot), z(\cdot))$ *are nonanticipative with respect to* $\{w_i^\alpha(\bar{\lambda}_i^a \cdot);\ \alpha = a, d,\ i \leq K\}$. *Write* (5.7) *in vector form as*

$$WL(t) = WL(0) + bt + \bar{w}(t) + z(t).$$

Suppose that the covariance of $\bar{w}(\cdot)$ *is positive definite. Then*

$$P(x, t, \partial G) = 0 \text{ for all } x \text{ and } t > 0, \quad (5.8)$$

and $WL(\cdot)$ *is a strong Feller process.*

Proof. The main problem concerns the tightness and convergence of $z^n(\cdot)$. The proof of Theorem 3.6.1 can be used for this case to show that $z^n(\cdot)$ is asymptotically continuous. (Note that this property does not hold for $y^n(\cdot)$ as soon as $x^n(\cdot)$ reaches an arbitrarily small neighborhood of the origin.) Thus (5.7) holds as asserted. The uniqueness and nonanticipativity assertions follow from Lemma 5.1.

The proof of Theorem 4.2.1 can be used to show that (5.8) holds and that the probability of being in a small neighborhood of the boundary is small if the process is stopped as soon as it reaches some arbitrarily small neighborhood of the origin. In particular, the proof of Theorem 4.2.1 yields that for any $\delta > 0$,

$$E \int_0^t I_{\{0\}}(WL_i(s)) I_{\{|WL(s)| \geq \delta\}} ds = 0,\ i \leq K. \quad (5.9)$$

Define $\tilde{WL}(\cdot) = \sum_i WL_i(\cdot)$. Then

$$\tilde{WL}(t) = \tilde{WL}(0) + \sum_i \left[b_i t + \bar{w}_i(t)\right] + \tilde{y}(t),$$

where $\tilde{y}(\cdot)$ is a reflection term at the origin. Applying Theorem 4.2.1 to $\tilde{WL}(\cdot)$ for $0 < t_0 < t_1 < \infty$ yields that

$$\lim_{\delta \to 0} \sup_{\tilde{WL}(0)} \sup_{t_0 \leq t \leq t_1} P(\tilde{WL}(0), t, N_\delta(0)\} = 0.$$

This implies that (5.9) holds with the right-hand indicator function deleted. The rest of the details are left to the reader. ■

Remarks. One can allow any current job to be completed before switching, provided that the proportions are adhered to as closely as possible. Parekh and Gallagher [205] propose the following version, which they call PGPS: packet by packet generalized processor sharing. If the service scheme used in the theorem were followed, then each arriving job (or packet) would have some known completion time. The PGPS scheme works on one job until it is completed, then switches to the job that would have the smallest completion time if the ideal method were used. If the $\Delta_l^{\alpha,n}$ are bounded, then [205] shows that the performance of this modification and the ideal system are arbitrarily close as $n \to \infty$. A similar conclusion follows by the heavy traffic analysis, without the boundedness assumption.

8

State Dependence

Chapter Outline

In Chapters 5–7 it was supposed that the arrival and service processes, the routing process, batch magnitudes, etc., did not depend on the state of the queues. State-dependent arrival and service processes would arise if the arrival and work processes were allowed to depend on the queue levels, as for example, if longer queues led to balking or leaving before service, or to faster processing. State-dependent service times also occur in applications to communications and computer networks when the service or arrival rates depend on the number in the system. Other examples are in [186]. State dependencies can easily be incorporated into the results of Chapters 5–7 with only minor changes in the proofs.

In Chapters 5–7 the various processes $w_i^{\alpha,n}(\cdot)$, $\alpha = a, d, r$, $i \leq K$, converged weakly to martingales that were actually Wiener processes. The primary difference in the development and results is that in this chapter the limit martingales are not always Wiener processes. But they will always be representable as stochastic integrals with respect to Wiener processes. Although we concentrate on the problem formulation of the network of Section 6.1, it should be apparent that there are analogous modifications of the proofs and conditions of the other models of Chapters 6 and 7.

Two basic types of state-dependent models will be considered. The first model, dealt with in Section 1, is an adaptation of the setup used in Theorem 6.1.1. The state dependence is small, but highly significant in its effect on the heavy traffic limit. In the second model, the service times and arrivals are defined in terms of state-dependent "rates." This is, essentially, a state-dependent form of the Poisson arrival and exponential service time assumption. It will be seen that this second model is actually a special case of the first. Arbitrary combinations of the two forms can be treated.

In subsequent work on the optimal control problem it is useful to know that the Wiener processes that are used to represent the limit martingales are actually weak-sense limits of processes that are representable in terms of the actual physical data. Such a result is given in Theorem 1.3. Vacations, where either the rate of starting depends on the current state or the distribution of the duration depends on the state at the start, are treated at the end of Section 2.

In most of the results, for specificity, $x_i^n(\cdot)$ is defined to be $1/\sqrt{n}$ times the queue size at real time nt. But the fast-arrival model of Section 6.6 can be used just as well, with the obvious changes. Section 3 concerns the so-called self-service problem, where each customer has its own server. The telephone network is one classical example. For such models, the population from which the arrivals are drawn is supposed to be large, and it is indexed by n, which is also the order of the rate of arrivals. Hence, for this case, $x_i^n(\cdot)$ is defined to be $1/\sqrt{n}$ times the (centered about a mean value) queue size at real time t. There will not be a reflection term in the limit.

The dynamical terms need not be continuous, provided that the system noise has a "smoothing" effect, and this is developed in Section 4. Discontinuities arise, for example, when a control is fixed a priori and is of the threshold or "switching curve" type. Two examples are given where the routing probabilities are based on a threshold-type function of the difference between values of the queues, hence are discontinuous.

More complicated state-dependent forms can also be treated, where the dependence is via the state and some underlying auxiliary correlated process. For such cases, the perturbed test function methods of [157, 177], that are used effectively for complicated models in stochastic approximation, are useful, although not treated here.

The methods can be adapted to the workload formulations of Chapter 6. If there are fixed priorities and the state dependence is on the workload itself, then there is little change in the development. If the state dependence is on the scaled queued numbers, then one needs to introduce the asymptotic relations such as in Theorems 6.4.1, 6.4.3, and 6.5.3.

An interesting problem arises when there are long vacations and the rate of termination of the vacation depends on the current state. Although this is not treated in this chapter, a closely related problem is dealt with in the control context in Chapter 10.

In Section 5 we return to the multiplexer problem of Chapter 1, with fixed controls. It is shown that the methods of Sections 1 and 2 are readily adapted to get the weak convergence, and the assumptions are easily verified. Section 6 applies the previous results to problems where there is balking or withdrawing, with possible retrials.

8.1 Marginal State Dependence

The model in this section is a natural state-dependent analogue of the setup used in Theorem 6.1.2. The state dependency will be "small," but it can have significant effects. What is meant by "small" is made precise in (A1.1) and (A1.3). Let $\mathcal{F}_{i,l}^{\alpha,n}$ and $E_{i,l}^{\alpha,n}$, $\alpha = a, d, r$, be defined as in Subsection 6.1.2.

In (A1.1) and (A1.2), x denotes the value of the state at the start of the $(l+1)$st (interarrival or service) interval at P_i. In (A1.3), x denotes the value just before the end of the lth service interval at P_i. The $o_{i,n}^{\alpha,n}(1/\sqrt{n})$ and $o_{ij,n}^{\alpha,n}(1/\sqrt{n})$ are of order $o(1/\sqrt{n})$ uniformly in n, i, j, l, in each compact x-set. The continuity conditions on the functions introduced in the assumptions will be weakened later.

A1.1. *For $\alpha = a, d$, $i \leq K$, there are constants $\bar{\Delta}_i^\alpha$ and continuous functions $\Delta_i^\alpha(\cdot)$, that have at most a linear growth in x and satisfy*

$$E_{i,l}^{\alpha,n} \Delta_{i,l+1}^{\alpha,n} \equiv \bar{\Delta}_{i,l+1}^{\alpha,n}(x) = \bar{\Delta}_i^\alpha \left(1 - \frac{1}{\sqrt{n}} \Delta_i^\alpha(x)\right) + o_{i,l}^{\alpha,n}(1/\sqrt{n}). \qquad (1.1)$$

A1.2. *There are bounded and continuous functions $\sigma_{\alpha,i}^2(\cdot)$, $\alpha = a, d$, $i \leq K$, such that in the mean and uniformly in α, l, i, and in x in each bounded set,*

$$E_{i,l}^{\alpha,n} \left[1 - \frac{\Delta_{i,l+1}^{\alpha,n}}{\bar{\Delta}_{i,l+1}^{\alpha,n}(x)}\right]^2 = \left[\sigma_{i,l+1}^{\alpha,n}\right]^2 \to \sigma_{\alpha,i}^2(x),$$

as $n \to \infty$.

A1.3. *There are constants $\bar{q}_{ij}$ and continuous functions $q_{ij}(\cdot)$ that have at most a linear growth in x and for $i, j \leq K$, satisfy*

$$P_{\mathcal{F}_{i,l}^{r,n}} \left\{I_{ij,l}^n = 1\right\} \equiv q_{ij,l}^n(x) = \bar{q}_{ij} + \frac{1}{\sqrt{n}} q_{ij}(x) + o_{ij,l}^{r,n}(1/\sqrt{n}),$$

where we recall that $I_{ij,l}^n$ denotes the indicator function of the event that the lth departure from P_i goes to P_j. The spectral radius of the matrix $Q = \{\bar{q}_{ij}; i, j\}$ is less than unity.

A1.4. *The heavy traffic condition is*

$$\frac{1}{\bar{\Delta}_i^a} = \frac{1}{\bar{\Delta}_i^d} - \sum_j \frac{\bar{q}_{ji}}{\bar{\Delta}_j^d},$$

or, equivalently, defining $\bar{\lambda}_i^\alpha = 1/\bar{\Delta}_i^\alpha$, $\alpha = a, d$,

$$\bar{\lambda}_i^a = \bar{\lambda}_i^d - \sum_j \bar{q}_{ji} \bar{\lambda}_j^d.$$

A1.5. $\left\{|\Delta_{i,l}^{a,n}|^2, |\Delta_{i,l}^{d,n}|^2; i, l, n\right\}$ *is uniformly integrable.*

Define the functions

$$\lambda_i^\alpha(x) = \bar{\lambda}_i^\alpha \Delta_i^\alpha(x), \quad \alpha = a, d,$$

and

$$b_i(x) = \lambda_i^a(x) + \sum_j \left[\bar{q}_{ji}\lambda_j^d(x) + \bar{\lambda}_j^d q_{ji}(x)\right] - \lambda_i^d(x). \tag{1.2}$$

The assumptions imply that the conditional distribution of the intervals or the routing, given the "past," depends mainly on the state at the start of the interval, or just before the routing is to take place, according to the case. Owing to the $o(\cdot)$ terms in (A1.1) and (A1.3), the dependence can be a little more general; for example, it can include the change in the state during the interval in question, as will be seen in Subsection 2.2.

Terminology. Henceforth, we will use the following notation. Let $x_{i,l}^{a,n}$ denote the sample value of the state at the start of the lth interarrival interval at P_i, and define $x_{i,l}^{d,n}$ analogously. Let $x_{i,l}^{r,n}$ denote the sample value of the state just before the end of the lth service interval at P_i. The terminology of Section 6.1 will be used unless mentioned otherwise. For $\alpha = a, d$, $i \leq K$, the following processes replace the $w_i^{\alpha,n}(\cdot)$ of Section 6.1. Define

$$m_i^{\alpha,n}(t) = \frac{1}{\sqrt{n}} \sum_{l=1}^{nt} \left[1 - \frac{\Delta_{i,l}^{\alpha,n}}{\bar{\Delta}_{i,l}^{\alpha,n}(x_{i,l}^{\alpha,n})}\right], \tag{1.3}$$

$$m_{ij}^{r,n}(t) = \frac{1}{\sqrt{n}} \sum_{l=1}^{nt} \left[I_{ij,l}^n - q_{ij,l}^n(x_{i,l}^{r,n})\right], \tag{1.4}$$

$$m_{ij}^{d,n}(t) = \bar{q}_{ij} w_i^{d,n}(t). \tag{1.5}$$

Also define

$$\begin{aligned} M_i^{\alpha,n}(t) &= m_i^{\alpha,n}(N_i^{\alpha,n}(t)), \\ M_{ij}^{r,n}(t) &= m_{ij}^{r,n}(N_i^{d,n}(t)), \qquad M_i^{r,n}(\cdot) = \{M_{ij}^{r,n}(\cdot),\ j \leq K\}, \\ M_i^n(t) &= M_i^{a,n}(t) + \textstyle\sum_j \bar{q}_{ji} M_j^{d,n}(t) - M_i^{d,n}(t) + \textstyle\sum_j M_{ji}^{r,n}(t). \end{aligned} \tag{1.6}$$

The processes in (1.3)–(1.5) are martingales, since the summands are centered about the conditional expectations given the "past."

The next theorem is the analogue of Theorem 6.1.1. It asserts the weak convergence of $x^n(\cdot)$ to the solution of a reflected stochastic differential equation (1.8). It also implies the convergence of the martingales (that represent the randomness) to stochastic integrals. Define $y_i^n(\cdot) = \bar{\lambda}_i^d T_i^{d,n}(t)$, where $T_i^{d,n}(t)$ again denotes $1/\sqrt{n}$ times the idle time at P_i by real time nt.

Theorem 1.1. *Let $x^n(0)$ converge weakly to $x(0)$ and assume* (A1.1)–(A1.5) *and buffers of sizes $\sqrt{n}B_i$, $B_i \leq \infty$. Then $x^n(\cdot)$ can be represented as*

$$x_i^n(t) = x_i^n(0) + \int_0^t b_i(x^n(s))ds + M_i^n(t) + z_i^n(t) + \epsilon_i^n(t), \; i \leq K, \tag{1.7}$$

where $\epsilon_i^n(\cdot)$ converges weakly to the "zero" process and

$$z^n(t) = [I - Q']\, y^n(t) - U^n(t).$$

Define

$$\Psi^n(\cdot) = (x^n(\cdot), M_i^{\alpha,n}(\cdot), N_i^{\alpha,n}(\cdot), M_i^{r,n}(\cdot), y^n(\cdot), U^n(\cdot);\; \alpha = a, d,\; i \leq K).$$

Then $\{\Psi^n(\cdot)\}$ is tight, and any weak-sense limit $\Psi(\cdot)$ satisfies

$$x(t) = x(0) + \int_0^t b(x(s))ds + M(t) + z(t), \tag{1.8}$$

where

$$z(t) = [I - Q']\, y(t) - U(t), \tag{1.9}$$

$y_i(0) = 0$, and $y_i(\cdot)$ is continuous and nondecreasing and can increase only at t where $x_i(t) = 0$, and analogously for $U_i(\cdot)$ at the upper boundary. Let $\{\mathcal{F}_t, t \geq 0\}$ denote the filtration engendered by $\Psi(\cdot)$. The weak-sense limits of the $N_i^{\alpha,n}(\cdot)$ have values $\bar{\lambda}_i^\alpha t$. The weak-sense limits of the $(M_i^{\alpha,n}(\cdot)$; $\alpha = a, d, r,\; i \leq K)$, denoted by $(M_i^\alpha(\cdot);\; \alpha = a, d, r,\; i \leq K)$, are continuous orthogonal $\mathcal{F}_t$-martingales, where $M_i^\alpha(\cdot)$, $\alpha = a, d$, are real-valued and $M_i^r(\cdot)$ is $I\!R^r$-valued. Define $M^\alpha(\cdot) = (M_i^\alpha(\cdot), i \leq K)$, $\alpha = a, d$. Then

$$M(t) = M^a(t) - [I - Q']\, M^d(t) + M^r(t), \tag{1.10}$$

where $M^r(\cdot) = \sum_i M_i^r(\cdot)$. The quadratic variations of the $M_i^\alpha(\cdot)$, $\alpha = a, d$, are

$$\bar{\lambda}_i^\alpha \int_0^t \sigma_{\alpha,i}^2(x(s))ds. \tag{1.11}$$

The $M_i^r(\cdot)$, $i \leq K$, are mutually independent $\mathcal{F}_t$-Wiener processes whose covariance matrices are $\bar{\lambda}_i^d$ times the matrix $\Sigma_{r,i}$ defined in (6.1.14).

Representation of the martingales as stochastic integrals. Theorem 1.1 and Theorem 2.2.1 imply that (with the possible augmentation of the probability space by the addition of a vector-valued Wiener process that is independent of the other weak-sense limit processes) there are mutually independent standard $\mathcal{F}_t$-Wiener processes $(w_i^\alpha(\cdot);\ \alpha = a, d,\ i \leq K)$ that are also independent of $(x(0), M_i^r(\cdot),\ i \leq K)$ and such that for $\alpha = a, d$,

$$M_i^\alpha(t) = \int_0^t \sigma_{\alpha,i}(x(s))dw_i^\alpha(\bar{\lambda}_i^\alpha s). \tag{1.12}$$

Furthermore, $x(\cdot)$ and $z(\cdot)$ are nonanticipative with respect to $(w_i^\alpha(\bar{\lambda}_i^\alpha \cdot)$, $M_i^r(\cdot);\ i \leq K,\ \alpha = a, d)$.

Proof. Most of the proof is an adaptation of the proof of Theorem 6.1.2. The main difference is that the Wiener processes in Theorem 6.1.2 are replaced by martingales, which are representable as stochastic integrals. If the functions $\Delta_i^\alpha(\cdot)$, $\alpha = a, d$, and $q_{ij}(\cdot)$ are bounded and the $\sqrt{n}o(1/\sqrt{n})$, $no(1/n)$ go to zero uniformly in x, then it will be shown that all solutions satisfy (1.8)–(1.11). Otherwise, we need to use a truncation argument of the type in Subsection 2.5.6. By the linear growth condition on $b(\cdot)$ and the boundedness of the $\sigma_{\alpha,i}^2(\cdot)$, all solutions to (1.8)–(1.11) satisfy (2.5.19) whether $b(\cdot)$ is bounded or not. Hence, as noted in Subsection 2.5.5, without loss of generality we can suppose that the functions of x are all bounded and the $o(1/\sqrt{n})$ and $o(1/n)$ terms are uniform in x.

Recall the proof of the weak convergence of $N^{\alpha,n}(\cdot)$ and $S^{\alpha,n}(\cdot)$ in Theorem 5.1.1. In the present case, when centering each term at the conditional expectation given the "past," the decomposition (5.1.20) takes the form, for $\alpha = a, d$ and each i,

$$\mathcal{T}_i^{\alpha,n}(t) = \frac{1}{n}\sum_{l=1}^{nt} \bar{\Delta}_{i,l}^{\alpha,n}(x_{i,l}^{\alpha,n}) + \frac{1}{n}\sum_{l=1}^{nt}\left[\Delta_{i,l}^{\alpha,n} - \bar{\Delta}_{i,l}^{\alpha,n}(x_{i,l}^{\alpha,n})\right]. \tag{1.13}$$

The last term on the right-hand side of (1.13) is a martingale and converges weakly to the zero process by (A1.2). By using the expansion of $\bar{\Delta}_{i,l}^{\alpha,n}(x)$ of (A1.1) in the first term on the right side of (1.13), we see that only the dominant part $\bar{\Delta}_i^\alpha$ is relevant asymptotically. Then the proof of Theorem 5.1.1 implies that $\mathcal{T}_i^{\alpha,n}(\cdot)$ converges weakly to the process with values $\bar{\Delta}_i^\alpha t$, that $N_i^{a,n}(\cdot)$ converges weakly to the process with values $\bar{\lambda}_i^a t$, and that $N_i^{d,n}(\cdot)$ is asymptotically Lipschitz continuous.

Equations (6.1.19)–(6.1.20) hold with $\bar{\Delta}_i^{\alpha,n}$ replaced by $\bar{\Delta}_{i,l}^{\alpha,n}(x_{i,l}^{\alpha,n})$. Thus, the last terms on the right sides of (6.1.19) and (6.1.20) are replaced by, for $\alpha = a, d$,

$$\frac{1}{\sqrt{n}}\sum_{l=1}^{nS_i^{\alpha,n}(t)} \frac{\Delta_{i,l}^{\alpha,n}}{\bar{\Delta}_i^\alpha(1 - \Delta_i^\alpha(x_{i,l}^{\alpha,n})/\sqrt{n}) + o_{i,l}^{\alpha,n}(1/\sqrt{n})}.$$

Under (A1.1) and (A1.5), for $\alpha = a, d$, the above expression can be written as, respectively,

$$\sqrt{n}\bar{\lambda}_i^a t + \frac{\bar{\lambda}_i^a}{n} \sum_{l=1}^{nS_i^{a,n}(t)} \Delta_{i,l}^{a,n} \bar{\Delta}_i^a(x_{i,l}^{a,n}) + \epsilon_i^{a,n}(t) + [n\, o(1/n)]t, \tag{1.14}$$

$$\bar{\lambda}_i^d \left[\sqrt{n}t - T_i^{d,n}(t)\right] + \frac{\bar{\lambda}_i^d}{n} \sum_{l=1}^{nS_i^{d,n}(t)} \Delta_{i,l}^{d,n} \bar{\Delta}_i^d(x_{i,l}^{d,n}) + \epsilon_i^{d,n}(t) + [n\, o(1/n)]t, \tag{1.15}$$

where the $\epsilon_i^{\alpha,n}(\cdot)$ converge weakly to the "zero" process. Using the asymptotic Lipschitz continuity of $S_i^{\alpha,n}(\cdot)$, $\alpha = a, d$ and (A1.5), we see that the set of processes defined by the second terms in (1.14) and (1.15) are tight and that all of their weak-sense limits are absolutely continuous.

Note the following for later use. Suppose that $x^n(\cdot)$ is asymptotically continuous (to be shown below). Then viewing $\Delta_{i,l}^{\alpha,n}/n$ as an interpolation time interval, the second terms in (1.14) and (1.15) interpolate (asymptotically) into integrals of $\bar{\Delta}_i^\alpha(x^n(s))$. More precisely, since $x_{i,l}^{a,n} = x_i^n(\mathcal{T}_i^{a,n}((l-1)/n))$, the presumed asymptotic continuity of $x^n(\cdot)$, the convergence of $\mathcal{T}_i^{a,n}(\cdot)$, and (A1.1) and (A1.2) imply that the difference between the second term in (1.14) and

$$\bar{\lambda}_i^a \sum_{l=1}^{n\bar{\lambda}_i^a t} \bar{\Delta}_i^a \left(x^n\left(\frac{(l-1)\bar{\Delta}_i^a}{n}\right)\right) \frac{\bar{\Delta}_i^a}{n}$$

converges weakly to the "zero" process. Similarly, the difference between the above expression and the integral

$$\bar{\lambda}_i^a \int_0^t \bar{\Delta}_i^a(x^n(s))ds$$

converges weakly to the "zero" process. If, in addition, $T_i^{d,n}(\cdot)/\sqrt{n}$ converges weakly to the "zero" process (to be shown below), then $N_i^{d,n}(\cdot)$ converges weakly to the process with values $\bar{\lambda}_i^d t$, and the analogous result holds for the second term in (1.15).

The analog of (6.1.21) is

$$\begin{aligned} D_{ij}^n(t) &= \frac{1}{\sqrt{n}} \sum_{l=1}^{nS_i^{d,n}(t)} I_{ij,l}^n = \frac{1}{\sqrt{n}} \sum_{l=1}^{nS_i^{d,n}(t)} \left[I_{ij,l}^n - q_{ij,l}^n(x_{i,l}^{r,n})\right] \\ &+ \frac{1}{\sqrt{n}} \sum_{l=1}^{nS_i^{d,n}(t)} q_{ij,l}^n(x_{i,l}^{r,n}) \left[1 - \frac{\Delta_{i,l+1}^{d,n}}{\bar{\Delta}_{i,l+1}^{d,n}(x_{i,l+1}^{d,n})}\right] \\ &+ \frac{1}{\sqrt{n}} \sum_{l=1}^{nS_i^{d,n}(t)} q_{ij,l}^n(x_{i,l}^{r,n}) \frac{\Delta_{i,l+1}^{d,n}}{\bar{\Delta}_{i,l+1}^{d,n}(x_{i,l+1}^{d,n})}. \end{aligned} \tag{1.16}$$

Using the expansions of $\bar{\Delta}^{d,n}_{i,l+1}(x)$ and $q^n_{ij,l}(x)$, and neglecting some terms of order $o(1/n)$, write the last term of (1.16) as

$$\frac{\bar{\lambda}^d_i}{\sqrt{n}} \sum_{l=1}^{nS^{d,n}_i(t)} \Delta^{d,n}_{i,l+1} \left[\bar{q}_{ij} + \frac{1}{\sqrt{n}} q_{ij}(x^{r,n}_{i,l}) \right] \left[1 + \frac{1}{\sqrt{n}} \Delta^d_i(x^{d,n}_{i,l+1}) \right]. \tag{1.17}$$

This can be written as (neglecting residual-time error and other terms that converge weakly to the "zero" process)

$$\bar{q}_{ij} \bar{\lambda}^d_i \left[\sqrt{n} t - T^{d,n}_i(t) \right] + \frac{\bar{\lambda}^d_i}{n} \sum_{l=1}^{nS^{d,n}_i(t)} \Delta^{d,n}_{i,l+1} \left[q_{ij}(x^{r,n}_{i,l}) + \bar{q}_{ij} \Delta^d_i(x^{d,n}_{i,l+1}) \right]. \tag{1.18}$$

Write the next to last term on the right side of (1.16) as

$$\frac{1}{\sqrt{n}} \sum_{l=1}^{nS^{d,n}_i(t)} \left[\bar{q}_{ij} + \frac{q_{ij}(x^{r,n}_{i,l})}{\sqrt{n}} + o^{r,n}_{i,l}\left(\frac{1}{\sqrt{n}}\right) \right] \left[1 - \frac{\Delta^{d,n}_{i,l+1}}{\bar{\Delta}^{d,n}_{i,l+1}(x^{d,n}_{i,l+1})} \right].$$

Rewrite the above expression as the sum of $\bar{q}_{ij} M^{d,n}_i(t)$ and a term that converges weakly to the "zero" process.

Define

$$\begin{aligned} B^n_i(t) = {} & \frac{\bar{\lambda}^a_i}{n} \sum_{l=1}^{nS^{a,n}_i(t)} \Delta^{a,n}_{i,l} \Delta^a_i(x^{a,n}_{i,l}) - \frac{\bar{\lambda}^d_i}{n} \sum_{l=1}^{nS^{d,n}_i(t)} \Delta^{d,n}_{i,l} \Delta^d_i(x^{d,n}_{i,l}) \\ & + \sum_j \frac{\bar{\lambda}^d_j}{n} \sum_{l=1}^{nS^{d,n}_j(t)} \Delta^{d,n}_{j,l+1} \left[q_{ji}(x^{r,n}_{j,l}) + \bar{q}_{ji} \Delta^d_j(x^{d,n}_{j,l+1}) \right]. \end{aligned} \tag{1.19}$$

Putting the pieces together, replacing $S^{\alpha,n}_i(\cdot)$ by $N^{\alpha,n}_i(\cdot)$ in the martingale terms, and using the heavy traffic condition (A1.4) yields the analog of (6.1.25):

$$x^n_i(t) = x^n_i(0) + B^n_i(t) + M^n(t) + z^n_i(t) + \epsilon^n_i(t), \tag{1.20}$$

where $\epsilon^n_i(\cdot)$ contains the residual-time error terms and the other terms that converge weakly to the zero process.

The set of all processes defined by the terms in (1.19) is tight with absolutely continuous weak-sense limits, as are the $\{N^{d,n}_i(\cdot), S^{d,n}_i(\cdot)\}$. Define $\bar{M}^n(\cdot) = (M^{\alpha,n}_i(\cdot);\ \alpha = a, d, r,\ i \le K)$. Then $\{\bar{M}^n(\cdot)\}$ is tight, and all weak-sense limits are continuous by the martingale properties of the processes in (1.3) and (1.4) and condition (A1.5), exactly analogous to the situation in Theorem 6.1.1. Now the method of Theorem 3.6.1 yields the tightness and asymptotic continuity of the reflection terms and of $\{x^n(\cdot)\}$. Thus, $T^{d,n}_i(\cdot)/\sqrt{n}$ converges weakly to the "zero" process, and $N^{d,n}_i(\cdot)$ converges weakly to the process with values $\bar{\lambda}^d_i t$. Given the continuity of the

$\Delta_i^\alpha(\cdot), \alpha = a, d$, and the tightness and asymptotic continuity of $x^n(\cdot)$ and the weak convergence of the $S_i^{\alpha,n}(\cdot)$, it is easy to see that the processes defined by the terms in (1.19) converge to integrals in the sense that the difference between (1.19) and

$$\int_0^t \lambda_i^a(x^n(s))ds - \int_0^t \lambda_i^d(x^n(s))ds + \sum_j \int_0^t \left[\bar{\lambda}_j^d q_{ji}(x^n(s)) + \bar{q}_{ji}\lambda_j^d(x^n(s))\right] ds \tag{1.21}$$

converges weakly to the "zero" process, where the functions $\lambda_i^\alpha(\cdot)$ are defined above (1.2).

Let

$$\big(x(\cdot), M_i^\alpha(\cdot), N_i^\alpha(\cdot), M_i^r(\cdot), y(\cdot), U(\cdot);\ \alpha = a, d,\ i \le K\big)$$

denote the weak-sense limit of a weakly convergent subsequence. Then by the weak convergence, $z(\cdot) = [I - Q']y(\cdot) - U(\cdot)$ is the reflection term, as in Theorems 6.1.1 and 6.1.2. Also, (1.8) and (1.10) hold.

We need only prove the asserted $\mathcal{F}_t$-martingale and quadratic variation properties of $M(\cdot)$. To do this, we follow the scheme used in Theorem 6.1.2, but use Theorem 2.8.2. Note that we used the definition (6.1.30) and seek to evaluate the limit of (6.1.31) as $n \to \infty$. Analogously to what was done in Theorem 6.1.2, let $\xi_{i,l}^{\alpha,n}$ denote the summands in the $m_i^{\alpha,n}(\cdot)$. Recall the definition of $t + s_{i,l}^{\alpha,n}$ from Theorem 6.1.2. Let $\tilde{x}_{i,l}^{\alpha,n}$, $\alpha = a, d$, denote the sample value of the state at the beginning of the lth interval starting at or after interpolated time t.

Without loss of generality, as in Theorem 6.1.2 we continue to suppose that the jump times are distinct. Proceeding as in Theorem 6.1.2, E_t^n acting on the bracketed term in (6.1.31) equals the sum of (6.1.34)–(6.1.40) plus a term that goes to zero in mean as $n \to \infty$. The first order terms go to zero in mean by (A1.1), as does (6.1.40). The expressions (6.1.37) and (6.1.38) are now approximated by (for $\alpha = a, d$, and where all approximations are in the mean)

$$K^{\alpha,n}(\tau) = \frac{1}{2n}\sum_i E_t^n \sum_{l=1}^{n\bar{\lambda}_i^\alpha \tau} F_{m_i^\alpha m_i^\alpha}(\bar{M}^n(t + s_{i,l}^{\alpha,n}))\sigma_{\alpha,i}^2(\tilde{x}_{i,l}^{\alpha,n}). \tag{1.22}$$

This, in turn, is approximated by

$$\frac{\bar{\lambda}_i^\alpha}{2}\sum_i E_t^n \int_t^{t+\tau} F_{m_i^\alpha m_i^\alpha}(\bar{M}^n(s))\sigma_{\alpha,i}^2(x^n(s))ds. \tag{1.23}$$

The expression (6.1.39) is still approximated by (6.1.44).

From this point, substituting the approximations back into (6.1.31) and using the weak convergence and Theorem 2.8.2 yields the assertions concerning the M-processes. ■

The Skorohod Problem model. Theorem 6.1.4 extended the results of Theorems 6.1.1 and 6.1.2 to the Skorohod problem model (6.1.45). The same extension holds when $w^n(\cdot)$ in (6.1.45) is replaced by a process of the type $M^n(\cdot)$ arising in Theorem 1.1, where $M^n(\cdot)$ is asymptotically continuous, and any weak-sense limit has the appropriate martingale property.

The "pre-Wiener" processes. According to Theorem 1.1, the processes $M_i^{r,n}(\cdot)$, $i \le K$, converge to mutually independent Wiener processes. The $M_i^{\alpha,n}(\cdot)$; $\alpha = a, d$, $i \le K$, converge to mutually orthogonal martingales that are orthogonal to the set $M_i^{r,n}(\cdot)$, $i \le K$. The $M_i^{\alpha,n}(\cdot)$; $\alpha = a, d$, $i \le K$, are not Wiener processes if their quadratic variations depend on the state. But if there is state dependence, they can always be represented in terms of standard Wiener processes as in (1.12). For the control problem, one wishes to prove that the limit of the optimal costs is an optimal cost for the weak-sense limit system. For such a proof it is often useful (as in [167, Section 10.4]) to construct "pre-Wiener" processes $w_i^{\alpha,n}(\cdot)$, $\alpha = a, d$, $i \le K$, from the physical data such that their weak-sense limits are the $w_i^\alpha(\cdot)$, $\alpha = a, d$, $i \le K$, in (1.12). The samples of the "pre–Wiener" processes are used to approximate the samples of the Wiener processes in (3.8.3) and (3.8.7). See Chapters 9 and 10 and [167, Section 10.4] for examples where such a method is employed.

In particular, we would like to show that $M_i^{\alpha,n}(\cdot)$, $\alpha = a, d$, $i \le K$, can be approximated in the form

$$M_i^{\alpha,n}(t) \approx \int_0^t \sigma_{\alpha,i}(x^n(s))dw_i^{\alpha,n}(\bar{\lambda}_i^\alpha s), \tag{1.24}$$

where $(w_i^{\alpha,n}(\cdot)$, $\alpha = a, d$, $i \le K)$, converges weakly to the set of Wiener processes used in (1.12). The approximations $w_i^{\alpha,n}(\cdot)$ can be readily constructed, and the details given in [167, Section 10.4] for a related limit problem can be easily adjusted to cover the present case.

Let $\tilde{w}_i^\alpha(\cdot)$, $i \le K$, $\alpha = a, d$, be mutually independent standard real-valued Wiener processes that are also independent of all of the processes driving the queues. In the next theorem, the probability space and filtrations $\{\mathcal{F}_t^n, t \ge 0\}$ are augmented to include the $\tilde{w}_i^\alpha(N_i^{\alpha,n}(\cdot))$. All square roots are taken to be nonnegative. Let $\xi_{i,l}^{\alpha,n}$ denote the lth summand in (1.3).

Recall the definition of $\sigma_{i,l}^{\alpha,n}$ from (A1.2). Define, for $\alpha = a, d$, $i \le K$,

$$\begin{aligned}\delta w_{i,l}^{\alpha,n} &= \frac{1}{\sqrt{n}} I_{\{\sigma_{i,l}^{\alpha,n} \neq 0\}} \left[\sigma_{i,l}^{\alpha,n}\right]^{-1} \xi_{i,l}^{\alpha,n} \\ &\quad + I_{\{\sigma_{i,l}^{\alpha,n} = 0\}} \left[\tilde{w}_i^{\alpha,n}(l/n) - \tilde{w}_i^{\alpha,n}(l/n - 1/n)\right],\end{aligned} \tag{1.25}$$

and

$$w_i^{\alpha,n}(t) = \sum_{l=1}^{nt} \delta w_{i,l}^{\alpha,n}.$$

Then we can write

$$m_i^{\alpha,n}(t) = \sum_{l=1}^{nt} \sigma_{i,l}^{\alpha,n} \delta w_{i,l}^{\alpha,n}. \tag{1.26}$$

Under (A1.2), we can write

$$M_i^{\alpha,n}(t) = \int_0^t \sigma_{\alpha,i}(x^n(s-))dw_i^{\alpha,n}(N_i^{\alpha,n}(s)) + \epsilon_i^{m,\alpha,n}(t),$$

where

$$\sup_{s\le t} E\left[\epsilon_i^{m,\alpha,n}(s)\right]^2 = O(t)\delta_n,$$

where $\delta_n \to 0$ as $n \to 0$. The error term $\epsilon_i^{m,\alpha,n}(\cdot)$ is due to the replacement of $\sigma_{i,l}^{\alpha,n}$ by $\sigma_{\alpha,i}(x_{i,l}^{\alpha,n})$.

Theorem 1.2. *Assume the conditions and notation of Theorem 1.1. Then*[1] $(\Psi^n(\cdot), w_i^{\alpha,n}(\cdot);\ \alpha = a, d,\ i \le K)$ *converges weakly to* $(\Psi(\cdot), w_i^\alpha(\cdot);\ \alpha = a, d,\ i \le K)$, *where the* $M_i^\alpha(\cdot)$ *have the representation* (1.12). *The* $w_i^\alpha(\cdot)$, $\alpha = a, d; M_i^r(\cdot)$, $i \le K$, *are mutually independent Wiener processes, independent of* $x(0)$, *and the* $w_i^\alpha(\cdot)$; $\alpha = a, d$, $i \le K$, *are standard. The pair* $(x(\cdot), z(\cdot))$ *is nonanticipative with respect to* $(w_i^\alpha(\bar{\lambda}_i^\alpha \cdot);\ \alpha = a, d,\ M_i^r(\cdot),\ i \le K)$.

Remark on the proof. The proof is an application of the method of Theorem 6.1.2 and the weak-convergence results of Theorem 1.1. We note only that the summands in (1.26) are martingale differences and

$$E_{i,l}^{\alpha,n}\left[\delta w_{i,l+1}^{a,n}\right]^2 = 1/n,$$

where the conditioning now includes the "past" of the $(\tilde{w}_i^{a,n}(\cdot), \tilde{w}_i^{d,n}(\cdot))$ as well.

Vector-valued processes in Theorem 1.2. Theorem 1.2 was stated as it was to accommodate the form of the problem in this (and in the next) section, where the martingales composed from the individual service and interarrival intervals are asymptotically orthogonal. Then one constructs the "pre-Wiener" process component by component. It can happen that some subset of the martingales are not asymptotically orthogonal. For example, one set of arrival times might be a subset of another. The general method in [167, Section 10.4] can be adapted to handle quite general situations, and no more will be said here.

Conditional means, variances, and routing probabilities depending on an auxiliary process. Suppose that there is a Euclidean-space-valued process $v^n(\cdot)$ (which might depend on $x^n(\cdot)$) such that if $v^n(t) = v$

---

[1] $\Psi^n(\cdot)$ was defined in Theorem 1.1.

at the start of an interarrival or service interval (respectively, just before the end of a service interval), then the $\Delta_i^\alpha(x)$ (respectively, $q_{ij}(x)$) in (A1.1)–(A1.3) are replaced by $\Delta_i^\alpha(x,v)$ (respectively, $q_{ij}(x,v)$). Suppose that these functions are bounded and continuous and that $v^n(\cdot)$ is asymptotically continuous if $x^n(\cdot)$ is. Further, suppose that the x-component of any weak-sense limit $(x(\cdot),v(\cdot))$ of $(x^n(\cdot),v^n(\cdot))$ satisfies (1.8)–(1.11), but with $b(x,v)$ and $\sigma_{\alpha,i}^2(x,v)$ used.

Since the proof of asymptotic continuity of $x^n(\cdot)$ in Theorem 1.1 continues to hold with no change, one can add such auxiliary processes $v^n(\cdot)$, and the theorem remains true with this change. The analogous result will hold for the model of Theorem 2.1.

8.2 Poisson-Type Input and Service Process

8.2.1 The Basic Model

One of the most popular and historical models in queueing theory takes the exogenous arrival processes to be Poisson, and the service times to be mutually independent and exponentially distributed [132]. Apart from the fact that such a model fits many applications, it has the very convenient property of being Markovian.

A Poisson exogenous arrival process can be viewed "locally," in the sense that the Poisson process is a consequence of the assumptions that the arrivals over disjoint intervals are mutually independent, that the probability of an arrival over a small time interval $[t,t+\delta)$ is $\mu\delta + o(\delta)$ for some $\mu > 0$, and that the probability of more than one arrival on such an interval is $o(\delta)$. In this sense, if the arrivals are drawn from many possible sources with each acting independently, then the Poisson assumption is reasonable. Similarly, the "exponential" assumption on the service time is equivalent to supposing that the event that a job in service at real time t will be completed in the interval $[t,t+\delta)$ is independent of the past history, and its probability is $O(\delta)$. The network analogue is the popular Jackson network [125].

We will consider the state-dependent form of such a network, where the arrival and service rates and the routing probabilities depend on the current state. Many examples are in [2, 163, 171, 186]. The terminology and network structure of the previous section will be retained unless otherwise stated. Thus, $x_i^n(t)$ continues to denote $1/\sqrt{n}$ times the content of the queue at P_i at real time nt. Recall the definition of $A_i^n(t)$ (respectively, $D_i^n(t)$) as $1/\sqrt{n}$ times the number of exogenous arrivals to (respectively, departures from) P_i by real time nt. They will be jump processes of the "state-dependent" Poisson type. Let $\mathcal{F}_t^n$ denote the minimal σ-algebra that measures all of the system data up to real time nt.

Assumptions. The assumptions concerning the arrival and service rates are stated for the original unscaled problem in real time, in the following way. All $o_i^{\alpha,n}(1/\sqrt{n})$ are $o(1/\sqrt{n})$, uniformly in x in each bounded set.

A2.1. *There are constants $\bar{\lambda}_i^a$, $i \leq K$, and continuous functions $\lambda_i^a(\cdot)$, with at most linear growth, such that the (conditional mean) exogenous arrival rate at P_i at real time nt, when $x^n(t) = x$, is*

$$\bar{\lambda}_i^{a,n}(x) = \bar{\lambda}_i^a + \lambda_i^a(x)/\sqrt{n} + o_i^{a,n}(1/\sqrt{n}). \tag{2.1}$$

That is, the process redefined by

$$M_i^{a,n}(t) = A_i^n(t) - \sqrt{n}\int_0^t \bar{\lambda}_i^{a,n}(x^n(s))ds \tag{2.2}$$

is an $\mathcal{F}_t^n$-martingale.

A2.2. *There are constants $\bar{\lambda}_i^d$, $i \leq K$, and continuous functions $\lambda_i^d(\cdot)$, with at most linear growth, such that the (conditional mean) service rate at P_i at real time nt when $x^n(t) = x$ and $x_i^n(t) > 0$ is*

$$\bar{\lambda}_i^{d,n}(x) = \bar{\lambda}_i^d + \lambda_i^d(x)/\sqrt{n} + o_i^{a,n}(1/\sqrt{n}). \tag{2.3}$$

That is, the process redefined by

$$M_i^{d,n}(t) = D_i^n(t) - \sqrt{n}\int_0^t I_{\{x_i^n(s)>0\}}\bar{\lambda}_i^{d,n}(x^n(s))ds \tag{2.4}$$

is an $\mathcal{F}_t^n$-martingale.

A2.3. *The probability, conditioned on $\mathcal{F}_t^n$, of any two or more events (arrival or service completion) in the system on the real time interval $[nt, nt+\delta]$ is $o(\delta)$, where $o(\delta)$ is uniform in n, t and in $x^n(t) = x$ for x in any compact set.*

A2.4. The heavy traffic condition is $\bar{\lambda}_i^a = \bar{\lambda}_i^d - \sum_j \bar{q}_{ji}\bar{\lambda}_j^d$,

where $\bar{q}_{ij}$ is defined in (A1.3).

Let $I_{ij}^{d,n}(t)$ denote the indicator of the event that there is a routing from P_i to P_j at real time nt. The process $M_{ij}^{r,n}(t)$ in (1.6) is redefined as

$$M_{ij}^{r,n}(t) = \int_0^t \left[I_{ij}^{d,n}(s) - q_{ij}^n(x^n(s-))\right] dD_i^n(s). \tag{2.5}$$

Recall the definition

$$y_i^n(t) = \bar{\lambda}_i^d\sqrt{n}\int_0^t I_{\{x_i^n(s)=0\}}ds = \bar{\lambda}_i^d T_i^{d,n}(t). \tag{2.6}$$

Using (2.2), (2.4), and (2.5), redefine the vectors

$$M^{\alpha,n}(\cdot) = (M_i^{\alpha,n}(\cdot),\ i \le K),\ \ \alpha = a, d,$$
$$M_i^{r,n}(\cdot) = (M_{ij}^{r,n}(\cdot),\ j \le K),\ \ M^{r,n}(\cdot) = \sum_i M_i^{r,n}(\cdot).$$

Set $\bar{M}^n(\cdot) = (M^{a,n}(\cdot), M^{d,n}(\cdot), M_i^{r,n}(\cdot), i \le K)$, and use the same notation, without the superscript n, for any weak-sense limit.

Theorem 2.1. *Let $x^n(0)$ converge weakly to $x(0)$. Assume buffers of sizes $\sqrt{n}B_i$, $B_i \le \infty$, (A1.3), and (A2.1)–(A2.4). Then $(x^n(\cdot), y^n(\cdot), U^n(\cdot), \bar{M}^n(\cdot))$ is asymptotically continuous. The weak-sense limit of any weakly convergent subsequence satisfies (1.8)–(1.10), where $b(\cdot)$ is defined by (1.2). The $M_i^\alpha(\cdot)$, $\alpha = a, d, r$, $i \le K$, are mutually independent Wiener processes. The $M_i^\alpha(\cdot)$ have variances $\bar{\lambda}_i^\alpha$, and the covariance of $M_i^r(\cdot)$ is $\bar{\lambda}_i^d$ times the matrix $\Sigma_{r,i}$ defined in (6.1.14) with $\bar{q}_{ij}$ replacing the q_{ij} there.*

Proof. The proof is very close to that of Theorem 1.1. To simplify the development, suppose that the $\lambda_i^\alpha(\cdot)$ are bounded, and that the $o(1/\sqrt{n})$ terms in (A2.1)–(A2.3) and (A1.3) are uniform in x. Otherwise, a truncation argument is used to get the same results. The terms in the equation

$$x_i^n(t) = x_i^n(0) + A_i^n(t) + \sum_j D_{ji}^n(t) - D_i^n(t) - U_i^n(t) \tag{2.7}$$

can be written as

$$A_i^n(t) = M_i^{a,n}(t) + \sqrt{n}\int_0^t \bar{\lambda}_i^{a,n}(x^n(s))ds, \tag{2.8}$$

$$D_i^n(t) = M_i^{d,n}(t) + \sqrt{n}\int_0^t I_{\{x_i^n(s)>0\}}\bar{\lambda}_i^{d,n}(x^n(s))ds, \tag{2.9}$$

$$D_{ji}^n(t) = \int_0^t I_{ji}^{d,n}(s)dD_j^n(s).$$

Rewrite the last expression as

$$\begin{aligned} D_{ji}^{r,n}(t) &= M_{ji}^{r,n}(t) + \int_0^t q_{ji}^n(x^n(s-))dD_j^n(s) \\ &= M_{ji}^{r,n}(t) + \int_0^t q_{ji}^n(x^n(s-))dM_j^{d,n}(s) \\ &\quad + \sqrt{n}\int_0^t I_{\{x_j^n(s)>0\}} q_{ji}^n(x^n(s))\bar{\lambda}_j^{d,n}(x^n(s))ds. \end{aligned} \tag{2.10}$$

The representations (2.1), (2.3) and the heavy traffic condition yield

$$\sqrt{n}\int_0^t \Big[\bar\lambda_i^a(x^n(s)) - I_{\{x_i^n(s)>0\}}\bar\lambda_i^d(x^n(s)) + \sum_j I_{\{x_j^n(s)>0\}} q_{ji}^n(x^n(s))\bar\lambda_j^d(x^n(s))\Big] ds$$
$$= \int_0^t \Big[\lambda_i^a(x^n(s)) - \lambda_i^d(x^n(s)) + \sum_j \left[\bar q_{ji}\lambda_j^d(x^n(s)) + \bar\lambda_j^d q_{ji}(x^n(s))\right]\Big] ds + y_i^n(t) - \sum_j \bar q_{ji} y_j^n(t) + \sqrt{n}o(1/\sqrt{n}) + \sum_j O(T_j^{d,n}(t)/\sqrt{n}). \tag{2.11}$$

The nonreflection terms on the last two lines of (2.11) are asymptotically continuous. The contribution of the martingale terms to (2.7) is

$$M_i^{a,n}(t) - M_i^{d,n}(t) + \sum_j M_{ji}^{r,n}(t) + \sum_j \int_0^t q_{ji}^n(x^n(s-))dM_j^{d,n}(s), \tag{2.12}$$

and the discontinuities in these terms are of order $1/\sqrt{n}$. Thus, if they are tight, they are asymptotically continuous. Tightness will be verified by evaluating the associated Doob–Meyer processes.

Using the amplitude scaling of $1/\sqrt{n}$, it is easily verified (see Subsection 2.4.2) that the Doob–Meyer process associated with $M_i^{a,n}(\cdot)$ is

$$\langle M_i^{a,n}\rangle(t) = \int_0^t \left[\bar\lambda_i^a + \frac{1}{\sqrt{n}}\lambda_i^a(x^n(s)) + o(1/\sqrt{n})\right] ds \tag{2.13}$$

in the sense that $[M_i^{a,n}(\cdot)]^2 - \langle M_i^{a,n}\rangle(\cdot)$ is an $\mathcal{F}_t^n$-martingale. The $o(1/\sqrt{n})$ term is uniform in time in any bounded interval and in the state value. Similarly, the Doob–Meyer process associated with $M_i^{d,n}(\cdot)$ is

$$\langle M_i^{d,n}\rangle(t) = \int_0^t I_{\{x_i^n(s)>0\}}\left[\bar\lambda_i^d + \frac{1}{\sqrt{n}}\lambda_i^d(x^n(s)) + o(1/\sqrt{n})\right] ds. \tag{2.14}$$

The Doob–Meyer processes associated with $M_{ij}^{r,n}(\cdot)$ are ($j \neq k$ in the second equation)

$$\begin{aligned}\langle M_{ij}^{r,n}\rangle(t) &= \int_0^t I_{\{x_i^n(s)>0\}} q_{ij}^n(x^n(s))\left[1 - q_{ij}^n(x^n(s))\right]\bar\lambda_i^{d,n}(x^n(s))ds,\\ \langle M_{ij}^{r,n}, M_{ik}^{r,n}\rangle(t) &= -\int_0^t I_{\{x_i^n(s)>0\}} q_{ij}^n(x^n(s)) q_{ik}^n(x^n(s))\bar\lambda_i^{d,n}(x^n(s))ds,\end{aligned} \tag{2.15}$$

Equations (2.13)–(2.15) and the criterion of Theorem 2.8.3 imply that the martingales are tight. This implies that the term $q_{ji}^n(x^n(s))$ in (2.12) can be replaced by $\bar q_{ji}$ without changing the weak-sense limits. Putting the above representations together, (1.7) holds (modulo $\sum_i O(T_i^{d,n}(\cdot)/\sqrt{n})$, which is asymptotically continuous) where $\epsilon_i^n(\cdot)$ converges weakly to the "zero" process. Thus, Theorem 3.6.1 implies that $y^n(\cdot)$ and $U^n(\cdot)$ are asymptotically continuous. Thus, so is $x^n(\cdot)$, and the above $O(T_i^{d,n}(\cdot)/\sqrt{n})$ terms converge weakly to the "zero" process. Also, the processes defined by $\int_0^t I_{\{x_i^n(s)=0\}}ds$ converge weakly to the "zero" process. Thus we can drop the indicator functions in (2.14) and (2.15) without changing their asymptotic values.

We need only characterize the weak-sense limits of the martingales. One could use Theorem 2.8.8 together with a representation of $F(\bar{M}(t+\tau)) - F(\bar{M}(t))$ via Itô's lemma. Alternatively, one could use Theorem 2.8.2 together with the fact that the jumps of $\bar{M}^n(\cdot)$ are $O(1/\sqrt{n})$. But for the sake of variety, we will do a direct proof and verify the conditions of Theorem 2.1.3.

Write $\Psi^n(t) = (x^n(t), \bar{M}^n(t), y^n(t), U^n(\cdot))$, and let $\Psi(\cdot) = (x(\cdot), \bar{M}(\cdot), y(\cdot), U(\cdot))$ denote the weak-sense limit (with the convergent subsequence also indexed by n). Let $\{\mathcal{F}_t, t \geq 0\}$ denote the filtration engendered by $\Psi(\cdot)$. It can be shown by a use of Itô's lemma (details left to the reader) that

$$E\left|\bar{M}^n(t)\right|^4 = O(t^2),$$

uniformly in n. Thus, for any $t > 0$,

$$\{|\bar{M}^n(s)|^2; s \leq t, n\} \text{ is uniformly integrable.} \tag{2.16}$$

Using the terminology of Theorem 2.1.3, we have

$$Eh(\Psi^n(s_i), i \leq p)E_t^n\left[\bar{M}^n(t+\tau) - \bar{M}^n(t)\right] = 0. \tag{2.17}$$

By the uniform integrability (2.16) and the weak convergence, (2.17) also holds if the n is dropped. Thus (2.1.6) holds, and $\bar{M}(\cdot)$ is an $\mathcal{F}_t$-martingale. Now turning to the verification of (2.1.7), we evaluate

$$Eh(\Psi^n(s_i), i \leq p)E_t^n\left[\bar{M}^n(t+\tau) - \bar{M}^n(t)\right]\left[\bar{M}^n(t+\tau) - \bar{M}^n(t)\right]'. \tag{2.18}$$

The Doob–Meyer processes

$$\begin{array}{ll} \langle M_i^{a,n}, M_j^{a,n}\rangle(\cdot), & \langle M_i^{d,n}, M_j^{d,n}\rangle(\cdot),\ i,j \leq K, \text{ for } i \neq j, \\ \langle M_i^{a,n}, M_j^{d,n}\rangle(\cdot), & \langle M_i^{a,n}, M_j^{r,n}\rangle(\cdot),\ i,j \leq K, \end{array}$$

are all zero, since the exogenous arrival and service completion times are distinct with probability one. The processes $\langle M_i^{d,n}, M_j^{r,n}\rangle(\cdot)$ are zero due to the distinct times of completion of services (for $i \neq j$) and to (A1.3) for $i = j$. Similarly, the $\langle M_i^{r,n}, M_j^{r,n}\rangle(\cdot)$ are zero for $i \neq j$.

Thus the conditional expectation in (2.18) can be written as a diagonal block matrix with blocks

$$E_t^n\left[M_i^{\alpha,n}(t+\tau) - M_i^{\alpha,n}(t)\right]^2 \to \bar{\lambda}_i^\alpha t, \quad i \leq K,\ \alpha = a, d, \tag{2.19}$$

and

$$E_t^n \left[M_i^{r,n}(t+\tau) - M_i^{r,n}(t)\right] [M_i^{r,n}(t+\tau) - M_i^{r,n}(t)]' \to \bar{\lambda}_i^d \Sigma_{r,i} t, \quad i \le K. \tag{2.20}$$

The convergence in (2.19) and (2.20) is in the mean and follows from the form of the Doob–Meyer processes in (2.13)–(2.15) with the indicator functions dropped. Thus, (2.1.7) holds for the asserted variances. ■

8.2.2 A Generalization: Relationships with Section 1

Theorem 2.1 is a special case of Theorem 1.1, and combinations of the assumptions of Theorems 1.1 and 2.1 can also be used. To see this, consider the model of Theorem 2.1 and suppose for simplicity in the discussion that the functions $\lambda_i^\alpha(\cdot)$, $\alpha = a, d$ and $q_{ij}(\cdot)$ are bounded. Suppose that the $(l+1)$st interval at P_i (either exogenous interarrival or service) starts at real time nt. The probability, conditioned on the data to t, that it has not ended by real time $nt + s$ is

$$E_t^n \exp\left[- \int_0^s \bar{\lambda}_i^{\alpha,n}(x^n(t+u/n))du \right].$$

The mean value of the interval, conditioned on the data up to real time nt, is

$$E_t^n \int_0^\infty s\bar{\lambda}_i^{\alpha,n}(x^n(t+s/n)) \left\{ \exp - \int_0^s [\bar{\lambda}_i^{\alpha,n}(x^n(t+u/n))du] \right\} ds,$$

which equals

$$\frac{1}{\bar{\lambda}_i^\alpha} + O\left(\frac{1}{\sqrt{n}}\right). \tag{2.21}$$

A similar computation shows that the conditional variance is $1/[\bar{\lambda}_i^\alpha]^2 + O(1/\sqrt{n})$, and that the fourth conditional moments are uniformly bounded. Thus, in the terminology of Section 1, and using $\bar{\Delta}_i^\alpha = 1/\bar{\lambda}_i^\alpha$,

$$\bar{\Delta}_{i,l}^{\alpha,n}(x) = \bar{\Delta}_i^a + O_{i,l}^{\alpha,n}\left(\frac{1}{\sqrt{n}}\right),$$

where $O_{i,l}^{\alpha,n}(1/\sqrt{n})$ is $O(1/\sqrt{n})$ uniformly in l, n, i, x, and the limit (as $n \to \infty$) of the conditional variance is $1/[\bar{\Delta}_i^\alpha]^2$. The above results are enough to get tightness and the continuity of the weak-sense limits via the method of Section 1. This is true, since the proof of tightness does not involve the joint properties of the M-processes, only that the means and variances of each are of the correct order and that the heavy traffic condition (A1.4) holds. Then the asymptotic continuity of $x^n(\cdot)$ and the expression (2.21) for the conditional mean imply that (A1.1) holds for a continuous function $\Delta_i^\alpha(\cdot)$ which can be obtained from the representation (2.3). Similarly, (A1.2) holds for $\sigma^2_{\alpha,i}(x) = 1$. In more detail, using the asymptotic continuity of $x^n(\cdot)$ and (2.3) , the conditional expectation of the length of the interval (the state at the start of the interval is x) can be written as

$$\frac{1}{\bar{\lambda}_i^a + \lambda_i^a(x)/\sqrt{n} + o(1/\sqrt{n})} = \frac{1}{\bar{\lambda}_i^\alpha}\left[1 - \frac{\lambda_i^\alpha(x)}{\sqrt{n}\bar{\lambda}_i^\alpha} + o(\frac{1}{\sqrt{n}})\right].$$

Using the definitions $\lambda_i^\alpha(x) = \bar{\lambda}_i^\alpha \Delta_i^\alpha(x)$ from Section 1, we see that the conditional mean length can be written as

$$\bar{\Delta}_i^\alpha \left[1 - \frac{\Delta_i^\alpha(x)}{\sqrt{n}} + o(\frac{1}{\sqrt{n}})\right],$$

which is just the form (A1.1) of Section 1.

By the above results, Theorem 2.1 is a special case of Theorem 1.1. From the point of view of modeling, the models of Theorems 1.1 and 2.1 can be combined, in that some exogenous arrival or service processes can be modeled by one form, and others by the other form. Theorem 1.1 would cover such a combination.

8.2.3 Vacations

All of the results concerning vacations can be adapted to the state-dependent case. State-dependent variations of the "short" vacations in Subsection 7.3.1 are similar to what has been done with the drift and variance terms in Theorem 1.1. The situation in Theorem 2.1 is more complicated. One needs to replace the indicator function in (2.4) by the indicator function of the event that both $x_i^n(s) > 0$ and there is no vacation at scaled time s and introduce a process for the vacations.

In Subsection 7.3.2, the "rate" of starting of the long vacations at P_i was $\bar{\lambda}_i^{s,n}/n$ in real time and $\bar{\lambda}_i^{s,n}$ in scaled time. Let us work in scaled time so that the result will include the fast arrivals of Section 6.6 as well. The comments apply to both Theorems 1.1 and 2.1, since we work on the intervacation and vacation sections separately, as in Theorem 7.3.2. Suppose that (in scaled time) the rate of starting vacations is $\lambda_i^{s,n}(x)$, when the state is x, where for each x, $\lambda_i^{s,n}(x) \to \lambda_i^s(x)$, a bounded and continuous function, uniformly on each compact set. Suppose that the probability of simultaneous vacations at different P_i is zero. Define $\bar{\lambda} = \sup_{x,i} \lambda_i^s(x)$. For the moment, keep the assumption (A7.3.4) concerning the durations of the vacations. Recall (7.3.16).

The form of the extension of Theorems 1.1 and 2.1 for the problem with long vacations is that of Theorem 7.3.2, but with a minor modification in the representation of the jumps, which will now be partially in terms of random Poisson measures (see Sections 2.3 and 3.7). Let $N_i^0(\cdot)$, $i \leq K$, be mutually independent Poisson random measures, each with rate (conditioned on the past data) $\bar{\lambda}$ and jumps uniformly distributed on $[0, 1]$. Define the function $q_i^0(x,\gamma_0)$ to equal unity if $\gamma_0 \in [0, \lambda_i^s(x)/\bar{\lambda}]$, with $q_i^0(x,\gamma_1) = 0$ otherwise. Let $\nu_{i,k}$ denote the jump times of $\int_0^t \int q_i^0(x(s-),\gamma_0) N_i^0(d\gamma_0\, ds)$. The jump value for the kth vacation at P_i is then obtained from (7.3.16). It is the solution to

$$\delta x(t) = x(\nu_{i,k}-) + \bar{b}^i t + [I - Q']\, \delta y(t) - \delta U(t), \tag{2.22}$$

at time $\tau_{i,k}^v$.

Now suppose that the conditional distribution of $\tau_{i,l}^{v,n}$, given the past, depends only on the value of the state x at the start of the vacation and (as $n \to \infty$) converges weakly to a distribution $\Pi_i^v(x,\cdot)$, uniformly in x in each compact set. Suppose that $\Pi_i^v(x,\cdot)$ is weakly continuous in x. Then use the solution to (2.22) at time $\tau_{i,k}^v$, which has the conditional distribution $\Pi_i^v(x(\nu_{i,k}-),\cdot)$.

8.3 Self-Service with Fast Arrivals

Networks such as telephone systems, where there are a large number of servers (lines), each one dedicated to a single user, and no queue are called *self-service*. We will consider the case where the number of servers is infinite, and the arrival rate of order n. This would be the case if the arrivals were drawn at random from a large population whose size is indexed by n, as for the telephone system.

A single-stage system. Suppose that there is a single arrival stream and a single processing stage that has an infinite number of independent servers. Let $Q^n(t)$ denote the number of servers that are busy at real time t and define $x^n(t) = (Q^n(t) - n\bar{\rho})/\sqrt{n}$, where $\bar{\rho} = \bar{\lambda}^a/\bar{\lambda}^d$, where $\bar{\lambda}^a$ is defined in (A3.1) and $\bar{\lambda}^d$ is defined in (A3.4). Let $A^n(t)$ (respectively, $D^n(t)$) denote $1/\sqrt{n}$ times the number of arrivals (respectively, departures) by real time t. Let $\Delta_l^{a,n}/n$, $l = 1, \ldots$, denote the interarrival intervals. In the model to be presented, the service rate depends only slightly on the state, analogous to the situation in Sections 1 and 2, but this dependence can have substantial effects. An alternative model of self-service arises in communications systems where there is a single channel to be shared equally by all users, as is the case of the internet. Here the service rate would be roughly inversely proportional to the number of users. See Section 1.5 and Chapter 10 for a controlled version of this problem. The terminology of Section 1 is used unless otherwise noted. The techniques are a combination of those used in Theorems 1.1 and 2.1.

The following conditions will be assumed, where x denotes the state at the start of the interval.

A3.1. *There is a constant $\bar{\Delta}^a = 1/\bar{\lambda}^a$ and a continuous function $\Delta^\alpha(\cdot)$, that has at most linear growth in x and satisfy*

$$E_l^{\alpha,n}\Delta_{l+1}^{\alpha,n} \equiv \bar{\Delta}_{l+1}^{\alpha,n}(x) = \bar{\Delta}^\alpha\left(1 - \frac{1}{\sqrt{n}}\Delta^\alpha(x)\right) + o_l^{\alpha,n}(1/\sqrt{n}). \quad (3.1)$$

A3.2. *There is a bounded and continuous function $\sigma_a^2(\cdot)$ such that in the mean and uniformly in l and in x in each bounded set,*

$$E_l^{a,n}\left[1 - \frac{\Delta_{l+1}^{a,n}}{\bar{\Delta}_{l+1}^{a,n}(x)}\right]^2 \to \sigma_a^2(x), \text{ as } n \to \infty.$$

A3.3. $\left\{|\Delta_l^{a,n}|^2; l, n\right\}$ *is uniformly integrable.*

The service-time model is the following form of what was used in Theorem 2.1.

A3.4. *There is a constant $\bar{\lambda}^d$ and a continuous function $\lambda^d(\cdot)$, with at most linear growth, such that the (conditional mean) service rate per busy server at real time t when $x^n(t) = x$ is*

$$\bar{\lambda}^{d,n}(x) = \left[\bar{\lambda}^d + \lambda^d(x)/\sqrt{n} + o^{d,n}(1/\sqrt{n})\right].$$

That is, the process defined by

$$M^{d,n}(t) = D^n(t) - \int_0^t \frac{Q^n(s)}{\sqrt{n}}\bar{\lambda}^{d,n}(x^n(s))ds \quad (3.2)$$

is an $\mathcal{F}_t^n$-martingale.

Note that there is no reflection term in (3.3), and there is no explicit heavy traffic condition.

Theorem 3.1. *Let $x^n(0)$ converge weakly to $x(0)$ and assume* (A3.1)–(A3.4). *Define*

$$b(x) = \bar{\lambda}^a\Delta^a(x) - \bar{\rho}\lambda^d(x) - x\bar{\lambda}^d,$$

and let $M^{a,n}(\cdot)$ be defined as in (1.6). Then $x^n(\cdot)$, $M^{a,n}(\cdot)$, and $M^{d,n}(\cdot)$ are asymptotically continuous, and the weak-sense limit of any weakly convergent subsequence satisfies

$$x(t) = x(0) + \int_0^t b(x(s))ds + M^a(t) - M^d(t). \quad (3.3)$$

Let $\mathcal{F}_t$ denote the minimal σ-algebra that measures $\{x(s), M^a(s), M^d(s), s \le t\}$. Then $M^a(t)$ is a continuous $\mathcal{F}_t$-martingale with quadratic variation process $\bar{\lambda}^a \int_0^t \sigma_a^2(x(s))ds$, and $M^d(\cdot)$ is an $\mathcal{F}_t$-Wiener process with variance $\bar{\lambda}^a$. Also, $M^a(\cdot)$ is orthogonal to $M^d(\cdot)$.

Proof. If the $\lambda^\alpha(\cdot)$, $\alpha = a, d$, are not uniformly bounded or the order $o(\cdot)$ is not uniform in x, then use the truncation method of Subsection 2.5.6. Similarly, the truncation method allows us to suppose that the $x^n(\cdot)$ are bounded. Expand $M^{d,n}(\cdot)$ as

$$D^n(t) - \int_0^t \left[\bar{\lambda}^d x^n(t) + \bar{\rho}\lambda^d(x^n(s)) + \sqrt{n}\bar{\rho}\bar{\lambda}^d + O(1/\sqrt{n}) + \sqrt{n}o(1/\sqrt{n})\right] ds.$$

The Doob–Meyer process associated with $M^{d,n}(\cdot)$ is

$$\frac{1}{n}\int_0^t Q^n(s)\bar{\lambda}^{d,n}(x^n(s))ds, \tag{3.4}$$

which equals $\bar{\lambda}^a t + O(1/\sqrt{n})t$. Write

$$D^n(t) = M^{d,n}(t) + \sqrt{n}\bar{\lambda}^a t + \int_0^t \left[\bar{\lambda}^d x^n(s) + \bar{\rho}\lambda^d(x^n(s)) + \epsilon^n(t)\right] ds,$$

where $\epsilon^n(\cdot)$ converges weakly to the "zero" process.

Let $S^{a,n}(t)$ denote $1/n$ times the number of arrivals by real time t, and $x_l^{a,n}$ the value of the state at the start of the lth interarrival interval. Then write the scaled arrival process as

$$A^n(t) = \frac{1}{\sqrt{n}} \sum_{l=1}^{nS^{a,n}(t)} \left[1 - \frac{\Delta_l^{a,n}}{\bar{\Delta}_l^{a,n}(x_l^{a,n})}\right] + \frac{1}{\sqrt{n}} \sum_{l=1}^{nS^{a,n}(t)} \frac{\Delta_l^{a,n}}{\bar{\Delta}_l^{a,n}(x_l^{a,n})}. \tag{3.5}$$

Then modulo a term that converges weakly to the "zero" process, the second term on the right of (3.5) is

$$\sqrt{n}\bar{\lambda}^a t + \bar{\lambda}^a \sum_{l=1}^{nS^{a,n}(t)} \frac{\Delta_l^{a,n}\Delta^a(x_l^{a,n})}{n}.$$

The remaining details are now readily completed and are left to the reader. ■

Networks. The result of Theorem 3.1 can readily be extended to networks. One needs a representation of the output process from any stage, and this can be obtained via the method used in Theorem 6.1.1. To illustrate the idea, consider a two-stage tandem network. If the output of the first stage is the only input to a second stage, then that input process is just the $D^n(\cdot)$

of Theorem 3.1. Now suppose that an output of stage 1 is an input to stage 2 only with probability (conditional on the past data) $q^n(x) = \bar{q} + q(x)/\sqrt{n}$, where $q(\cdot)$ is bounded and continuous and x is the value of the state just before the time of the service completion. Rewrite $D^n(\cdot)$ as $D_1^n(\cdot)$, and let $D_{12}^n(t)$ denote $1/\sqrt{n}$ times the actual number of outputs of stage 1 going to stage 2 by time t.

Let $I^{r,n}(t)$ denote the indicator function of the event that a completed service at time t will go to stage 2. Then $q^n(x^n(t-))$ is the probability of this event, conditioned on the past data. Write

$$D_{12}^n(t) = \int_0^t \left[I^{r,n}(s) - q^n(x^n(s-))\right] dD_1^n(s) + \int_0^t q^n(x^n(s-)) dD_1^n(s). \quad (3.6)$$

The second term in (3.6) equals (modulo a term that converges weakly to the "zero" process)

$$\sqrt{n}\bar{q}\bar{\lambda}^a t + \int_0^t \left[\bar{q}\left(\bar{\lambda}^d x^n(s) + \bar{p}\lambda^d(x^n(s))\right) + q(x^n(s))\bar{\lambda}^a\right] ds + \bar{q}M^{d,n}(t).$$

The first term in (3.6) converges weakly to a Wiener process that is independent of the others and that has variance $\bar{\lambda}^a\bar{q}(1-\bar{q})$.

The references [22, 25, 84, 106, 202, 251] deal with related problems, sometimes with more general service times. See, in particular, [84], which contains much useful information.

8.4 Discontinuous Dynamics

Continuity of the drifts, covariances, and routing probabilities was assumed for simplicity of the proofs. However, the dynamical terms will be discontinuous whenever threshold or similar "switching-type" controls are used. The following general result uses Theorem 2.5.4 to extend Theorems 1.1 and 2.1 to a large class of discontinuous dynamical terms.

Theorem 4.1. *Assume the conditions of Theorems* 1.1 *or* 2.1, *but replace continuity of the* $\Delta_i^\alpha(\cdot), \lambda_i^\alpha(\cdot)$, $\alpha = a, d$, $q_{ij}(\cdot), i, j$, *by measurability. Then* $\Psi^n(\cdot)$ *in Theorem* 1.1 *is asymptotically continuous. Suppose that for each* t, $\alpha = a, d$, *and* $i \le K$, *the functions whose values are*

$$\begin{aligned} &\int_0^t \Delta_i^\alpha(\phi(s))ds, \quad \int_0^t \lambda_i^\alpha(\phi(s))ds, \\ &\int_0^t q_{ij}(\phi(s))ds, \quad \int_0^t \sigma_{i,\alpha}^2(\phi(s))ds \end{aligned} \quad (4.1)$$

are continuous on $D(\mathbb{R}; 0, \infty)$ *with probability one with respect to the measure induced by any weak-sense limit process* $x(\cdot)$. *Then the conclusions of*

Theorems 1.1 and 2.1 hold. Let G_d denote the set of points in G where at least one of the integrand functions in (4.1) is discontinuous. If for each $\delta > 0$ and $t > 0$,

$$\lim_{\epsilon \to 0} P\{\text{time } x(\cdot) \text{ spends in } N_\epsilon(G_d) \text{ on } [0,t] \text{ is } \geq \delta\} = 0 \tag{4.2}$$

holds for any weak-sense limit process $x(\cdot)$, then the desired continuity with probability one of (4.1) holds.

Comments on the proof. For concreteness, we work with the model of Theorem 1.1. The proofs of asymptotic continuity of the set $\Psi^n(\cdot)$, the fact that the weak-sense limits of the $M_i^{\alpha,n}(\cdot)$ are martingales, and the facts that the weak-sense limits of the processes defined by the $B^n(\cdot)$ in (1.19) and the Doob–Meyer processes of the limit martingales are Lipschitz continuous do not require continuity of the dynamical terms. The only issue is the characterization of the weak-sense limits of the $B^n(\cdot)$ and of the Doob–Meyer processes of the limit martingales. The theorem follows from Theorem 2.5.4 and the (with probability one) continuity of the functions defined in (4.1). The continuity (with probability one) of the functions defined in (4.1) is implied by (4.2).

Condition (4.2) can also be used directly without recourse to Theorem 2.5.4. To do this, fix $\epsilon > 0$ small and define the following stopping times recursively. Let $\tau_0^n = 0$, and, for $k > 0$,

$$\begin{aligned} \sigma_k^n &= \min\{t \geq \tau_{k-1}^n : x^n(t) \in N_{\epsilon/2}(G_d)\}, \\ \tau_k^n &= \min\{t > \sigma_k^n : x^n(t) \notin N_\epsilon(G_d)\}. \end{aligned}$$

Divide time into intervals

$$[\tau_0^n, \sigma_1^n),\ [\sigma_1^n,\ \tau_1^n), \ldots.$$

Asymptotically, for any bounded time segment, the probability that the path $x^n(\cdot)$ can jump from $N_{\epsilon/2}(G_d)$ to $G - N_\epsilon(G_d)$ or in reverse in one step goes to zero. Also, for any $t > 0, \epsilon > 0$, the probability that more than N of the above intervals are needed to cover $[0,t]$ goes to zero as $N \to \infty$, uniformly in n. Approximate the first term in (1.19) by a sum of the type

$$\sum_{k=0}^{\infty} \sum_{l=nS_i^{a,n}(t\wedge\tau_k^n)}^{nS_i^{a,n}(t\wedge\sigma_{k+1}^n)} + \sum_{k=1}^{\infty} \sum_{l=nS_i^{a,n}(t\wedge\sigma_k^n)+1}^{nS_i^{a,n}(t\wedge\tau_k^n)} .$$

Consider the second sum. By (4.2), the weak convergence and continuity of the limits, its contribution can be made as small as desired as $n \to \infty$ by making ϵ small. Now consider the first sum. The argument of Theorem 1.1 can be applied to each of the summands $k = 0, 1, \ldots$. Repeat for the other terms in (1.19).

The condition (4.2) is satisfied if the $M(\cdot)$ in (1.8) is nondegenerate on some open set that contains $\bar{G}_d$. By nondegeneracy it is meant that $\sigma(x)\sigma'(x)$ is is uniformly positive definite in such a set. More generally, $M(\cdot)$ need only be nondegenerate in any direction that forces the process out of any arbitrarily small neighborhood of G_d.

Examples of discontinuous routing. Consider the particular form of the model of Theorem 6.1.1, where there are two mutually independent input streams A_i^n, and two processors P_i, $i = 1, 2$, each with its own queue. The outputs leave the system. The input stream A_i^n goes to P_i unless rerouted according to the following policy. The (conditional on past data) probability that a job from A_i^n goes to P_j, $j \neq i$, is $q_{ij}(x)/\sqrt{n}$, where x is the state just before the time of arrival of that job. Suppose that $\sigma_{i,a}^2 + \sigma_{i,d}^2 > 0$ for each i. The heavy traffic condition is that there are b_i such that

$$\lim_n \sqrt{n} \left[\frac{1}{\bar{\Delta}_i^{a,n}} - \frac{1}{\bar{\Delta}_i^{d,n}} \right] = b_i. \tag{4.3}$$

Case 1. Let $C > 0$ and let $q_{ij}(x)$ equal zero if $x_i - x_j < C$ and equal $\bar{q} > 0$ otherwise. Define

$$q_{ii}(x) = 1 - \frac{q_{ij}(x)}{\sqrt{n}}, \quad j \neq i.$$

Let $A_{ij}^n(t)$ denote $1/\sqrt{n}$ times the number from A_i^n that were directed to P_j by real time nt. Then for $i \neq j$,

$$x_i^n(t) = x_i^n(0) + A_{ii}^n(t) + A_{ji}^n(t) - D_i^n(t), \tag{4.4}$$

where $A_{ii}^n(t) = A_i^n(t) - A_{ij}^n(t)$. Theorem 4.1 applies and yields that the weak-sense limit process satisfies, for $i \neq j$,

$$\begin{aligned} x_i(t) &= x_i(0) + w_i^a(\bar{\lambda}_i^a t) - w_i^d(\bar{\lambda}_i^a t) + b_i t + y_i(t) \\ &\quad - \bar{\lambda}_i^a \int_0^t q_{ij}(x(s))ds + \bar{\lambda}_j^a \int_0^t q_{ji}(x(s))ds, \end{aligned} \tag{4.5}$$

where $y_i(\cdot)$ is the reflection term at the origin.

Case 2. Now let all of A_i^n be directed to P_j, $j \neq i$, when $x_i^n(t) - x_j^n(t) > C$. Rewrite (4.4) as $(i \neq j)$

$$\begin{aligned} x_i^n(t) &= x_i^n(0) + w_i^{a,n}(S_i^{a,n}(t)) - w_i^{d,n}(S_i^{d,n}(t)) \\ &\quad + b_i^n t + y_i^n(t) - A_{ij}^n(t) + A_{ji}^n(t) + \epsilon_i^n(t), \end{aligned}$$

where $\epsilon_i^n(\cdot)$ converges weakly to the "zero" process. The processes $A_{ij}^n(\cdot)$, $i \neq j$, function as reflection terms acting on and normal to the shifted diagonals. If, $x_i^n(t) - x_j^n(t) > C$ at the moment of an arrival to P_i, then there is an immediate rerouting. Theorem 3.6.1 can be used to show that the $A_{ij}^n(\cdot)$, $i \neq j$, are asymptotically continuous. The weak-sense limit processes satisfy

$$x_i(t) = x_i(0) + w_i^a(\bar{\lambda}_i^a t) - w_i^d(\bar{\lambda}_i^a t) + b_i t + y_i(t) - A_{ij}(t) + A_{ji}(t), \quad (4.6)$$

where $A_{ij}(\cdot)$ are reflection terms on the shifted diagonals. The solution to (4.6) is unique in the strong sense.

8.5 An Application to the Multiplexer–Buffer System

8.5.1 Introduction and Problem Description

The methods of the previous sections will be illustrated on the class of problems in communications systems that was discussed in Section 1.4. In various forms such problems are ubiquitous. There is a data transmission system with n independent sources, each generating data in a bursty way and competing for a single shared channel. The data are multiplexed into a single stream and then buffered, until being transmitted. The buffer is large but finite, being $\sqrt{n}B$, where $0 < B < \infty$. Inputs are processed immediately if the server is idle. Arrivals to a full buffer are rejected. The $\sqrt{n}$ factor in the scaling is suggested by heavy traffic analysis. As usual in heavy traffic analysis, if the buffer size were $o(\sqrt{n})$, then (asymptotically) it would always be full, and a large part of the input would be lost. If the buffer size were of order larger than $O(\sqrt{n})$, it would play no role in the limit.

The general idea in Section 1.4 covered many possible types of input processes. All that was required is that when suitably scaled by an appropriate function of n, they superimpose into a stream with a well defined limit as $n \to \infty$. In this section one particular realization will be treated. There might be fixed controls, but the general control problem is deferred to Section 9.3. The first issue is the selection of the mechanism by which each source generates its own bursty data stream. The "Markov modulated" model of [5, 70, 136] will be used. In the simplest form of this model, each source alternates between *on* and *off* periods, and can send messages only during the *on* periods. It is supposed that there is only a single type of source, so that they are statistically identical. More general cases are in [169, 171]. In particular, the *on* and *off* intervals are mutually independent and exponentially distributed with the *on* (respectively, *off*) period having mean value $1/\mu$ (respectively, $1/\lambda$). The intervals are also independent of the initial condition and the service process.

Now we need to fix the data-creation scheme when the source is *on*. The data are assumed to be divided into very small units, or "cells." Two extreme possibilities (and combinations of them) will be allowed. The first is the so-called fluid model, where the amount of data (cells) created is just ν times the length of the *on* time. If, as is usually the case, the cell size is small relative to the durations of the *on* periods, then the possibility that the number of cells sent is nonintegral is unimportant. The "fluid" model for the server is simply that each cell requires the same amount of time, and this is very small, so that one supposes that the number of cells transmitted is the channel capacity times [current time minus idle time]. The second model for the creation of data during the *on* periods supposes that it is Poisson with rate ν and independent of the initial condition and service process. The data-creation processes are assumed to be independent of the other driving random variables. The traffic is heavy in the usual sense that the capacity of the transmitter is only "slightly" larger than the stationary rate of creation of messages.

More generally, the source need not be off during an *off* period. It could simply transmit at a lower rate. The development would then be nearly identical to what will be done. In addition, there might be many statistically distinct classes of source (say, with various types of data, voice, video, etc.). The class of any such source can be fixed for all time, or there might be Markov switching between them. Many such variations are in [169, 171] and we will concentrate on the basic single-class models.

Among the key issues in the design of such systems are the delay or other similar measure of performance, and the loss of cells due to buffer overflow. These are dealt with via appropriate sizing of the buffer and control over the input, as discussed in Section 9.3, together with numerical data. The method is easily adapted to other types of control problems. For example, there might be an alternative route that can be used at an additional cost. A routing control can be chosen to, say, minimize the additional cost plus the cost of waiting on the primary queue.

To formally describe the input model, let $q_i(t)$ denote the indicator function of the event that source i is *on* at time t, and let $a_i(t)$ denote the total number of outputs provided by source i up to time t. Then following the above description, $(q_i(\cdot), a_i(\cdot))$ is a continuous-time Markov process, under either the Markov modulated "fluid" or the Markov modulated Poisson data creation model. Let $\Delta_l^{d,n}/n$ denote the sequence of service times. We will suppose either (A5.1a) or (A5.1b) (the "fluid" service model). But keep in mind that the same development works with some sources being Markov modulated "fluid" and others being Markov modulated Poisson.

A5.1a. *For each n, the $\Delta_l^{d,n}$ are mutually independent and identically distributed with mean value $\bar{\Delta}^{d,n}$ and*

$$\left\{ \left| \frac{\Delta_1^{d,n}}{\bar{\Delta}^{d,n}} \right|^2 ; n \right\} \text{ is uniformly integrable.}$$

Also,

$$E\left[1-\frac{\Delta_1^{d,n}}{\bar{\Delta}^{d,n}}\right]^2 \to \sigma_d^2$$

as $n \to \infty$. The service times are independent of the initial condition and the interarrival intervals and states of the sources.

A5.1b. *There are constants $\bar{\Delta}^{d,n}$ and $\bar{\Delta}^d$ such that $\bar{\Delta}^{d,n} \to \bar{\Delta}^d$ and the number of cells served by real time t is*

$$\frac{n}{\bar{\Delta}^{d,n}}\,[t - \text{idle time by } t.]$$

We will assume that the system is in heavy traffic, in the sense that the server works at a rate that is close to the *stationary* arrival rate of messages from all sources, that is $n\nu\lambda/(\lambda+\mu)$. Define $C_n = n(\bar{\Delta}^{d,n})^{-1}$, the effective "mean" service rate. The fundamental heavy traffic assumption says that the mean service rate is "slightly" larger than the mean (stationary) arrival rate. In particular, we make the following assumption.

A5.2. *There is a real number b such that*

$$C_n = \frac{n\nu\lambda}{\lambda+\mu} + \sqrt{n}b.$$

Controls. Suppose that the buffer content and possibility of overflow are controlled in a state-dependent way by selectively deleting cells before they reach the buffer, either at the sources or at the entrance to the multiplexer or buffer. The following general scheme will be used. Let $F^n(t)$ denote $1/\sqrt{n}$ times the number of cells deleted by t. The next assumption specifies a restriction on the control. The constant $\bar{u}$ in (A5.3) is a limitation on the maximum mean rate at which the controller can delete incoming cells. It is related to the quality of service and is discussed further below. One must select the control to find a good balance between various criteria of good service. The subject of controls for this problem is discussed further in Section 9.3.

A5.3. *The rule for the cell deletion can depend only on the system data that are available to date, namely, the initial condition and the history of arrivals and departures. There is a bound on the instantaneous rate of cell deletion in the sense that there is a $\bar{u}$ and processes $u^n(\cdot)$ and $\epsilon^{u,n}(\cdot)$ such that $\epsilon^{u,n}(\cdot)$ converges weakly to the "zero" process, $0 \le u^n(t) \le \bar{u}$, and*

$$F^n(t) = \int_0^t u^n(s)ds + \epsilon^{u,n}(t).$$

It was shown in [169] that this form can give excellent performance. See the data and further discussion in Section 9.4. Note that the control is viewed in an "aggregated" form in that the exact mechanism for deleting the cells is not specified. The value of $\bar{u}$ in this example is a decision of the designer. The larger its value, the quicker the controller can respond to the threat of potential overload and the smaller the total average overflow losses will be. As $\bar{u}$ decreases, the controller must anticipate overload even sooner, and the overall overflow and control losses will be greater. On the other hand, the larger $\bar{u}$ is, the greater the degradation in the quality of service at those times when the maximum control is exercised. The data in Section 9.4 and in [169] illustrate the interactions between the value of $\bar{u}$, the form of the controls, and the tradeoffs among the types of losses. In principle, one could set $\bar{u} = \infty$, in which case there is no limit on the rate at which the controller can delete incoming cells. This leads to a singular control problem (Chapter 11). Our aim is to illustrate the role of heavy traffic analysis and there are many possibilities within this framework.

In motivating the form of the control in (A5.3), it is important to keep in mind the difference between a desired or ideal control policy and the actual effects of its realization. Most frequently, optimization methods yield a desired policy, without regard to issues of realization. In this example we call $F^n(\cdot)$ the control function, since that is what we aim to choose as best as possible, and it is the weak-sense limit of $F^n(\cdot)$ that appears as the well-defined control in the weak-sense limit system. The form in which the control is written in (A5.3) was determined by the simple fact that in a large and possibly decentralized system one cannot always ensure that the precise desired instantaneous effects of the chosen policy will be realized, especially in view of the possibility that what is called the control might be the cumulative effect of many individual actions, each of which are directed independently by the chosen policy. But in a large and suitably scaled system, the overall effects of the individual actions would average out to be nearly what is the ideal given by the optimization procedure. See Chapter 9 for more detail.

8.5.2 Convergence

Representation of the input process. In this section we obtain asymptotic results for the appropriately centered and normalized input process. To facilitate the weak convergence proofs, we introduce a martingale decomposition for the "input" processes $q_i(\cdot)$ and $a_i(\cdot)$. This will allow us to represent these processes in a form that resembles a reflected diffusion process analogous to the development in Section 2. Let $\mathcal{F}_t^n$ denote the σ-algebra generated by all of the system data by real time t. The $q_i(\cdot)$ will be assumed to be stationary. The reference [171] treats the nonstationary case, with little difference in the development. Theorem 5.1 gives the weak-convergence result for the scaled and centered number of *on* sources. Then in Theorem 5.2 the weak convergence of the scaled buffer content process will be proved.

Martingale decompositions. Define the processes $\tilde{q}_i(\cdot)$ and $\tilde{a}_i(\cdot)$ by

$$\begin{aligned} dq_i(t) &= \left[-\mu q_i(t) + \lambda(1 - q_i(t))\right] dt + d\tilde{q}_i(t), \\ da_i(t) &= \nu q_i(t)\, dt + d\tilde{a}_i(t), \end{aligned} \tag{5.1}$$

and $\tilde{q}_i(0) = \tilde{a}_i(0) = 0$. For the "fluid" input model, $\tilde{a}_i(t) = 0$ in (5.1). Since

$$\begin{aligned} E\left[q_i(t+s) - q_i(t)\big|\mathcal{F}_t^n\right] &= [-\mu q_i(t) + \lambda(1 - q_i(t))]s + o(s), \\ E\left[a_i(t+s) - a_i(t)\big|\mathcal{F}_t^n\right] &= \nu q_i(t)s + o(s), \end{aligned}$$

the processes $\tilde{q}_i(\cdot)$ and $\tilde{a}_i(\cdot)$ are $\mathcal{F}_t^n$-martingales. The following second moments can be easily computed by standard Markov chain arguments:

$$\begin{aligned} &E\left[(\tilde{q}_i(t+s) - \tilde{q}_i(t))^2\big|\mathcal{F}_t^n\right] = [\mu q_i(t) + \lambda(1 - q_i(t))]s + o(s), \\ &E\left[(\tilde{a}_i(t+s) - \tilde{a}_i(t))^2\big|\mathcal{F}_t^n\right] = \nu q_i(t)s + o(s), \\ &E\left[(\tilde{q}_i(t+s) - \tilde{q}_i(t))(\tilde{a}_i(t+s) - \tilde{a}_i(s))\big|\mathcal{F}_t^n\right] = o(s). \end{aligned} \tag{5.2}$$

The mean values of the $o(s)$ terms are also $o(s)$. Throughout, the various $o(\cdot)$ terms will have this property. For the fluid arrival process, $\tilde{a}_i(t) = 0$. The Doob–Meyer processes for the martingales are just the integrals of the coefficients of s on the right side of (5.2).

Define the averages

$$Q^n(t) = \frac{1}{n}\sum_{i=1}^{n} q_i(t) \quad \text{and} \quad \tilde{Q}^n(t) = \frac{1}{n}\sum_{i=1}^{n} \tilde{q}_i(t).$$

Then (5.1) yields

$$dQ^n(t) = [-(\lambda + \mu)Q^n(t) + \lambda]\, dt + d\tilde{Q}^n(t), \tag{5.3}$$

where $\tilde{Q}^n(\cdot)$ is an $\mathcal{F}_t^n$-martingale, and (5.2) yields

$$E\left[(\tilde{Q}^n(t+s) - \tilde{Q}^n(t))^2\big|\mathcal{F}_t^n\right] = \frac{1}{n}[\mu Q^n(t) + \lambda(1 - Q^n(t))]s + o(s).$$

Define the *centered* and *normalized* process

$$v^n(t) = \sqrt{n}\left[Q^n(t) - \frac{\lambda}{\lambda + \mu}\right].$$

Then

$$dv^n(t) = -(\lambda + \mu)v^n(t)\, dt + dw^{v,n}(t), \tag{5.4}$$

where

$$w^{v,n}(t) = \frac{1}{\sqrt{n}} \sum_{i=1}^{n} \tilde{q}_i(t) = \sqrt{n}\tilde{Q}^n(t).$$

The next result is needed in Theorem 5.2.

Theorem 5.1. $(v^n(\cdot), w^{v,n}(\cdot))$ *converges weakly to a solution of*

$$dv(t) = -(\lambda + \mu)v(t)\,dt + dw^v(t), \tag{5.5}$$

where $w^v(\cdot)$ *is a Wiener process with variance* $2\lambda\mu/(\lambda+\mu)$. *Furthermore, the process* $v(\cdot)$ *is stationary.*

Proof. The $v^n(\cdot)$ are stationary, so any weak-sense limit process must be stationary. By the first line of (5.2), the Doob–Meyer process associated with the martingale $w^{v,n}(\cdot)$ is

$$\int_0^t [\lambda + (\mu - \lambda)Q^n(s)]\,ds.$$

Since $Q^n(\cdot)$ converges weakly to the "constant" process with values $\lambda/(\lambda+\mu)$, the above integral converges weakly to the function defined by $2\lambda t\mu/(\lambda+\mu)$. The set $\{v^n(0)\}$ is tight. This can be seen by applying Itô's lemma (2.4.4) to $[v^n(\cdot)]^2$, which yields

$$d[v^n(t)]^2 = -2(\lambda+\mu)[v^n(t)]^2dt + \left[\int \gamma^2 \lambda^{v,n}(t, d\gamma)\right] dt + dM^n(t),$$

where $\lambda^{v,n}(t, d\gamma)$ is the jump-rate measure of $w^{v,n}(\cdot)$ at t and $M^n(\cdot)$ is a martingale. Since $\int \gamma^2 \lambda^{v,n}(t, d\gamma)$, the derivative of the Doob–Meyer process, is bounded by some constant C_0 uniformly in n, t, via stationarity we have

$$E[v^n(0)]^2 = \lim_t E[v^n(t)]^2 \le \frac{C_0}{2(\lambda+\mu)},$$

hence the tightness of $\{v^n(0)\}$.

The convergence of the Doob–Meyer process, the fact that the discontinuities of $w^{v,n}(\cdot)$ are $O(1/\sqrt{n})$, and the proof of Theorem 2.1 show that $w^{v,n}(\cdot)$ converges weakly to the cited Wiener process. It follows from this and the form (5.4) that $\{v^n(\cdot)\}$ is tight and converges weakly to the solution of (5.5). ■

The queue length process. Let $x^n(t)$ denote $1/\sqrt{n}$ times the number of cells in the buffer at time t, let $U^n(t)$ denote $1/\sqrt{n}$ times the number of cells lost by time t due to buffer overflow, and define $y^n(t)$ to be $\sqrt{n}/\bar{\Delta}^{d,n}$ times the idle time by t. Let $A^n(t)$ (respectively, $D^n(t)$) denote $1/\sqrt{n}$ times the total number of arrivals (respectively, departures) by time t. Then

$$x^n(t) = x^n(0) + A^n(t) - D^n(t) - F^n(t) - U^n(t). \tag{5.6}$$

Let $S^{d,n}(t)$ denote $1/n$ times the number of departures by time t. Following the usual expansion procedure, modulo an asymptotically negligible error,

$$\begin{aligned} D^n(t) &= \frac{1}{\sqrt{n}} \sum_{l=1}^{nS^{d,n}(t)} 1 = w^{d,n}(S^{d,n}(t)) + \frac{1}{\sqrt{n}} \sum_{l=1}^{nS^{d,n}(t)} \frac{\Delta_l^{d,n}}{\bar{\Delta}^{d,n}} \\ &= w^{d,n}(S^{d,n}(t)) + \frac{\sqrt{n}t}{\bar{\Delta}^{d,n}} - y^n(t). \end{aligned}$$

For the fluid service model, $w^{d,n}(t) \equiv 0$. By (5.1),

$$A^n(t) = \nu \int_0^t v^n(s)ds + \frac{\nu\lambda\sqrt{n}t}{\lambda+\mu} + w^{a,n}(t), \tag{5.7}$$

where

$$w^{a,n}(t) = \frac{1}{\sqrt{n}} \sum_{i=1}^{n} \tilde{a}_i(t) = \frac{1}{\sqrt{n}} \sum_{i=1}^{n} \left[a_i(t) - \nu \int_0^t q_i(s)\,ds \right].$$

Putting together the above representations and using the definition of C_n yields, modulo an asymptotically negligible error,

$$\begin{aligned} x^n(t) = x^n(0) &+ \nu \int_0^t v^n(s)ds - bt - F^n(t) \\ &+ w^{a,n}(t) - w^{d,n}(S^{d,n}(t)) + y^n(t) - U^n(t). \end{aligned} \tag{5.8}$$

The convergence theorem.

Theorem 5.2. *Assume* (A5.1)–(A5.3) *and that* $\{x^n(0)\}$ *is tight. Then* $S^{d,n}(\cdot)$ *converges weakly to the process with values* $\bar{\lambda}^d t = \nu\lambda t/(\lambda+\mu)$. *The sequence* $\{x^n(\cdot), v^n(\cdot), F^n(\cdot), w^{v,n}(\cdot), w^{a,n}(\cdot), w^{d,n}(\cdot), y^n(\cdot), U^n(\cdot)\}$ *is tight, and any weak-sense limit* $\big(x(\cdot), v(\cdot), F(\cdot), w^v(\cdot), w^a(\cdot), w^d(\cdot), y(\cdot), U(\cdot)\big)$ *satisfies*

$$\begin{aligned} dx(t) &= [\nu v(t) - b]dt - dF(t) + dw^a(t) - dw^d(\bar{\lambda}^d t) + dy(t) - dU(t), \\ dv(t) &= -(\lambda+\mu)v(t)\,dt + dw^v(t), \end{aligned} \tag{5.9}$$

where $0 \le x \le B$ *and*

$$F(t) = \int_0^t u(s)ds, \quad 0 \le u(t) \le \bar{u}.$$

The $w^a(\cdot)$, $w^d(\bar{\lambda}^d\cdot)$, $w^v(\cdot)$ *are mutually independent Wiener processes, and* $v(\cdot)$, $x(\cdot)$, $F(\cdot)$, $y(\cdot)$, $U(\cdot)$ *are nonanticipative with respect to them. The process* $v(\cdot)$ *is stationary. The variances of* $w^v(\cdot)$ *and* $w^d(\cdot)$ *are, respectively,* $2\lambda\mu/[\lambda+\mu]$ *and* σ_d^2. *For the Poisson input model, the variance of* $w^a(\cdot)$ *is*

$\nu\lambda/[\lambda+\mu]$, *and it is zero for the fluid input model. For the fluid service model,* $\sigma_d^2 = 0$.

Proof. By (A5.3), $\{F^n(\cdot)\}$ is tight and its weak-sense limits are Lipschitz continuous with Lipschitz constant no greater than $\bar{u}$. The process $S^{d,n}(\cdot)$ is easily (via the method of Theorem 5.1.1) shown to be asymptotically Lipschitz continuous, with Lipschitz constant no greater than $\bar{\lambda}^d$. The martingales are asymptotically continuous. Then the asymptotic smoothness of the nonreflection terms in (5.8) implies the asymptotic continuity of the reflection terms via Theorem 3.6.1. Then as in Theorem 5.1.1, $S^{d,n}(\cdot)$ converges as asserted. The verification of the assertions concerning the asymptotic properties of the martingales and the nonanticipativity follows the lines of Theorem 6.2.1 and is left to the reader. ■

A control determined by a switching curve. In typical applications the control is determined by a switching curve. Namely, the state space is split into two parts by a continuous curve such that the controller deletes cells at the maximum rate of $\bar{u}$ when (x, v) is above the curve, does not delete at all if (x, v) is below the curve, and can do anything if (x, v) is on the curve. One common form (see Section 9.3) uses the line $a_0 + a_1 x + a_2 v = 0$, $a_1 > 0$, $a_2 > 0$, to divide the regions. Thus, cells are deleted at the maximum rate if $a_0 + a_1 x + a_2 v > 0$, etc. On average, the maximum rate of deletion for any one source is $\bar{u}/\sqrt{n}$.

With this threshold control,

$$u^n(t) = \bar{u} I_{\{a_0 + a_1 x^n(t) + a_2 v^n(t) > 0\}}.$$

Theorem 5.2 still holds, and the process $u(\cdot)$ still exists (after the selection of a weakly convergent subsequence). If the covariance matrix of the noise $(w^a(1) + w^d(\bar{\lambda}^d), w^v(1))$ is positive definite, then Theorem 4.1 implies that (almost everywhere)

$$u(t) = \bar{u} I_{\{a_0 + a_1 x(t) + a_2 v(t) > 0\}}, \tag{5.10}$$

and (via the Girsanov transformation method) the solution to (5.9) under this control is weak-sense unique. Furthermore, the results in Section 4.3 imply that there is a unique invariant measure and that the transition function $P(x, v; t, \cdot)$ converges strongly to the invariant measure as $t \to \infty$. The results of Section 4.3 require that G be bounded, but it is $[0, B] \times I\!R$ here. However, owing to the stability properties of the $v(\cdot)$ process and the fact that the control is only in the x-component, the results can be carried over.

It is always true that $\sigma_v^2 > 0$. But suppose that $\sigma_a^2 = \sigma_d^2 = 0$. Then the Girsanov transformation method cannot be used. Theorem 5.2 still holds, except possibly for the form of the derivative of $F(\cdot)$ on the discontinuity. The property (4.2) and Theorem 4.1 hold, which implies that $F(t)$ has the form $\int_0^t u(s)ds$ for the $u(\cdot)$ of (5.10) and that there is a unique weak-sense solution to (5.9) under this control.

8.6 Balking, Withdrawing, and Retrials

Balking. Consider the model and assumptions of Theorem 1.1 or 2.1, but allow the possibility that an exogenous arrival decides not to join the queue, perhaps because it is too long: in other words, it balks. Let the probability of balking for an exogenous arrival at P_i at scaled time t, conditioned on the past data, be $p_i^b(x^n(t-))/\sqrt{n}$, where $p_i^b(\cdot)$ is continuous and has at most linear growth in x. Theorems 1.1 and 2.1 continue to hold, with weak-sense limit model

$$x_i(t) = x_i(0) + \int_0^t b_i(x(s))ds + M_i(t) - \bar{\lambda}_i^a \int_0^t p_i^b(x(s))ds + z_i(t), \quad i \le K. \tag{6.1}$$

The proof involves only a minor change. Let $I_{i,l}^{b,n}$ denote the indicator function of the event that the lth exogenous arrival to queue i does not join. We need to subtract

$$\frac{1}{\sqrt{n}} \sum_{l=1}^{nS_i^{a,n}(t)} I_{i,l}^{b,n} \tag{6.2}$$

from $A_i^n(t)$. Rewrite (6.2) as

$$\frac{1}{\sqrt{n}} \sum_{l=1}^{nS_i^{a,n}(t)} \left[I_{i,l}^{b,n} - p_i^b(x^n(\mathcal{T}_i^{a,n}(l/n)-))/\sqrt{n}\right] + \frac{1}{n} \sum_{l=1}^{nS_i^{a,n}(t)} p_i^b(x^n(\mathcal{T}_i^{a,n}(l/n)-)).$$

The first term converges weakly to the "zero" process. The tightness and asymptotic continuity of $x^n(\cdot)$, $M^n(\cdot)$, $z^n(\cdot)$ continue to hold, and the right-hand term converges weakly to

$$\bar{\lambda}_i^a \int_0^t p_i^b(x(s))ds. \tag{6.3}$$

Withdrawing before service. Now suppose that, in addition to the balking of the last paragraph, any customer can voluntarily leave the system at any time if its patience runs out. Let $p_i^w(\cdot)$, $i \le K$, be continuous functions on $[\mathbb{R}^+]^r \times \mathbb{R}^+$. Suppose that the probability (conditioned on all past data) that a customer in the lth position in queue i will leave the system in the scaled time interval $[t, t+\delta)$ is

$$\frac{1}{\sqrt{n}} p_i^w(x^n(t), l/\sqrt{n})\delta + \frac{o(\delta)}{\sqrt{n}},$$

where the order $o(\delta)$ is uniform on any bounded x-set and time interval. Each customer acts independently of the others. Then the conditional rate at which the scaled queue i decreases due to withdrawing is

$$\bar{p}_i^w(x^n(t)) = \frac{1}{\sqrt{n}} \sum_{l=1}^{\sqrt{n}x_i^n(t)} p_i^w(x^n(t), l/\sqrt{n}).$$

Suppose that $\bar{p}_i^w(\cdot)$ is continuous and has at most linear growth in x. Then Theorems 1.1 and 2.1 continue to hold with weak-sense limit system

$$x_i(t) = x_i(0) + \int_0^t b_i(x(s))ds + M_i(t) - \int_0^t \left[\bar{\lambda}_i^a p_i^b(x(s)) + \bar{p}_i^w(x(s))\right] ds + z_i(t). \tag{6.4}$$

To modify the proofs of Theorems 1.1 and 1.2, write the scaled number of withdrawn customers at P_i by scaled time t as $D_i^{w,n}(t)$ and factor it into a compensator and martingale. The martingale converges weakly to the "zero" process, and the compensator yields the addition to the drift term.

Balking and Retrials. Now consider the problem where there is only a single processor and a finite buffer of size $B\sqrt{n}$. Arriving customers might balk, as in the first part of the section, whose terminology will be reused without the subscript i. Thus, an arriving customer will not join the queue if it balks or if the buffer is full when it arrives. Suppose that an arriving customer at scaled time t who has balked enters a "retrial" queue with probability (conditioned on the past) $p^r(x^n(t-))$, where $p^r(\cdot)$ is continuous and has at most linear growth in x. Also suppose that a customer who arrives when the buffer is full will join the retrial queue with probability (conditioned on the past) p_1. Let $x_r^n(t)$ denote $1/\sqrt{n}$ times the number of customers in the retrial queue at scaled time t. Each customer in the retrial queue attempts entry into the main queue independently (and independently of the past data) and at rate (in scaled time) $\bar{\lambda}_r$ and exits the retrial queue at random (again choosing independently) without making such an attempt with rate (in scaled time) $\bar{\lambda}_e$. Retrials can balk and reenter the retrial queue and retry in the same way as do newly arriving exogenous customers. Let $I_l^{f,n}$ (respectively, $I_l^{b,n}$) denote the indicator function of the event that the lth arriving customer is rejected due to a full buffer (respectively, balks), and let $I_l^{r,n}$ denote the indicator function that such a customer joins the retrial queue. Let $D^{r,n}(t)$ (respectively, $D^{e,n}(t)$) denote $1/\sqrt{n}$ times the number in the retrial queue that have actually joined the main queue (respectively, exited the system) by scaled time t. Then

$$x_r^n(t) = x_r^n(0) + \frac{1}{\sqrt{n}} \sum_{l=1}^{nS^{a,n}(t)} I_l^{b,n} I_l^{r,n} + \frac{1}{\sqrt{n}} \sum_{l=1}^{nS^{a,n}(t)} I_l^{f,n} I_l^{r,n} - D^{e,n}(t) - D^{r,n}(t). \tag{6.5}$$

The third term on the right of (6.5) converges weakly to $p_1 U(t)$, and the second to $\bar{\lambda}_i^a \int_0^t p^b(x(s))p^r(x(s))ds$. Next, decompose $D^{e,n}(\cdot)$ and $D^{r,n}(\cdot)$ into the sum of a martingale and compensator. Only the compensators affect the limit. These are (modulo asymptotically negligible terms) $\bar{\lambda}_e \int_0^t x_r^n(s)ds$ and $\bar{\lambda}_r \int_0^t x_r^n(s)ds$, respectively. The drift term $\bar{\lambda}_r \int_0^t x_r(s)ds$ needs to be added to (6.1). An interesting application of balking and retrials to the modeling of service centers is in [185].

9
Bounded Controls

Chapter Outline

This is the first chapter that is concerned with optimal control. The general heavy traffic problem that leads to the controlled reflected stochastic differential equation with bounded controls is treated, with both the discounted and the ergodic cost criteria. The model is the heavy traffic analogue of the classical control problem where the controls are bounded, and the weak-sense limit system is (1.1) or (1.2). We will use the canonical model (1.3) for the physical system, and assumptions of a general nature are made. It is supposed that the dynamical equation for the application can be manipulated into this general form. The assumptions are tailored to what would be typical in applications, so that they do fit a large number of cases. Random jumps can be added to (1.1) and (1.2) and vacations (or other jump-causing phenomena) incorporated in (1.3) with little change in the development, as noted at the end of Section 1. It will be shown that the optimal costs for the physical problem converge to the optimal cost for the weak-sense limit system and that (loosely speaking) the optimal physical processes converge to the optimal weak-sense limit process. Also, "nice" nearly optimal controls for the weak-sense limit system are nearly optimal for the physical system, under heavy traffic. Such results justify using the weak-sense limit system for the design and analysis of nearly optimal physical systems under heavy traffic.

A word needs to be said concerning the definition of admissible control for the physical problem. In the limit problem (or in the "ideal" model), the control term $u(\cdot)$ will take values in a compact set $\mathcal{U}$. The control term might also be restricted in this way for the physical problem, as when the control is a service rate that one might be able to choose at will, but subject to magnitude constraints. Then the perturbations of the set $\mathcal{U}$ that are to be allowed in the definition of admissible control in (A1.6) or (A1.8) will be zero. For many physical problems the definition of admissible control needs to be more flexible. The admissible controls for the physical system are, by definition, just the set of physically possible control terms that can appear in the evolution equation for the state, and they are often the (perhaps random) consequence of actions that are taken according to some general control rule. Then in defining admissible control one needs to account for the way that the control rule might be physically realized, and we might not be able to restrict the values of the control term to lie completely in the desired set $\mathcal{U}$. It might actually take values in a slightly larger set, which converges to $\mathcal{U}$ as $n \to \infty$. This discrepancy, if it occurs, is due to the structure of the actual physical mechanism that realizes the control term, and we take a broad point of view to cover such applications.

Consider the following example, where the actual control term, as it appears in the evolution equation for the physical system, is determined by a random sampling mechanism. Consider the model of Section 5.1. Suppose that the control rule at (scaled) time t says that $p^n(t)/\sqrt{n}$ percent of the exogenous inputs must be deleted, where $p^n(t)$ is $\mathcal{U}$-valued and continuous. Suppose that the control is realized by choosing at random with this probability. The control term that appears in the actual evolution equation is ($1/\sqrt{n}$ times) the actual number deleted and will be random. But for large n, $1/\sqrt{n}$ times the quantity deleted in scaled time $[0,t]$ will differ little from $\bar{\lambda}^a \int_0^t p^n(s)ds$. In this case, the control rule specifies $p^n(\cdot)$. The weak-sense limit of $\int_0^t p^n(s)ds$ will appear in the limit equation as an integral $\int_0^t p(s)ds$, where $p(t)$ is $\mathcal{U}$-valued.

For this example the control term for the physical model (the admissible control) is defined as the actual outcome of the sampling, but it is the effect of the control rule in the limit that appears as the control in the limit equation. Such considerations lead to a definition of admissible control for the physical problem as a data-dependent process that differs from a desired $\mathcal{U}$-valued data-dependent process by a (perhaps random) perturbation that goes to zero as $n \to \infty$. This is formalized in (A1.6) and (A1.8).

The convergence proofs are simplest when the control appears in an additive and linear form in the dynamics and cost function. We start the development with this case, which also covers many applications in communications networks, where the control is often over admissions or routing. The more general nonadditive or nonlinear case requires the use of the relaxed control representation of the control process, as used in Chapters 3 and 4. The relaxed control representation is used for theoretical purposes only. It is necessary for the convergence results for the general nonlinear problem, since in general one cannot show that the control processes $u^n(\cdot)$ themselves converge in any way. But, it is hard to find a problem of practical interest where the optimal control is a relaxed control and not an ordinary control.

Section 1 deals with the discounted cost criterion and Section 2 with the ergodic cost criterion. In Section 3 we return to the multiplexer problem of Section 8.5 and show that the general conditions of Sections 1 and 2 are satisfied. The numerical data and discussion for this problem in Section 4 (see also [169]) show the great advantages of the heavy traffic–optimal control approach for design and analysis.

9.1 Discounted Cost

9.1.1 A Canonical Model and Assumptions

In this section we will prove general convergence results for a canonical model. The model is canonical in that it fits numerous applications where the instantaneous control forces are bounded. The model and assumptions will be given in a general form, which will need to be validated in particular applications. As noted in the Chapter Outline, the set of possible control values $\mathcal{U}$ for the weak-sense limit process is compact. The set of possible control values for the physical problem will be defined slightly differently below, taking into account how the control might actually be realized. Let $\{\mathcal{F}_t, t \geq 0\}$ be a filtration, and $w(\cdot)$ a standard $\mathcal{F}_t$-Wiener process. For appropriate functions $b(\cdot)$ and $\sigma(\cdot)$, the weak-sense limit systems will be of the form

$$dx(t) = b(x(t), u(t))dt + dM(t) + dz(t), \tag{1.1}$$

or in relaxed control form

$$dx(t) = \int_{\mathcal{U}} b(x(t), \alpha) r_t(d\alpha)dt + dM(t) + dz(t), \tag{1.2}$$

where $M(\cdot)$ is a stochastic integral of the form

$$M(t) = \int_0^t \sigma(x(s))dw(s),$$

and the dimension of $x(t)$ is K.

Let us recall the definition of admissible controls for (1.2), that were given in Section 3.1: The control process $u(\cdot)$ in (1.1) is said to be admissible if $u(t) \in \mathcal{U}$ and $u(\cdot)$ is $\mathcal{F}_t$-adapted. Otherwise said, $(w(\cdot), u(\cdot))$ is an admissible pair, or $u(\cdot)$ is admissible with respect to $w(\cdot)$. The relaxed control $r(\cdot)$ in (1.2) is said to be admissible if $r(t, \cdot)$ is a measure on $\mathcal{B}(\mathcal{U})$, with $r(t, \mathcal{U}) = t$, and $r(\cdot, B)$ is $\mathcal{F}_t$-adapted for each $B \in \mathcal{B}(\mathcal{U})$. The derivative $r_t(\cdot)$ (of $r(\cdot)$) is defined so that it is $\mathcal{F}_t$-adapted and is a probability measure on $\mathcal{B}(\mathcal{U})$. When we speak of a solution $(x(\cdot), z(\cdot))$, it is assumed that it is also $\mathcal{F}_t$-adapted.

The general assumptions that are to be used cover typical applications, and will be verified for some particular cases of current interest in Section 3. It is supposed that the physical problem can be put into the form of the Skorohod problem

$$x^n(t) = x^n(0) + B^n(t) + M^n(t) + z^n(t) + \epsilon^n(t) \tag{1.3}$$

where $z^n(\cdot)$ satisfies (A1.4) and $\epsilon^n(\cdot)$ converges weakly to the "zero" process. The conditions (A3.5.1)–(A3.5.3) on the state space and reflection directions will be used. Under our conditions, $M^n(\cdot)$ will converge weakly to either a Wiener process or a stochastic integral.

Controls appearing nonlinearly. Two classes of models will be dealt with. In the first, the control can appear in a nonlinear way in the dynamics and cost function. Then we suppose that $B^n(\cdot)$ is representable either in terms of an ordinary control as in

$$B^n(t) = \int_0^t b(x^n(s), u^n(s))ds, \tag{1.4}$$

or, more generally, in terms of a relaxed control as in

$$B^n(t) = \int_0^t \int_{\mathcal{U}} b(x^n(s), \alpha) r^n(d\alpha\, ds) = \int_0^t \int_{\mathcal{U}} b(x^n(s), \alpha) r^n_s(d\alpha) ds. \tag{1.5}$$

The actual controls for the physical problem will usually be physically realizable ordinary (not relaxed) controls. Nevertheless, in dealing with convergence questions it is often convenient to represent them in relaxed control form. The restrictions that will be put on the $u^n(\cdot)$ and $r^n(\cdot)$ in (1.4) and (1.5) will be discussed above (A1.1) and stated in (A1.6) or (A1.8). Subject to these restrictions, let us suppose that there is a well-defined class of admissible controls for (1.3) and that the conditions (A1.2)–(A1.5) hold for the controls in this class. The initial part of any heavy traffic analysis involves putting the problem into a form similar to (1.3), and many examples appear elsewhere in this book. The method of getting forms such as (1.3) might be problem-dependent but the convergence analysis starts only when such a form has been obtained.

Let $\beta > 0$ and $c_i \geq 0$. The cost function in this section will take the discounted forms, for $x^n(0) = x$, $x(0) = x$,

$$W_\beta(x, r) = E_x^r \int_0^\infty e^{-\beta t} \left[\int_{\mathcal{U}} k(x(t), \alpha) r_t(d\alpha) dt + c' dy(t) \right], \tag{1.6}$$

$$W_\beta^n(x, u^n) = E_x^{u^n} \int_0^\infty e^{-\beta t} \left[k(x^n(t), u^n(t)) dt + c' dy^n(t) \right]. \tag{1.7a}$$

If $r^n(\cdot)$ is the relaxed control representation of $u^n(\cdot)$, then write

$$W_\beta^n(x, r^n) = E_x^{r^n} \int_0^\infty \int_{\mathcal{U}} e^{-\beta t} \left[k(x^n(t), \alpha) r_t^n(d\alpha) dt + c' dy^n(t) \right]. \quad (1.7b)$$

Controls appearing linearly. In the second class of models, the control occurs linearly and additively in the dynamics and cost. This class is not quite a subclass of the model described in the previous paragraph, since the definition of admissible control will be slightly more general. For some matrix D and vector $\bar{c}$, the physical system and cost function take the form

$$x^n(t) = x^n(0) + \int_0^t b_0(x^n(s)) ds + DF^n(t) + M^n(t) + z^n(t) + \epsilon^n(t), \quad (1.8)$$

$$W_\beta^n(x, F^n) = E_x^{F^n} \int_0^\infty e^{-\beta t} \left[k_0(x^n(t)) dt + \bar{c}' dF^n(t) + c' dy^n(t) \right], \quad (1.9)$$

where $F^n(\cdot)$ is the control term, and is further specified in (A1.8). The weak-sense limit and cost will have the forms

$$dx(t) = b_0(x(t)) dt + Du(t) dt + dM(t) + dz(t), \quad (1.10)$$

$$W_\beta(x, u) = E_x^u \int_0^\infty e^{-\beta t} \left[k_0(x(t)) dt + \bar{c}' u(t) dt + c' dy(t) \right], \quad (1.11)$$

where $u(t) \in \mathcal{U}$.

Let $\mathcal{F}_t^n$ denote the minimal σ-algebra that measures all of the system's data that are available to the controller at scaled time t, and let E_t^n denote the associated conditional expectation. The definition of $\mathcal{F}_t^n$ might be slightly different in different applications. It will always measure $\{x^n(s), M^n(s), z^n(s), u^n(s), s < t\}$. In particular applications it might measure more detail, for example, the arrivals or departures up to scaled time t, provided that they are not affected by the control action at t. By definition, admissibility of $u^n(\cdot)$ implies that $u^n(t)$ is $\mathcal{F}_t^n$-measurable. Actually, all that matters concerning $\mathcal{F}_t^n$ is that (1.13) hold.

Assumptions. Subsets of the assumptions (A1.1)–(A1.12) will be used for our general model. The conditions are deliberately general, so that they cover a wide variety of applications. If there were no control, then (in Chapter 6) under some smoothness conditions on $b(\cdot)$, the weak convergence of $\epsilon^n(\cdot)$ to the zero process and of $M^n(\cdot)$ to a Wiener process was sufficient to guarantee that any subsequence of the processes in (1.3) had a further subsequence that converged weakly to a solution of (1.2). Two new issues arise in the control problem. The first concerns the nonanticipativity of the (weak-sense limit) control and of $(x(\cdot), z(\cdot))$ with respect to the (weak-sense limit) Wiener process. Without a control, but with a Lipschitz condition on $b(\cdot)$ and constant $\sigma(\cdot)$, the nonanticipativity of $(x(\cdot), z(\cdot))$ follows from the fact that there is a unique strong-sense solution. To prove the nonanticipativity for the controlled problem or when $\sigma(\cdot)$ is not constant, we need to know that the weak-sense limit pairs $(M(\cdot), r(\cdot))$ are admissible. This is the point of (1.13). It is given in the form that is a usual and convenient way of verifying this nonanticipativity in particular applications. In Chapter 8, $\sigma(\cdot)$ was allowed to depend on x, and the summands that defined $M^n(\cdot)$ were martingale differences. The required nonanticipativity of the weak-sense limit solution was a consequence of this martingale property. Condition (1.13) covers the analogous cases for the control problem and is weaker. The bursty noise model of Section 5.5 satisfies (A1.3), as do the various correlated service-time models of Chapter 7. The process $M^n(\cdot)$ in (A1.3) is often represented as a scaled sum, in which case (A1.3) is satisfied under quite weak mixing conditions on the summands. See, for example, the state-dependent noise results of [157, 160], where conditions such as (A1.3) are verified for a wide variety of such processes via perturbed test function methods.

The second new issue arising in the control problem concerns the cost function. In order to verify that the costs converge (as well as the processes in (1.3)) we need a uniform integrability condition on those parts of the cost function (1.7) that might not be bounded. This is the reason for (A1.1), (A1.2), (1.12), and (A1.5). The purpose of the terms involving t^p is to ensure that the various components of the cost grow slowly enough with respect to the discount factor $e^{-\beta t}$. The factor t^p is used simply because $\lim_{T\to\infty} \int_T^\infty e^{-\beta t} t^p dt = 0$. Condition (A1.3) holds for the M-processes in Sections 8.1–8.3, with $p = 2$. Concerning (A1.5), in many applications $b(\cdot)$ is bounded or the control can only reduce its average value.

Pathwise costs. Even without (A1.2), (1.12) and (A1.5) (but still assuming that $\epsilon^n(\cdot)$ converges weakly to the "zero" process), the pathwise cost process defined by $\int_0^t e^{-\beta t}[\text{cost rate}]\, dt$ will still converge weakly to that for the limit process. Confining attention to pathwise convergence would greatly simplify the analysis.

Admissible controls. The form (1.3) is a representation of the physical system in terms of an $\mathcal{F}_t^n$-adapted process $u^n(\cdot)$ which is called the control, and represents the effects of the deliberate attempts to manage the system. In applications it is not always possible to restrain the values of $u^n(t)$ to any particular set $\mathcal{U}$ w.p.1 or even to ensure that it does not contain delta functions, as noted in the introduction and seen from the following example. Consider a one-dimensional example where $b(x, u) = b_0(x) + u$. Suppose that the control action is simply that of admitting or not admitting exogenous arrivals to the queue, as in the problem of Section 3. Let

$\int_0^t u^n(s)ds$ denote $1/\sqrt{n}$ times the number of exogenous arrivals denied admission to the system by scaled time t. Owing to the discrete nature of the process, $u^n(t)$ is either zero or a delta function of value $1/\sqrt{n}$. Suppose that one wishes (as the "ideal" control) to have the maximum scaled "rate" of denied admissions be bounded by some constant $\bar{u}$ in the sense that there is a $\mathcal{U} = [0, \bar{u}]$-valued function $\bar{u}^n(\cdot)$ such that the scaled number of denied admissions is $\int_0^t \bar{u}^n(s)ds$, modulo $1/\sqrt{n}$. In any particular application, the actual mechanism that is used to control the admissions might not be able to ensure such a representation w.p.1. For example, the deletion might be done by random sampling or in some decentralized way so that the goal is approximated in some statistical sense, or only asymptotically. In the random sampling approach, if the control rule is to not admit an arrival at scaled time t with probability $\bar{u}^n(t)/[\sqrt{n}\bar{\lambda}^a]$, then one can write $\int_0^t u^n(s)ds = \int_0^t \bar{u}^n(s)ds + \epsilon^{u,n}(t)$, where $\epsilon^{u,n}(\cdot)$ is an "error" process that is due to the random sampling, and it converges weakly to zero as $n \to \infty$. Alternatively, if one could centralize the control such that $\int_0^t \bar{u}^n(s)ds$ is realized as closely as possible, pathwise, then $\epsilon^{u,n}(\cdot) = O(1/\sqrt{n})$.

Generally, the values taken by the physical control process depend on whatever constraints there are on the physical mechanisms that are used to realize it. Thus, with the "random sampling" mechanism, a $\mathcal{U}$-valued control is just an idealization, and we seek to realize it as well as possible given the physical mechanisms. On the other hand, for other problems we might be able to select the control values at will to take values in a desired set $\mathcal{U}$, as for example if the control were a service rate, in which case the $\hat{u}^n(\cdot)$ that is introduced in (A1.6) and the $\epsilon^{u,n}(\cdot)$ in (A1.8) are zero. We need to ensure that any weak-sense limit of the physically realizable controls (in relaxed control form) is an admissible relaxed control for the limit problem. Also, the admissible controls must be arbitrarily close to the ideal ones for large n. These comments motivate (A1.6) and (A1.8) (according to the case), the form of which fits many common applications. Indeed, (A1.6) (or (A1.8), according to the case) can be taken to be the definition of an admissible control. See Section 3 for an example of its verification.

For the multiplexer problem of Section 3 there are several ways of realizing a control, where (A1.8) is to hold, and the actual method that is used is determined by the system design. Given any desired $\mathcal{U}$-valued adapted process $\bar{u}^n(\cdot)$, one method of approximately realizing it is by denying admission to an arrival at time t at random with an appropriate probability as above. Alternatively, one could deny admissions in a deterministic way (either at the buffer entrance or coordinated among the sources) such that the scaled number of deletions by time t is as close to $\int_0^t \bar{u}^n(s)ds$ as possible. If the total input is the superposition of inputs from many independent sources, as in the multiplexer example, and the control is localized at the sources, then it is hard to synchronize the deletions among all the sources, and one must resort to some mechanism that "works on the average." Our definition of admissible control is a consequence of these problems of physical realization. See the comments below (A8.5.3) for further motivation. In examples such as those arising in communications systems, and where the control consists in nonadmission or rerouting, the control term often appears linearly in the dynamics and cost function. In such cases, the discrete nature of the number not admitted or rerouted leads to the representation of (A1.8) and Theorem 1.2.

Assumptions. As noted above, the conditions are phrased so that they cover a large variety of applications. Assumptions (A1.2), (A1.3), and (A1.5) are to hold for any sequence of admissible controls, and the value of p is chosen independently of the controls. The point of (A1.5) is to ensure that the cost functionals are well-defined, as well as the uniform integrability of the reflection term and cost rate on any finite time interval. The condition obviously holds if $b(\cdot)$ is bounded. In the example of Subsection 3.2, $b(\cdot)$ has two components, one bounded and the other obviously bounded in mean square.

A1.1.
$$\sup_n E\left|x^n(0)\right|^2 < \infty.$$

A1.2. *There is a* $p \geq 0$ *such that*

$$\limsup_n \sup_t \frac{E \sup_{s\leq t} |\epsilon^n(s)|}{1+t^p} = 0.$$

A1.3. $M^n(\cdot)$ *is asymptotically continuous, and for* $p \geq 0$ *and under* (A1.1),

$$\limsup_n \sup_t \frac{E \sup_{s\leq t} |M^n(s)|^2}{1+t^p} = O(1). \tag{1.12}$$

If $z^n(\cdot)$ *is asymptotically continuous, then for all* $t, \tau \geq 0$ *and arbitrary real-valued* $f(\cdot)$ *with compact support that is continuous together with its partial derivatives up to second order,*

$$E_t^n \left[f(M^n(t+\tau)) - f(M^n(t)) - \frac{1}{2}\int_t^{t+\tau} \operatorname{trace}\left[f_{mm}(M^n(s))\Sigma(x^n(s))\right] ds \right] \to 0 \tag{1.13}$$

in probability, where $\Sigma(x) = \sigma(x)\sigma'(x)$ *and* $\sigma(\cdot)$ *is bounded and Lipschitz continuous.*

A1.4. $z^n(t) = \sum_i y_i^n(t) d_i$. *There are real* $\epsilon_n \to 0$ *such that* $y_i^n(\cdot)$ *can increase only at* t *where* $x^n(t)$ *is within a distance of* ϵ_n *from* ∂G_i. *For* *each $T > 0$ and $\epsilon > 0$,*

$$\lim_n P\left\{\sup_{s\le T}|z^n(s) - z^n(s-)| \ge \epsilon\right\} = 0. \tag{1.14}$$

Typically, $z^n(\cdot)$ is either continuous (when it is scaled idle time), or its discontinuities are of order $O(1/\sqrt{n})$.

A1.5. *$b(\cdot)$ is continuous and $b(\cdot, u)$ is Lipschitz continuous, uniformly in $u \in \mathcal{U}$, a compact set. There is $p \ge 0$ such that*

$$\limsup_n \sup_t \frac{E|b(x^n(t), u^n(t))|^2}{1 + t^p} = O(1),$$

or in relaxed control notation,

$$\limsup_n \sup_t \frac{E\left|\int_{\mathcal{U}} b(x^n(t), \alpha) r_t^n(d\alpha)\right|^2}{1 + t^p} = O(1).$$

A1.6. *An admissible control $u^n(\cdot)$ has the representation $u^n(\cdot) = \bar{u}^n(\cdot) + \hat{u}^n(\cdot)$, where $\bar{u}^n(\cdot)$ and $u^n(\cdot)$ are $\mathcal{F}_t^n$-adapted, $\bar{u}^n(t)$ takes values in $\mathcal{U}$, and for each t and any sequence of admissible $u^n(\cdot)$,*

$$\lim_n E\int_0^t |\hat{u}^n(s)| ds = 0. \tag{1.15}$$

We will also have use for the following, when we need to ensure that $z(\cdot)$ defines $y(\cdot)$ uniquely.

A1.7. *Either* (a) *or* (b) *holds:* (a) *The covariance $\Sigma(x)$ in* (1.13) *is nondegenerate for each x.* (b) *For each i such that $c_i > 0$ (so that there is a positive cost associated with boundary face ∂G_i), at each edge or corner that involves ∂G_i, the set of reflection directions on the adjoining faces are linearly independent.*

Weakening the conditions. The bounds on the expectations in (1.12), (A1.2), and (A1.5) will be used to guarantee that

$$\lim_{T\to\infty} \sup_n \int_T^\infty e^{-\beta t} E\left[|k(x^n(t), u^n(t))| dt + c' dy^n(t)\right] = 0$$

and that for each T, $\{k(x^n(t), u^n(t)), y^n(T), t \le T, n\}$ is uniformly integrable. But $k(\cdot)$ might not depend on all of the components of x, and some of the c_i might be zero. One needs these bounds to hold only for the components of $\epsilon^n(\cdot)$, $M^n(\cdot)$, $b(x^n(\cdot), u^n(\cdot))$, and $\hat{u}^n(\cdot)$ that affect these interests, and the conditions in Theorems 1.1–1.5 can be changed to reflect this.

Define $V_\beta^n(x) = \inf_{u^n} W_\beta^n(x, u^n)$, where the infimum is taken over the admissible controls. For the model (1.8), (1.9), define $V_\beta^n(x) = \inf_{F^n} W_\beta^n(x, F^n)$. Define $V_\beta(x) = \inf_r W_\beta(x, r)$, where the infimums over the admissible relaxed controls, as defined in Section 3.2. The following facts will be useful to bound the growth rate of $y^n(t)$ in t.

Theorem 1.1. *Assume* (A3.5.1)–(A3.5.3), (1.12), (A1.1), (A1.2), (A1.4), *and* (A1.5). *Then for any sequence of admissible controls there are* $\bar{y}^n(\cdot)$ *and* $\epsilon^{y,n}(\cdot)$ *such that we can write* $|y^n(\cdot)| \le |\bar{y}^n(\cdot)| + |\epsilon^{y,n}(\cdot)|$, *where*

$$E\left|\bar{y}^n(t)\right|^2 = O(t^{p+2}), \quad \limsup_n \sup_t \frac{E|\epsilon^{y,n}(t)|}{1+t^p} = 0. \tag{1.16}$$

Proof. By the Lipschitz condition in Theorem 3.5.1, there is $C < \infty$ such that

$$|y^n(t)| \le C \sup_{s \le t} \left| x^n(0) + \int_0^s b(x^n(\tau), u^n(\tau))d\tau + M^n(s) + \epsilon^n(s) \right|.$$

Now split the last expression into two parts, the second part being $|\epsilon^{y,n}(\cdot)|$ and due to $\epsilon^n(\cdot)$. The theorem now follows from the bounds in the assumptions. ■

9.1.2 Controls Appearing Linearly

In order to illustrate the general ideas of the convergence theorems with minimal encumbrance, we will start with the simpler forms where the control appears in the additive and linear way of (1.8) and (1.9). The following alternative to (A1.6) is used.

A1.8. *Let* $F^n(\cdot), n < \infty$, *be admissible control terms for* (1.8). *Then, for each* n, *there are* $\mathcal{F}_t^n$*-adapted* $\bar{u}^n(\cdot)$ *with values in* $\mathcal{U}$, *and* $\mathcal{F}_t^n$*-adapted* $\epsilon^{u,n}(\cdot)$ *satisfying* (A1.2) *and of bounded variation w.p.1 on each bounded time interval, and the control term has the form*

$$F^n(t) = \int_0^t \bar{u}^n(s)ds + \epsilon^{u,n}(t).$$

In this subsection we will suppose that $\mathcal{U}$ is convex as well as compact. This is a common form in applications, and it holds for the multiplexer example in Section 3. The combination of linearity and convexity allows us to illustrate the method with minimal effort, since we can obtain a convergence theorem without the need of relaxed controls.

The main result of the next theorem is (1.17), which is one-half of the ultimate convergence result of interest, namely, $V_\beta^n(x^n(0)) \to V_\beta(x(0))$. The proof of the "reverse inequality" $\limsup_n V_\beta^n(x^n(0)) \le V_\beta(x(0))$ is similar to that given for the nonlinear control problem in Theorem 1.4. Define $\Psi^n(\cdot) = (x^n(\cdot), F^n(\cdot), M^n(\cdot), z^n(\cdot))$.

Theorem 1.2. *Let $c_i \ge 0$, and let $k_0(\cdot)$ be continuous with at most a linear growth rate. Assume the model* (1.8) *and* (1.9), *and let $x^n(0)$ converge weakly to $x(0)$. Assume* (A3.5.1)–(A3.5.3), (A1.1)–(A1.5), (A1.7), and (A1.8), *with convex $\mathcal{U}$. Then $\{\Psi^n(\cdot)\}$ is tight and for any weak-sense limit $\Psi(\cdot) = (x(\cdot), F(\cdot), M(\cdot), z(\cdot))$ there is a $\mathcal{U}$-valued process $u(\cdot)$ such that* (1.10) *holds, where $z(\cdot)$ is the reflection term. There is a standard Wiener process $w(\cdot)$ such that $M(t) = \int_0^t \sigma(x(s))dw(s)$, and $x(\cdot), u(\cdot), z(\cdot)$, are nonanticipative with respect to $w(\cdot)$. Along the weakly convergent subsequence, where the expectation is over the (possibly random) initial condition,*

$$EW_\beta^n(x^n(0), F^n) \to EW_\beta(x(0), u).$$

Also,

$$\liminf_n EV_\beta^n(x^n(0)) \ge EV_\beta(x(0)). \tag{1.17}$$

Proof. First, consider the weak convergence. In the weak-convergence argument, if $b_0(\cdot)$ is not bounded, then use the truncation method of Section 2.5, which allows us to suppose (without loss of generality) that $b_0(\cdot)$ is bounded. The sequence $M^n(\cdot)$ is asymptotically continuous by hypothesis, and $F^n(\cdot)$ is asymptotically uniformly Lipschitz continuous. Since all of the nonreflection processes on the right of (1.8) are asymptotically continuous, Theorem 3.6.1 implies that $z^n(\cdot)$ is asymptotically continuous. Thus, $x^n(\cdot)$ is asymptotically continuous. Extract a weakly convergent subsequence of $\Psi^n(\cdot)$, indexed also by n, and with weak-sense limit denoted by $\Psi(\cdot)$. If $\sigma(\cdot)$ is constant, then (1.13) and Theorem 2.8.2 yield that there is a standard Wiener process $w(\cdot)$ such that $M(t) = \sigma w(t)$, as well as the nonanticipativity. If $\sigma(\cdot)$ is not constant, then the existence of $w(\cdot)$, and the form of $M(\cdot)$ and the nonanticipativity follow from (1.13) and Theorems 2.8.2 and 2.2.1.

By the convexity and compactness of $\mathcal{U}$, for any $t, \delta \ge 0$, $\int_t^{t+\delta} \bar{u}^n(s)ds \in \mathcal{U}\,\delta$. Thus, there is a $\mathcal{U}$-valued process $u(\cdot)$, which we can take to be nonanticipative as well, such that $F(t) = \int_0^t u(s)ds$. Hence, the weak-sense limit satisfies (1.10).

Now let us turn our attention to the convergence of the costs. By (A1.1), (A1.2), (1.12), (A1.4), (A1.5), the assumptions on $\bar{u}^n(\cdot)$ and $\epsilon^{u,n}(\cdot)$ in (A1.8), and Theorem 1.1 and the Lipschitz condition in Theorem 3.5.1, there are $\bar{x}^n(\cdot)$ and $\epsilon^{x,n}(\cdot)$ such that $|x^n(t)| \le |\bar{x}^n(t)| + |\epsilon^{x,n}(t)|$, where (there is a similar construction in Theorem 1.1)

$$E|\bar{x}^n(t)|^2 = O(t^{p+2}), \quad \limsup_n \sup_t \frac{E|\epsilon^{x,n}(t)|}{1+t^p} = 0. \tag{1.18}$$

Define

$$C^n(T_1, T_2) = \int_{T_1}^{T_2} e^{-\beta t} \left[k_0(x^n(t))dt + \bar{c}' dF^n(t) + c' dy^n(t) \right]. \tag{1.19}$$

The convergence of the costs follows from the weak convergence, the uniform integrability of $\{C^n(0,T)\}$ for any T, and the fact that

$$\lim_{T\to\infty} \sup_n E|C^n(T,\infty)| = 0.$$

The last assertion of the theorem follows from the fact that the $F^n(\cdot)$ are arbitrary: In particular, for $\epsilon > 0$, let $F^n(\cdot)$ be ϵ-optimal controls, and selecting a subsequence if necessary,

$$\epsilon + EV_\beta^n(x^n(0)) \ge EW_\beta^n(x^n(0), F^n) \to EW_\beta(x(0), u) \ge EV_\beta(x(0)). \tag{1.20}$$

Then (1.17) follows due to the arbitrariness of ϵ. ■

Remark. The assumptions in Theorem 1.2 were chosen to illustrate the method of proof for a simple problem. The additive way that the control entered and the convexity of $\mathcal{U}$ allowed us to avoid using relaxed controls. If we replace the additive form by nonlinear forms, but suppose that $(b(x,\mathcal{U}), k(x,\mathcal{U}))$ is convex for each x, then one can still do the analysis without the introduction of relaxed controls, as seen in [153] or [156, Chapter 9], but it is no easier.

9.1.3 Controls Appearing Nonlinearly

We now turn to the general model of (1.3)–(1.6). The following conditions will be used.

A1.9. $b(x,u) = b_0(x) + b_c(x,u)$, *where* $b_0(\cdot)$ *is Lipschitz continuous, and* $b_c(\cdot,u)$ *is bounded and Lipschitz continuous uniformly in* $u \in \mathcal{U}$, *with* $b_c(x,0) = 0$.

A1.10. $k(x,u) = k_0(x) + k_c(x,u)$, *where* $k_0(\cdot)$ *is nonnegative and has at most linear growth, and* $k_c(\cdot,u)$ *is bounded and continuous, with* $k_c(x,0) = 0$. *Also,* $c_i \ge 0$ *in the cost function.*

A1.11. *Let* $\delta > 0$ *and let* $\mathcal{U}_0$ *be an arbitrary finite subset of* $\mathcal{U}$. *For arbitrary* $k \ge 0$, *let* $u^n(\cdot) = \bar{u}^n(\cdot) + \hat{u}^n(\cdot)$ *be an admissible control on* $[0, k\delta)$, *with*

$\bar{u}^n(t)$ being $\mathcal{U}_0$-valued and constant on each interval $[i\delta, i\delta+\delta), i < k$. Let u_k^n be $\mathcal{F}_{k\delta}^n$-measurable and $\mathcal{U}_0$-valued. Then there is an admissible control $u^n(\cdot) = \bar{u}^n(\cdot) + \hat{u}^n(\cdot)$ on $[k\delta, k\delta+\delta)$, where $\bar{u}^n(t) = u_k^n$, $t \in [k\delta, k\delta+\delta)$, and

$$\lim_n E \int_{k\delta}^{k\delta+\delta} |\hat{u}^n(s)| ds = 0.$$

The following is the form of (A1.11) when the control appears linearly as in (1.8) and (1.9).

A1.12. *Let $\delta, \mathcal{U}_0$ be as in* (A1.11). *For arbitrary $k \geq 0$, let*

$$F^n(t) = \int_0^t \bar{u}^n(s) ds + \epsilon^{u,n}(t) \tag{1.21}$$

be an admissible control on $[0, k\delta)$, where $\bar{u}^n(t)$ is $\mathcal{U}_0$-valued and constant on each interval $[i\delta, i\delta+\delta)$, $i < k$. Let u_k^n be $\mathcal{F}_{k\delta}^n$-measurable and $\mathcal{U}_0$-valued. Then there is an admissible control of the form (1.21) *on $[k\delta, k\delta+\delta)$, where $\bar{u}^n(t) = u_k^n$, $t \in [k\delta, k\delta+\delta)$, and $\epsilon^{u,n}(\cdot)$ satisfies* (A1.2).

Redefine $\Psi^n(\cdot) = (x^n(\cdot), r^n(\cdot), M^n(\cdot), z^n(\cdot))$, where $r^n(\cdot)$ is the relaxed control representation of $u^n(\cdot)$. Let $\bar{r}^n(\cdot)$ and $\hat{r}^n(\cdot)$ denote the relaxed control representations of $\bar{u}^n(\cdot)$ and $\hat{u}^n(\cdot)$, respectively, where $\hat{u}^n(\cdot)$ and $\bar{u}^n(\cdot)$ are defined as in (A1.6). The next theorem is the nonlinear analogue of Theorem 1.2.

Theorem 1.3. *Assume* (A3.5.1)–(A3.5.3), (A1.1)–(A1.7), (A1.9), *and* (A1.10). *Let $x^n(0)$ converge weakly to $x(0)$ and let $u^n(\cdot)$ be admissible. Then $\{\Psi^n(\cdot)\}$ is tight, and any weak-sense limit process $\Psi(\cdot)$ satisfies* (1.2), *where $z(\cdot)$ is the reflection term. There is a standard Wiener process $w(\cdot)$ such that $M(t) = \int_0^t \sigma(x(s))dw(s)$, $(x(\cdot), r(\cdot), z(\cdot))$ is nonanticipative with respect to $w(\cdot)$, and $r(\cdot)$ is an admissible relaxed control. Also, along the chosen weakly convergent subsequence,*

$$EW_\beta^n(x^n(0), r^n) \to EW_\beta(x(0), r) \tag{1.22}$$

and (1.17) *holds.*

Proof. The setup differs from that of Theorem 1.2 mainly in the more general way that the control appears, which requires the use of relaxed controls. If $b_0(\cdot)$ is not bounded, then use the truncation method and the uniqueness of the solution to (1.2) to get the tightness and weak convergence of some subsequence. Thus, suppose (without loss of generality) that $b_0(\cdot)$ is bounded. Write $u^n(\cdot) = \bar{u}^n(\cdot) + \hat{u}^n(\cdot)$ as in (A1.6). Factor

$$\int b_c(x^n(t), \alpha) r_t^n(d\alpha) = \int_{\mathcal{U}} b_c(x^n(t), \alpha) \bar{r}_t^n(d\alpha) + \int b_c(x^n(t), \alpha) \hat{r}_t^n(d\alpha).$$

Factor $\int k_c(x^n(t),\alpha)r_t^n(d\alpha)$ analogously. By (A1.5), (A1.6), (A1.9), and (A1.10), the influence of $\hat{r}^n(\cdot)$ on both the dynamics and the cost function is asymptotically negligible, and it will be neglected henceforth.

The sequence of relaxed controls $\{\bar{r}^n(\cdot)\}$ is tight. The sequence $\{z^n(\cdot)\}$ is asymptotically continuous by Theorem 3.6.1 and the asymptotic continuity of the other driving processes in (1.3). Extract a weakly convergent subsequence (indexed by n also) of $\Psi^n(\cdot)$, and denote the weak-sense limit by $\Psi(\cdot) = (x(\cdot), r(\cdot), M(\cdot), z(\cdot))$. By the asymptotic continuity of the $z^n(\cdot)$, (1.13), and Theorems 2.8.2 and 2.2.1, there is standard Wiener process $w(\cdot)$ such that $M(\cdot)$ can be written as $\int_0^t \sigma(x(s))dw(s)$ and $(x(\cdot), r(\cdot), z(\cdot))$ is nonanticipative with respect to $w(\cdot)$.

Let us use the Skorohod representation (Theorem 2.5.5) henceforth, so that all convergences are w.p.1 in the metric on the path space. With this representation,

$$\int_0^t \int_{\mathcal{U}} b_c(x^n(s),\alpha)\bar{r}^n(d\alpha\, ds) - \int_0^t \int_{\mathcal{U}} b_c(x(s),\alpha)\bar{r}^n(d\alpha\, ds)$$

goes to zero. Also, by the convergence of $r^n(\cdot)$ to $r(\cdot)$,

$$\int_0^t \int_{\mathcal{U}} b_c(x(s),\alpha)\bar{r}^n(d\alpha\, ds) - \int_0^t \int_{\mathcal{U}} b_c(x(s),\alpha)\bar{r}(d\alpha\, ds)$$

goes to zero. It follows from this and the convergence $\Psi^n(\cdot) \to \Psi(\cdot)$ that (1.2) holds. Analogously, the above expressions hold with $k_c(\cdot)$ replacing $b_c(\cdot)$. The bounds on $|y^n(\cdot)|$ that were obtained in Theorem 1.1 still hold. Hence, as in Theorem 1.3, the set $\{C^n(0,T)\}$ is uniformly integrable for each T and $\lim_{T\to\infty} \sup_n E|C^n(T,\infty)| = 0$. This implies that we need concern ourselves only with $C^n(0,T)$ for arbitrary T. Now the uniform integrability and weak convergence of $C^n(0,T)$ for each T imply (1.22) and (1.17). ∎

The next theorem completes the proof of the convergence (1.23).

Theorem 1.4. *Assume the conditions of Theorem* 1.3 *and* (A1.11). *Then*

$$EV_\beta^n(x^n(0)) \to EV_\beta(x(0)). \tag{1.23}$$

Now assume the model (1.8) *and* (1.9), *with* (A1.8) *replacing* (A1.6) *and* (A1.12) *replacing* (A1.11). *Then* (1.23) *continues to hold.*

Proof. The proof under (A1.6) and (A1.11) will be given, since the problem with the control appearing linearly is a special case. By (1.17), it remains only to prove

$$\limsup_n EV_\beta^n(x^n(0)) \le EV_\beta(x(0)). \tag{1.24}$$

Suppose that for each $\epsilon > 0$ there is a smooth $\mathcal{U}$-valued feedback control $u^\epsilon(\cdot)$ that is ϵ-optimal for the system (1.1) with cost (1.6). In many cases, such a control can be applied to the physical system directly in that $u^\epsilon(x^n(\cdot))$ is an admissible control for (1.3), (1.7). If it is not, due to issues of physical realizability, suppose for the moment that there is an $\mathcal{F}_t^n$-adapted process $\hat{u}^n(\cdot)$ that satisfies (A1.6) and such that the control $u^n(t) = u^\epsilon(x^n(t)) + \hat{u}^n(t)$ is admissible. Then apply this control to the physical system, and use the fact that it is not necessarily optimal for the physical system, but is ϵ-optimal for the *limit system,* to get (1.24). The main difficulty is in showing that there is such a smooth ϵ-optimal feedback control for the limit system. In order to avoid this problem, a slightly indirect approach will be taken.

The proof will be carried out in the following way. First, note that given any admissible pair $(w(\cdot), r(\cdot))$, (1.2) has a unique strong-sense solution. First suppose that σ is constant, so that $M(t) = \sigma w(t)$ in (1.2). Suppose that for each $\epsilon > 0$, there is an admissible control $u^n(\cdot)$ with relaxed control representation $r^n(\cdot)$ such that $(M^n(\cdot), r^n(\cdot))$ converges weakly to an admissible pair $(M(\cdot), r(\cdot))$ for (1.2), where $r(\cdot)$ is the relaxed control representation of an ordinary (but not a feedback) control $u^\epsilon(\cdot)$ that is ϵ-optimal for (1.1) (which is equivalent to (1.2) in this case) with initial condition $x(0)$. The limit process $M(\cdot)$ will always be a Wiener process with covariance $\sigma\sigma'$. Then

$$EV_\beta^n(x^n(0)) \le EW_\beta^n(x^n(0), u^n) \to EW_\beta(x(0), u^\epsilon) \le EV_\beta(x(0)) + \epsilon, \tag{1.25}$$

and this implies (1.24). The actual control $u^n(\cdot)$ that is used need not be practical. Any sequence guaranteeing (1.25) will do.

Theorems 3.8.1 and 3.8.3 constructed a finite-valued and piecewise constant ϵ-optimal control that can be adapted to our needs. The Wiener process $w^\epsilon(\cdot)$ and solution $(x^\epsilon(\cdot), z^\epsilon(\cdot))$ were indexed by ϵ in those theorems. In Theorem 3.8.1, $(w^\epsilon(\cdot), u^\epsilon(\cdot))$ was an admissible pair, and the piecewise constant $u^\epsilon(\cdot)$ was ϵ-optimal. The control $u^\epsilon(\cdot)$ was randomized and its (conditional on the past) probability law was defined by (3.8.3). The conditional probability of value of the control to be used on an interval $[k\delta, k\delta + \delta)$ depended on samples of the driving Wiener process $w^\epsilon(\cdot)$ over $[0, k\delta)$. By Theorem 3.8.3, we can use samples of $M^\epsilon(\cdot) = \int_0^t \sigma(x^\epsilon(s))dw^\epsilon(s)$ instead. For our purposes, it is simpler to use the $M^\epsilon(\cdot)$. Let $r^\epsilon(\cdot)$ denote the relaxed control representation of this ϵ-optimal $u^\epsilon(\cdot)$.

An admissible adaptation of $u^\epsilon(\cdot)$ for the physical model (1.3) will be defined and called $u^n(\cdot)$, with $r^n(\cdot)$ denoting its relaxed control representation. This control will have the property that the sequence $\Psi^n(\cdot) = (x^n(\cdot), r^n(\cdot), M^n(\cdot), z^n(\cdot))$ converges weakly to a set of processes $\Psi(\cdot) = (x(\cdot), r(\cdot), M(\cdot), z(\cdot))$ that solves (1.2). Furthermore, $r(\cdot)$ will be the relaxed control representation of an ordinary admissible control $u(\cdot)$ that is equivalent to $u^\epsilon(\cdot)$ in that $(M(\cdot), u(\cdot))$ has the distribution of $(M^\epsilon(\cdot), u^\epsilon(\cdot))$. From this point on, the uniqueness of the solution to (1.2) under the admissible pair $(M^\epsilon(\cdot), u^\epsilon(\cdot))$ implies (1.25). The control $u^n(\cdot)$ will be constructed by adapting the method of [167, Chapter 10]. The adaptation is a little more complicated than what was done in the reference due to the problem of "physical realizability" that was discussed earlier in the chapter.

Using the rule (3.8.3), which defines the functions $q_k(\cdot)$, define random variables $\bar{u}^n(k\delta)$, $k = 0, 1, \ldots$, recursively via the conditional probability law

$$\begin{aligned} P\left\{\bar{u}^n(k\delta) = \alpha \middle| x^n(0), \bar{u}^n(i\delta), i < k, M^n(s), \epsilon^n(s), s < k\delta\right\} \\ = q_k\left(\alpha; x^n(0), \bar{u}^n(i\delta), i < k, M^n(p\theta), p\theta < k\delta\right). \end{aligned} \tag{1.26}$$

By Theorems 3.8.1 and 3.8.3, we can take the functions $q_k(\cdot)$ to be continuous in the M and $x(0)$ arguments for each value of the other arguments. Use (1.26) on the intervals $[k\delta, k\delta + \delta)$ recursively. Generally, the process $M^n(\cdot)$ depends on the actual control that is used. First, define $\bar{u}_0^n$ by drawing at random with the law (1.26) with $k = 0$. By (A1.11), there is an admissible control $u^n(\cdot)$ on $[0, \delta)$ which is close to $\bar{u}_0^n$ in the sense that for $k = 0$,

$$\lim_n E \int_{k\delta}^{k\delta+\delta} |u^n(s) - \bar{u}_0^n(s)| ds = 0, \tag{1.27}$$

where $\bar{u}_0^n(s) = \bar{u}_0^n$ on $[0, \delta)$. Now with $u^n(\cdot)$ available on $[0, \delta)$, so is $M^n(\cdot)$. Continuing, choose $\bar{u}_\delta^n$ by drawing at random via the conditional probability law (1.26) with $k = 1$. Define $\bar{u}^n(s) = \bar{u}_\delta^n$ on $[\delta, 2\delta)$. By (A1.11), there is admissible $u^n(\cdot)$ on $[\delta, 2\delta)$ such that (1.27) holds for $k = 1$. Continue for all intervals $[k\delta, k\delta + \delta)$, $k = 2, 3, \cdots$.

Let $\bar{u}^n(\cdot)$ denote the piecewise constant process with values $\bar{u}_{k\delta}^n$ on $[k\delta, k\delta + \delta)$. Let $\bar{r}^n(\cdot)$ and $r^n(\cdot)$ denote the relaxed control representations of $\bar{u}^n(\cdot)$ and $u^n(\cdot)$, respectively. By (A1.11), the difference $r^n(\cdot) - \bar{r}^n(\cdot)$ converges weakly to the "zero" process. By the method of construction of $\bar{u}^n(\cdot)$ and the continuity of the functions $q_k(\cdot)$ in the M and $x(0)$ arguments for each value of the other arguments, the probability law of $(M^n(\cdot), r^n(\cdot))$ converges to that of $(M^\epsilon(\cdot), r^\epsilon(\cdot))$ as $n \to \infty$. Let $\Psi(\cdot) = (x(\cdot), r(\cdot), M(\cdot), z(\cdot))$ denote the weak-sense limit of a weakly convergent subsequence. By the method of construction of $u^n(\cdot)$ and the uniqueness of the solution to (1.2), $(x(\cdot), r(\cdot), M(\cdot), z(\cdot))$ must have the distribution of $(x^\epsilon(\cdot), r^\epsilon(\cdot), M^\epsilon(\cdot), z^\epsilon(\cdot))$. Hence the weak-sense limit does not depend on the subsequence, which completes the proof for the constant $\sigma(\cdot)$ case. If $\sigma(\cdot)$ is not a constant, then use Theorem 3.8.3 in lieu of Theorem 3.8.1. ■

9.1.4 Extensions: State-Dependent Variance, Discontinuous Dynamics, and Vacations

Extensions and comments; workload, x-dependent $\mathcal{U}$, and vacations. The model (1.3) could have been obtained from any of the networks of the previous chapters, open or closed or even "fluid." There could be batch arrivals or processing, and short vacations. It could represent the workload equation as well. There is one slight change in the proof of Theorem 1.4 that would be required for the workload case. The values of the $M^n(t)$ that are obtained in the workload representation in Section 6.4 involve the future service times of the jobs that are in the system at scaled time t, and the control cannot be based on such data if it not available until the processing of those jobs is completed. This problem affects only the construction of the control $\bar{u}^n(\cdot)$ in Theorem 1.4, since $\bar{u}^n(\cdot)$ must be $\mathcal{F}_t^n$-adapted. But, owing to the use of $p\theta < kd$ (note the strict inequality) in (1.26), this problem does not create a difficulty for the proof. Just use whatever part of $M^n(\cdot)$ is available when using the construction in (1.26) and note that, asymptotically, the values $M^n(p\theta)$, $p\theta < k\delta$, will be known at scaled time $k\delta$, with a probability that goes to unity as $n \to \infty$.

We have supposed that the set of control values $\mathcal{U}$ does not depend on the state, mainly for simplicity. A similar development yields the following, which we state without proof. Let $\mathcal{U}(x)$ be compact and continuous in x, and contained in a compact set $\bar{\mathcal{U}}$. Let the $r_t(\cdot)$ in (1.2) be probability measures on $\mathcal{B}(\mathcal{U}(x(t))$ and let the $\bar{u}^n(t)$ part of admissible $u^n(\cdot)$ take values in $\mathcal{U}(x^n(t))$. Then under the other conditions, all of the previous results continue to hold.

Discontinuous dynamics and switching controls. Given a continuous optimal or ϵ-optimal feedback control $u(\cdot)$ for the limit system under which it has a weak-sense unique solution, it is commonly true either that it can be applied directly to the physical system in that $u(x^n(\cdot))$ is admissible or that there is an adaptation of the form $u^n(t) = u(x^n(t)) + \hat{u}^n(t)$ that is admissible, where $\hat{u}^n(\cdot)$ is small in that it satisfies (A1.6). One can make an analogous claim for the system (1.8), (1.9), where (A1.8) replaces (A1.6).

Let the control have q components. Suppose that the control appears linearly in the dynamics and cost function, and the constraint $u(t) \in \mathcal{U}$ takes the form $a_i \le u_i \le b_i$, $i \le q$. Then whenever the control problem has been solved either analytically or numerically, the optimal feedback control $u(\cdot)$ for the limit system has been of the switching curve type. That is, there are continuous surfaces $\phi_i(\cdot)$ each of which divides G into two parts. In one part $u_i(x) = a_i$, and it equals b_i in the other part, with either value being optimal on the boundary. Whether such controls can be adapted for use on the physical system and be nearly optimal under heavy traffic needs to be examined case by case. However, if there is a weak-sense unique solution to the limit system under this control and $(x^n(\cdot), \int_0^{\cdot} u(x^n(s))ds)$ converges weakly to $(x(\cdot), \int_0^{\cdot} u(x(s))ds)$, then the control is nearly optimal for the physical system under heavy traffic, assuming that the conditions of Theorem 1.2 hold and there is $\epsilon^{u,n}(\cdot)$ satisfying (A1.2) such that $F^n(t) = \int_0^t u(x^n(s))ds + \epsilon^{u,n}(t)$ is admissible. (In this case, the feedback control $u(\cdot)$

can be used as suggested at the beginning of the proof of Theorem 1.4, even though it is not smooth.) See also the discussion in Section 8.4 and at the end of Section 8.5.

Long vacations. Suppose that the system (1.3) has the network origin of the problem of Section 7.3.2, which was concerned with the so-called long and infrequent vacations. The vacations were called "long," since the effect of the service interruptions was long enough to cause a jump in the state as $n \to \infty$. Controls can readily be added, in several ways. The simplest is the model used in Theorems 1.1–1.4, but with vacations of the type used in Section 7.3.2 added. The physical model is then the "adjusted" (1.3), namely,

$$x^n(t) = x^n(0) + B^n(t) + M^n(t) + J^n(t) + z^n(t) + \epsilon^n(t), \tag{1.28}$$

where $J^n(t)$ is the scaled correction for the effects of the services which were not done at the processors that were on vacation. Augment (1.12) by supposing that for some $p \geq 0$ and any sequence of admissible controls,

$$\limsup_n \frac{E \sup_{s \leq t} |J^n(s)|^2}{1 + t^p} = O(1), \tag{1.29}$$

and this would have to be verified in any application. Assume, in addition, the conditions on the vacation and intervacation intervals of Theorem 7.3.2, and that the jump due to the kth vacation of processor P_i converges weakly to the solution of (7.3.16) at $\tau^v_{i,k}$. Augment $\mathcal{F}^n_t$ to include the past data on the vacations. Then under their other conditions, the conclusions of Theorems 1.1–1.4 continue to hold, except that weak convergence is understood to refer to the set of intervacation sections, and (1.2) is replaced by

$$dx(t) = \int_{\mathcal{U}} b(x(t), \alpha) r_t(d\alpha) dt + dM(t) + dJ(t) + dz(t). \tag{1.30}$$

With this model, the distribution function of the jumps in the weak-sense limit model is not affected by the state. One could allow the distribution of the vacation interval to depend on the state at its start. See, for example, the comments in Subsection 8.2.3. One could also let the rate (conditional, given the past data) at which vacations start be dependent on the state and control. Even with these modifications, the distribution of the jumps still does not depend on the control, and the methods of proof are essentially what we have been using. One could let the jump distribution depend on the control just before the jump started. An example of such a problem that arises naturally in controlled polling when the connections between the queues and server are subject to random breakdown is given in Chapter 11. Then the method of proof of (1.23) involves some new considerations. For the singular control problem of Chapter 10, one can control during a jump so as to affect the value of the jump, and this will be dealt with in that chapter. See also [174].

9.2 The Ergodic Cost Problem

The average cost per unit time problem of Section 4.4 is one of the most common in applications to queueing networks. The system model will be either (1.3)–(1.5) or (1.8). For the model (1.3)–(1.5) and an admissible control $u^n(\cdot)$ with relaxed control representation $r^n(\cdot)$, define the *pathwise average* cost over $[0, T]$,

$$C_T^n(x, r^n) = \frac{1}{T}\int_0^T \left[k_0(x^n(t)) + \int_{\mathcal{U}} k_c(x^n(t), \alpha) r_t^n(d\alpha)\right] dt + \frac{c'y^n(T)}{T} \tag{2.1a}$$

and define $\gamma_T^n(x, r^n) = E_x^{r^n} C_T^n(x, r^n)$. Then the cost of interest is

$$\gamma^n(x, r^n) = \limsup_T \gamma_T^n(x, r^n). \tag{2.2}$$

The weak-sense limit system is (1.2) with cost (4.4.1).

For the model (1.8), define the pathwise average cost to be

$$C_T^n(x, F^n) = \frac{1}{T}\int_0^T k_0(x^n(t))dt + \frac{\bar{c}'F^n(T) + c'y^n(T)}{T} \tag{2.1b}$$

and define $\gamma_T^n(x, F^n) = E_x^n C_T^n(x, F^n)$. The cost is

$$\gamma^n(x, F^n) = \limsup_T \gamma_T^n(x, F^n).$$

The weak-sense limit system is (1.10), with cost (4.4.1), where $k_c(x, \alpha) = \bar{c}'\alpha$.

The problem will be set up in the terminology of Section 4.7, via the functional occupation measure. The development will be for the model (1.3)–(1.5). The procedure for (1.8) is similar. Recall the definitions of the shifted function $\phi_t(\cdot)$ and of the shifted and centered function $\Delta_t\phi(\cdot)$ given above (4.7.5). The assumptions below are analogues of what was used in the previous section. They differ in that here they hold for the processes starting at an arbitrary time and not just at time zero. By (4.7.5), the system (1.3)–(1.5) with the initial time shifted to $t + a$ for arbitrary nonnegative t, a is

$$\begin{aligned} x_{t+a}^n(s) &= x^n(t+a) + \int_0^s b_0(x_{t+a}^n(v))dv \\ &\quad + \int_0^s \int_{\mathcal{U}} b_c(x_{t+a}^n(v), \alpha)(\Delta_{t+a} r^n)_v(d\alpha)dv \\ &\quad + \Delta_{t+a}M^n(s) + \Delta_{t+a}z^n(s) + \Delta_{t+a}\epsilon^n(s). \end{aligned}$$

Recall, in particular, that $(\Delta_{t+a}r^n)_v(d\alpha)$ is the (left hand) time derivative of $\Delta_{t+a}r^n(d\alpha) = r^n(t + a + \cdot, d\alpha) - r^n(t + a, \delta\alpha)$ for fixed $t + a$.

Assumptions. Assumptions (A2.2)–(A2.4) are to hold for any sequence of admissible controls. Condition (A2.3) replaces (A1.3).

A2.1. *G is compact, $b(\cdot)$ is continuous, $b(\cdot, u)$ and $\sigma(\cdot)$ are Lipschitz continuous, uniformly in $u \in \mathcal{U}$, a compact set, and $\sigma(\cdot)$ is bounded.*

A2.2.
$$\limsup_n \sup_t E \sup_{s \leq 1} |\epsilon^n(t+s) - \epsilon^n(t)| = 0.$$

A2.3. *$\Delta_t M^n(\cdot)$ is asymptotically continuous as $n, t \to \infty$ in any way at all. Also,*
$$\limsup_n \sup_t E \sup_{s \leq 1} |\Delta_t M^n(s)|^2 = O(1).$$
The matrix $\Sigma(x) = \sigma(x)\sigma'(x)$ is uniformly positive definite. If $z^n(\cdot)$ is asymptotically continuous, then for each $s, \tau \geq 0$, the following holds:
$$E_t^n \Big[f(\Delta_t M^n(s+\tau)) - f(\Delta_t M^n(s)) \\ - \frac{1}{2} \int_s^{s+\tau} \text{trace}\left[f_{ww}(\Delta_t M^n(v)) \Sigma(x_t^n(v)) \right] dv \Big] \to 0$$
in the sense of probability as $n, t \to \infty$, where $f(\cdot)$ is an arbitrary real-valued function with compact support that is continuous together with its partial derivatives up to second order.

A2.4.
$$\limsup_n \sup_t E|b(x^n(t), u^n(t))|^2 = O(1).$$

A2.5. *Let $u^n(\cdot)$ be an admissible control. Then $u^n(t)$ has the representation $u^n(\cdot) = \bar{u}^n(\cdot) + \hat{u}(\cdot)$, where $\bar{u}^n(\cdot)$ and $\hat{u}(\cdot)$ are $\mathcal{F}_t^n$-adapted and $\bar{u}^n(t)$ takes values in $\mathcal{U}$. Also, $\lim_n \sup_t E \int_0^1 |\hat{u}^n(t+\tau)| d\tau = 0$.*

The next assumption is the analogue of (A1.8) and will replace (A2.5) when the control appears linearly in the dynamics and cost function, as in (1.8) and (1.9).

A2.6. *Let $F^n(\cdot)$ be an admissible control term for (1.8). Then there are $\mathcal{F}_t^n$-adapted $\bar{u}^n(\cdot)$ and $\epsilon^{u,n}(\cdot)$ such that $\bar{u}^n(t)$ is $\mathcal{U}$-valued and*
$$F^n(t) = \int_0^t \bar{u}^n(s) ds + \epsilon^{u,n}(t),$$
where $\epsilon^{u,n}(\cdot)$ satisfies (A2.2).

A2.7. $c_i \geq 0$ *and* $k(\cdot)$ *is continuous.*

Theorem 2.1 shows that the limit of the optimal pathwise average costs for the physical problem cannot be better than the ergodic cost for the heavy traffic limit in the sense that (2.3) and (2.4) hold. The assertions in the second and third paragraphs of Theorem 2.1 are used to get the converse Theorem 2.2. Theorems 2.1 and 2.2 together imply that a nice nearly optimal feedback control for the limit system will be nearly optimal for the physical system under heavy traffic. Furthermore, for that feedback control, the pathwise average costs on $[0,T]$ for the physical system converge weakly to the ergodic costs for the limit system, as $n, T \to \infty$.

Theorem 2.1. *Assume the model* (1.3)–(1.5) *and conditions* (A3.5.1)–(A3.5.3), (A4.7.2), (A2.1)–(A2.5), *and* (A2.7). *Let* $x^n(0)$ *be arbitrary. Suppose that no more than* K *constraints (where* K *is the dimension of* $x^n(t)$*, and which are assumed to be linearly independent) are active at any boundary point. Then for any sequence of admissible relaxed controls* $r^n(\cdot)$

$$\liminf_{n,T\to\infty} \gamma_T^n(x^n(0), r^n) \geq \bar{\gamma}, \tag{2.3}$$

where $\bar{\gamma}$ *is the infimum of the costs for the limit system. Furthermore, for any* $\epsilon > 0$,

$$\lim_{n,T\to\infty} P\left\{C_T^n(x^n(0), r^n) \leq \bar{\gamma} - \epsilon\right\} = 0. \tag{2.4}$$

The lim *and* lim inf *are taken as* $n, T \to \infty$ *in any way at all.*

Let $\bar{u}(\cdot)$ *be a measurable* $\mathcal{U}$*-valued feedback control that is continuous almost everywhere with respect to Lebesgue measure and such that there is* $\hat{u}^n(\cdot)$ *satisfying* (A2.5) *and* $u^n(\cdot) = \bar{u}(x^n(\cdot)) + \hat{u}^n(\cdot)$ *is an admissible control. Then* $C_T^n(x^n(0), r^n)$ *converges in mean to* $\gamma(\bar{u})$ *(defined in* (4.4.2b)*), as* $n, T \to \infty$.

Now suppose that (1.8) *holds and let* (A2.6) *replace* (A2.5). *Then the analogues of* (2.3) *and* (2.4) *hold. Let* $\bar{u}(\cdot)$ *be as in the above paragraph, and let there be* $\epsilon^{u,n}(\cdot)$ *satisfying* (A2.6) *for which* $F^n(t) = \int_0^t \bar{u}(x^n(s))ds + \epsilon^{u,n}(t)$ *is an admissible control term. Then* $C_T^n(x^n(0), F^n)$ *converges in mean to* $\gamma(\bar{u})$.

Proof. The proof will be given under (A2.5). Let $u^n(\cdot)$ be any sequence of admissible controls with relaxed control representations $r^n(\cdot)$. Recall the definition of $\Psi_t^n(\cdot)$ given above (4.7.6). Under our hypotheses the sequence of occupation measures $\{\mathcal{Q}_T^n\}$ that was introduced in Section 4.7 is tight, $\hat{u}^n(\cdot)$ has no effect on the weak-sense limits, and Theorem 4.7.1 holds. Let $n, T \to \infty$ index a weakly convergent subsequence of $\{\mathcal{Q}_T^n\}$, with weak-sense limit $\mathcal{Q}$. By Theorem 4.7.1, the representation (4.7.11) holds for the processes induced by the sample values $\mathcal{Q}^{\omega'}$ of $\mathcal{Q}$ (for almost all ω'). By our hypotheses, the cost can be represented in terms of a function $\bar{F}(\cdot)$ as in

Section 4.7, where $\bar{F}(\cdot)$ is continuous w.p.1 with respect to the measure induced by any stationary solution to (1.2), and $\{\bar{F}(\Psi_t^n(\cdot)); n, t\}$ is uniformly integrable. Thus, by Theorem 4.7.1 (and in the notation of that theorem) as $n, T \to \infty$ along the convergent subsequence

$$C_T^n(x^n(0), r^n) \Rightarrow \int \bar{F}(\tilde{\psi}(\cdot))dQ(\tilde{\psi}(\cdot)). \tag{2.5}$$

By Theorem 4.4.3, there is an optimal control for (1.2) in the class of relaxed feedback controls, and Theorem 4.6.1 implies that this control is optimal with respect to all admissible controls. This implies that the right-hand side of (2.5) is no less than $\bar{\gamma}$ and yields (2.3) and (2.4). The assertion in the second paragraph of the theorem is proved in the same way. ■

Theorem 2.2. *Assume the model and conditions in the first three sentences of Theorem 2.1. Suppose that for each $\epsilon > 0$ there is an ϵ-optimal feedback control $\bar{u}^\epsilon(\cdot)$ for (1.2) that satisfies the condition in the second paragraph of Theorem 2.1. Let $r^n(\cdot)$ denote the relaxed control representation of the admissible adaptation of $\bar{u}^\epsilon(\cdot)$ to the physical system. Then*

$$C_T^n(x^n(0), r^n) \Rightarrow \gamma(\bar{u}^\epsilon) \leq \bar{\gamma} + \epsilon. \tag{2.6}$$

Let $r^n(\cdot)$ be ϵ_n-optimal relaxed controls for the physical system, where $\epsilon_n \to 0$. Then

$$C_T^n(x^n(0), r^n) \Rightarrow \bar{\gamma}. \tag{2.7}$$

The analogous conclusions hold for the model (1.8).

Remark on the control $\bar{u}^\epsilon(\cdot)$ in Theorem 2.2; convergence of optimal costs and nearly optimal feedback controls. The question of admissibility for the physical system was discussed at length in Section 1, where the definition took "physical realizability" into account. This definition required either that we specify the physical control mechanism so that the precise requirements are clear or that we simply make a general assumption that could be validated in a large set of applications. The second approach was taken for the sake of generality, with the validation being left for specific examples, as in Section 3.

Theorem 2.2 says that if for each $\epsilon > 0$ there is an ϵ-optimal control $\bar{u}^\epsilon(\cdot)$ for (1.1) that satisfies the almost everywhere continuity condition of Theorem 2.1 and can be adapted to the physical system in the sense of the theorem, then (2.6) and (2.7) hold. The theorems are concerned with the convergence of costs, and they do not guarantee the existence of the $\bar{u}^\epsilon(\cdot)$. However, it is important to note that in applications to date the numerically or analytically obtained optimal controls for the model (1.1) have satisfied the conditions in the second or third paragraph of Theorem 2.1.

The sets $(b_c(x,\mathcal{U}), k_c(x,\mathcal{U}))$ are continuous in x. If they are also convex for each x, then Theorem 4.4.5 says that arbitrarily smooth ϵ-optimal feedback controls exist for (1.1). Thus, if such a control can be adapted for use on the physical system, then Theorems 2.1 and 2.2 together imply that the optimal costs for the physical problem converge to the optimal for the limit problem. The convexity condition covers the bulk of applications in controlled queues to date.

When the above convexity condition does not hold or if we cannot guarantee the existence of the ϵ-optimal controls satisfying the conditions of Theorem 2.2, then the problem of proving the convergence of the optimal costs for the physical problem to the optimal for the limit is more complicated. For purposes of the convergence proof, we are stuck with the problem of adapting ϵ-optimal relaxed controls for use on the physical system. (Recall the method of proof in Theorem 1.4.) This was the reason for the more complicated setup in Section 4.8, which implies the following result. While the statement of Theorem 4.8.2 might seem complicated, it is has not been hard to verify in applications to date.

Theorem 2.3. *Assume the model and conditions in the first three sentences of Theorem* 2.1. *Let* $\delta > 0$ *and let* $\mathcal{U}_0$ *be a finite subset of* $\mathcal{U}$. *Let* $\bar{u}^n(\cdot)$ *be a* $\mathcal{U}_0$*-valued and* $\mathcal{F}_t^n$*-adapted process that is constant on the intervals* $[k\delta, k\delta + \delta)$. *Suppose that there are admissible controls* $u^n(\cdot)$ *such that* (A2.5) *holds. Then* (2.7) *holds. The analogous result holds for the model* (1.8) *if* (A2.6) *replaces* (A2.5).

9.3 Examples

There are many examples of current importance where the conditions of Sections 1 and 2 are satisfied. A detailed demonstration for the multiplexer problem of Section 8.5 will be given below. The problem of assigning or reassigning incoming jobs to one of a bank of processors, where the required work depends on the assignment, will also fit the model, if only the assignment of a small part can be controlled [166]. Under heavy traffic, such control can have a substantial effect on the performance. Another example is that of service rate or (marginal) power control, as follows.

9.3.1 Service Rate and Assignment Controls

Service rate control. Consider the following simple form of the setup of Section 8.2, where the marginal service rates can be controlled so that $\lambda_i^d(x)$ is replaced by continuous $\lambda_i^d(x,u)$, where the control parameter u takes values in some compact set $\mathcal{U}$. We assume that the service rate at any time can be selected and physically realized at will, provided that it depends only on the past data. Then the term $\hat{u}^n(\cdot)$ in (A1.6) does not appear and (A1.11) holds. Let the functions $\lambda_i^a(\cdot)$ and $\lambda_i^d(\cdot, u)$ be Lipschitz continuous uniformly in $u \in \mathcal{U}$ and bounded. Then under (A8.1.3) and (A8.2.1)–(A8.2.4) (where we suppose that the $o(1/\sqrt{n})$ terms are of this order uniformly in (x, u)), all of (A1.2)–(A1.5) hold.

In some applications, the service rate is controlled by controlling the probability that a job will have to be redone. For example, in a wireless communications system, the control parameter u might correspond to the power level of the transmitted signal, and then the probability that a message will either not be received or not received correctly will depend on that power level.

9.3.2 The Multiplexer Problem: Convergence

The conditions of Sections 1 and 2 will be verified for the multiplexer problem of Section 8.5. Refer to that section for a discussion of the form chosen for the problem (where $\mathcal{U} = [0, \bar{u}]$) and the relationship of the bound $\bar{u}$ to the quality of service. The equations for the physical system are (8.5.4) and (8.5.8), and the limit equation is (8.5.9). With the definitions of $w^{\alpha,n}(\cdot)$, $\alpha = v, a, d$, from Section 8.5, we have $M^n(\cdot) = (w^{v,n}(\cdot), w^{a,n}(\cdot) - w^{d,n}(S^{a,n}(\cdot)))$. The system state and the data-creation processes are as in Section 8.5, and are independent of the initial condition and service intervals. The condition (A8.5.2), the first sentence of (A8.5.3), and either the fluid model of (A8.5.1b) or the following special case of (A8.5.1a) will be assumed, as well as that $v^n(\cdot)$ is stationary. Define $\xi_l^{d,n} = 1 - \Delta_l^{d,n}/\bar{\Delta}^{d,n}$. If the service process is not fluid, then suppose that for each n the $\{\Delta_l^{d,n}\}$ are mutually independent and identically distributed, with $E[\xi_l^{d,n}]^2 \to \sigma_d^2$ and $\{|\Delta_l^{d,n}/\bar{\Delta}^{d,n}|^2; n, l\}$ uniformly integrable. For the fluid service model, $\sigma_d^2 = 0$. The control mechanism will be specified in a way that guarantees the second part of (A8.5.3), (A1.8) and (A1.12).

Discounted cost criterion. Recall that $F^n(t)$ denotes $1/\sqrt{n}$ times the number of cells from all sources that have been rejected by time t. We start with the discounted cost functions

$$W_\beta^n(x, u^n) = E_x^{u^n} \int_0^\infty e^{-\beta t} \left[k_0(x^n(t))dt + \bar{c}dF^n(t) + cdU^n(t)\right] \tag{3.1}$$

and

$$W_\beta(x, u) = E_x^u \int_0^\infty e^{-\beta t} \left[(k_0(x(t)) + \bar{c}u(t))\, dt + cdU(t)\right], \tag{3.2}$$

with $x^n(0) = x$, $c > 0, \bar{c} > 0$. The $k_0(\cdot)$ penalizes delay, the $\bar{c}u$-term penalizes the control losses and the cU-term penalizes buffer overflow. The function $k_0(\cdot)$ might be nonlinear. For example, if small delays (or queue sizes) are not a concern, but large queue sizes are to be discouraged, then let $k_0(x) = 0$ for small x and set $k_0(x)$ large for large x. Such nonlinearities can be used as surrogates for constraints that might be difficult to handle otherwise; for example, we might wish to minimize some combination of the control and overflow losses, subject to a constraint on the probability that $x^n(t)$ is greater than some constant $\bar{x}$. Then experiment with the form of $k_0(\cdot)$ to approximately attain this goal. An example will be given when the data are discussed in the next section.

The relation (1.13) holds under the given conditions where Σ is diagonal with entries given by Theorems 8.5.1 and 8.5.2. The condition (1.12) is easy to show for $w^{v,n}(\cdot)$ and $w^{a,n}(\cdot)$ (with $p = 1$) by using (8.5.2) and the values for the Doob–Meyer processes of these martingales. If the service is "fluid," then $w^{d,n}(\cdot)$ is zero. Otherwise, (1.12) follows for $p = 1$ from the following theorem and the conditions on the $\Delta_l^{d,n}$.

Theorem 3.1. *For each n, let $\{\Delta_l^n\}$ be a sequence of real-valued random variables, let $\{\hat{\mathcal{F}}_l^n, l \geq 0\}$ denote the filtration that they engender and write $\hat{E}_l^n$ for the expectation conditioned on $\hat{\mathcal{F}}_l^n$. Define $\bar{\Delta}_l^n = E_l^n \Delta_{l+1}^n$, and suppose that for constants $C_i > 0$, $C_1 \leq \inf_n \bar{\Delta}_l^n \leq \sup \bar{\Delta}_l^n \leq C_2$. Suppose that $\{|\Delta_l^n/\bar{\Delta}_l^n|^2; n, l\}$ is uniformly integrable and that $\sup_n \hat{E}_l^n |\Delta_{l+1}^n / \bar{\Delta}_{l+1}^n|^2 \leq C_3$. Define the $\hat{\mathcal{F}}_l^n$-martingale differences $\xi_l^n = [1 - \Delta_l^n/\bar{\Delta}_l^n]$. Define $N^n(t) = \min\{m : \sum_{l=1}^m \Delta_l^n \geq nt\}/n$ and the process*

$$w^n(t) = \frac{1}{\sqrt{n}} \sum_{l=1}^{nt} \xi_l^n.$$

Then

$$\sup_n E \max_{s \leq N^n(t)} |w^n(s)|^2 = O(t), \tag{3.3}$$

and for any $T > 0$,

$$\limsup_n \sup_t E \sup_{nN^n(t) < l \leq nN^n(t+T)} \left| \frac{\Delta_l^n}{\sqrt{n}\bar{\Delta}_l^n} \right| = 0. \tag{3.4}$$

Proof. By (2.1.3),

$$E \max_{s \leq N^n(t)} |w^n(s)|^2 \leq \sup_l 4E|\xi_l^n|^2 EN^n(t). \tag{3.5}$$

To help evaluate $EN^n(t)$, write

$$[N^n(t)]^2 \leq \frac{2}{n} \left[\frac{1}{\sqrt{n}} \sum_{l=1}^{nN^n(t)-1} 1 \right]^2 + \frac{2}{n}.$$

The first term on the right can be written as

$$\frac{2}{n}\left[w^n(N^n(t)-1/n)+\frac{1}{\sqrt{n}}\sum_{l=1}^{nN^n(t-1/n)}\frac{\Delta_l^n}{\bar{\Delta}_l^n}\right]^2. \tag{3.6}$$

The term with the sum in (3.6) is bounded by $O(1)\sqrt{nt}$. Then using (3.5) we have

$$E[N^n(t)]^2 \le O(1)[EN^n(t)/n+t^2],$$

and this yields that $E[N^n(t)]^2 = O(t^2)$, which (together with (3.5)) yields (3.3). A similar analysis yields

$$\limsup_n \sup_t E[N^n(t+T)-N^n(t)] = O(T), \tag{3.7a}$$

$$\limsup_n \sup_t E \sup_{N^n(t)<s\le N^n(t+T)} |w^n(s)|^2 = O(T). \tag{3.7b}$$

By (3.7a) and the hypothesis concerning the uniform integrability of the squares of the martingale differences, for any $\epsilon > 0$ and $T > 0$,

$$\limsup_n \sup_t P\left\{\sup_{nN^n(t)<l\le nN^n(t+T)}\left|\frac{\Delta_l^n}{\sqrt{n}\bar{\Delta}_l^n}\right| \ge \epsilon\right\} = 0, \tag{3.8}$$

and

$$\limsup_n \sup_t E \sup_{nN^n(t)<l\le nN^n(t+T)}\left|\frac{\Delta_l^n}{\sqrt{n}\bar{\Delta}_l^n}\right|^2 < \infty. \tag{3.9}$$

Finally, (3.8) and (3.9) imply (3.4). ■

Now to verify (1.12), identify $N^n(\cdot) = N^{d,n}(\cdot)$ and $\Delta_l^n = \Delta_l^{d,n}$. Then equation (3.3) implies (1.12) with $p = 1$. Assumption (A1.2) is a condition on the residual-time error terms for the service and interarrival times. It is easy to verify for the arrival processes if they are either Markov modulated Poisson or Markov modulated "fluid" (see Section 8.5 for the definitions), and also for the service process if it is "fluid." It will next be demonstrated for the more general nonfluid service process. The component $\epsilon^{d,n}(\cdot)$ of $\epsilon^n(\cdot)$ that is due to the service process is bounded by $1/[\bar{\Delta}^{d,n}\sqrt{n}]$ times the length of the quantity $\Delta_l^{d,n}$ associated with the service interval that covers t. But (3.4) implies that for any $T < \infty$, $\sup_t E \sup_{t<s\le t+T} |\epsilon^{d,n}(s)| \to 0$ as $n \to 0$, which yields (A1.2).

Condition (A1.4) is obvious, as are (A1.9) and (A1.10). The boundedness of $\sup_{n,t} E|v^n(t)|^2$ (recall that $v^n(\cdot)$ is defined by (8.5.4) and is stationary) implies (A1.5).

The physical realization of the control. The verification of (A1.8) and (A1.12) requires that we specify the control mechanism more precisely.

Recall that $F^n(t)$ denotes $1/\sqrt{n}$ times the number of cells deleted by time t, and that the current problem formulation entails specifying a constant $\bar{u}$ that bounds $\bar{u}^n(t)$, the main component of the control in (A1.8). Suppose that the input is the Markov modulated fluid process. Then the cells are essentially infinitely divisible, and for any $\mathcal{F}_t^n$-adapted and $\mathcal{U}$-valued process $\bar{u}^n(\cdot)$, the value of the scaled number not admitted can be kept arbitrarily close to $\int_0^t \bar{u}^n(s)ds$. Hence, we can suppose that the process $\epsilon^{u,n}(\cdot)$ in (A1.8) is zero and that (A1.12) holds.

Next, suppose that the input process for each source is Markov modulated Poisson. The rate of the total (over all sources) unscaled arrival process is $n\nu\lambda/[\lambda+\mu] + \sqrt{n}\nu v^n(t)$. The deleted cells can be chosen by many means. But for specificity, suppose that the control is done at the individual sources and, while the choices made by the various individual sources will depend on the current information, they are otherwise not synchronized. For specificity, let the sources control by deleting via random sampling. Given a $\mathcal{U}$-valued and $\mathcal{F}_t^n$-adapted process $\bar{u}^n(\cdot)$, delete an arrival at any source at time t with probability (conditioned on $\mathcal{F}_t^n$) $p^n(t)/\sqrt{n} \le 1$, where $p^n(\cdot)$ satisfies

$$\frac{p^n(t)}{\sqrt{n}}\left[\frac{\sqrt{n}\nu\lambda}{\lambda+\mu} + \nu v^n(t)\right] = \bar{u}^n(t). \tag{3.10}$$

There will be a $p^n(t)$ such that $p^n(t)/\sqrt{n} \le 1$, except when $v^n(t)$ is of the order $-\sqrt{n}$ (or more negative, and then set $p^n(t) = 0$); the fraction of time that $v^n(t)$ is so negative goes to zero as $n \to \infty$. In fact, there will not be any control when $v^n(t)$ is very negative. Also, one can simplify by dropping the $v^n(t)$ from (3.10), with little asymptotic effect. With this control rule, the unscaled number of deleted cells is a Poisson process with rate $\sqrt{n}\bar{u}^n(t)$ at time t. Thus, the mean (times $1/\sqrt{n}$) number of deletions by time t is $\int_0^t \bar{u}^n(s)ds$, and the Doob–Meyer process associated with the centered process $F^n(t) - \int_0^t \bar{u}^n(s)ds$ is $\int_0^t \bar{u}^n(s)ds/\sqrt{n}$. From this point on, it is straightforward to show that (A1.8) and (A1.12) hold, and the rest of the details are omitted. Keep in mind that the random sampling approach was chosen for specificity. There are alternatives that allow greater flexibility, including correlated sampling, or sampling from a subset of cells or sources.

The ergodic cost problem. Here the costs are

$$\gamma^n(x,u^n) = \limsup_T \frac{1}{T} E\left[\int_0^T k_0(x^n(t))dt + \bar{c}F^n(T) + cU^n(T)\right], \tag{3.11}$$

$$\gamma(x,u) = \limsup_T \frac{1}{T} E\left[\int_0^T [k_0(x(t)) + \bar{c}u(t)]\,dt + cU(t)\right]. \tag{3.12}$$

The mathematics of the ergodic cost problem, as developed in Chapter 4 and Section 2, requires that the driving vector-valued Wiener process in the limit equation (8.5.9) have a positive definite covariance matrix. The variance of $w^v(\cdot)$ is always positive. The variance of $w^a(\cdot)$ will be positive if there is some randomness in the cell generation process when a source is *on*. While we have assumed either a Markov modulated fluid or a Markov modulated Poisson process for the data generation, any combination of these two can be used, and any positive combination yields a positive variance for $w^a(\cdot)$. Suppose that the variance of the Wiener process $w^a(\cdot) - w^d(\bar{\lambda}^d\cdot)$ in (8.5.9) is positive. Then the problem is nondegenerate in that the matrix $\Sigma(x)$ (which is constant in this problem) is positive definite. If, in addition, the $v(\cdot)$ variable were bounded, then the state space G would be bounded, and the results of Section 2 could be applied immediately. The restriction in Chapter 4 that the state space G be compact was made to simplify the problem of the nature of the convergence of the transition probabilities to the invariant measure, as $t \to \infty$. Owing to the stability of the process $v(\cdot)$ in the present case, Theorem 2.1 remains valid. Indeed, the original work [156] on the maximum principle allowed unbounded state spaces, under a "uniform" stability condition. While [156] dealt with the unreflected process, all of the results carry over to this example, owing to the strong stability properties of the one-dimensional Gauss–Markov process $v(\cdot)$.

The computations that were done in the first part of the section to verify the assumptions for the discounted cost case also imply the conditions in Section 2 for the ergodic cost case, except for the compactness of G. It is worth noting that the numerical data showed that the ergodic cost problem is also very well behaved for the degenerate fully fluid case where $w^a(\cdot)$ and $w^d(\cdot)$ are both zero.

9.4 Data for the Multiplexer Problem

One class of sources. Some of the numerical data from [169] will be discussed, since they illustrate some of the advantages of the heavy traffic approximations for control and analysis and the fundamental role played by the numerical methods for optimization problems when used as an exploratory tool. The data will be given only for the case where all sources are statistically identical, since the development in Section 8.5 and in the previous section was confined to that case. However, analogous results are readily obtained for any number of classes (subject to computational limitations as the dimension grows), as seen in [169, 171]. The data for the multiclass case are quite interesting, since the control of any one source type affects the environment for the other source types and they can interact in rather surprising ways. The data for the discounted cost problem are nearly identical to those for the ergodic cost problem for all cases, when β

is small, and the cost functions are normalized by multiplying them by β. All of the data below are for the ergodic cost function.

To get the numerical data, the optimization problems for the appropriate heavy traffic limit equations (8.5.9) and ergodic cost function (3.12) are solved numerically via the Markov chain approximation method [156, 159, 167]. The method is fully described in [167]. Convergence theorems exist for all the problems of concern in current heavy traffic analysis. The basic idea of the numerical problem is to approximate the controlled process (8.5.9) (actually, a rescaled form such that $\nu = 1$) by a suitable controlled finite-state discrete-time Markov chain on a state space that is a "discretization" of the state space of (x, v) and for which the computation can be readily carried out. If the chain and the approximating cost function are chosen suitably, then the value of the associated cost approximates that of the original problem arbitrarily well.

The numerical results depend on the discretization level h. The exact solution to the optimization problem for the limit system and that for the physical problem converge to each other as $n \to \infty$ and $h \to 0$. But, one is also interested in the level of discretization that yields the best results for a system of some given (large) size. Such issues are discussed in [169, 172], which provide useful heuristic rules. No matter what the (small) discretization level, the key qualitative features are retained. In an application the channel capacity, the size of the system, and the basic parameters would be given or estimated, and then the value of the "excess" capacity parameter b would be computed from (A8.5.2). A primary interest concerns the possible tradeoffs between the losses due to the controller action (the number of cells deleted by the controller) and the losses due to buffer overflow. Because of this, in the presented data the function $k_0(\cdot)$ in (3.11) and (3.12) is set to zero. Typically, under the optimal operating conditions the buffers were usually nearly empty. Naturally, buffer overflow losses are weighed more heavily than control losses, since one can select which cells are to be rejected by the controller and would select the lower priority ones, where possible.

The optimal cost by itself is not usually the most interesting result. What is of greater interest are the various components of the cost, say the individual buffer overflow and control losses, since they tell us what are the possible tradeoffs under optimal controls. In addition, we want to know the dependence of the control functions on the parameters of the system. Suppose that we have solved for the numerical approximation for the optimal control (called $u^h(\cdot)$) and cost (called $\bar{\gamma}^h$). Then calculate the numerical approximations (denoted by $\bar{\gamma}^h_U$ and $\bar{\gamma}^h_u$, respectively.) to the stationary expectation of the components $EU(1)$ and $E\int_0^1 u^h(s)ds$ under the computed optimal control. These are the estimates of the buffer overflow and control losses, respectively. Let $EU^n(1)$ denote the optimal stationary buffer overflow loss for the physical system. By the limit theorems,

$$\bar{\gamma}_U^h \approx EU^n(1) = \frac{\text{E[\# cells lost due to buffer overflow per unit time]}}{\sqrt{n}},$$

and this value is also approximately what one gets if the optimal control for the limit problem is applied to the physical problem for large n. With a rescaling such that $\nu = 1$, the total mean cell creation rate is $n\lambda/(\lambda+\mu)$. Thus, the ratio of buffer overflow loss to the total number of cells can be approximated by $\bar{\gamma}_U^h[\lambda+\mu]/[\sqrt{n}\lambda]$, which gives the average probability of cell loss. The tables and graphs give the $\log_{10}$ of either the total costs or of the two loss components $\bar{\gamma}_U^h$ and $\bar{\gamma}_u^h$.

The data sets. Unless otherwise noted, the data are for the Markov modulated "fluid" arrival and "fluid" service case. Figure 4.1 plots the minimal cost and associated buffer overflow loss component as a function of the scaled buffer size B for the case $n = 400, \mu = 1$, $\lambda = .4$, $c = 200$, $\bar{c} = 1$, $b = .95$ and $\bar{u} = .4$. The value of n is used as a guide to the appropriate level of discretization and to compute the effective traffic intensity, which is the ratio of the total mean input rate to the channel capacity in (A8.5.2). The discretization level $h = 1/\sqrt{n}$ gave good results.

The upper line in Figure 4.1 gives the numerical approximation to $\log_{10} \bar{\gamma}^n$. The lower line gives the numerical approximation to $\log_{10} EU^n(1)$. Note that the curves in Figure 4.1 are virtually straight lines. The asymptotic (as buffer size increases) linearity is not obvious a priori, due to the nonlinearity of the optimal control problem. The linearity in the figure suggests some type of large deviations result, but to date there are no proofs of such a fact. The linearity of the curves is important, since it facilitates extrapolation to larger buffer sizes, with the corresponding smaller losses. Figure 4.1 is for the fluid model. If $w^a(\cdot) - w^d(\bar{\lambda}^d\cdot)$ in (8.5.9) is not zero, then a similar almost linearity is observed, but it commences at slightly larger B-values. The probability of buffer overflow can increase quite a bit as the variance of this Wiener process increases. For example, if the variance of this Wiener process increases to 0.05 from zero, then the buffer losses increase by an order of magnitude. Standard controllers such as the so-called leaky bucket or token bank type systems, that do not depend on feedback data, act (asymptotically) as though the deletion takes place continuously, essentially analogous to $u(\cdot)$ being constant.

Figure 4.2 plots the optimal controls for two different values of $\bar{u}$, for the case $\nu = \mu = 1$, $n = 75$, $\lambda = .2$, $b = 0.48$, which yields a traffic intensity $\rho = .75$. The scaled buffer size is $B = 16$ in all of the data from this point on. The optimal controls are determined by a switching line, with the control working at a maximum rate above the line and being inactive below it. The switching curves are linear, and parallel. Again, there are no current proofs of these numerically observed facts. The fact that the switching line drops when $\bar{u}$ decreases is intuitively reasonable. With a smaller maximum mean deletion rate, the controller must act earlier to reduce potential buffer overflow. Figure 4.3 illustrates the effect of burstiness on the control. We say that the burstiness of a source increases if the probability of being *on* is unchanged, but the duration of the periods increases. With burstier sources, the current number of *on* sources does not change fast, so that if the number is above average, the controller will be more cautious and control "earlier." Figure 4.4 illustrates the effect of a small amount of noise in the x-equation. The switching curve moves rapidly downward as x approaches B. This is also intuitively reasonable. Owing to the greater randomness, the likelihood of buffer overflow can increase dramatically, and the control must act sooner in anticipation of the possibility of overflow.

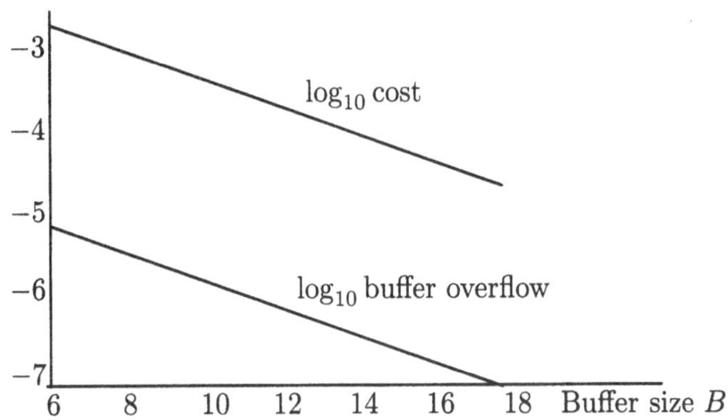

Figure 4.1. Losses as a function of scaled buffer size.

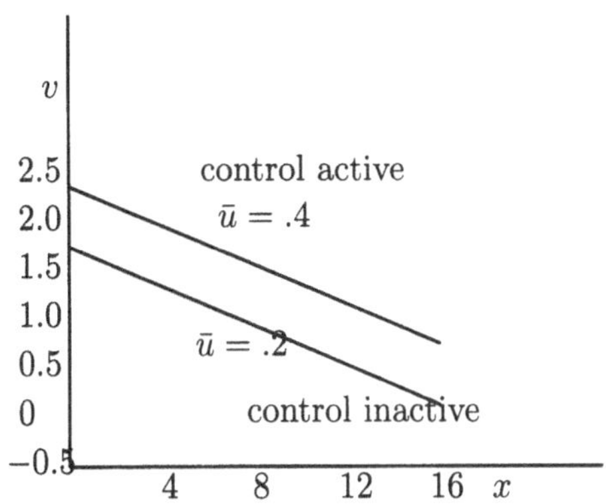

Figure 4.2. The control switching curves as function of $\bar{u}$.

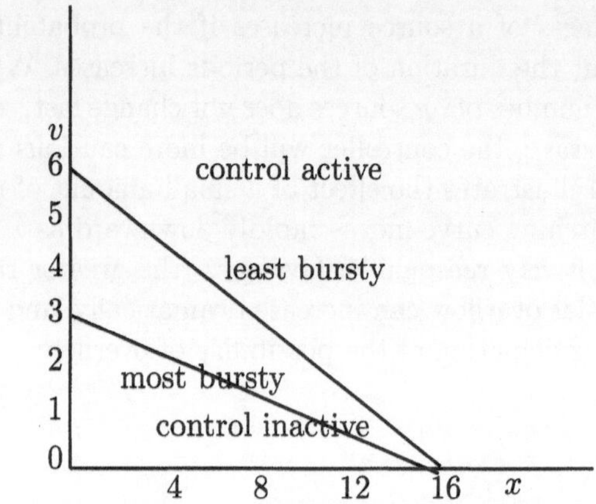

Figure 4.3. The switching lines as a function of source burstiness.

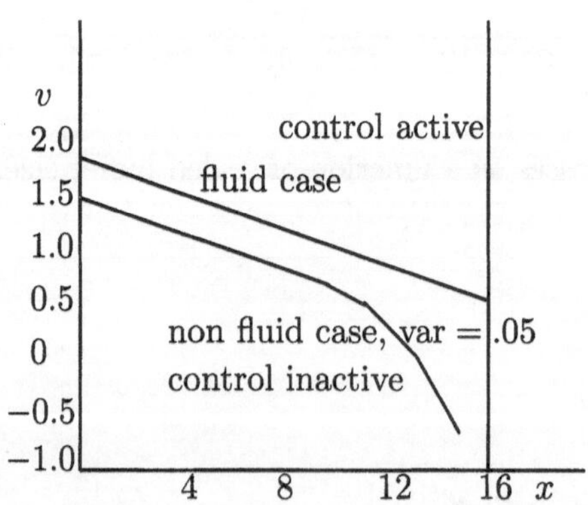

Figure 4.4. Switching curves for the fluid and non fluid cases.

Optimal controls. We next demonstrate the improvement under state feedback controls. Refer to Table 4.1, a fluid model case, where we give the $\log_{10}$ of both the optimal costs and the loss components (the buffer overflow and control losses). Various values of $\bar{u}$ and $c, \bar{c}$ are used.

$n = 75$, $\lambda = 1$, $\mu = .2$, Buffer=$16\sqrt{n}$, fluid case				
$\bar{u}$	$c, \bar{c}$	$\log_{10} \bar{\gamma}^h$	buffer overflow loss	control loss
0.0	n.a.	−4.82	−4.82	n.a.
0.2	5,1	−4.42	−5.35	−4.81
0.2	50,1	−3.77	−5.64	−4.27
0.2	200,1	−3.36	−5.81	−3.90
0.4	5,1	−4.61	−5.86	−4.75
0.4	50,1	−4.22	−6.28	−4.47
0.4	200,1	−3.93	−6.54	−4.24
0.6	5,1	−4.68	−6.42	−4.72
0.6	50,1	−4.53	−6.85	−4.64
0.6	100,1	−4.47	−7.0	−4.58
0.6	200.1	−4.35	−7.15	−4.51

Table 4.1.Optimal losses and components.

Table 4.1 shows the possible tradeoffs between the controller and buffer overflow losses, by varying the weights on the cost components and the value of $\bar{u}$, and using the associated optimal controls. Tables such as this are useful for design. Typically, one wishes to do the best possible subject to whatever constraints are on the problem. The upper limit $\bar{u}$ in this problem is a type of quality of service constraint in the sense that it governs the maximum mean rate at which cells can be deleted. One might wish to control so that the probability of buffer overflow is no greater than some given level ϵ, and minimize the control loss with this constraint. This constrained problem is hard to solve, but we can get to the same end by solving a sequence of optimal control problems with varying cost weights and using the results (such as those in Table 4.1) to get a policy that realizes the desired goals. The table tells us what the (optimal) tradeoffs are between the buffer overflow loss and the other quantities of interest. We can choose between the pairs ($\bar{u}$, control loss) to get the system that best realizes the goal of ϵ buffer overflow loss. Such "constrained" formulations are an important application of optimal control techniques. In general, one would solve a sequence of optimal control problems in order to experiment with good tradeoffs between the various costs that are of interest. Given a desired buffer overflow loss, the reduction in control loss under optimal feedback control can be 2 to 3 orders of magnitude better than that for constantly acting controls. Tables such as this can also be used to estimate the gains from marginal increments in the channel capacity.

The average buffer content for the various cases in Table 4.1 with $\bar{u} = .4$ is $0.04 \times \sqrt{n}$. With the normalization $\nu = 1$, it is just 0.346, out of a maximum (scaled) size of 16. The probability that the buffer is nonempty is approximately 0.2. These numbers are quite insensitive to the cost coefficients, corroborating the fact that the buffer overflow bursts are very rare.

10
Singular Controls

Chapter Outline

The singular control problem of Section 3.3 is dealt with in this chapter. Recall that the control $F(\cdot)$ in (1.1) is called singular if there is no integrable process $u(\cdot)$ such that $F(t) = \int_0^t u(s)ds$. Such control processes arise, for example, when we have either direct or indirect control over admission, routing, or idle time and there is no a priori bound on the number not admitted. Following the approach set out in Chapter 9, we start with a canonical model and phrase the assumptions in a way that they cover many cases of interest. The description of the canonical model and the main convergence theorem are in Section 1. Part of the proof that the optimal costs for the physical model converge to the optimal for the limit problem uses the time transformation method of Section 2.6. For the other part, it is first shown that the singular control for the weak-sense limit model can be arbitrarily well approximated by a differentiable control, and then the methods of Chapter 9 are applied. Such an approximation has independent interest in the study of singular controls.

Typically, the controls are determined by a switching surface, where the control is active on one side and not on the other. Under heavy traffic, good approximations to this surface are obtained by solving the optimal control problem for the weak-sense limit. Although it is not treated, very similar results can be obtained for the impulsive control problem of Section 3.3. Impulsive controls occur, for example, when a processor or input can be turned off and on, but at a cost that does not vanish as $n \to \infty$. Interesting and nonstandard control problems then arise, as seen in [174].

The work is confined to the discounted cost function. Little of any generality is known concerning the ergodic cost for singular controls, and the results of Chapter 4 have not been extended to this case.

An interesting form of the singular control problem arises when there are *long* vacations or processor interruptions of the type in Subsection 7.3.2, and when one *can control* during the vacations. Consider the situation where the control is either the rejection of exogenous inputs or rerouting, which might be particularly important to do during vacations. The behavior of the physical problem becomes complicated owing to the interaction of the effects of both the vacations and the controls. But the heavy traffic limit retains its simplicity. The weak-sense limit problem is nonclassical, however: While the effects of the vacation reduce to an *instantaneous* jump in the limit, the value of the jump depends on the arrivals as well as the controls *during* the physical vacation.

For the problem where there can be control during the vacations, the part of the process between vacations and the behavior during vacations are treated separately. The sequence of sections between vacations converges weakly to the sections of a singularly controlled diffusion. The approach of Subsection 7.3.2 is used to handle the convergence of the vacation sections. In Subsection 7.3.2 time during the vacations was rescaled, and what was called the *local fluid time scale* was used, in which the duration of the lth vacation of P_i became $\tau_{i,l}^{v,n}$, and a mean or fluid approximation to the inputs, outputs, and routings was justified. Then the processes defined by (7.3.16) gave the values of the jumps in the limit, although there was no control. A similar analysis for the current problem yields (7.3.16), but with a control term added.

The convergence analysis also yields the nonclassical structure of the correct control problem and state space for the weak-sense limit. In the part between vacations, the model is just the singularly controlled diffusion, as noted, and the state is just $x(t)$, the limit of the scale queue sizes. What is nonclassical in the limit model is the proper state space description for the effects of the control during the vacations. To get this, we need to see what information is necessary for the controller "during" the vacations. The value of the controlled jump during a vacation is determined by a controlled version of (7.3.16). Consider the jump due to the lth vacation of P_i and use the local fluid time scaling so that the duration is $\tau_{i,l}^{v,n}$. During this time the mean or fluid approximation can be used to represent the exogenous inputs, completed services, and routings, as in Section 7.3. If, for example, the control is over the admission of the exogenous inputs, then (in the limit, and using the local fluid time scale) the control at P_i is a process whose derivative is bounded by the rate of arrival $\bar{\lambda}_i^a$ at P_i, and analogously if the control is over the routings. In scaled time, the limit of the effects of the vacations is just a jump, whether there is or is not control during the vacations. However, from the point of view of the controller for the limit problem, the evolution during the "vacation" must be viewed in terms of the limit in the local fluid time scale. The correct state description is actually the *pair* (queue state, time since the vacation started), all in the limit of the local fluid time scale. Unless the limit $\tau^v_{i,l}$ is exponentially distributed, it is necessary to keep track of the elapsed time since the start of the vacation, since the control actions might depend on the prediction of the conditional distribution of its remaining time, given its duration to date. See Section 2.

Section 3 is concerned with the modeling and convergence proof of the problem of admission control for the ISDN (integrated systems data network) system that was introduced in Section 1.5. It is shown that the problem fits the canonical model of Section 1 and that the conditions are verified. Many alternative forms of this problem are in [2].

Other forms of the singularly controlled problem will be addressed in the context of the workload (and state space collapse) examples in Chapters 11 and 12. Section 4 briefly outlines some other models that are closer to the format of this chapter. The first concerns the problem where an a priori routing is given, but a job can be rerouted at a cost. The second arises in admission and routing control for trunk-line-type systems such as the long-distance telephone network. In the simple form dealt with here, one can reroute only if the assigned buffer is full. Then the control takes a different form, being an integral with respect to the boundary overflow process. The examples chosen are intended to be illustrative of general techniques and are certainly not exhaustive.

The basic model of Section 1 allows state dependence. Thus, phenomena such as balking and withdrawing can be handled, and these can be incorporated into the models of Sections 3 and 4. Any combination (for discounted cost) of the singular and ordinary control problems of this and of Chapter 9 can be handled simply by combining their assumptions and methods.

10.1 A Canonical Model

In Chapter 9 we worked with a canonical model that covered numerous problems of practical interest and showed how to verify the conditions for a class of applications. An analogous procedure will be followed in this Chapter. Let $\{\mathcal{F}_t, t \geq 0\}$ be a filtration and $w(\cdot)$ a standard $\mathcal{F}_t$-Wiener process. The weak-sense limit system will have the form

$$\begin{aligned} x(t) &= x + \int_0^t b_0(x(s))ds + M(s) + DF(s) + z(s), \\ z(t) &= \sum_i d_i y_i(t), \end{aligned} \tag{1.1}$$

where D is a matrix and, as in Chapter 9, $M(t) = \int_0^t \sigma(x(s))dw(s)$. If $\sigma(\cdot)$ is a constant, then $M(\cdot)$ and $w(\cdot)$ are used interchangeably. The control $F(\cdot)$ has paths in $D(I\!R^k; 0, \infty)$ for some k, and its components $F_i(\cdot)$ are nondecreasing with $F_i(0) = 0$. As in Section 3.3, $F(\cdot)$ is called admissible (alternatively, $(w(\cdot), F(\cdot))$ is an admissible pair) if $F(\cdot)$ is $\mathcal{F}_t$-adapted. Such functions are called *singular controls* because (when considered as measures on $[0, \infty)$) they might be singular with respect to Lebesgue measure in that they cannot necessarily be represented in the form $F(t) = \int_0^t u(s)ds$ for some integrable function $u(\cdot)$ [179, 189, 225, 231]. For $x(0) = x$, the discounted cost function of interest for (1.1) is

$$W_\beta(x, F) = E_x^F \int_0^\infty e^{-\beta t} \left[k_0(x(t))dt + \bar{c}'dF(t) + c'dy(t)\right]. \tag{1.2}$$

The canonical model for the physical problem is taken to be

$$x^n(t) = x^n(0) + \int_0^t b_0(x^n(s))ds + M^n(s) + DF^n(s) + z^n(s) + \epsilon^n(s), \tag{1.3}$$

where it will be assumed that $\epsilon^n(\cdot)$ converges weakly to the "zero" process, and $z^n(\cdot)$ is the reflection term. The control $F^n(\cdot)$ has paths in $D(I\!R^k; 0, \infty)$, and the components $F_i^n(\cdot)$ are nondecreasing and satisfy $F_i^n(0) = 0$. The σ-algebra $\mathcal{F}_t^n$ is defined as in Section 9.1, and $F^n(\cdot)$ is $\mathcal{F}_t^n$-adapted. For $x^n(0) = x$, the cost function for (1.3) is

$$W_\beta^n(x, F^n) = E_x^{F^n} \int_0^\infty e^{-\beta t} \left[k_0(x^n(t))dt + \bar{c}'dF^n(t) + c'dy^n(t)\right]. \tag{1.4}$$

Recall the definitions $V_\beta^n(x) = \inf_{F^n} W_\beta^n(x, F^n)$ and $V_\beta(x) = \inf_F W_\beta(x, F)$.

The system (1.3) arises when the control consists in the rejection of inputs or in rerouting [2, 189]. In this sense, it is a very common model, particularly in applications to telecommunications.

Assumptions

A1.1. *$k_0(\cdot)$ is Lipschitz continuous and nonnegative, and $c_i \geq 0$, $\bar{c}_i > 0$.*

A1.2. *$b_0(\cdot)$ and $\sigma(\cdot)$ are Lipschitz continuous, with $\sigma(\cdot)$ bounded.*

A1.3. *Let $\sup_n E|x^n(0)|^2 \leq C_0$ for some $C_0 < \infty$. Then there is $p \geq 0$ such that, uniformly in the control,*

$$\limsup_n \sup_t \frac{E\left|b_0(x^n(t))\right|^2}{1 + t^p} = O(1).$$

A1.4. *Let* $E|x(0)|^2 \leq C_0$ *for some* $C_0 < \infty$. *Then there is* $p \geq 0$ *such that, uniformly in the control,*

$$\sup_t \frac{E\left|b_0(x(t))\right|^2}{1+t^p} = O(1).$$

The Lipschitz conditions in (A1.2) together with (A3.5.1)–(A3.5.3) imply that for any admissible pair, there is a strong-sense unique solution to (1.1), as noted at the end of Section 3.5.

Weakening the conditions. Recall the comments below (A9.1.7). The analogous weakening holds in this chapter. One needs the bounds on the expectations in (9.1.12), (A9.1.2), and in (A1.3) and (A1.4) to hold only for whatever components are necessary to assure that

$$E\int_T^\infty e^{-\beta t}[k_0(x^n(t))dt + c'dy^n(t)]$$

goes to zero uniformly in n as $T \to \infty$ and $\{|k_0(x^n(s))|, y^n(s); s \leq T, n\}$ is uniformly integrable for each T, and the conditions in the theorems can be changed to reflect this.

The next theorem establishes the boundedness of the optimal costs.

Theorem 1.1. *Assume the model* (1.3) *and the conditions* (A1.1)–(A1.3), (A9.1.1), (A9.1.2), (A9.1.4), (9.1.12), *and* (A3.5.1)–(A3.5.3). *Then*

$$\sup_n EV_\beta^n(x^n(0)) < \infty,$$

where the expectation is over the possibly random initial condition.

Proof. Let $F^n(t) = 0$ for all t. Then the costs for (1.3) are bounded uniformly in n by Theorem 9.1.1 and the bounds in the assumptions. ∎

The next result is used to show that we can restrict attention to controls that do not increase after some finite time when proving the limit theorems for the costs.

Theorem 1.2. *Assume the model* (1.1) *and the conditions* (A3.5.1)–(A3.5.3), (A1.1), (A1.2), *and* (A1.4). *Then for any* $C_0 < \infty$, *all* $x(0)$ *satisfying* $E|x(0)|^2 \leq C_0$, *and each* $\epsilon > 0$, *there is a* $T_\epsilon < \infty$ *and* ϵ*-optimal controls for* (1.1) *that do not increase after* T_ϵ. *Also, for the given class of* $x(0)$,

$$\sup_{x(0)} EV_\beta(x(0)) < \infty. \tag{1.5}$$

The expectation is over the possibly random initial condition.

Proof. Fix C_0 and restrict $x(0)$ such that $E|x(0)|^2 \leq C_0 < \infty$. The controls $F(\cdot)$ can depend on $x(0)$. The inequality (1.5) can be established by showing that $\sup_{x(0)} EW_\beta(x(0), 0) < \infty$ and the details are left to the reader. Let $F(\cdot)$ be admissible for (1.1) and satisfy

$$\sup_{x(0)} EW_\beta(x(0), F) < \infty. \tag{1.6}$$

For $T > 0$, define $F^T(\cdot) = F(t \wedge T)$, and let $(x^T(\cdot), z^T(\cdot), y^T(\cdot))$ denote the associated solution to (1.1). The Lipschitz conditions and Theorem 3.5.1 ensure that we can construct a solution for each admissible pair. Thus, it can be supposed that the same Wiener process is used, no matter what T is. Note that

$$\int_0^\infty e^{-\beta t} dF(t) = F(t)e^{-\beta t}\Big|_0^\infty + \beta \int_0^\infty e^{-\beta t} F(t)dt,$$

and similarly if $y(\cdot)$ replaces $F(\cdot)$. For bounded costs, the first term on the right must be zero.

Define $B(t) = \int_0^t b_0(x(s))ds$, and let $B^T(\cdot)$ and $M^T(\cdot)$ denote the use of $F^T(\cdot)$. By (A1.1), there is a constant C such that for any T and $t \geq T$,

$$\begin{aligned} \left|k_0(x(t)) - k_0(x^T(t))\right| \leq\ & C\left|F(t) - F(T)\right| + C\left|B(t) - B^T(t)\right| \\ & + C\left|y(t) - y^T(t)\right| + C\left|M(t) - M^T(t)\right|. \end{aligned}$$

By (A1.1) and the fact that (3.4.13) holds by Theorem 3.5.1, there is a constant C_1 such that, for $t \geq T$,

$$\begin{aligned} \left|y(t) - y^T(t)\right| \leq\ & C_1\left|F(t) - F(T)\right| \\ & + \sup_{T \leq s \leq t} C_1\left|B(s) - B^T(s)\right| + \sup_{T \leq s \leq t} C_1\left|M(s) - M^T(s)\right|. \end{aligned}$$

Thus, for some $C_2 < \infty$,

$$\begin{aligned} \left|k_0(x(t)) - k_0(x^T(t))\right| \leq\ & C_2\left|F(t) - F(T)\right| \\ & + \sup_{T \leq s \leq t} C_2\left|B(s) - B^T(s)\right| + \sup_{T \leq s \leq t} C_2\left|M(s) - M^T(s)\right|. \end{aligned}$$

By the boundedness of $\sigma(\cdot)$,

$$E \sup_{T \leq s \leq t} \left|M(s) - M^T(s)\right|^2 = O(t).$$

By (A1.4),

$$E \sup_{T \leq s \leq t} \left|B(s) - B^T(s)\right|^2 = O(1)t \int_0^t O(s^p)ds.$$

Hence, there is $C_3 < \infty$ that does not depend on T or $F(\cdot)$ or on $x(0)$ in the chosen class such that

$$E\int_T^\infty e^{-\beta t}[k_0(x^T(t)) + |y^T(t)|]dt \le C_3 E\int_T^\infty e^{-\beta t}[|F(t)| + O(t^{p+1})]dt. \tag{1.7}$$

Fix $x(0)$ in the chosen class. By the dominated convergence theorem and the finiteness of the cost under $F(\cdot)$, the right side goes to zero as $T \to \infty$. Hence, under $F^T(\cdot)$ the contribution to the cost on $[T,\infty)$ goes to zero as $T \to \infty$. Take $F(\cdot)$ to be an $\epsilon/2$-optimal control. Then we lose at most $\epsilon/2$ in the cost by taking T suitably large.

Now suppose that given $\epsilon > 0$ there is no single value of T that can be used for all initial conditions in the chosen class. Then for each small $\epsilon > 0$ there are $T_k \to \infty$ and $x_k(0)$, $k = 1,\ldots,$ satisfying $E|x_k(0)|^2 \le C_0$ and such that $x_k(0)$ converges weakly to some $x(0)$ with $E|x(0)|^2 \le C_0$, and such that for *any* sequence of associated ϵ-optimal controls $F_k(\cdot)$,

$$EW_\beta(x_k(0), F_k^{T_k}) - EW_\beta(x_k(0), F_k) \ge \rho$$

for some $\rho > 0$ and all k. Let $F_0(\cdot)$ be an $\epsilon/4$-optimal control for $x(0)$, that is constant after some finite time T. Apply it to the systems with the other initial conditions $x_k(0)$. That is, the same pair $(w(\cdot), F_0(\cdot))$ is used for all $x_k(0)$. The following two facts establish a contradiction:

1. $EW_\beta(x_k(0), F_0) \to EW_\beta(x(0), F_0) \le EV_\beta(x(0)) + \epsilon/4$,
2. $\liminf_k EW_\beta(x_k(0), F_k) \ge EV_\beta(x(0))$.

Thus, $F_0(\cdot)$ is ϵ-optimal for initial condition $x_k(0)$ and large k. ■

Theorem 1.3. *Assume the model* (1.3), *the conditions* (A1.1)–(A1.3), (A9.1.1)–(A9.1.4), (A9.1.7), (A3.5.1)–(A3.5.3), *and that* $x^n(0)$ *converges weakly to* $x(0)$. *Then*

$$\liminf_n EV_\beta^n(x^n(0)) \ge EV_\beta(x(0)). \tag{1.8}$$

Proof. For simplicity in the proof let $\sigma(\cdot)$ be a constant, called σ. For the general case, we need only show that the limit martingale can be represented as a stochastic integral. For arbitrary $\epsilon > 0$, let $F^n(\cdot)$ be ϵ-optimal controls. The main new difficulty in the proof over that of Theorem 9.1.2 arises from the possibility that $\{F^n(\cdot)\}$ will not be tight. The time transformation method of Section 2.6 is designed to deal with this problem. See [167, Chapter 11] for another effective use of this method.

Define $T^n(t) = t + \sum_i F_i^n(t)$, analogously to what was done in (2.6.7). As in (2.6.8), define the "inverse"

$$\hat{T}^n(t) = \inf\{s : T^n(s) > t\}, \tag{1.9}$$

and define the "hat" processes as $\hat{x}^n(t) = x^n(\hat{T}^n(t))$, etc. The time transformation $\hat{T}^n(\cdot)$ stretches time so that all of the sequences of concern are asymptotically continuous. We need only consider $F^n(\cdot)$ such that for each $T < \infty$,

$$\sup_n E^n_{x^n(0)} F^n_i(T) < \infty. \tag{1.10}$$

Then (2.6.9) holds with $M^n(\cdot)$ replacing $w^n(\cdot)$. By the construction of the time transformation, the sequences $\{\hat{F}^n(\cdot)\}$ and $\{\hat{T}^n(\cdot)\}$ are asymptotically continuous. In fact, they are asymptotically Lipschitz continuous with Lipschitz constant no greater than unity. The $\hat{M}^n(\cdot)$ are still asymptotically continuous and $\hat{\epsilon}^n(\cdot)$ converges weakly to the "zero" process.

For purposes of characterizing the correct limit process, we can suppose that $b_0(\cdot)$ is bounded and use the truncation method of Subsection 2.5.5 and the strong-sense uniqueness properties of the solution to (1.1). Then the process defined by $\int_0^t b_0(x^n(s))ds$ is asymptotically continuous, and this implies the asymptotic continuity of the process defined by $\int_0^t b_0(\hat{x}^n(s))d\hat{T}^n(s)$. Then by Theorem 3.6.1, $\hat{z}^n(\cdot)$ is asymptotically continuous, hence $\hat{x}^n(\cdot)$ is. Now extract a weakly convergent subsequence with weak-sense limit denoted by $\hat{x}(\cdot), \ldots, \hat{T}(\cdot)$, and proceed as in Section 2.6. The weak-sense limit satisfies (2.6.10) with $\hat{w}(\cdot)$ replaced by $\hat{M}(\cdot)$. By (1.10), the inverse function $T(t) = \inf\{s : \hat{T}(s) > t\}$ exists w.p.1. Define $x(t) = \hat{x}(T(t))$, etc. Then as in Section 2.6, (1.1) holds.

We have not proved the integrability properties that would ensure the analogue of the convergence of the costs as in (2.6.12), but (letting n index the weakly convergent subsequence) Fatou's lemma implies that

$$\begin{aligned} &\liminf_n EW^n_\beta(x^n(0), F^n) \\ &\qquad = \liminf_n E\int_0^\infty e^{-\beta\hat{T}^n(t)}\left[k_0(\hat{x}^n(t))d\hat{T}^n(t) + \bar{c}' d\hat{F}^n(t) + c' d\hat{y}^n(t)\right] \\ &\qquad \geq E\int_0^\infty e^{-\beta\hat{T}(t)}\left[k_0(\hat{x}(t))d\hat{T}(t) + \bar{c}' d\hat{F}(t) + c' d\hat{y}(t)\right]. \end{aligned} \tag{1.11}$$

As in Section 2.6, the right side of (1.11) equals $EW_\beta(x(0), F)$. ■

Approximations of controls for (1.1). Given $R \geq 0$ and an admissible control $F(\cdot)$ for (1.1) that does not increase after a time T, let $F^R(\cdot)$ denote the admissible control whose components $F^R_i(\cdot)$ have derivatives that are bounded by R and are the best pointwise approximations to $F_i(\cdot)$ that satisfy $F^R_i(t) \leq F_i(t)$ for all t. The approximations that are given in Theorems 1.4 will be used to get the"converse" to (1.8) in Theorem 1.5. They are of independent interest in singular control theory, since they allow us to replace singular by classical controls for many theoretical purposes.

Theorem 1.4. *Let $C_0 < \infty$ and confine attention to $x(0)$ such that $E|x(0)|^2 \leq C_0$ for some $C_0 < \infty$. Assume the conditions of Theorem 1.2 and let*

$\epsilon > 0$. Then there are $T < \infty$ and ϵ-optimal admissible controls $F(\cdot)$ for (1.1) with $\sup_{x(0)} EW_\beta(x(0), F) < \infty$, and $F(t)$ is constant for $t \geq T$. The $F(\cdot)$ might depend on $x(0)$. By using the bounded control $F_A(\cdot)$, defined by its components $F_{A,i}(t) = F_i(t) \wedge A$, the cost is increased by at most ϵ (uniformly in $x(0)$) for large A. Furthermore, there is $R < \infty$ and differentiable $F_A^R(\cdot)$ such that $\dot{F}_A^R(t) \leq R, t \leq T$, $\dot{F}_A^R(t) = 0$, $t > T$, and

$$EW_\beta(x(0), F_A^R) \leq EW_\beta(x(0), F_A) + \epsilon \tag{1.12}$$

for all $x(0)$, where the expectation is over the possibly random initial condition $x(0)$. Write $F_A^R(t) = \int_0^t u^R(s)ds$. Without loss of generality, it can be supposed that $u^R(\cdot)$ is piecewise constant, and takes values in a finite set, where the intervals and set of possible values do not depend on $x(0)$.

Proof. Let $F(\cdot)$ be $\epsilon/4$-optimal. By Theorem 1.2 we can suppose that there is $T < \infty$ such that $F(\cdot)$ is constant after T for all considered $x(0)$. Given $A < \infty$, define $T_A = \min\{t : F_i(t) \geq A$, for some $i\} \wedge T$ and set $F_{A,i}(t) = F_i(t) \wedge A$. Let the subscript A denote quantities associated with the use of $F_A(\cdot)$. For the assertion concerning bounded controls, we need only show that, uniformly in $x(0)$ in the chosen class,

$$\limsup_{A\to\infty} E \int_{T_A}^{\infty} e^{-\beta t} \left[k_0(x_A(t)) + |F_A(t)| + |y_A(t)|\right] dt \leq \epsilon/2. \tag{1.13}$$

By the computations leading to (1.7), for some $C_1 < \infty$ the left side is bounded by

$$C_1 \limsup_{A\to\infty} E \int_{T_A}^{\infty} e^{-\beta t} \left[1 + t^{p+1} + |F(t)|\right] dt.$$

Now fix $x(0)$, hence $F(\cdot)$. Then the dominated convergence theorem implies that the last expression equals zero. The uniformity in $x(0)$ in the chosen class is shown by an argument by contradiction, and the details are omitted.

Now moving to the assertion (1.12), let us suppose that $F(T) = F(\infty)$ is bounded by some constant A and show that $F(\cdot)$ can be approximated by a differentiable control with the derivatives bounded by some $R < \infty$. Given $R > 0$, let $F_i^R(\cdot)$ be the best pointwise approximation to $F_i(\cdot)$ that is no greater than $F_i(\cdot)$ and that is differentiable with derivative no larger than R. Then for large R, $F^R(\cdot)$ will be arbitrarily close to $F(\cdot)$, except in a set whose Lebesgue measure is bounded by some δ^R, that goes to zero uniformly in $x(0)$ as $R \to \infty$. Let $x^R(\cdot)$, etc., denote the associated solution.

The time transformation method that was employed in Theorem 1.3 can be used. Use $T^R(t) = t + \sum_i F_i^R(t)$ with inverse $\hat{T}^R(t) = \inf\{s : T^R(s) > t\}$. The sequence $\left\{\hat{x}^R(\cdot), \hat{z}^R(\cdot), \hat{w}^R(\cdot), \hat{F}^R(\cdot), \hat{T}^R(\cdot)\right\}$ converges weakly to a weak-sense limit denoted by $(\hat{x}(\cdot), \hat{z}(\cdot), \hat{w}(\cdot), \hat{F}(\cdot), \hat{T}(\cdot))$. The inverse $T(t) = \inf\{s : \hat{T}(s) > t\}$ exists, and the process $(\hat{w}(T(\cdot)), \hat{F}(T(\cdot)))$ has the same probability law as the original $(w(\cdot), F(\cdot))$. The inverse process $(\hat{x}(T(\cdot)), \hat{z}(T(\cdot)), \hat{w}(T(\cdot)), \hat{F}(T(\cdot)))$ satisfies (1.1). By the strong-sense uniqueness of the solution to (1.1) for each admissible pair $(w(\cdot), F(\cdot))$, and the boundedness of $F(\cdot)$, $EW_\beta(x(0), F^R) \to EW_\beta(x(0), F)$ uniformly in $x(0)$ in the chosen class. Thus, the cost changes by an arbitrarily small amount using $F^R(\cdot)$ for large enough R.

The assertion that the differentiable $F^R(\cdot)$ can be replaced by a piecewise constant control with arbitrarily little penalty is straightforward and is left to the reader. ∎

Theorem 1.5. *Assume the model* (1.3), *let* $x^n(0)$ *converge weakly to* $x(0)$, *and assume the conditions* (A1.1)–(A1.4), (A9.1.1)–(A9.1.4), (A9.1.7), (A9.1.12), *and* (A3.5.1)–(A3.5.3). *Then*

$$EV_\beta^n(x^n(0)) \to EV_\beta(x(0)). \tag{1.14}$$

Proof. By Theorem 1.3, we need only prove that

$$\limsup_n EV_\beta^n(x^n(0)) \le EV_\beta(x(0)). \tag{1.15}$$

The approach used in Theorem 9.1.4, which used the "comparison control" defined by the conditional probability law (9.1.26), can be used. Let $\epsilon > 0$. By Theorem 1.4, we can suppose that there is an $\epsilon/2$-optimal control for (1.1) that has the form $F(t) = \int_0^t u(s)ds$, where $u(\cdot)$ is bounded, piecewise constant, and finite-valued, and equals zero after some finite time T. Thus, we need only consider such controls. From this point on use (A9.1.12) and the method of Theorem 9.1.4. ∎

10.2 Vacations

Recall the problem of Subsection 7.3.2, where the vacations were "rare," but were long enough to cause jumps in the state in the limit. Suppose that one can control *during* as well as between the vacations. For example, consider the problem described by Figure 7.3.1, but suppose that the exogenous inputs can be refused entry at any time if desired, and that there are buffers of finite capacity $B_i\sqrt{n}$ and any buffer overflow is lost. Exercising such control during the vacations changes the value of the jump and leads to a nonclassical weak-sense limit problem.

We will consider such a problem for a general network and continue to use the scaling and general structure of the network of Subsection 7.3.2, where the rate at which a vacation started in real time was $O(1/n)$ and the duration was $O(\sqrt{n})$. But the reader should keep in mind that the fast scaling of Section 6.6 could also be used.

The new problems concern the effects of the control actions during a vacation. The control between vacations is treated as in Section 1. Consider the kth vacation of P_i. During this vacation and without control, the queue at P_i would increase, and the queues that P_i feeds will decrease. The changes can be managed by controlling the admissions. But whether to control the admissions or not at any time during the vacation would depend on the information that is then available concerning the (possibly a priori unknown) duration of that vacation. If $\tau_{i,k}^{v,n}$ is exponentially distributed, then no information can be obtained on its residual value until the vacation has ended, owing to the "memoryless" property. But if $\tau_{i,k}^{v,n}$ has a general distribution, then the conditional distribution of the time to go (given the current system data) will depend on the elapsed time since the vacation started. In this case of a nonexponential distribution, whether to control or not during a vacation would depend on this elapsed time. The possibility of such control requires a closer look at the behavior during a vacation, and the use of a more detailed model for the limit process.

The analysis involves a combination of the methods that were used in Theorem 7.3.2 and in Section 1, since there can be control both during and between the vacations. Let $F_i^n(\cdot)$ (respectively, $U_i^n(t)$) denote $1/\sqrt{n}$ times the total number of exogenous arrivals not admitted to (respectively, buffer overflow at) P_i by real time nt. Let $F_i^{s,n}(t)$ (respectively, $U_i^{s,n}(t)$) denote $1/\sqrt{n}$ times the number of jobs refused entry (respectively, lost due to buffer overflow) at P_i by real time nt *during the intervacation periods.*

Define $F_{i,l}^{v,n}(t) = (F_{ij,l}^{v,n}(t), j \leq K)$, where $F_{ij,l}^{v,n}(t)$ is $1/\sqrt{n}$ times the number of exogenous arrivals not admitted to P_j during the first $\sqrt{n}t$ time units of the lth vacation of P_i (where $t \leq \tau_{i,l}^{v,n}$). Define the buffer overflow term $U_{ij,l}^{v,n}(t)$ and $U_{i,l}^{v,n}(t) = (U_{ij,l}^{v,n}(t), j \leq K)$ analogously. Thus, these processes are defined in the local fluid time scale.

The state equation is, for $i = 1, \ldots, K$,

$$x_i^n(t) = x_i^n(0) + A_i^n(t) + \sum_j D_{ji}^n(t) - D_i^n(t) - F_i^n(t) - U_i^n(t). \tag{2.1}$$

Rewrite (2.1) in the terminology of Section 6.1 as

$$x_i^n(t) = x_i^n(0) + b_i^n t + M_i^n(t) - F^n(t) + z_i^n(t) + \epsilon_i^n(t), \tag{2.2}$$

where $\epsilon^n(\cdot)$ is a residual-time error term and we recall the definitions

$$\begin{aligned} M_i^n(t) = {} & w_i^{a,n}(S_i^{a,n}(t)) - w_i^{d,n}(S_i^{d,n}(t)) \\ & + \sum_j q_{ji}^n w_j^{d,n}(S_j^{d,n}(t)) + \sum_j w_{ji}^{r,n}(S_j^{d,n}(t)), \end{aligned} \tag{2.3}$$

$$b_i^n = \sqrt{n}\left[\frac{1}{\bar{\Delta}_i^{a,n}} + \sum_j \frac{q_{ji}^n}{\bar{\Delta}_j^{d,n}} - \frac{1}{\bar{\Delta}_i^{d,n}}\right],$$

$$z^n(t) = [I - Q_n']\, y^n(t) - U^n(t).$$

For initial condition $x^n(0) = x$, and $c_i > 0$, $\bar{c}_i > 0$, $i \le K$, let the cost function for the physical system be

$$W_\beta^n(x, F^n) = E_x^{F^n} \int_0^\infty e^{-\beta t}\left[k_0(x^n(t))dt + \bar{c}' dF^n(t) + c' dU^n(t)\right]. \quad (2.4)$$

The weak-sense limit system and its cost function are defined in Theorem 2.1 and in the discussion after it. The notation of Theorem 7.3.2 will be used, with the addition of what is required from Section 1 to deal with the controls. The analysis of the intervacation intervals adapts the methods of Theorem 1.3, and the analysis of the effects of the vacations uses the methods of Theorem 7.3.2. Assumption (A7.3.3) implies that the probability is $O(1/\sqrt{n})$ that any two vacations will overlap or abut on any finite time interval. Thus, for the purposes of the weak convergence analysis, we can suppose that the set of vacations of all of the processors do not overlap or abut. Recall (see below (7.3.13)) that $\bar{\nu}_{l-1}^n$ (respectively, ν_l^n) is the scaled time of the start (respectively, end) of the lth intervacation interval.

Because of the problem of proving tightness due to the singular control, the time transformation method was used in Theorem 1.3 for the weak convergence analysis. The time transformation needs to be modified, owing to the presence of vacations. The procedure for getting the correct limit process works in sections. The sequence of processes that are stopped at the beginning of the first vacation are treated as in Theorem 1.3. Then the set of possible jumps caused by the first vacation is shown to be tight, and the weak-sense limits are characterized. Thus, the set of initial conditions at the beginning of the second intervacation interval is tight, and the sequence of processes defined on the time interval between the end of the first vacation up to the beginning of the second is dealt with as in Theorem 1.3, etc. Since we work with the intervacation sections of the processes, the time transformation will depend on the index of the section. Thus, define

$$T_l^n(t) = t + \sum_i \left[F_i^{s,n}((\bar{\nu}_{l-1}^n + t) \wedge \nu_l^n) - F_i^{s,n}(\bar{\nu}_{l-1}^n)\right] \quad (2.5)$$

and

$$\hat{T}_l^n(t) = \inf\{s : T_l^n(s) > t\},$$

the time transformation that is used to deal with the lth intervacation section of the processes. The time transformation of the scaled queue paths in the lth intervacation interval is defined by

$$\hat{x}_{i,l}^n(t) = x_i^n((\bar{\nu}_{l-1}^n + \hat{T}_l^n(t)) \wedge \nu_l^n). \quad (2.6)$$

Define

$$M_{i,l}^n(t) = M_i^n((\bar{\nu}_{l-1}^n + t) \wedge \nu_l^n) - M_i^n(\bar{\nu}_{l-1}^n).$$

Analogously to (2.6), define the time transformed process during the lth intervacation interval:,

$$\hat{M}_{i,l}^n(t) = M_{i,l}^n((\bar{\nu}_{l-1}^n + \hat{T}_l^n(t)) \wedge \nu_l^n) - M_{i,l}^n(\bar{\nu}_{l-1}^n),$$

and define $y_{i,l}^n(\cdot), U_{i,l}^n(\cdot)$, $\epsilon_{i,l}^n(\cdot)$, and $\hat{y}_{i,l}^n(\cdot), \hat{U}_{i,l}^n(\cdot)$, $\hat{\epsilon}_{i,l}^n(\cdot)$ analogously. Set

$$\hat{\Theta}_{i,l}^n(t) = \left(\hat{x}_{i,l}^n(t), \hat{M}_{i,l}^n(t), \hat{y}_{i,l}^n(t), \hat{U}_{i,l}^n(t), \hat{\epsilon}_{i,l}^n(t)\right).$$

Owing to the length of the next theorem, it will be confined to mutually independent interarrival and service intervals. The length of the statement is due to the fact that it attempts to completely characterize the correct model which serves as the limit. Conditions (A2.1) and (A6.1.2) could be replaced by (A9.1.3) and the assumption that the spectral radius of $\{|q_{ij}|; i,j\}$ is less than unity, where $q_{ij}^n \to q_{ij}$. Theorem 2.2 continues to hold with these changes in the conditions, provided that (A9.1.2) and (2.11) also hold, so that we have the required uniform integrability. Condition (A9.1.2) holds under (A2.1), by Theorem 9.3.1.

A2.1. *For each n, the random variables $\{\Delta_{i,l}^{\alpha,n};\ l,i,\alpha\}$ are mutually independent, $E\Delta_{i,l}^{n,\alpha} = \bar{\Delta}_i^{\alpha,n} \to \bar{\lambda}_i^\alpha$, and $E[\Delta_{i,l}^{\alpha,n}/\bar{\Delta}_i^{\alpha,n}|^2 = [\sigma_{\alpha,i}^n]^2 \to \sigma_{\alpha,i}^2$. The set $\{|\Delta_{i,l}^{\alpha,n}/\bar{\Delta}_i^{\alpha,n}|^2;\ n,l,\alpha,i\}$ is uniformly integrable.*

Theorem 2.1. *Assume the network form and vacation model that were used in Theorem 7.3.2, but with admission control. Assume* (A2.1), (A6.1.0), (A6.1.2), (A6.1.3), (A7.3.3), (A7.3.4), *and suppose that no source is on vacation at time zero. Let $x^n(0)$ converge weakly to $x(0)$. Let $k_0(\cdot) \geq 0$ be continuous with at most linear growth, and let $\sup_n E|x^n(0)|^2 < \infty$. Fix the controls $F^n(\cdot)$ such that the costs are bounded. For each l and $i \leq K$, the set*

$$\left\{\bar{\nu}_{l-1}^n, \nu_l^n, \tau_{i,l}^{v,n}, \tau_{i,l}^{s,n}, \nu_{i,l}^n, \hat{\Theta}_{i,l}^n(\cdot), \hat{T}_l^n(\cdot), F_{i,l}^{v,n}(\cdot), U_{i,l}^{v,n}(\cdot)\right\} \tag{2.7}$$

is tight, as is the set of jumps caused by the vacations. Fix a weakly convergent subsequence for all l, i and index it by n. Denote the weak-sense limit by dropping the superscript n. Define the inverse transformations $T_l(t) = \inf\{s : \hat{T}_l(s) > t\}$, set $x_{i,l}(t) = \hat{x}_{i,l}(T_l(t))$, and similarly define the inverse transformed processes $M_{i,l}(\cdot)$, etc. We have $\nu_i = \bar{\nu}_i$. Define the process $x_i(\cdot)$ by $x_i(t) = x_{i,l}(t)$ for $\nu_{l-1} \leq t < \nu_l$, and similarly define $M_i(\cdot)$, etc. Then

$$x(t) = x(0) + bt + M(t) + J(t) - F^s(t) + [I - Q']y(t) - U^s(t), \tag{2.8}$$

where $y_i(\cdot), U_i(\cdot)$, $i \le K$, *are the reflection terms at the lower and upper boundaries, respectively. The* $M(\cdot)$ *is the Wiener process* $w(\cdot)$ *of (6.1.18b).*

The control components $F^v_{ij,l}(\cdot)$, *which are the weak-sense limits of* $F^{v,n}_{ij,l}(\cdot)$, *are differentiable, with* $0 \le d[F^v_{ij,l}(t)]/dt \le \bar{\lambda}^a_j$. *Analogously to the situation in Theorem 7.3.2,* $J(\cdot) = \sum_i J_i(\cdot)$, *where*

$$J_i(t) = \sum_{l:\nu_{i,l} \le t} \delta x(\nu_{i,l}), \quad \nu_{i,l} = \sum_{k \le l} \tau^s_{i,k},$$

and $\delta x(\nu_{i,l})$ *is the solution at time* $\tau^v_{i,l}$ *to the Skorohod problem*

$$\delta x(t) = x(\nu_{i,l}-) + \bar{b}^i t - F^v_{i,l}(t) + [I - Q']\,\delta y(t) - \delta U(t), \tag{2.9}$$

where $\dot{F}^v_{ij,l}(t) \le \bar{\lambda}^a_j$, $\bar{b}^i$ *is defined below (7.3.13), and* $F^v_{i,l}(\cdot) = (F^v_{ij,l}(\cdot), j \le K)$. *The solution* $\delta U(\cdot)$ *in (2.9) equals* $U^v_{i,l}(\cdot) = (U^v_{ij,l}(\cdot), j \le K)$, *the weak-sense limit of* $U^{v,n}_{i,l}(\cdot)$. *The nonanticipativity and other assertions of Theorem 7.3.2 hold, with the control included.*

Define

$$F(t) = F^s(t) + \sum_i \sum_{l:\nu_{i,l} \le t} F^v_{i,l}(\tau^v_{i,l}),$$

and define $U(t)$ *analogously. Let* E^F_x *denote the expectation of functionals of the processes in (2.8) under initial condition* x *and with controls* $(F^s(\cdot), F^v_{i,l}(\cdot), i \le K, l < \infty)$ *used. Then*

$$\begin{aligned}
\liminf_n W^n_\beta(x^n(0), F^n) &\ge W_\beta(x(0), F) \\
&= E^F_{x(0)} \int_0^\infty e^{-\beta t} \left[k_0(x(t))dt + \bar{c}' dF^s(t) + c' dU^s(t)\right] \\
&\quad + E^F_{x(0)} \sum_{i,l} e^{-\beta \nu_{i,l}} \left[\bar{c}' F^v_{i,l}(\tau^v_{i,l}) + c' U^v_{i,l}(\tau^v_{i,l})\right] \\
&= E^F_{x(0)} \int_0^\infty e^{-\beta t} \left[k_0(x(t))dt + \bar{c}' dF(t) + c' dU(t)\right].
\end{aligned} \tag{2.10}$$

Remarks on the proof. Recall that in the current case $\epsilon^n(t)$ is a residual-time error term and is bounded by $1/\sqrt{n}$ times the sums of the interarrival and service intervals that cover real time nt. It converges weakly to the "zero" process by the asymptotic continuity of the $w^{\alpha,n}_i(S^{\alpha,n}_i(\cdot))$. The proof of the weak convergence of the intervacation sections of the processes is as in Theorem 1.3, assuming tightness of $\{x^n(\bar{\nu}^n_{l-1})\}$ for each $l = 1, \ldots$. The set of initial conditions is tight. Then (A7.3.3) implies that the set $\{x^n(\nu^n_1)\}$, the values at the start of the first vacation, is tight. The main difference with Theorem 1.3 concerns the behavior during the vacations. This is handled analogously to what was done in Theorem 7.3.2, and yields tightness of $\{x^n(\bar{\nu}^n_l)\}$ if $\{x^n(\nu^n_l)\}$ is tight.

Consider the lth vacation of P_k. Now (7.3.17) is modified by subtracting the exogenous arrivals that were not admitted owing to control action during the vacation. Let $A^{v,n}_{kj,l}(t)$ denote $1/\sqrt{n}$ times the number of exogenous arrivals to P_j in the scaled time interval $[\nu^n_{k,l}, \nu^n_{k,l} + (t \wedge \tau^{v,n}_{k,l})/\sqrt{n})$. Then in the sense of a measure on the real line, $F^{v,n}_{kj,l}(\cdot)$ is absolutely continuous with respect to $A^{v,n}_{kj,l}(\cdot)$, uniformly in n. Since $A^{v,n}_{kj,l}(\cdot)$ converges weakly to the process defined by $\bar{\lambda}^a_k t$ (Theorem 5.6.4), the sets $\{F^{v,n}_{kj,l}(\cdot), n < \infty\}$ are tight for each j, k and l and any weak-sense limit $F^v_{kj,l}(\cdot)$ must be differentiable with $0 \leq d[F^v_{kj,l}(t)]/dt \leq \bar{\lambda}^a_j$, for $t \leq \tau^v_{k,l}$. Then the form of (2.9) follows from the development in Theorem 7.3.2.

The form of the weak-sense limit control problem: A nonclassical control problem. Since the effects of the control during a vacation appear in the weak-sense limit system, the limit control problem is more complicated than that in Section 1. Once a vacation starts, one must keep track of the elapsed time, since the control actions will in general depend on a prediction of how long that vacation will continue. The mathematical description of the limit process is given by (2.8) between vacations and by (2.9) at the vacation. Equation (2.9) gives the values associated with the lth vacation of P_i. Keep in mind that the time scale that is used in (2.9) is the limit of the local-fluid time scale, and it is valid for $t \leq \tau^v_{i,l}$. Thus, for the limit problem, the state description is different for the intervacation intervals and the vacation intervals.

Let $F(\cdot)$ denote the full control policy, both between and during the vacations. Suppose that no processor is on vacation at the initial time, and the initial condition is x. Then define the infimal cost $V_\beta(x) = \inf_F W_\beta(x, F)$. If some processor is on vacation at the initial time (as noted above, in the limit we need consider only the situation where at most one processor is on vacation at a time), then the state is (initial value of the state, index of the processor which is on vacation, time since that vacation began)$= (x, v, \tau)$. Then let $W_\beta(x, v, \tau, F)$ denote the cost under control $F(\cdot)$, and define $V_\beta(x, v, \tau)$ as the infimal cost. If the duration of the vacations is exponential, then τ is redundant and can be dropped.

The next theorem follows from the ideas of Theorems 1.5 and 2.1, modified to handle the control during the vacations, and the details of proof are omitted, except for those showing (2.11). An analysis of a related problem, where the control can shut down a processor, is in [174].

Theorem 2.2. *Assume the conditions of Theorem* 2.1, *and that there is* $p \geq 0$ *such that, uniformly in the control and initial condition,*

$$\limsup_n \sup_t \frac{E|J^n(t)|^2}{1+t^p} = O(1). \tag{2.11}$$

Then

$$EV^n_\beta(x^n(0)) \to EV_\beta(x(0)). \tag{2.12}$$

(2.11) *holds if* $\{\Delta^{a,n}_{i,l}\}$ *satisfies the independence and uniform integrability part of* (A2.1) *and* $\sup_{k,i,n} E[\tau^{v,n}_{i,k}]^2 < \infty$.

Proof of (2.11). Let $A^{v,n}_{ji,k}$ denote $1/\sqrt{n}$ times the number of exogenous arrivals to P_i during the kth vacation of P_j. We need to evaluate

$$E\left|\sum_{k:\nu^n_{j,k}\le t} A^{v,n}_{ji,k}\right|^2.$$

First, we bound $\sup_{n,j,i,k} E|A^{v,n}_{ji,k}|^2$. Fix i, j. We can write

$$A^{v,n}_{ji,k} \le \frac{1}{\sqrt{n}} \sum_{l=nS^{a,n}_i(\nu^n_{j,k})+2}^{nS^{a,n}_i(\nu^n_{j,k}+\tau^{v,n}_{j,k}/\sqrt{n})} 1 + \frac{1}{\sqrt{n}}. \tag{2.13}$$

Define $\xi^{a,n}_{i,l} = [1 - \Delta^{a,n}_{i,l}/\bar{\Delta}^{a,n}_i]$. The main term on the right side of (2.13) equals

$$\frac{1}{\sqrt{n}} \sum_{l=nS^{a,n}_i(\nu^n_{j,k})+2}^{nS^{a,n}_i(\nu^n_{j,k}+\tau^{v,n}_{j,k}/\sqrt{n})} \frac{\Delta^{a,n}_{i,l}}{\bar{\Delta}^{a,n}_i} + \frac{1}{\sqrt{n}} \sum_{l=nS^{a,n}_i(\nu^n_{j,k})+2}^{nS^{a,n}_i(\nu^n_{j,k}+\tau^{v,n}_{j,k}/\sqrt{n})} \xi^{a,n}_{i,l}.$$

The first term of the last expression is bounded by $\tau^{v,n}_{j,k}/\bar{\Delta}^{a,n}_i$. Following the proof in Theorem 9.3.1, the mean square value of the second term is bounded by $O(1)E[\tau^{v,n}_{j,k}]/\sqrt{n} + O(1/\sqrt{n})$.

The effect of any one vacation is a function of the increased idle time at the processor on vacation and the number of exogenous arrivals not admitted at each processor during that time. Thus, owing to the independent routing and spectral radius assumptions in (A6.1.2), the mean square value $E|x^n(\nu^n_{j,l} + \tau^{v,n}_{j,l}/\sqrt{n}) - x^n(\nu^n_{j,l})|^2$ is bounded by some constant times

$$\sup_{n,j,i,k} E|A^{v,n}_{ji,k}|^2 + \sup_{n,i,k} E|\tau^{v,n}_{i,k}|^2.$$

Now use the fact that the vacations start at random and ignore the scaled time gaps $\tau^{v,n}_{j,l}/\sqrt{n}$ to stochastically bound the number of vacations that start on the scaled time interval $[0, t]$ by a random variable L with a Poisson distribution with rate $\sum_i \bar{\lambda}^{s,n}_i$. Then use the fact that L is independent of the arrival, vacation intervals, and routing processes and has mean square value $O(t^2)$ to show that (2.11) holds with $p = 2$. ■

Another point of view to estimating the effect of a vacation is in [3, Section 4].

10.3 Controlled Admission in a Multiservice System

In this section we return to the type of ISDN system that was introduced in Section 1.5. Some of the discussion and assumptions will be repeated. Two forms of the problem will be developed. In all cases, all of the GP users in the system share the available bandwidth equally. The BE requests are always admitted, unless some very large limit is reached. In the model of the next subsection, the BE users use *all* of the available bandwidth; i.e., the available capacity is the only limit on the rate at which work can be done on the BE users. This is the situation where each BE arrival is the entire work for a "session," which arrives essentially instantaneously, and is buffered and then fed to the channel at any rate up to the channel capacity.

The second subsection is concerned with the case where there is a maximum rate at which data can be fed to the channel, irrespective of the available bandwidth. In this latter case, if there are not many BE customers in the system, then not all of the available bandwidth will be used. This changes the dynamics and the scaling, but the analysis is essentially the same. Other variations, including a stability analysis and a consideration of channel errors, are in [2].

10.3.1 Introduction: The Basic System

The model. The subscripts b and g, respectively, are used to denote variables associated with the BE and GP users. In heavy traffic, the mean service capacity is slightly greater than the mean demand. The system is parameterized by the scale or size parameter n, which is the order of the mean number of GP users and the arrival rates for the BE and GP users. The system capacity, excess capacity (over what is required for the average demands) and demand grow as $n \to \infty$. The bandwidth is normalized so that each GP user gets one unit of bandwidth. The "fast" scaling of Section 5.7 will be used. Unless otherwise noted we will use the definitions and notation of Section 1.5, although the assumptions will be repeated. Recall that t denotes real time, $\Delta_{\alpha,l}^{a,n}/n$ denotes the lth interarrival interval for class $\alpha = b, g$, that we allow at most $\sqrt{n}B_b$ of the BE users in the system at any one time, $U_g^n(t)$ denotes $1/\sqrt{n}$ times the number of GP users denied admission by t due to the entire channel being occupied by GP, and $U_b^n(t)$ equals $1/\sqrt{n}$ times the number of BE users denied admission due to there already being $\sqrt{n}B_b$ in the system on their arrival.

For each n and $\alpha = b, g$, the $\Delta_{\alpha,l}^{a,n}$ are assumed to be mutually independent and identically distributed with means $\bar{\Delta}_\alpha^a = 1/\bar{\lambda}_\alpha^a$. The set

$$\left\{ \left| \frac{\Delta_{\alpha,l}^{a,n}}{\bar{\Delta}_\alpha^a} \right|^2 ; n, l, \alpha = b, g \right\}$$

is assumed to be uniformly integrable and $E\left[1 - \Delta_{\alpha,l}^{a,n}/\bar{\Delta}_\alpha^a\right]^2 \to \sigma_{a,\alpha}^2$. The arrival processes are mutually independent and independent of the initial condition and the service times for the GP users. The service times for the members of class GP are exponentially distributed with rate $\bar{\lambda}_g^d$. The set of interarrival times and service times for the GP are assumed to be mutually independent and independent of the initial condition. Any GP arrival that is denied admission disappears from the system.

The service time requirements for the BE depend on the history of the available bandwidth during their stay in the system and are defined as follows. Suppose that $B(t)$ is the total bandwidth that is unused by the GP in the system at time t, and there are $N(t) > 0$ of the BE customers in the system. Then there is $\bar{\lambda}_b^d > 0$ such that the probability (conditioned on the data up to t) that any particular BE user will depart on the interval $[t, t+\delta)$ is $\bar{\lambda}_b^d B(t)\delta/N(t) + o(\delta)$. The (conditional) probability that there will be only one departure in that interval is $\bar{\lambda}_b^d B(t)\delta + o(\delta)$, and the (conditional) probability of more than one departure is $o(\delta)$.

Define $x_b^n(t)$ to be $1/\sqrt{n}$ times the number of BE users in the system at time t, and set (see Section 1.5 for a discussion of the definition)

$$x_g^n(t) = \frac{1}{\sqrt{n}} \left[\text{number of } GP \text{ users in system at } t - n\frac{\bar{\lambda}_g^a}{\bar{\lambda}_g^d} \right]. \tag{3.1}$$

For the heavy traffic condition, suppose that there is a constant b such that the channel capacity is defined by

$$C_n = n\left[\frac{\bar{\lambda}_g^a}{\bar{\lambda}_g^d} + \frac{\bar{\lambda}_b^a}{\bar{\lambda}_b^d} \right] + b\sqrt{n}. \tag{3.2}$$

Let $\mathcal{F}_t^n$ denote the minimal σ-algebra that measures the data that are available at time t, before the control actions at that time are taken. The control process $F_g^n(\cdot)$ is assumed to be $\mathcal{F}_t^n$-adapted, where $F_g^n(t)$ is $1/\sqrt{n}$ times the number of GP not admitted by t. The evolution equations are (1.5.4), namely,

$$x_b^n(t) = x_b^n(0) + A_b^n(t) - D_b^n(t) - U_b^n(t), \tag{3.3a}$$

$$x_g^n(t) = x_g^n(0) + A_g^n(t) - D_g^n(t) - F_g^n(t) - U_g^n(t). \tag{3.3b}$$

Representation of the input processes. Recall that $S_\alpha^{a,n}(t)$ is $1/n$ times the number of arrivals of class α by time t. For $\alpha = b, g$, write

$$A_\alpha^n(t) = \frac{1}{\sqrt{n}} \sum_{l=1}^{nS_\alpha^{a,n}(t)} 1 = \frac{1}{\sqrt{n}} \sum_{l=1}^{nS_\alpha^{a,n}(t)} \left[1 - \frac{\Delta_{\alpha,l}^{a,n}}{\bar{\Delta}_\alpha^a}\right] + \frac{1}{\sqrt{n}} \sum_{l=1}^{nS_\alpha^{a,n}(t)} \frac{\Delta_{\alpha,l}^{a,n}}{\bar{\Delta}_\alpha^a}. \tag{3.4}$$

Note that

$$\sum_{l=1}^{nS_\alpha^{a,n}(t)} \frac{\Delta_{\alpha,l}^{a,n}}{n}$$

equals t minus the time since the last arrival before or at t. Thus, the last term on the right of (3.4) equals $-\sqrt{n}t/\bar{\Delta}_\alpha^d + \epsilon_\alpha^n(t)$, where $\epsilon_\alpha^n(t)$ is $\sqrt{n}/\bar{\Delta}_\alpha^d$ times the real time between t and the last arrival of class α before or at t. This will converge to the "zero" process as $n \to \infty$. Write

$$A_\alpha^n(t) = w_\alpha^{a,n}(S_\alpha^{a,n}(t)) + \sqrt{n}\bar{\lambda}_\alpha^a t + \epsilon_\alpha^n(t). \tag{3.5}$$

Representation of the departure processes. The next step is to decompose the departure process $D_g^n(\cdot)$ into "drift" and "random" components, analogously to (3.5). Owing to the fact that the service times are mutually independent and exponentially distributed, by the results in Section 2.3, $D_g^n(\cdot)$ can be decomposed into the sum of the integral of the *conditional mean rate* at which $D_g^n(\cdot)$ increases (the compensator) and a martingale process, called $w_g^{d,n}(\cdot)$. To do this, first note that the conditional mean rate of increase of $D_g^n(\cdot)$ at t is $\bar{\lambda}_g^d/\sqrt{n}$ times the number of class GP customers in the system at t. Thus,

$$D_g^n(t) = \bar{\lambda}_g^d \int_0^t \left[x_g^n(s) + \sqrt{n}\frac{\bar{\lambda}_g^a}{\bar{\lambda}_g^d}\right] ds + w_g^{d,n}(t). \tag{3.6}$$

The Doob–Meyer process (see Section 2.4) associated with the martingale $w_g^{d,n}(\cdot)$ is just $1/\sqrt{n}$ times the first term on the right of (3.6), namely,

$$\langle w_g^{d,n}\rangle(t) = \frac{\bar{\lambda}_g^d}{\sqrt{n}} \int_0^t \left[x_g^n(s) + \sqrt{n}\frac{\bar{\lambda}_g^a}{\bar{\lambda}_g^d}\right] ds. \tag{3.7}$$

The factor $1/\sqrt{n}$ appears, since $D_\alpha^n(t)$ is $1/\sqrt{n}$ times the number of departures of class α by time t. Similar decompositions were used in [163, 171].

The departure process for class BE is similarly decomposed into a sum of the integral of the "conditional mean rate" at which $D_b^n(\cdot)$ increases, and a martingale $w_b^{d,n}(\cdot)$. In preparation for this, let $I_b^n(t)$ denote the indicator function of the event that there are BE users in the system at t, and note that the available bandwidth per class BE customer at time t is

$$\frac{C_n - \left[\sqrt{n}x_g^n(t) + n\bar{\lambda}_g^a/\bar{\lambda}_g^d\right]}{\sqrt{n}x_b^n(t)} I_b^n(t),$$

which equals

$$\frac{n\bar{\lambda}_b^a/\bar{\lambda}_b^d + \sqrt{n}b - \sqrt{n}x_g^n(t)}{\sqrt{n}x_b^n(t)} I_b^n(t).$$

The conditional mean rate at which $D_b^n(\cdot)$ increases at t is $\bar{\lambda}_b^d/\sqrt{n}$ times the above expression times the number of BE users in the system at t and is

$$\bar{\lambda}_b^d \left[\sqrt{n}\frac{\bar{\lambda}_b^a}{\bar{\lambda}_b^d} + b - x_g^n(t)\right] I_b^n(t).$$

Hence,

$$D_b^n(t) = \bar{\lambda}_b^d \int_0^t \left[\sqrt{n}\frac{\bar{\lambda}_b^a}{\bar{\lambda}_b^d} + b - x_g^n(s)\right] I_b^n(s)ds + w_b^{d,n}(t).$$

Rewrite the last expression as

$$D_b^n(t) = \bar{\lambda}_b^d \int_0^t \left[\sqrt{n}\frac{\bar{\lambda}_b^a}{\bar{\lambda}_b^d} + b - x_g^n(s)\right] ds + w_b^{d,n}(t) - y_b^n(t). \tag{3.8}$$

The term $y_b^n(\cdot)$ is a reflection term. It corrects for the effects of the $I_b^n(\cdot)$. It can increase only when $x_b^n(t) = 0$ and ensures that the $x_b^n(t)$ are non-negative. The Doob–Meyer process associated with $w_b^{d,n}(\cdot)$ is

$$\langle w_b^{d,n}\rangle(t) = \bar{\lambda}_b^d \int_0^t \left[\frac{\bar{\lambda}_b^a}{\bar{\lambda}_b^d} + \frac{b}{\sqrt{n}} - \frac{x_g^n(s)}{\sqrt{n}}\right] I_b^n(s)ds. \tag{3.9}$$

Now putting all of the above representations together into (3.3) and canceling the $\pm\sqrt{n}\bar{\lambda}_\alpha^a t$ terms yields the forms

$$\begin{aligned} x_b^n(t) = x_b^n(0) - \bar{\lambda}_b^d \int_0^t g_0(x^n(s))ds + w_b^{a,n}(S_b^{a,n}(t)) \\ - w_b^{d,n}(t) + y_b^n(t) - U_b^n(t) + \epsilon_b^n(t), \end{aligned} \tag{3.10a}$$

$$\begin{aligned} x_g^n(t) = x_g^n(0) - \bar{\lambda}_g^d \int_0^t x_g^n(s)ds + w_g^{a,n}(S_g^{a,n}(t)) \\ - w_g^{d,n}(t) - F_g^n(t) - U_g^n(t) + \epsilon_g^n(t), \end{aligned} \tag{3.10b}$$

where we define

$$g_0(x) = b - x_g. \tag{3.11}$$

The values of $x_b^n(t)$ are constrained to $[0, B_b]$. The limit processes will have the form

$$x_b(t) = x_b(0) - \bar{\lambda}_b^d \int_0^t g_0(x(s))ds + w_b^a(\bar{\lambda}_b^a t) - w_b^d(t) + y_b(t) - U_b(t), \tag{3.12a}$$

$$x_g(t) = x_g(0) - \bar{\lambda}_g^d \int_0^t x_g(s)ds + w_g^a(\bar{\lambda}_g^a t) - w_g^d(t) - F_g(t), \tag{3.12b}$$

with $x_b(t) \in [0, B_b]$. The various w-processes are Wiener processes, whose properties will be given in Theorem 3.1. The $F_g(\cdot)$ in (3.12b) is a singular control.

Let $\beta > 0$, $c > 0$ and let $k_0(\cdot)$ be a continuous function with at most linear growth in x and $k_0(0) = 0$. We will use the discounted cost function, for $x^n(0) = x$,

$$W_\beta^n(x, F_g^n) = E_x^{F_g^n} \int_0^\infty e^{-\beta t} \left[k_0(x_b^n(t))dt + cdF_g^n(t)\right]. \tag{3.13}$$

The cost for the weak-sense limit system with control $F_g(\cdot)$ and initial condition $x(0) = x$ is

$$W_\beta(x, F_g) = E_x^{F_g} \int_0^\infty e^{-\beta t} \left[k_0(x_b(t))dt + cdF_g(t)\right]. \tag{3.14}$$

The dF_g terms penalize the rejections. We do not penalize the loss $U_g^n(\cdot)$, since it is zero in the limit as $n \to \infty$ no matter what the controls are. If $k_0(\cdot)$ is linear, then it simply penalizes the waiting time for the BE users. It can be nonlinear. For example, it might be zero for small values of the argument (if the delays at small values are considered to be unimportant), or it might increase faster for large x_b to discourage long delays. Define $V_\beta^n(x) = \inf_{F_g^n} W_\beta^n(x, F_g^n)$, where the inf is over the admissible controls. Define $V_\beta(x)$ analogously. Define $M_\alpha^n(\cdot) = w_\alpha^{a,n}(S_\alpha^{a,n}(\cdot)) - w_\alpha^{d,n}(\cdot)$.

Theorem 3.1. *Let* $\sup_n E|x^n(0)|^2 < \infty$, *suppose that* $x^n(0)$ *converges weakly to* $x(0)$, *and assume the other conditions of this subsection. The conclusions of Theorems* 1.1–1.5 *hold, with limit system* (3.12). *Let* $\{\mathcal{F}_t, t \geq 0\}$ *denote the filtration engendered by the processes in* (3.12). *Then* $w_\alpha^a(\bar{\lambda}_\alpha^a \cdot)$, $w_\alpha^d(\cdot), \alpha = b, g$, *are mutually independent* $\mathcal{F}_t$*-Wiener processes with* $w_\alpha^a(\bar{\lambda}_\alpha^a \cdot)$ *having variance* $\sigma_{a,\alpha}^2 \bar{\lambda}_\alpha^a$ *and* $w_\alpha^d(\cdot)$ *having variance* $\bar{\lambda}_\alpha^a$. *Also,* $0 \leq x_b(t) \leq B_b$. *The process* $y_b(\cdot)$ *is the reflection term for* $x_b(\cdot)$ *at zero, and* $U_b(\cdot)$ *is the reflection term at the upper bound. The sequence* $U_g^n(\cdot)$ *converges weakly to the "zero" process.*

Proof. Let $F_g^n(\cdot)$ be a sequence of controls with bounded costs. The conditions on the interarrival intervals imply that the residual-time error terms converge weakly to the "zero" process. Define

$$\mathcal{T}_\alpha^{a,n}(t) = \frac{1}{n} \sum_{l=1}^{nt} \Delta_{\alpha,l}^{a,n}.$$

Modulo an asymptotically negligible error,

$$\mathcal{T}_\alpha^{a,n}(S_\alpha^{a,n}(t)) = t, \quad S_\alpha^{a,n}(\mathcal{T}_\alpha^{a,n}(t)) = t.$$

As in Theorem 5.1.1, these equations imply that $\mathcal{T}_\alpha^{a,n}(\cdot)$ converges weakly to the process with values $\bar{\Delta}_\alpha^a t$, and that $S_\alpha^{a,n}(\cdot)$ converges weakly to the process with values $\bar{\lambda}_\alpha^a t$. Hence the $w_\alpha^{a,n}(S_\alpha^{a,n}(\cdot))$ converge weakly to Wiener processes with variances $\sigma_{a,\alpha}^2 \bar{\lambda}_\alpha^a$.

Since $\bar{\lambda}_g^d/\sqrt{n}$ times the integrand in (3.7) is always bounded, Theorem 2.8.3 implies that $\{w_g^{d,n}(\cdot)\}$ is tight and has continuous weak-sense limits. Furthermore, since $\sup_n EF_g^n(T) < \infty$ for each T, $x_g^n(\cdot)$ is bounded in probability on any finite interval. Thus, the effect of $x_g^n(\cdot)$ on the integrand in (3.7) disappears as $n \to \infty$, and Theorem 2.8.2 implies that $w_g^{d,n}(\cdot)$ converges weakly to a Wiener process with variance $\bar{\lambda}_g^a$.

The sequence $\{w_b^{d,n}(\cdot)\}$ is tight. The tightness of $\{y_b^n(\cdot)\}$ follows from Theorem 6.6.1 and implies that $I_b^n(\cdot)$ can be nonzero only on a set whose Lebesgue measure goes to zero as $n \to \infty$. Hence, $I_b^n(\cdot)$ has no asymptotic influence on the Doob–Meyer process associated with the martingale $w_b^{d,n}(\cdot)$. These facts, the representation (3.9), and Theorem 2.8.2 imply that $w_b^{d,n}(\cdot)$ converges weakly to a Wiener processes with variance $\bar{\lambda}_b^a$.

Recall the comments concerning weakening the conditions that were made below (A1.4). Since $y^n(\cdot)$ does not appear in the cost equation and $k_0(\cdot)$ is bounded, (A1.3), (A1.4), (A9.1.2), and the other conditions concerned with bounds on the expectations are not needed. The sequence $U_g^n(\cdot)$ has "zero" weak-sense limits, since the "upper boundary" for x_g^n gets "pushed" to infinity as $n \to \infty$. The remaining details, including the mutual independence of the weak-sense-limit Wiener processes and the verification of (A9.1.12) (which is needed for the proof of Theorem 1.5), are left to the reader. ■

10.3.2 Upper Limit to the Bandwidth for the BE-Sharing Customers

In the model of the previous subsection, the BE users shared the available bandwidth and were able to use all of it. It is not always possible for all of the available bandwidth to be used. For example, depending on the way that the data arrive and are buffered, there might be restrictions on the rate at which data can enter the channel. This problem changes the model only slightly, and a similar analysis can be used. The few required adjustments for this and other variations of the basic model can be found in [2] and we give only the weak-sense limit system.

Suppose that the *maximum* bandwidth that any individual BE user can use is C_b. The main difference in the development concerns the departure process for the BE users. Redefine

$$x_b^n(t) = \frac{\text{number of } BE \text{ users at } t - n\bar{\lambda}_b^a/[C_b\bar{\lambda}_b^d]}{\sqrt{n}}. \tag{3.15}$$

Now $x_b^n(t)$ is a scaled deviation from the mean that would hold if each BE user used exactly C_b units of bandwidth and all BE were admitted. In the previous subsection the number of BE users in the system at any time was $O(\sqrt{n})$, and $x_b^n(t)$ measured that actual number, scaled by $1/\sqrt{n}$. Now due to the bandwidth limitation, the number in the system at any time is $O(n)$. That this is the correct order follows from the fact that the $x_b^n(\cdot)$ defined by (3.15) converge weakly to a well-defined process. We suppose that BE users are rejected if $x_b^n(t) \geq B_b$, where $B_b < \infty$. This will have negligible effect if B_b is large. The changes in the previous formulation and the end result will be described.

The martingale decompositions of the service processes that were given in the previous subsection still hold, with the appropriate modification for the current case for the conditional mean departure rate for the BE users. The conditional mean rate at which $D_b^n(\cdot)$ increases at time t is

$$\begin{aligned} I_b^n(t)\frac{\bar{\lambda}_b^d}{\sqrt{n}} &\times \min\{\text{available BW at } t,\ C_b(\#\text{ of } BE \text{ in system at } t)\} \\ &= I_b^n(t)\bar{\lambda}_b^a\sqrt{n} + I_b^n(t)\bar{\lambda}_b^d g_1(x^n(t)), \end{aligned} \tag{3.16}$$

where we define (b is defined in (3.2))

$$g_1(x) = \min\{b - x_g, C_b x_b\}. \tag{3.17}$$

Now analogously to the procedure in the previous subsection, decompose $D_b^n(\cdot)$ as

$$D_b^n(t) = \bar{\lambda}_b^a\sqrt{n}t + \int_0^t \bar{\lambda}_b^d g_1(x^n(s))ds + w_b^{d,n}(t) - y_b^n(t),$$

where the Doob–Meyer process associated with the martingale $w_b^{d,n}(\cdot)$ is now

$$\langle w_b^{d,n}\rangle(t) = \int_0^t \left[\bar{\lambda}_b^a + \frac{\bar{\lambda}_b^d g_1(x^n(s))}{\sqrt{n}}\right] I_b^n(s)ds.$$

The state equation is (3.10b), and (replacing (3.10a))

$$\begin{aligned} x_b^n(t) = x_b^n(0) &- \bar{\lambda}_b^d \int_0^t g_1(x^n(s))ds \\ &+ w_b^{a,n}(S_b^{a,n}(t)) - w_b^{d,n}(t) + y_b^n(t) - U_b^n(t), \end{aligned} \tag{3.18}$$

and the limit equation is (3.12b) and (replacing (3.12a))

$$x_b(t) = x_b(0) - \bar{\lambda}_b^d \int_0^t g_1(x(s))ds + w_b^a(\bar{\lambda}_b^a t) - w_b^d(t) - U_b(t). \tag{3.19}$$

Theorem 2.1 holds with these modifications.

10.4 Rerouting and Singular Control

Two forms of the rerouting problem will be discussed to illustrate other ways that singular controls can appear in the limit.

10.4.1 Rerouting with Penalty

Consider the following rerouting problem, where assignment is made on *arrival,* as opposed to the problems of Chapter 12, where assignment is made only when processing starts. The problem to be described was, in fact, the subject of the first actual derivation of the singular control problem as a heavy traffic limit, starting with the physical model [189]. The results can be extended to a multistage network.

There are K mutually independent exogenous input streams and K processors. Stream i is a priori assigned to processor P_i, but can be reassigned on arrival if desired. The reassignment is the control, and the limit singularly controlled process is (4.1). Completed jobs leave the system. The notation that is not defined or redefined here is the same as used in Chapter 6. Let $\Delta_{ij,l}^{d,n}$, with centering constants $\bar{\Delta}_{ij}^{d,n} \to \bar{\Delta}_{ij}^{d} = 1/\bar{\lambda}_{ij}^{d}$, denote the time required for the lth job from input stream i that is processed at P_j. Let the set of all work requirements be mutually independent, have uniformly bounded variances, and be independent of the initial condition and the arrival process. Suppose that the distribution of $\Delta_{ij,l}^{d,n}$ does not depend on l. The set

$$\left\{ \left| \frac{\Delta_{ii,l}^{d,n}}{\bar{\Delta}_{ii}^{d,n}} \right|^2 ; i, l, n \right\}$$

is assumed to be uniformly integrable and $E[1 - \Delta_{ii,l}^{d,n}/\bar{\Delta}_{ii}^{d,n}]^2 \to \sigma_{d,i}^2$. The analogous assumptions will be made on the interarrival intervals. They are mutually independent, independent of the initial condition, $E[1 - \Delta_{i,l}^{a,n}/\bar{\Delta}_{i}^{a,n}]^2 \to \sigma_{a,i}^2$ and the set

$$\left\{ \left| \frac{\Delta_{i,l}^{a,n}}{\bar{\Delta}_{i}^{a,n}} \right|^2 ; i, l, n \right\}$$

is uniformly integrable.

Let $I_{ij,l}^{r,n}$ denote the indicator function of the event that the lth exogenous arrival from stream i is rerouted on arrival from P_i to P_j, $j \neq i$. Let $S_i^{d,n}(t)$ (respectively, $S_{ij}^{d,n}(t)$) denote $1/n$ times the number of stream i jobs that were processed at P_i (respectively, at P_j, $j \neq i$) by real time nt, write $D_{ij}^n(t) = \sqrt{n} S_{ij}^{d,n}(t)$, and use $D_i^n(t) = D_{ii}^n(t)$. Let $F_{ij}^n(t)$ denote $1/\sqrt{n}$ times the number rerouted from P_i to P_j by real time nt and define

$$w_{ij}^{d,n}(t) = \frac{1}{\sqrt{n}} \sum_{l=1}^{nt} \left[1 - \frac{\Delta_{ij,l}^{d,n}}{\bar{\Delta}_{ij}^{d,n}} \right], \quad w_i^{d,n}(t) = w_{ii}^{d,n}(t).$$

Suppose that there are buffers of capacity $\sqrt{n}B_i, B_i < \infty$, that the decision to reassign does not depend on the work requirements of the job (which might not be known on arrival), and that there are real numbers b_i such that

$$\sqrt{n} \left[\frac{1}{\bar{\Delta}_i^{a,n}} - \frac{1}{\bar{\Delta}_{ii}^{d,n}} \right] = b_i^n \to b_i.$$

The modifications that are required when the work is known on arrival are minor and can be found in [166].

The system equations are, for $i \leq K$,

$$x_i^n(t) = x_i^n(0) + A_i^n(t) - D_i^n(t) - \sum_{j:j\neq i} D_{ji}^n(t) + \sum_{j:j\neq i} \left[F_{ji}^n(t) - F_{ij}^n(t) \right] - U_i^n(t),$$

where $U_i^n(t)$ is $1/\sqrt{n}$ of the number denied admission by real time nt due to a full buffer. The scaled number of departures from P_i by nt can be written as

$$\frac{1}{\sqrt{n}} \sum_{l=1}^{nS_i^{d,n}(t)} \left[1 - \frac{\Delta_{ii,l}^{d,n}}{\bar{\Delta}_{ii}^{d,n}} \right] + \sum_{j:j\neq i} \frac{1}{\sqrt{n}} \sum_{l=1}^{\sqrt{n}D_{ji}^{d,n}(t)} \left[1 - \frac{\Delta_{ji,l}^{d,n}}{\bar{\Delta}_{ji}^{d,n}} \right]$$
$$+ \frac{1}{\sqrt{n}} \sum_{l=1}^{nS_i^{d,n}(t)} \frac{\Delta_{ii,l}^{d,n}}{\bar{\Delta}_{ii}^{d,n}} + \sum_{j:j\neq i} \frac{1}{\sqrt{n}} \sum_{l=1}^{\sqrt{n}D_{ji}^{d,n}(t)} \frac{\Delta_{ji,l}^{d,n}}{\bar{\Delta}_{ji}^{d,n}}.$$

Now suppose that for all $i, j \neq i$, $\bar{\Delta}_{ii}^{d,n} = \bar{\Delta}_{ji}^{d,n}$. (Otherwise, the workload formulation that will be given below is more convenient to use.) Then following the usual procedure, the system equations can be rewritten as (modulo an asymptotically negligible residual-time error term)

$$\begin{aligned} x_i^n(t) &= x_i^n(0) + w_i^{a,n}(S_i^{a,n}(t)) - w_i^{d,n}(S_i^{d,n}(t)) + b_i^n t \\ &+ \sum_{j:j\neq i} \left[F_{ji}^n(t) - F_{ij}^n(t) \right] + y_i^n(t) - U_i^n(t) - \sum_{j:j\neq i} w_{ji}^{d,n}(D_{ji}^n(t)/\sqrt{n}). \end{aligned}$$

For $\beta > 0$, $c_i > 0$, $\bar{c}_{ij} > 0$, $j \neq i$, $x^n(0) = x$ and a continuous function $k_0(\cdot)$, let the cost function be

$$W_\beta^n(x, F^n) = E_x^{F^n} \int_0^\infty e^{-\beta t} \left[k_0(x^n(t))dt + \sum_{j,i:j\neq i} \left(\bar{c}_{ij} dF_{ij}^n(t) + c_i dU_i^n(t) \right) \right].$$

Let $F^n(\cdot)$ be a sequence of admissible controls with uniformly bounded costs. Then $\sup_n E|F^n(T)| < \infty$ for each T, which implies that $D_{ij}^n(\cdot)/\sqrt{n}$ converges weakly to the "zero" process. Due to the fact that the cost of rerouting is positive, Theorems 1.1 to 1.5 can be shown to hold with the limit system and cost being

$$x_i(t) = x_i(0) + w_i^a(\bar{\lambda}_i^a t) - w_i^d(\bar{\lambda}_i^a t) + b_i t + \sum_{j:j\neq i} [F_{ji}(t) - F_{ij}(t)] + y_i(t) - U_i(t), \tag{4.1}$$

$$W_\beta(x, F) = E_x^F \int_0^\infty e^{-\beta t} \left[k_0(x(t))dt + \sum_{j,i:j\neq i} (\bar{c}_{ij} dF_{ij}(t) + c_i dU_i(t)) \right].$$

A workload formulation. Now drop the assumption that $\bar{\Delta}_{ii}^{d,n} = \bar{\Delta}_{ij}^{d,n}$ for all $i, j \neq i$. Then the formulation in terms of workload is more convenient. Define

$$\bar{w}_{ij}^{r,n}(t) = \frac{1}{\sqrt{n}} \sum_{l=1}^{nt} I_{ij,l}^{r,n} \left[\Delta_{ii,l}^{d,n} - \bar{\Delta}_{ii}^{d,n} \right], \quad i \neq j,$$

$$\hat{w}_{ij}^{r,n}(t) = \frac{1}{\sqrt{n}} \sum_{l=1}^{nt} I_{ij,l}^{r,n} \left[\Delta_{ij,l}^{d,n} - \bar{\Delta}_{ij}^{d,n} \right], \quad i \neq j.$$

The scaled work scheduled by time nt to be done at P_i can be represented as

$$\frac{1}{\sqrt{n}} \sum_{l=1}^{nS_i^{a,n}(t)} \Delta_{ii,l}^{d,n} + \sum_{j:j\neq i} \frac{1}{\sqrt{n}} \sum_{l=1}^{nS_i^{a,n}(t)} I_{ji,l}^{r,n} \Delta_{ji,l}^{d,n} - \sum_{j:j\neq i} \frac{1}{\sqrt{n}} \sum_{l=1}^{nS_i^{a,n}(t)} I_{ij,l}^{r,n} \Delta_{ii,l}^{d,n}.$$

Suppose that the buffers are in terms of workload and not job numbers. The controller has available all of the system information up to the time that the routing decision is made. This will always be the full data for the jobs already arrived or processed, the past rerouting data, and the number of jobs of the various classes that are still in the various queues. It will include the work queued as well, if that is available. Asymptotically, the work in queue i is approximated by $\sum_j x_{ji}^n(t) \bar{\Delta}_{ji}^d$, where $x_{ji}^n(t)$ is the scaled number of jobs from stream j that are queued at P_i at nt. The approximation is in the sense that the difference converges weakly to the "zero" process.

The system workload equations can be reduced to (for $i \le K$, and modulo an asymptotically negligible residual-time error term)

$$\begin{aligned} WL_i^n(t) &= WL_i^n(0) + \bar{\Delta}_i^{d,n}\left[w_i^{a,n}(S_i^{a,n}(t)) - w_i^{d,n}(S_i^{a,n}(t)) + b_i^n t\right] \\ &+ \sum_{i,j:j\neq i}\left[\bar{\Delta}_{ji}^{d,n}F_{ji}^n(t) - \bar{\Delta}_{ii}^{d,n}F_{ij}^n(t)\right] \\ &- \sum_{i,j:j\neq i}\left[\bar{w}_{ij}^{r,n}(S_i^{a,n}(t)) - \hat{w}_{ji}^{r,n}(S_j^{a,n}(t))\right] + y_i^n(t) - U_i^n(t). \end{aligned} \tag{4.2}$$

The term $U_i^n(\cdot)$ now represents the process of lost work due to insufficient room in the buffer, and it is assumed that the entire work of a job is lost if there is insufficient room for all of it. For $WL^n(0) = WL$ and $\bar{c}_{ij} > 0$, $c_i > 0$, an appropriate cost function is

$$W_\beta^n(WL, F^n) = E_{WL}^{F^n}\int_0^\infty e^{-\beta t}\left[k_0(WL^n(t))dt + \sum_{j,i,j\neq i}\left(\bar{c}_{ij}dF_{ij}^n(t) + c_i dU_i^n(t)\right)\right]. \tag{4.3}$$

When $\sup_n E|F^n(t)| < \infty$ for each t, it turns out that the $\bar{w}_{ij}^{r,n}(S_i^{a,n}(\cdot))$ and $\hat{w}_{ij}^{d,n}(S_i^{a,n}(\cdot))$ converge weakly to the zero process. In fact (and with a similar estimate for $\hat{w}_{ij}^{r,n}(S_i^{a,n}(\cdot))$),

$$E\sup_{s\le t}\left|\bar{w}_{ij}^{r,n}(S_i^{a,n}(s))\right|^2 \le \frac{O(1)}{\sqrt{n}}E|F^n(t)|.$$

The conclusions of Theorems 1.1–1.5 hold with the limit system and cost being (4.2) and (4.3), respectively.

10.4.2 Rerouting Only When a Queue Is Full

Suppose that in the problem of the previous subsection the rerouting control is possible only if the primary queue is full, and then there is also the option of rejection. This is a simple case of the trunk line rerouting problem, as exemplified in the long-distance telephone network. That system consists of a number of nodes, each of which is connected to some others by a trunk line. When a call arrives at node i and is destined for node j, the system first attempts to route it directly on the trunk connecting i and j (if there is such a trunk). If that is full, then the system must make a choice. Either the call is rejected (giving a busy signal) or it is routed via some intermediate node k, thus using the trunks connecting i, k and k, j [124, 143, 161, 163, 175, 176, 198, 199, 214, 215]. The use of two trunks for the single call increases the possibility that future losses will be larger than the one call saved by rerouting. An analysis of various forms of the problem under heavy traffic conditions is in [161, 163, 175, 176, 214, 215], with

[175, 176] giving a method that gave good results even for systems with very many nodes. In [163], where the state is the vector of scaled available capacities, it is seen that the control problem for the general trunk line network is actually one over the reflection directions. For such problems, there is not usually a cost assigned to the rerouting itself.

In the example of the last paragraph, if a call is rejected, then the reflection is normal to the boundary face of the state space which corresponds to zero available capacity for the direct trunk. If the job is rerouted, then the reflection is at an angle of 45°, since the spare capacities of two trunks are decreased on that same boundary face.

Only a few comments will be made, in the context of the special model of the last subsection, concerning the problem of being able to reroute (or reject) only when the buffer is full. This might be called controlled processor sharing. Let $\bar{\Delta}_{ii}^{d,n} = \bar{\Delta}_{ji}^{d,n}$ for all i, j, such that $j \neq i$ (otherwise, use the workload formulation). Let $J_{ij}^n(t)$ denote the indicator function of the event that a job from input stream i is rerouted to queue $j \neq i$ at scaled time t. Overflows that are not rerouted are assumed to be lost to the system. Since the rerouting of an arrival to queue i can take place only if the buffer is full, such an arrival is rejected from queue i and

$$F_{ij}^n(t) = \int_0^t J_{ij}^n(s)dU_i^n(s).$$

The cost should weigh at least the content of the buffer and the number of jobs lost. We will weigh the number of jobs rerouted a well. Thus, for $c_i > 0, c_{ij} > 0$ and $x^n(0) = x$, let the cost function be

$$\begin{aligned} W_\beta^n(x, F^n) &= E_x^{F^n} \int_0^\infty e^{-\beta t} k_0(x^n(t))dt \\ &+ E_x^{F^n} \int_0^\infty e^{-\beta t} \left[\sum_i c_i \left(dU_i^n(t) - \sum_{j:j\neq i} dF_{ij}^n(t) \right) + \sum_{j,i:j\neq i} c_{ij} dF_{ij}^n(t) \right]. \end{aligned}$$

The term in the cost function containing $c_{ij}F_{ij}^n(\cdot)$, which charges for rerouting, "regularizes" the problem. The sum of the control and reflection terms in the equation for $x_i^n(\cdot)$ is

$$\sum_{j:j\neq i} F_{ji}^n(t) + y_i^n(t) - U_i^n(t).$$

The weak-sense limit has the form

$$x_i(t) = x_i(0) + b_i t + w_i(t) + \sum_{j:j\neq i} F_{ji}(t) + y_i(t) - U_i(t), \tag{4.4}$$

where $w_i(\cdot)$ is a Wiener process and $F_{ij}(\cdot)$ is absolutely continuous with respect to $U_i(\cdot)$. In particular, there are adapted $[0, 1]$-valued processes

$J_{ij}(\cdot)$ (which are the actual control processes for the limit problem) such that

$$F_{ij}(t) = \int_0^t J_{ij}(s)dU_i(s), \quad \sum_{j:j\neq i} J_{ij}(t) \leq 1.$$

The proof of (1.8) is similar to that given in Theorem 1.3. The approximation of the control by a differentiable process, as done in Theorem 1.4 is no longer possible, since the control is a priori restricted to be active only when a buffer is full. Nevertheless, (1.15), hence (1.14), can still be shown to hold.

11
Polling and Control of Polling

Chapter Outline

We return to the scheduling problems discussed in Section 1.3. Consider a queueing system with several queues and a single server. An original form of the problem arose in the management of time-shared computer systems, where there was a single processor and many users, each of which created a sequence of jobs at random and competed for the attention of the server. Typically, each user queued its work separately, and there was a mechanism that connected the chosen source to the central processor for a period of time, then connected another, etc. When a queue is chosen to be worked on, it is said to be *polled.* The polling problem arises whenever there is a single resource (for example, server or processor) that must be shared among competing queues. For linear holding costs, the fixed-priority policy known as the $c\mu$-rule (and other rules closely related to it) has been shown to be optimal under a variety of statistical assumptions and cost structures [1, 4, 46, 243]. Due to Little's rule [132], this policy also minimizes the overall average weighted waiting time in the system. Owing to its importance in applications, the polling problem has been the subject of a large literature; see, for example, [27, 181, 229, 230, 238] for discussions of many applications and modeling matters

The order in which the queues are polled might be deterministic and fixed in advance. This is referred to as *cyclic polling.* It might be determined by the actual work in the queues, via feedback control, or even chosen at random. (A model of a packet radio system as a randomly switched polling system is in [181, Section 4].) In an equivalent model in manufacturing, there are K part types, each type requiring a different type of processing, the material for each is queued separately, and there is a single processor that must be scheduled to work on all of the queues in some order. The polling might be *exhaustive* in the sense that a queue was polled until there was no work left in it. By *gated polling* we mean that the source is polled until completion of all the jobs that were queued at the time that the most recent polling of that queue started. These are the main forms used in applications at this time. Alternatively, one could poll until some fixed (or queue-dependent) number of jobs are completed. In some cases the time required for the server to switch between queues is nearly zero and is ignored. In other cases, there is a nonignorable *nonzero switchover time* (perhaps dependent on the identities of the current and next queue being polled). In the manufacturing problem, the switchover time is also called *setup time,* when the processor must be "reconfigured" or "resetup" when the part type being worked on changes. While negligible in some applications, this time can be significant in others, particularly when switching entails retooling. The completed work does not always leave the system. For example, in the manufacturing problem completed work from one queue might join another queue for further processing with a different setup.

Nonlinear cost functions are important for the general problem. Suppose that with linear costs and the $c\mu$-rule used, queue 1 would have priority and it is being polled at time t_0. If there were many arrivals to queue 1 while it was being polled, then those jobs would be served before jobs that were already at the other queues at t_0. Furthermore, the delay at queues with small arrival rates would be seriously affected by work at queues with high arrival rates. Due to the principle of fairness, one might wish to have a nonlinear weighing of delays among the queues. This leads to either gated policies or to policies where one stops polling at some (queue-dependent) point before exhaustion. We will be concerned with heavy traffic analysis with both fixed and optimal or nearly optimal polling policies.

This chapter deals with two classes of problems. Sections 1 to 4 are concerned with the type of problem discussed in Section 1.3, and Sections 5 to 9 are concerned with a polling problem where the connections between the servers can break down, as might happen in applications to mobile communications. The techniques that are developed are of much broader applicability, since they are concerned with general issues of modeling and optimization. In all cases, it is supposed for simplicity that there are two queues, although the analyses can be carried over to an arbitrary number.

As noted in Section 1.3 in connection with Figure 1.3.1, under heavy traffic the workloads in the individual queues change fast in scaled time, but the total workload is well approximated by a reflected diffusion process. For a broad class of polling policies, it is shown in Section 1 that a "uniform" fluid approximation to the behavior of the individual queues is valid. This is exploited in Theorem 1.2 to prove a general form of the averaging formula (1.3.6), that plays a fundamental role in analysis and optimization. Formulas for the rate of switching are developed in Section 2 as is the limit result when there are nonnegligible switchover times. It is shown in Section 3 that the so-called gated polling rule fits the framework of Sections 1 and 2. Section 4 is concerned with the optimal control problem. The results of the previous sections are combined to show that the limits of optimal processes are optimal, and that a well-defined limit control problem can be used to get nearly optimal controls for the physical problem under heavy traffic. This is done for nonzero switchover costs or time.

Let $WL_i^n(t)$ (the workload or individual workload at queue i) denote $1/\sqrt{n}$ times the real time that the server must work to complete all of the jobs that are in queue i at *real time* nt. Define the *total workload* or simply the *workload* $WL^n(t) = \sum_i WL_i^n(t)$. The polling problem with "long" vacations is introduced in Section 5. As with other problems with long vacations, the limit model is a jump diffusion. Consider the behavior during a vacation of queue 2 that starts at scaled time t_0, when the total workload is $WL^n(t_0)$. If the duration of the vacation is short enough so that the content of queue 1 does not go to zero, then the vacation will have no effect on the equation for the total workload. On the other hand, if the vacation is long, then the server will empty queue 1. Then it can work only on the new arrivals to queue 1, and there can be significant idle time until the vacation ends. This idle time causes a jump in the total workload. The size of the jump depends on $WL_1^n(t_0)$, which in turn is affected by the polling policy. Since the time of the jump is not known, the jump is not controlled directly, but only via what "happens" to be the content of the nonvacationing queue when the other goes on vacation. This effect of the control on the jump leads naturally to a limit problem where the jump is controlled, a problem that has not received attention to date. There are new issues in the modeling, that require a generalization of the concept of Poisson measure. The relevant theory is developed in Section 9, and is of basic importance for problems where the jump is controlled.

The vacation problem is formulated in Section 5, and a heavy traffic analysis under a fixed sequence of controls is done in Section 6. This is developed further in Section 7, where the ideas of Section 9 are exploited to get the full optimal limit results. Various extensions and special cases are discussed in Section 8. In some special cases, the optimal control can be determined analytically, and it is piecewise linear. Numerical data are briefly discussed. The scaling where time is compressed by a factor of n is used for specificity. But all of the results can readily be converted for the "fast" scaling of Sections 5.7 and 6.6.

11.1 An Averaging Principle for the Individual Queues

The section is concerned with the polling problem for two queues. The first subsection supposes that the switching policy is continuous and monotonic and derives an averaging principle for the behavior of the individual queues that is basic to the formulation of the control problem and the computation of the switching rates. The restrictions on the policy are weakened in the second subsection.

11.1.1 Continuous Switching Policies

We will consider the problem of polling when there are two sources, each with its own queue. The approach is similar irrespective of the number of sources. The terminology of Subsection 5.3.2 will be used unless otherwise noted. Then $\Delta_{i,l}^{a,n}$ (respectively, $\Delta_{i,l}^{d,n}$) denotes the interarrival (respectively, service) intervals for the sources $i = 1, 2$. With centering constants $\bar{\Delta}_i^{\alpha,n}$, that converge to $\bar{\Delta}_i^{\alpha} = 1/\bar{\lambda}_i^{\alpha}$, $\alpha = a, d$, $i = 1, 2$, respectively, define the processes $w_i^{\alpha,n}(\cdot)$ and $S_i^{\alpha,n}(\cdot)$ as in Section 5.3.2: In particular, recall that

$$w_i^{\alpha,n}(t) = \frac{1}{\sqrt{n}} \sum_{l=1}^{nt} \left[1 - \frac{\Delta_{i,l}^{\alpha,n}}{\bar{\Delta}_i^{\alpha,n}} \right],$$

and that $S_i^{\alpha,n}(t), \alpha = a, d$, denotes $1/n$ times the number of arrivals (respectively, services) from source i by *real time nt*. Let $x_i^n(t)$ denote $1/\sqrt{n}$ times the number of jobs in queue i at *real time nt*, including the one in service, if any. Unless mentioned otherwise, the switchover time is zero. There is never idling if there is work to be done. Thus, the polling policy is the rule for switching between the queues.

A class of polling policies. The initial results will use switching policies that depend continuously on the total queued workload, and it is assumed that the controller knows the individual workloads. Both of these restrictions will be relaxed later. In practice, the controller would not usually know the precise number of jobs or work queued in the source not being polled. It will be seen that if the controller knows only the total queued work at the start of a polling cycle (or even at the the start of a recent cycle), then the results will be the same asymptotically.

The class of policies to be considered first will now be described loosely. Suppose that the current total workload is $WL = w > 0$, and that it changes *very slowly*. Then the polling policy to be considered in this subsection is such that the workload vector $(WL_1^n(\cdot), WL_2^n(\cdot))$ oscillates between two distinct (and w-dependent) points $(w_1^a(w), w_2^a(w))$ and $(w_1^b(w), w_2^b(w))$, where $w_1^a(w) < w_1^b(w)$ and $w_2^b(w) < w_2^a(w)$. Define the functions $\theta_i(\cdot)$,

$i = 1, 2$, assumed to be continuous until further notice, such that $\theta_1(w) = w_1^a(w), \theta_2(w) = w_2^b(w)$, and suppose that $\theta_i(0) = 0$. Thus,

$$\sum_i \theta_i(w) < w, \qquad \text{for all } w > 0. \tag{1.1}$$

The functions $\theta_i(\cdot)$ can be considered to be the controls. Equation (1.1) implies that there is a "gap" between the endpoints, so that the (scaled) polling interval does not become zero unless $w = 0$. See Figure 1.1 for an an illustration. The actual set of allowed polling strategies will be spelled out in more detail below. In all cases, if the queue to be switched to is empty, polling continues in the current queue.

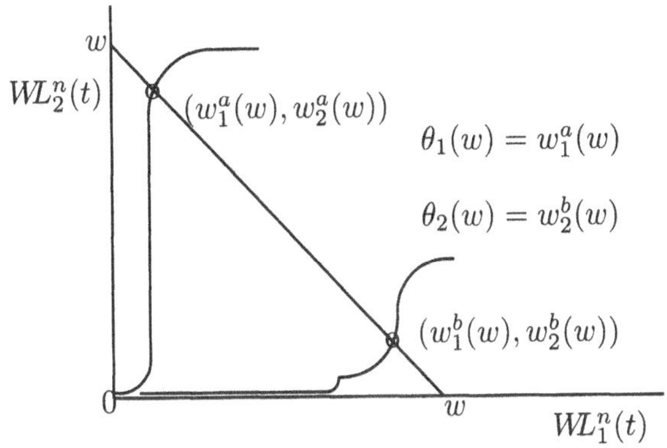

Figure 1.1. A class of polling strategies.

We will now define a precise and ideal (and versatile, as will be seen) class of strategies in terms of given functions $\theta_i(\cdot)$, $i = 1, 2$. For expository simplicity, suppose that $WL_2^n(0) \le \theta_2(WL^n(0))$ so that we start by polling queue 1. Define $\tau_0^n = 0$. Queue 1 is polled until the time

$$\sigma_1^n = \min\{t \ge 0 : WL_1^n(t) \le \theta_1(WL^n(t)),\ WL_2^n(t) > 0\}. \tag{1.2a}$$

Then queue 2 is polled until

$$\tau_1^n = \min\{t > \sigma_1^n : WL_2^n(t) \le \theta_2(WL^n(t)),\ WL_1^n(t) > 0\}. \tag{1.2b}$$

Define τ_i^n, σ_i^n, $i > 1$, recursively in the same way. Thus, queue 1 is polled on the real time intervals $(n\tau_k^n, n\sigma_{k+1}^n]$ and queue 2 on the alternate intervals.

The procedure, as stated above, is in terms of workload and might entail the stopping of a job before completion. It can be modified by allowing the completion of any job that was started before switching. Otherwise, we simply suppose that if a stopped job is restarted, then the remaining work is just the residual work. Under the conditions of Theorem 1.1 and from an asymptotic point of view, these are equivalent, and we will ignore the subtleties that differentiate them. Additionally, one can over- or undershoot by some *scaled* amount of work that goes to zero as $n \to \infty$.

This control procedure is defined in terms of workload. But as suggested by (5.3.12), the procedure can be rewritten in an asymptotically equivalent way in terms of job numbers. The stopping rule (1.2) for each segment requires keeping track of the current total workload. But since $WL^n(\cdot)$ varies slowly in comparison with the polling speed, one can use the simpler rule defined by

$$\sigma_1^n = \min\{t \geq 0 : WL_1^n(t) \leq \theta_1(WL^n(\tau_0^n)),\ WL_2^n(t) > 0\}. \tag{1.3a}$$

Then queue 2 is polled until

$$\tau_1^n = \min\{t > \sigma_1^n : WL_2^n(t) \leq \theta_2(WL^n(\sigma_1^n)),\ WL_1^n(t) > 0\}, \tag{1.3b}$$

etc., with the same asymptotic results. The allowed class of policies is quite broad, under conditions of heavy traffic.

Such polling schemes have the appearance of a process where at any time one of the queues is on vacation, with the vacations alternating between them. The difference between the setup in Subsection 5.3.2 and that in the present case is that here each source has its own queue. But the expression for the total workload is the same, since there is no switchover time or idling (to be relaxed later) if there is work to be done.

Assumptions. Define the traffic intensities $\rho_i^n = \bar{\Delta}_i^{d,n}/\bar{\Delta}_i^{a,n}$, $\rho_i = \bar{\Delta}_i^d/\bar{\Delta}_i^a = \bar{\Delta}_i^d\bar{\lambda}_i^a > 0$. The following assumptions will be used.

A1.0. *For each n, $(x_i^n(0), WL_i^n(0),\ i = 1,2)$ is independent of all of the "future" driving random variables.*

A1.1. *For $\alpha = a, d$, $i = 1, 2$, $\bar{\Delta}_i^{\alpha,n} \to \bar{\Delta}_i^\alpha = 1/\bar{\lambda}_i^\alpha$. There is a real b such that*

$$\sqrt{n}\left[\sum_i \rho_i^n - 1\right] = b^n \to b.$$

A1.2. *$w_i^{\alpha,n}(\cdot)$, $\alpha = a, d$, $i = 1, 2$, converge weakly to mutually independent Wiener processes with variances $\sigma_{\alpha,i}^2$, $i = 1, 2$, respectively, at least one of which is positive.*

The positivity is to ensure that the weak-sense limit Wiener process is not degenerate. Let $\mathcal{F}_t^n$ denote the minimal σ-algebra that measures all of the systems data to scaled time t. We will sometimes use the following stronger version of (A1.2), which is equivalent to (9.1.13) in (A9.1.3). Also, (A1.1) implies that $1 = \sum_i \rho_i$.

A1.2.′ *For each t, the conditional (given $\mathcal{F}_t^n$) probability law of the processes*

$$w_i^{\alpha,n}(S_i^{a,n}(t+\cdot)) - w_i^{\alpha,n}(S_i^{a,n}(t)),\ \alpha = a, d,\ i = 1, 2,$$

converges to that of the mutually independent Wiener processes $w_i^\alpha(\bar{\lambda}_i^a \cdot)$, *with variance parameters* $\bar{\lambda}_i^a \sigma_{\alpha,i}^2$, *at least one of which is positive.*

A1.3. *The* $\theta_i(\cdot)$, $i = 1, 2$, *are continuous and nondecreasing,* $WL - \sum_i \theta_i(WL) > 0$ *for* $WL > 0$, *and* $\theta_i(0) = 0$.

The expression (5.3.10) holds, which in the present terminology is

$$WL^n(t) = WL^n(0) + \sum_i \frac{1}{\sqrt{n}} \sum_{l=1}^{nS_i^{a,n}(t)} \Delta_{i,l}^{d,n} - t\sqrt{n} + z^n(t), \tag{1.4}$$

where $z^n(\cdot)$ is $1/\sqrt{n}$ times the idle time by real time nt. Theorem 5.3.2 also holds, with the limit process

$$WL(t) = WL(0) + bt + w(t) + z(t), \; WL(t) \geq 0, \tag{1.5}$$

where the variance of the Wiener process

$$w(t) = \sum_i \bar{\Delta}_i^d \left[w_i^a(\bar{\lambda}_i^a t) - w_i^d(\bar{\lambda}_i^a t) \right] \tag{1.6}$$

is

$$\sum_i \left[\bar{\Delta}_i^d \right]^2 \bar{\lambda}_i^a \left[\sigma_{a,i}^2 + \sigma_{d,i}^2 \right]. \tag{1.7}$$

The processes in (1.5) do not depend on the actual polling policy, since the polling does not produce extra idle time. Note that whatever the dimension of the original physical problem, the weak-sense limit process is one dimensional. This is is another example of what is called "state space collapse" [30, 206, 208, 209, 254] in the heavy traffic literature.

The local-fluid time scale. The duration of each polling cycle is of order $O(\sqrt{n})$ in real time (since the unscaled queue is $O(\sqrt{n})$) and $O(1/\sqrt{n})$ in the scaled time. Since we will need to analyze the path within a cycle, it is convenient to reintroduce the "local-fluid" time scale in which a cycle takes $O(1)$ units of time, and that was used in the representation (5.6.18) and in Theorem 7.3.2 to analyze the behavior of a queue during a vacation.

Thus, reparametrize the random switching times by defining, for $k \geq 1$,

$$\begin{aligned} \delta\sigma_k^n &= \sqrt{n}[\sigma_k^n - \tau_{k-1}^n], \\ \delta\tau_k^n &= \sqrt{n}[\tau_k^n - \sigma_k^n]. \end{aligned}$$

Consider the real time interval $(n\tau_k^n, n\sigma_{k+1}^n]$, $k \geq 0$, when queue 1 is being polled. Define

$$A_{i,k}^{1,n}(t) = \frac{1}{\sqrt{n}} \sum_{l=nS_i^{a,n}(\tau_k^n)+1}^{nS_i^{a,n}(\tau_k^n + (t \wedge \delta\sigma_{k+1}^n)/\sqrt{n})} \Delta_{i,l}^{d,n}. \tag{1.8a}$$

Note that $A_{i,k}^{1,n}(t)$ is $1/\sqrt{n}$ times the work arriving to queue i during the real time interval $(n\tau_k^n, n\tau_k^n + \sqrt{n}t]$, for $t \le \delta\sigma_{k+1}^n$. It is defined in the local-fluid time scale.

On the alternative real time intervals $(n\sigma_k^n, n\tau_k^n]$, when queue 2 is being polled, the workload arrival process to queue i in the local-fluid time scale is

$$A_{i,k}^{2,n}(t) = \frac{1}{\sqrt{n}} \sum_{l=nS_i^{a,n}(\sigma_k^n)+1}^{nS_i^{a,n}(\sigma_k^n+(t\wedge\delta\tau_k^n)/\sqrt{n})} \Delta_{i,l}^{d,n}. \tag{1.8b}$$

There is an ambiguity concerning the assignment of arrivals at the precise moments at which an interval begins and ends. But we take (1.8) to be the definitions. The assignment at those times has no effect asymptotically. Theorem 1.1 implies that, asymptotically, the path can be supposed to move between the upper and lower limits and back again at the constant fluid rate between switches.

Theorem 1.1. *Let $WL^n(0)$ converge weakly to $WL(0)$, assume* (A1.0)–(A1.3), *the polling rule* (1.2), *and let $T > 0$ be arbitrary. Then $\{\delta\sigma_k^n, \delta\tau_k^n; n, k\}$ is tight in the sense that*

$$\lim_{K\to\infty} P\left\{ \sup_{k:\tau_k^n \le T} [\delta\sigma_k^n + \delta\tau_k^n] \ge K \right\} = 0. \tag{1.9}$$

The processes defined in the local-fluid time scale by

$$\delta A_{i,k}^{1,n}(t) = A_{i,k}^{1,n}(t) - \rho_i(t \wedge \delta\sigma_{k+1}^n) \tag{1.10a}$$

and

$$\delta A_{i,k}^{2,n}(t) = A_{i,k}^{2,n}(t) - \rho_i(t \wedge \delta\tau_k^n) \tag{1.10b}$$

converge weakly to the zero process. Furthermore, the weak convergence is uniform in k in the sense that for any $\epsilon > 0$ and i, j,

$$\lim_n P\left\{ \sup_{k:\tau_k^n \le T} \sup_{s<\infty} \left|\delta A_{i,k}^{j,n}(s)\right| \ge \epsilon \right\} = 0. \tag{1.11}$$

Also, for $k \ge 1$ and when queue 1 is being polled, the sequence of processes defined by

$$\delta WL_{1,k}^{1,n}(t) = WL_1^n(\tau_k^n + (t \wedge \delta\sigma_{k+1}^n)/\sqrt{n}) - WL_1^n(\tau_k^n) - (\rho_1 - 1)(t \wedge \delta\sigma_{k+1}^n) \tag{1.12}$$

converges weakly to the "zero" process, also uniformly in k in the sense of (1.11), *and analogously when queue 2 is being polled. Finally,*

$$\delta\sigma_{k+1}^n \approx \frac{1}{1-\rho_1}\left[WL^n(\tau_k^n) - \sum_i \theta_i(WL^n(\tau_k^n)) \right], \tag{1.13a}$$

$$\delta\tau_k^n \approx \frac{1}{1-\rho_2}\left[WL^n(\sigma_k^n) - \sum_i \theta_i(WL^n(\sigma_k^n))\right] \tag{1.13b}$$

in the sense that $\sup_{k:\tau_k^n \le T}$ *of the absolute difference between the left and right-hand sides goes to zero in probability as* $n \to \infty$.

Proof. As noted above, Assumptions (A1.0)–(A1.2) and Theorem 5.3.2 imply that whatever the polling policy, provided that there is no idling if there is work to do, $WL^n(\cdot)$ converges weakly to the continuous process $WL(\cdot)$ on $[0,\infty)$ satisfying (1.5).

Following the usual procedure, center the $A_{i,k}^{1,n}(t)$ to get

$$\begin{aligned} A_{i,k}^{1,n}(t) &= \frac{1}{\sqrt{n}} \sum_{l=nS_i^{a,n}(\tau_k^n)+1}^{nS_i^{a,n}(\tau_k^n+(t\wedge\delta\sigma_{k+1}^n)/\sqrt{n})} \left[\Delta_{i,l}^{d,n} - \bar\Delta_i^{d,n}\right] \\ &+ \frac{1}{\sqrt{n}} \sum_{l=nS_i^{a,n}(\tau_k^n)+1}^{nS_i^{a,n}(\tau_k^n+(t\wedge\delta\sigma_{k+1}^n)/\sqrt{n})} \bar\Delta_i^{d,n}. \end{aligned} \tag{1.14}$$

Write the second term as

$$\begin{aligned} &\frac{\bar\Delta_i^{d,n}}{\sqrt{n}} \sum_{l=nS_i^{a,n}(\tau_k^n)+1}^{nS_i^{a,n}(\tau_k^n+(t\wedge\delta\sigma_{k+1}^n)/\sqrt{n})} \left[1 - \frac{\Delta_{i,l}^{a,n}}{\bar\Delta_i^{a,n}}\right] \\ &+ \frac{\bar\Delta_i^{d,n}}{\sqrt{n}\bar\Delta_i^{a,n}} \sum_{l=nS_i^{a,n}(\tau_k^n)+1}^{nS_i^{a,n}(\tau_k^n+(t\wedge\delta\sigma_{k+1}^n)/\sqrt{n})} \Delta_{i,l}^{a,n}. \end{aligned} \tag{1.15}$$

The second term of (1.15) is (modulo a residual-time error term, called $\epsilon_{i,k}^n(\cdot)$)

$$\frac{\bar\Delta_i^{d,n}}{\bar\Delta_i^{a,n}}(t \wedge \delta\sigma_{k+1}^n).$$

The function $\epsilon_{i,k}^n(t)$ is bounded by $1/\sqrt{n}$ times the service and interarrival intervals that cover the beginning or the end of the real time interval $(n\tau_k^k, n\tau_k^n + \sqrt{n}(t \wedge \delta\sigma_{k+1}^n)]$.

Fix $T > 0$, arbitrarily. Until further notice, let us modify the polling policy so that the $\delta\tau_k^n, \delta\sigma_k^n$ are no larger than $\sqrt{n}$. It will be seen that (1.9) holds under this restriction. Hence, the restriction always holds for large n. Under this restriction, the polling cycles that start by scaled time T involve the $w_i^{a,n}(S_i^{a,n}(\cdot))$ and $WL^n(\cdot)$ on at most the scaled interval $[0, T+2]$. Thus, by (A1.2), $\sup_{k:\tau_k^n \le T} |\epsilon_{i,k}^n(\cdot)|$ converges weakly to the zero process. Define

$$\bar\delta^n = \sup_{t<\infty, k:\tau_k^n \le T} [\text{sum of absolute values of first terms of (1.14) and (1.15)}].$$

By the assumed weak convergence in (A1.2), the weak convergence of $S_i^{a,n}(\cdot)$, and the $\sqrt{n}$ bounds on $\delta\tau_k^n$ and $\delta\sigma_k^n$, $\{\bar{\delta}^n\}$ is tight.

Consider the polling of queue 1 that starts at scaled time $\tau_k^n \le T$. Define

$$z_k^{1,n}(t) = \frac{1}{\sqrt{n}} \times \left[\text{idle time on real time interval } \left(n\tau_k^n, n\tau_k^n + \sqrt{n}(t \wedge \delta\sigma_{k+1}^n)\right]\right]$$

and

$$\bar{z}^{1,n} = \sup_{t<\infty, k:\tau_k^n \le T} z_k^{1,n}(t).$$

Then for $k \ge 0$,

$$\begin{aligned} &WL_1^n(\tau_k^n + (t \wedge \delta\sigma_{k+1}^n)/\sqrt{n}) \\ &\quad = WL_1^n(\tau_k^n) + A_{1,k}^{1,n}(t \wedge \delta\sigma_{k+1}^n) - (t \wedge \delta\sigma_{k+1}^n) + z_k^{1,n}(t). \end{aligned} \tag{1.16}$$

Since $z^n(\cdot)$ is tight with a continuous weak-sense limit and the $\delta\sigma_k^n$ and $\delta\tau_k^n$ are no larger than $\sqrt{n}$, $\{\bar{z}^{1,n}\}$ is tight. The sum of the second and third terms on the right of (1.16) are bounded above by

$$\bar{\delta}^n + \epsilon_{1,k}^n(t) + \frac{\bar{\Delta}_1^{d,n}}{\bar{\Delta}_1^{a,n}}(t \wedge \delta\sigma_{k+1}^n) - (t \wedge \delta\sigma_{k+1}^n). \tag{1.17}$$

Also,

$$\left\{\sup_{t \le T+2} |WL^n(t)|, n < \infty\right\} \tag{1.18}$$

is tight. Now the representations (1.16) and (1.17), the tightness of $\{\bar{z}^{1,n}, \bar{\delta}^n\}$, the cited convergence of the residual-time error terms, and (1.18) (together with a similar development for the alternate intervals) imply (1.9). The expression (1.9) implies that the $\sqrt{n}$ upper bound can be dropped.

The expression (1.9) and the convergence of $z^n(\cdot)$ to a continuous process imply that $\bar{z}^{1,n}$ converges weakly to zero. By (1.9) and (A1.2), $\bar{\delta}^n \to 0$. We can now conclude that the assertions concerning (1.10) and (1.11) hold, and hence (1.12) holds. The polling of queue 1 ends at the first t such that either $t \ge \sqrt{n}$ or (1.16) equals $\theta_1(WL^n(\tau_k^n + t/\sqrt{n}))$ and $WL_2^n(\tau_k^n + t/\sqrt{n}) > 0$. Using these facts together with (1.9), the assertion concerning the convergence of (1.12), and the weak convergence of $WL^n(\cdot)$ to a continuous limit, and for n, k such that $\tau_k^n \le T$, (1.16) yields that for some sequence δ_k^n for which $\sup_{k:\tau_k^n \le T} |\delta_k^n|$ can be assumed to converge weakly to zero,

$$\delta\sigma_{k+1}^n = \min\left\{t : \left[1 - \bar{\Delta}_1^{d,n}/\bar{\Delta}_1^{a,n}\right] t \ge WL_1^n(\tau_k^n) - \theta_1(WL^n(\sigma_{k+1}^n)) + \delta_k^n\right\}.$$

There is an analogous result for the alternate intervals. Modulo an asymptotically negligible error, for n, k such that $\tau_k^n \le T$,

$$WL_1^n(\tau_k^n) = WL^n(\tau_k^n) - \theta_2(WL^n(\tau_k)) \text{ and } WL^n(\sigma_{k+1}^n) = WL^n(\tau_k^n).$$

Thus, (1.13) and the associated uniformity assertion hold. ■

"Uniform" distribution of the individual workloads. Suppose that $WL^n(t) \approx w > 0$. Then Theorem 1.1 essentially implies that for small s, $WL_1^n(t+s)$ is nearly uniformly distributed on the interval $[\theta_1(w), w-\theta_2(w)]$ and $WL_2^n(t)$ is nearly uniformly distributed on $[\theta_2(w), w-\theta_1(w)]$. By (1.1), these intervals have positive length. This is formalized in the following extension of an important result from [42, 43]. Define the vector-valued functions $\phi_1(w) = (\theta_1(w), w-\theta_1(w))$ and $\phi_2(w) = (w-\theta_2(w), \theta_2(w))$.

Theorem 1.2. *Let $f(\cdot)$ be a continuous real-valued function on $[0,\infty)$, and assume the conditions of Theorem 1.1. Then the process defined by*

$$\int_0^t f\left(WL_1^n(s)\right) ds$$

converges weakly to the process with values

$$\int_0^t ds \int_0^1 f\left(v\left(WL(s) - \theta_2(WL(s))\right) + (1-v)\theta_1(WL(s))\right) dv. \tag{1.19}$$

If $WL_2^n(s)$ replaces $WL_1^n(s)$, interchange the indices 1 and 2 in (1.19). More generally, if $f(\cdot)$ depends on both coordinates, then the limit is

$$\int_0^t ds \int_0^1 f\left(\phi_1(WL(s)) + v\left(\phi_2(WL(s)) - \phi_1(WL(s))\right)\right) dv. \tag{1.20}$$

Proof. We work with the first case, for the others are treated identically. Modulo an asymptotically negligible "end" term, we can write

$$\begin{aligned} &\int_0^t f(WL_1^n(s))ds \\ &\quad = \sum_{k:\tau_k^n \le t} \left[\int_{\tau_k^n}^{\tau_k^n + \delta\sigma_{k+1}^n/\sqrt{n}} f(WL_1^n(s))ds + \int_{\sigma_{k+1}^n}^{\sigma_{k+1}^n + \delta\tau_{k+1}^n/\sqrt{n}} f(WL_1^n(s))ds \right]. \end{aligned} \tag{1.21}$$

Rewrite a typical term as

$$\int_0^{\delta\sigma_{k+1}^n/\sqrt{n}} f\left(WL_1^n\left(\tau_k^n + s\right)\right) ds = \frac{\delta\sigma_{k+1}^n}{\sqrt{n}} \int_0^1 f\left(WL_1^n\left(\tau_k^n + \frac{\delta\sigma_{k+1}^n}{\sqrt{n}} v\right)\right) dv. \tag{1.22}$$

By Theorem 1.1, we can approximate (for $0 \le v \le 1$)

$$\begin{aligned} &WL_1^n\left(\tau_k^n + \frac{\delta\sigma_{k+1}^n}{\sqrt{n}} v\right) \\ &\qquad \approx \left[WL^n(\tau_k^n) - \theta_2(WL^n(\tau_k^n))\right] + \left[\sum_i \theta_i(WL^n(\tau_k^n)) - WL^n(\tau_k^n)\right] v. \end{aligned} \tag{1.23}$$

This is illustrated in Figure 1.2. The approximation is in the sense that for any $T > 0$, $\sup_{k:\tau_k^n \le T, 0 \le v \le 1}$ of the absolute value of the difference between the two sides of (1.23) goes to zero as $n \to \infty$. The analogous result holds for the terms due to the second integral in (1.21). Thus, we can approximate, in the sense that the $\sup_{k:\tau_k^n \le T}$ of the difference goes to zero in probability as $n \to \infty$,

$$\int_0^1 f\left(WL_1^n\left(\tau_k^n + \frac{\delta\sigma_{k+1}^n}{\sqrt{n}}v\right)\right) dv$$
$$\approx \int_0^1 f\left(v\left(WL^n(\tau_k^n) - \theta_2(WL^n(\tau_k^n)\right) + (1-v)\theta_1\left(WL^n(\tau_k^n)\right)\right) dv, \tag{1.24}$$

with the analogous result for the terms in the second integral of (1.21). By the weak convergence of $WL^n(\cdot)$ and the continuity of the weak-sense limit process, (1.24) also approximates (in the above sense)

$$\int_0^1 f\left(WL_1^n\left(\sigma_k^n + \frac{\delta\tau_k^n}{\sqrt{n}}v\right)\right) dv.$$

Now use the weak convergence of $WL^n(\cdot)$ and the continuity of the weak-sense limit process to get the weak-sense limit (1.19). ■

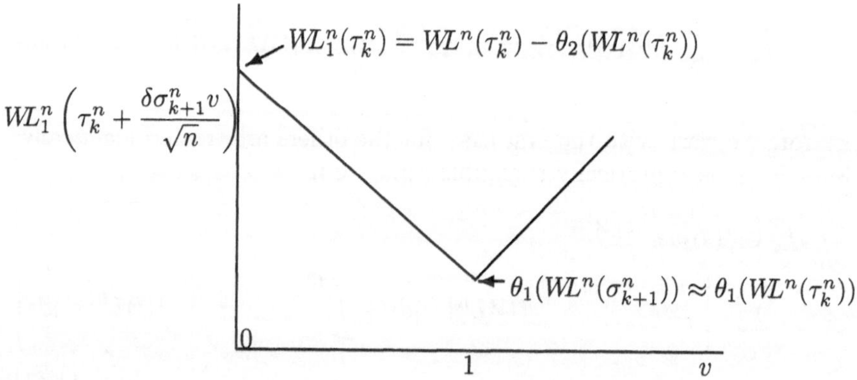

Figure 1.2. Approximation to $WL_1^n(\cdot)$ in the local-fluid time scale; polling queue 1 first.

The next theorem follows from Theorems 1.1, 1.2 and 5.3.3.

Theorem 1.3. *Assume the conditions of Theorem 1.1, except use the polling scheme where the switching rule is modified so that the next switching time depends on the workload at the time of the start of the current polling cycle or at the time of starting the polling of the current queue, or on the current workload in the queue being polled and the value of the other workload at the start of the cycle, all subject to the condition that we do* *not switch if the other queue is empty. Then the conclusions of Theorems 1.1 and 1.2 continue to hold.*

If the functions $\theta_i(WL^n(t))$ are replaced by

$$\frac{1}{\bar{\Delta}_i^d}\theta_i\left(\sum_i \bar{\Delta}_i^d x_i^n(t)\right), \tag{1.25}$$

and the switching decisions use either the value of $x^n(\cdot) = (x_1^n(\cdot), x_2^n(\cdot))$ (in lieu of the workloads) at the start of the current polling cycle or queue, its current value, or the current value of the component being polled and the value of the other component at the start of the current polling, and we do not switch if the other queue is empty, then the conclusions continue to hold.

11.1.2 Lower Thresholds and Relaxed Switching Policies

A lower threshold before switching. If the workload of queue 2 at time σ_k^n is small, then one might not want to switch to queue 2 immediately, and analogously if the workload of queue 1 is small at τ_k^n. It is reasonable to insist on a lower threshold for the workload in the other queue before switching. Return to the setup of Theorem 1.1 and modify the switching policy as follows.

A1.4. *Let $d > 0$ and let $\theta_i(\cdot)$ be continuous and nondecreasing functions on $[d, \infty)$ that satisfy $\theta_i(d) = 0$ and $\inf_{d \le WL < \infty}[WL - \sum_i \theta_i(WL)] > 0$. We do not switch unless the workload in the other queue is at least d. Otherwise, continue polling any arrivals to the current queue. Modify the definitions of $\tau_k^n, \sigma_k^n, \delta\tau_k^n, \delta\sigma_k^n$ in (1.2) in the obvious way.*

Theorem 1.4. *Assume* (A1.0)–(A1.2), (A1.4), *and let $WL^n(0) \ge d$ converge weakly to $WL(0)$. Then the conclusions of Theorems* 1.1 *and* 1.2 *continue to hold, but the limit system* (1.5) *is replaced by*

$$WL(t) = WL(0) + bt + w(t) + z(t), \quad WL(t) \ge d, \tag{1.26}$$

which has a reflection at $WL = d$ rather than at $WL = 0$. The extensions in the first paragraph of Theorem 1.3 also continue to hold, subject to the condition that we do not switch if the workload in the other queue is less than d. The same results hold if we do not switch if $\bar{\Delta}_i^{d,n} x_i^n(t) < d$ for the other queue.

Now modify (A1.4) *by supposing that there are thresholds $d_i, i = 1, 2$, such that switching from queue 1 to queue 2 will not occur if $WL^n(t) < d_1$, and switching from queue 2 to queue 1 will not occur if $WL^n(t) < d_2$. Let $\theta_i(\cdot)$ be defined on $[d_i, \infty)$, with $\theta_i(d_i) = 0$. Then the conclusions hold, with $d = \max\{d_1, d_2\}$.*

Remark on the proof. We will remark only on the proof that, for the case of the first paragraph, $WL^n(\cdot)$ converges weakly to the process defined by (1.26). Equation (1.4) still holds, where $z^n(\cdot)$ is the scaled idle-time process for the current case. We can write, modulo an error that converges weakly to the zero process,

$$WL^n(t) = WL^n(0) + b^n t + w^n(t) + z^n(t), \tag{1.27}$$

where

$$w^n(t) = \sum_i \bar{\Delta}_i^{d,n} \left[w_i^{a,n}(S_i^{a,n}(t)) - w_i^{d,n}(S_i^{a,n}(t)) \right] \tag{1.28}$$

and the process $z^n(\cdot)$ can increase only when $WL^n(t) \in [0, d]$.

Suppose that $z^n(\cdot)$ is not asymptotically continuous. Then there are $T > 0$ and $\epsilon, \rho > 0$ such that

$$\limsup_n P \left\{ \inf_{t \le T} WL^n(t) \le d - \epsilon \right\} \ge \rho.$$

Polling of a queue does not start unless the workload in that queue is at least d. Then the uniformity of the convergence of (1.10) and (1.12) in the sense of (1.11) and the fact that idle time does not decrease the workload yields a contradiction. Now the proof follows from that of Theorem 1.1.

Extensions. The following remarks are in preparation for the extensions in Theorem 1.5.

(a) Discontinuous $\theta_i(\cdot)$. An examination of the previous proofs shows that one needs only that the $\theta_i(\cdot)$ be nondecreasing, but not necessarily continuous. The proofs of the extensions in Theorem 1.5 are left to the reader. The first assertion uses the fact that due to the nondegeneracy of $w(\cdot)$, for any $\epsilon > 0$ and $0 < t \le T$ the probability is zero that $WL(t)$ takes values in the set of points where the discontinuity in either of the $\theta_i(\cdot)$ is greater than ϵ.

(b) General (not necessarily feedback) control functions. The control functions in Theorems 1.1–1.4 were given in feedback form. While this might be appropriate for the weak-sense Markovian limit problem, one might want more flexibility for the physical problem. The proof of Theorem 1.1 depended on the fact that the values of the individual workload sizes at the beginning and end points of the polling intervals changed slowly in time (asymptotically, and except possibly arbitrarily close to some random number of time points, as would be required for (a) above). In fact, under (A1.3) they were continuous or asymptotically continuous functions of the current workload. The $\theta_i(\cdot)$ in Theorems 1.1–1.4 can be replaced by a variety of conditions, while maintaining the asymptotic continuity in the queue sizes at the beginning and end of the polling intervals for almost all end points. Let $\delta u_i^n(\cdot)$, $i = 1, 2$, be $\mathcal{F}_t^n$-adapted processes that converge weakly to the "zero" process, and satisfy $0 \leq \theta_i(WL^n(t)) + \delta u_i^n(t) \leq WL^n(t)$. Then Theorems 1.1–1.4 remain valid if $\theta_i(WL^n(\cdot)) + \delta u_i^n(\cdot)$ replaces $\theta_i(WL^n(\cdot))$ in the definitions of σ_k^n and τ_k^n. More generally, the proofs work with minor modifications if the end points are defined by arbitrary processes that are asymptotically only piecewise continuous. This requires the comments in the next paragraph and leads to the second assertion of Theorem 1.5.

(c) Tightness in the Skorohod topology. Let Φ be a set of paths in $D(I\!R; 0, \infty)$. Then a necessary and sufficient condition [15, Theorem 14.3] for tightness of Φ is that for each $T > 0$,

$$\sup_{\phi \in \Phi} \sup_{s \leq T} |\phi(s)| < \infty$$

and

$$\lim_{\delta \to 0} \sup_{\phi \in \Phi} \inf_{\{t_i\}} \sup_{t_i \leq s < t_{i+1}} |\phi(s) - \phi(t_i)| = 0,$$

where $t_{i+1} \leq T$ and $t_{i+1} - t_i \geq \delta$. Thus, for each $T < \infty$ and $\epsilon > 0$ there is an integer k such that for each $\phi(\cdot) \in \Phi$, time can be divided into at most k half open intervals $\{[t_i, t_{i+1}),\ i \leq k\}$ and $\phi(\cdot)$ can change by at most ϵ on each interval. This characterization of tightness is useful in proving the second assertion of the theorem.

Theorem 1.5. *Let $WL^n(0)$ converge weakly to $WL(0)$, and assume* (A1.0)–(A1.3), *but drop the condition that the $\theta_i(\cdot)$ are continuous. Then under their conditions and polling schemes, the conclusions of Theorems* 1.1–1.3 *continue to hold. Let $u_i^n(\cdot)$, $i = 1, 2$, be $\mathcal{F}_t^n$-adapted processes such that $\{u_i^n(\cdot),\ n < \infty\}$ is tight for $i = 1, 2$. Suppose that for each $T > 0, \epsilon > 0$, and each n,*

$$\inf_{t \leq T : WL^n(t) \geq \epsilon} \left[WL^n(t) - \sum_i u_i^n(t) \right] > 0 \qquad (1.29)$$

with probability one. For any weakly convergent subsequence and any $\epsilon > 0$, let the weak-sense limit processes satisfy

$$\inf_{t \leq T : WL(t) \geq \epsilon} [WL(t) - \sum_i u_i(t)] > 0 \qquad (1.30)$$

with probability one. Then the conclusions of Theorems 1.1–1.3 *continue to hold with $u_i^n(\cdot)$ (respectively, $u_i(\cdot)$) replacing $\theta_i(WL^n(\cdot))$ (respectively, $\theta_i(WL(\cdot))$).*

For the case of Theorem 1.4, *in* (1.29) *and* (1.30) *replace* $\inf_{t \leq T : WL^n(t) \geq \epsilon}$ *by* $\inf_{t \leq T : WL^n(t) \geq d}$, *and* $\inf_{t \leq T : WL(t) \geq \epsilon}$ *by* $\inf_{t \leq T : WL(t) \geq d}$. *Then the conclusions of Theorem* 1.4 *hold.*

If (A1.2′) *replaces* (A1.2), *then* $(WL(\cdot), z(\cdot), u(\cdot))$ *is nonanticipative with respect to* $w(\cdot)$.

11.2 Rate of Switching and Nonzero Switching Time

The asymptotic values of the switching rate can be obtained from the results in the last section, and this is done in the first subsection. These, in turn, are needed to get the weak-sense limit process when there is a nonzero switching time, as in the second subsection, as well as for the control problem for nonzero switching time or cost, which is dealt with in Section 4.

11.2.1 Switching Rate

Define ($1/n$ times the total real time required for the first $\sqrt{n}t$ polling cycles)

$$T^{s,n}(t) = \frac{1}{\sqrt{n}} \sum_{k=1}^{\sqrt{n}t} [\delta\sigma_k^n + \delta\tau_k^n] \tag{2.1}$$

and let $S^{s,n}(t)$ denote $1/\sqrt{n}$ times the number of completed polling cycles by real time nt. The asymptotic *rate of switching* is defined to be the derivative of the weak-sense limit of $S^{s,n}(\cdot)$ and is given by Theorem 2.1.

Theorem 2.1. *Let* $WL^n(0)$ *converge weakly to* $WL(0)$ *and assume* (A1.0)–(A1.2) *and* (A1.4). *Then* $(T^{s,n}(\cdot), S^{s,n}(\cdot))$ *converges weakly to* $(T^s(\cdot), S^s(\cdot))$ *satisfying*

$$T^s(t) = \frac{1}{\rho_1\rho_2} \int_0^t \left[WL(T^s(s)) - \sum_i \theta_i(WL(T^s(s))) \right] ds, \tag{2.2}$$

$$S^s(t) = \rho_1\rho_2 \int_0^t \frac{ds}{[WL(s) - \sum_i \theta_i(WL(s))]}. \tag{2.3}$$

The theorem remains true if the more general controls $u_i^n(\cdot)$, $i = 1, 2$, *of Theorem 1.5 are used, subject to the lower threshold* $d > 0$. *If* (A1.2′) *replaces* (A1.2), *then the weak-sense limit solution* $WL(\cdot), z(\cdot)$ *and controls* $u_i(\cdot)$, $i = 1, 2$, *are nonanticipative with respect to* $w(\cdot)$.

Proof. Only (2.2) and (2.3) will be shown. The expression (1.9) implies that $T^{s,n}(\cdot)$ is asymptotically Lipschitz continuous on each interval $[0, T]$. Now the asymptotic continuity of $T^{s,n}(\cdot)$, Theorem 1.4, and the weak convergence of $WL^n(\cdot)$ to a continuous weak-sense limit driven by a nondegenerate

Wiener process imply that

$$
\begin{aligned}
&T^{s,n}(t) \\
&\approx \frac{1}{\sqrt{n}} \sum_{k=1}^{\sqrt{n}t} \left[\frac{WL^n(\tau_k^n) - \sum_i \theta_i(WL^n(\tau_k^n))}{1-\rho_1} + \frac{WL^n(\sigma_k^n) - \sum_i \theta_i(WL^n(\sigma_k^n))}{1-\rho_2} \right] \\
&\approx \frac{1}{\sqrt{n}\rho_1\rho_2} \sum_{k=1}^{\sqrt{n}t} \left[WL^n(\tau_k^n) - \sum_i \theta_i(WL^n(\tau_k^n)) \right] \\
&\approx \frac{1}{\sqrt{n}\rho_1\rho_2} \sum_{k=1}^{\sqrt{n}t} \left[WL^n\left(T^{s,n}\left(\frac{k}{\sqrt{n}}\right)\right) - \sum_i \theta_i\left(WL^n\left(T^{s,n}\left(\frac{k}{\sqrt{n}}\right)\right)\right) \right] \\
&\approx \frac{1}{\rho_1\rho_2} \int_0^t \left[WL^n(T^{s,n}(s)) - \sum_i \theta_i(WL^n(T^{s,n}(s))) \right] ds.
\end{aligned}
\tag{2.4}
$$

The expression (2.4) holds in the sense that the difference of the processes defined for the various sides converges weakly to the zero process. The weak-sense limit $T^s(\cdot)$ of any weakly convergent subsequence clearly satisfies (2.2).

By the definitions, modulo an asymptotically negligible error (bounded by the scaled time required by the cycle covering scaled time t),

$$T^{s,n}(S^{s,n}(t)) = t - z^n(t)/\sqrt{n}. \tag{2.5}$$

The integrand in (2.2) is continuous with probability one at almost all t. For purposes of the proof, it can be supposed that the integrand is bounded below by some positive number, which implies that the asymptotic slope of $T^{s,n}(\cdot)$ is bounded below by a positive number. This, (2.5), and the weak convergence of $z^n(\cdot)$ to a continuous limit imply that $\{S^{s,n}(\cdot)\}$ is tight, with absolutely continuous weak-sense limits and that for any weak-sense limit,

$$T^s(S^s(t)) = t, \tag{2.6}$$

with both $T^s(t)$ and $S^s(t)$ going to infinity as $t \to \infty$. Now by (2.6), for almost all (ω, t),

$$\frac{d}{dt} T^s(S^s(t)) = 1 = \left. \frac{dT^s(u)}{du} \right|_{u=S^s(t)} \frac{dS^s(t)}{dt}$$

and

$$
\begin{aligned}
\left. \frac{dT^s(u)}{du} \right|_{u=S^s(t)} &= \frac{1}{\rho_1\rho_2} \left[WL(T^s(S^s(t)) - \sum_i \theta_i(WL(T^s(S^s(t))) \right] \\
&= \frac{1}{\rho_1\rho_2} \left[WL(t) - \sum_i \theta_i(WL(t)) \right],
\end{aligned}
$$

which implies (2.3). ■

11.2.2 Nonzero Switching Time

Up to this point in the chapter the time required for the server to switch from one queue to another was assumed to be zero. In many applications with fast switches the time is negligible. The adjustments to the previous results when the switchover time is not negligible are minor and will be covered in this subsection. Let $\Delta_{i,l}^{s,n}$ denote the real time required for the lth switch of the server from queue i to queue $j \neq i$.

A2.1. *The switchover times are independent of the other driving random variables. For each i, there are centering constants $\bar{\Delta}_i^{s,n} \to \bar{\Delta}_i^s > 0$ such that*

$$w_i^{s,n}(t) = \frac{1}{\sqrt{n}} \sum_{l=1}^{\sqrt{n}t} \left[\Delta_{i,l}^{s,n} - \bar{\Delta}_i^{s,n} \right] \tag{2.7}$$

converges weakly to the "zero" process.

Define the scaled total switchover time in the first $\sqrt{nt}$ polling cycles:

$$G^n(t) = \frac{1}{\sqrt{n}} \sum_{l=1}^{\sqrt{n}t} \left[\Delta_{1,l}^{s,n} + \Delta_{2,l}^{s,n} \right].$$

Then (1.27) is changed to (modulo an asymptotically negligible error)

$$WL^n(t) = WL^n(0) + b^n t + w^n(t) + G^n(S^{s,n}(t)) + z^n(t) + \epsilon^n(t), \tag{2.8}$$

where $w^n(\cdot)$ is defined by (1.28). The $\epsilon^n(t)$ is a residual-time error term, and it has two parts, the first due to the residual service and interarrival times at nt as usual, and the second due to the residual switchover time of the current cycle, if any. The next theorem follows from Theorems 1.1–1.5 and 2.1.

Theorem 2.2. *Let $WL^n(0)$ converge weakly to $WL(0)$ and assume* (A1.0), (A1.1), (A1.2′), (A1.4), *and* (A2.1). *Then under their polling schemes, the conclusions of Theorems* 1.1–1.4 *and* 2.1 *hold with weak-sense limit system*

$$WL(t) = WL(0) + bt + w(t) + \int_0^t b_1(WL(s), \theta) ds + z(t), \tag{2.9}$$

where $WL(t) \geq d$ and

$$b_1(WL(t), \theta) = \frac{\left[\bar{\Delta}_1^s + \bar{\Delta}_2^s\right] \rho_1 \rho_2}{\left[WL(t) - \sum_i \theta_i(WL(t))\right]}. \tag{2.10}$$

The conclusions continue to hold under the more general assumptions on the control policy that were used in Theorem 1.5, *assuming the lower threshold $d > 0$. The weak-sense limit solution $WL(\cdot), z(\cdot)$, and controls $u_i(\cdot)$, $i = 1, 2$, are nonanticipative with respect to the Wiener process.*

Remark on the proof. The proof differs from those given previously only in the necessity for accounting for the forced idle time due to the switching between the queues. This is (modulo an asymptotically negligible error) $G^n(S^{s,n}(t))$. Now center the $\Delta_{i,l}^{s,n}$ about $\bar{\Delta}_i^{s,n}$ and use (A2.1) and the weak convergence of the $S^{s,n}(\cdot)$.

Remark on weak-sense uniqueness to (2.9). Since the variance of $w(\cdot)$ is positive, the Girsanov transformation method can be used to show that the nonanticipative solution to (2.9) is unique in the weak sense. Under (A1.2′), any weak-sense limit solution is nonanticipative with respect to the weak-sense limit Wiener process $w(\cdot)$. If the general nonanticipative controls of Theorem 1.5 replace the feedback controls for the physical system, then the weak-sense limit solution and controls are nonanticipative with respect to $w(\cdot)$, and the distribution of $(u_i(\cdot),\ i = 1, 2, w(\cdot))$ determines that of $(WL(\cdot), u_i(\cdot),\ i = 1, 2, w(\cdot))$.

11.3 Gated Polling

The gated polling scheme that was described in the Chapter Outline can be included in the format of Theorems 1.1–1.4 and 2.1–2.2 by appropriate selection of the functions $\theta_i(\cdot)$. Only a rough description will be given.

The polling procedure will be modified slightly, but the end result will be a good approximation to gated polling under heavy traffic. By the definition of gated polling, we are allowed to serve only those jobs that were in the queue at the moment that polling of that queue started. Violations of this policy will be allowed for the first polling cycle, and asymptotically small violations will be allowed after the first cycle. To determine the $\theta_i(\cdot)$ we will use the linear approximations determined by (1.10) and (1.12) to the behavior during the segments of (real time) lengths $\sqrt{n}\delta\sigma_k^n$ (respectively, $\sqrt{n}\delta\tau_k^n$) when queue 1 (respectively, queue 2) is being polled.

Let us start as for the procedure of Theorem 1.1 and poll up to (scaled time) σ_1^n as done there. In fact, to simplify the discussion and with no loss of generality, suppose that $WL_1^n(0) = WL^n(0) - \theta_2(WL^n(0))$. During the first polling period, where queue 1 is being polled, gated polling devotes $WL_1^n(0)\sqrt{n}$ units of real time to queue 1. Thus, $\delta\sigma_1^n = WL_1^n(0)$. Now using the linear approximations determined by (1.10) and (1.12), define the function $\theta_1(\cdot)$ such that $\theta_1(WL^n(\sigma_1^n))$ equals, asymptotically, the work that arrived at queue 1 during that polling time. Thus, asymptotically,

$$\theta_1(WL) \approx \rho_1(WL - \theta_2(WL)).$$

Analogously,

$$\theta_2(WL) \approx \rho_2(WL - \theta_1(WL)).$$

This leads to

$$\theta_i(WL) = \frac{\rho_i^2 WL}{1 - \rho_1\rho_2}.$$

This rule will approximate gated polling.

The definition of gated polling says that we can work only on what was in a queue at the start of its polling. More generally, the $\theta_i(\cdot)$ can be chosen such that only a desired fraction of those initial jobs will be served. The methods of the theorems in Sections 1 and 2 can be used to deal with such variations.

11.4 The Control Problem: Switching Cost and Time

We now turn our attention to the control problem where the switching times are to be controlled to minimize a cost function, both without and with a nonzero switchover time.

11.4.1 Switching Cost

Consider the control problem where *the switchover times are zero* but the cost penalizes the number of switches. In all cases, there is given a $d > 0$ (perhaps very small) such that no switching occurs if the work in the other queue is less than d.

Admissible controls. The control policy determines the timing of the beginning and end of each polling of the queues. If this policy is allowed to be arbitrary, it can be difficult to define the switching rate, whose computation in Theorem 2.1 depended on continuous or piecewise continuous (in state or time dependence) switching functions. Because of this, the class of allowable control policies for the physical system will be restricted to policies whose weak-sense limits are continuous or piecewise continuous. This restriction is done for mathematical reasons, but it does not appear to be harmful to the goal of minimizing the cost function, and is large enough to include all the controls that would be implemented in practice. The allowable rules are defined by the rules of Theorem 1.5, but with the possibility of controlled idling allowed as well. Let $F^n(t)$ denote $1/\sqrt{n}$ times the idle time that occurs when $WL^n(t) > d$, due to the controller's decision to idle. Thus, an admissible control for (4.1) is defined to be any $\mathcal{F}_t^n$-adapted set $(u_i^n(\cdot), i = 1, 2, F^n(\cdot))$ where $u^n(\cdot) = (u_i^n(\cdot), i = 1, 2)$ satisfies the conditions in Theorem 1.5.

The dynamics can be represented as

$$WL^n(t) = WL^n(0) + b^n t + w^n(t) + F^n(t) + z^n(t) + \epsilon^n(t) \tag{4.1}$$

and

$$WL(t) = WL(0) + bt + w(t) + F(t) + z(t), \tag{4.2}$$

where $w^n(\cdot)$ is defined in (1.28), $\epsilon^n(\cdot)$ is a residual-time error term, and $\epsilon^n(t)$ is bounded by $1/\sqrt{n}$ times the sum of the interarrival and service intervals that cover real time nt.

The cost function. Let $\beta > 0$ and $c > 0$ and let $k_i(WL)$ be continuous, nonnegative, and nondecreasing functions that go to infinity as $WL \to \infty$, but no faster than linearly. The cost function for the physical problem is

$$W_\beta^n(WL, u^n, F^n) = E_{WL}^{u^n,F^n} \int_0^\infty e^{-\beta t} \left[\sum_i k_i(WL_i^n(t))dt + cdS^{s,n}(t) \right], \tag{4.3}$$

where $(u^n(\cdot), F^n(\cdot))$ denotes the control policy, and $WL = WL^n(0)$ the initial condition. By Theorems 1.2 and 2.1, for control functions $\theta_i(\cdot)$ the natural form of the cost function for the weak-sense limit is, where $u_i(t) = \theta_i(WL(t))$,

$$W_\beta(WL, u, F) = E_{WL}^{u,F} \int_0^\infty e^{-\beta t} \left[\sum_i f_i(WL(t), \theta) + c\dot{S}^s(t) \right] dt, \tag{4.4}$$

where

$$f_1(WL, \theta) = \int_0^1 k_1 \left(v \left(WL - \theta_2(WL) \right) + (1-v)\theta_1(WL) \right) dv, \tag{4.5}$$

$f_2(\cdot)$ is obtained by interchanging the indices 1 and 2 in (4.5), and

$$\dot{S}^s(t) = \frac{\rho_1 \rho_2}{[WL(t) - \sum_i \theta_i(WL(t))]}. \tag{4.6}$$

Let us make the following assumption. There are unique continuous and nondecreasing minimizers, called $\bar{\theta}_i(WL)$, of the cost rate

$$k(WL, \theta) = \sum_i f_i(WL, \theta) + \frac{\rho_1 \rho_2}{[WL - \sum_i \theta_i(WL)]}.$$

Define $\bar{k}(WL) = k(WL, \bar{\theta})$. Suppose that $\bar{k}(WL)$ is continuous, nondecreasing, and has at most a linear growth in x. Consider the singular control problem with model (4.2), where $z(\cdot)$ is the reflection term at $WL = d$, and cost

$$\bar{W}_\beta(WL, F) = E \int_0^\infty e^{-\beta t} \bar{k}(WL(t))dt.$$

Let $V_\beta^n(WL)$ and $V_\beta(WL)$ denote the infimums of the cost functions over the admissible controls.

The Bellman equation. Let $\mathcal{L}$ denote the differential generator of the uncontrolled and unreflected process defined by $dx(t) = bdt + dw(t)$. We expect that there is a threshold $\bar{d} \geq d$ such that (in the limit) the optimal control component $\bar{F}(\cdot)$ acts as a reflection at $W\!L = \bar{d}$. Then the Bellman equation is [77]

$$\min_{\theta} \left[\mathcal{L}V_\beta(W\!L) + \bar{k}(W\!L) - \beta V_\beta(W\!L)\right] = 0, \; W\!L > \bar{d}, \tag{4.7}$$

$$V_{\beta,w}(W\!L) = 0 \text{ at } W\!L = \bar{d}, \tag{4.8}$$

where $V_{\beta,w}(\cdot)$ is the derivative of $V_\beta(\cdot)$ with respect to $W\!L$. The boundary condition (4.8) holds, since there is no boundary cost. In fact, the Bellman equation is the quasivariational inequality

$$\min\left\{\left[\mathcal{L}V_\beta(W\!L) + \bar{k}(W\!L) - \beta V_\beta(W\!L)\right],\, V_{\beta,w}(W\!L)\right\} = 0. \tag{4.9}$$

Since the control does not enter the dynamics, the problem reduces to a simple singular control problem with a cost that is a function of $W\!L$ and does not explicitly depend on the control. This two-step procedure, first computing the optimal $\theta_i(\cdot)$ and then the threshold $\bar{d}$, was used effectively in [211]. The function $V_\beta(\cdot)$ is twice continuously differentiable (see, for example, [193, 225]) and there is a $\bar{d} \geq d$ such that the optimal control $\bar{F}(\cdot)$ corresponds to a reflection at $\bar{d}$. Also (see the verification theorem for singular controls in [78]), $\bar{F}(\cdot)$ and $\theta_i(\cdot)$, $i = 1, 2$, are optimal with respect to all nonanticipative controls.

A sufficient condition for (4.10a) is given by Theorem 9.3.1. We restrict attention to controls satisfying the following property: For each $\epsilon > 0$. there are ϵ-optimal controls $(u_i^n(\cdot), i = 1, 2, F^n(\cdot))$ where $u_i^n(\cdot)$, $i = 1, 2$, satisfies any of the conditions of Theorem 1.5.

Theorem 4.1. *Let* $W\!L^n(0) = W\!L$. *Assume* (A1.0), (A1.1) *and* (A1.2′). *Assume the condition on* $\bar{k}(\cdot)$ *stated below* (4.6) *and that on admissible controls in the last paragraph. Suppose that, for some* $p \geq 1$ *and* $\alpha = a, d$, $i = 1, 2$,

$$\sup_n E \sup_{s \leq t} |w_i^{\alpha,n}(S_i^{a,n}(s))|^2 = O(t^p), \tag{4.10a}$$

$$\sup_n \sup_t \frac{E \sup_{s \leq t} |\epsilon^n(s)|}{1 + t^p} = 0. \tag{4.10b}$$

Then

$$\lim_n V_\beta^n(W\!L) = V_\beta(W\!L). \tag{4.11}$$

Proof. Consider the following control policy. Do not switch if $W\!L^n(t) \leq d$. Otherwise, do exhaustive polling. Then we can suppose that all the idle time is accounted for by $z^n(\cdot)$, and $F^n(t) = 0$. The bounds in (4.10) imply that $\sup_n Ez^n(t) = O(t^p)$. Also, $S^{s,n}(t) \leq t/d$. Thus, there is a policy with costs that are bounded uniformly in n.

Now fix an optimal or ϵ-optimal sequence $u^n(\cdot)$. The sequence $F^n(\cdot)$ might not be tight. But the time transformation method and Fatou's lemma, as used in Theorem 10.1.3, can be employed to show that

$$\liminf_n V_\beta^n(WL) \geq V_\beta(WL). \tag{4.12}$$

The reverse inequality

$$\limsup_n V_\beta^n(WL) \leq V_\beta(WL) \tag{4.13}$$

needs to be shown. Again, the procedure is analogous to what was done in Section 10.1: Apply the rule determined by $\bar{\theta}_i(\cdot)$, $i = 1, 2$, and $\bar{F}(\cdot)$ to (4.1). The $\{\bar{\theta}_i(WL^n(\cdot)), i = 1, 2\}$ will always be tight. ■

11.4.2 Nonzero Switching Time

We now turn to the case where there is a nonzero switchover time, according to the model used in Subsection 2.2, and continue to use the class of controls considered in Theorem 4.1. Then adding the idling time control to (2.8) yields the following dynamical equation for the physical model:

$$WL^n(t) = WL^n(0)+b^n t+w^n(t)+F^n(t)+G^n(S^{s,n}(t))+z^n(t)+\epsilon^n(t). \tag{4.14}$$

The weak-sense limit equation is, in terms of feedback controls $\theta_i(\cdot)$, $i = 1, 2$,

$$WL(t) = WL(0) + b(t) + w(t) + F(t) + \int_0^t b_1(WL(s), \theta)ds + z(t), \tag{4.15}$$

where $b_1(\cdot)$ is defined in (2.10). Suppose that there is no switchover cost, that (A2.1) holds, and that the switchover times are mutually independent and have uniformly bounded variances. Then under its other conditions, Theorem 4.1 holds. There is nothing new in the proof of (4.12). The idea in Theorem 4.1 was to use the optimal control for the limit problem to prove (4.13). But in the present case, at this time we do not know whether the solution to the Bellman equation is smooth. We can simply assume that there are continuous or piecewise continuous optimal switching functions for the limit system and that the optimal $F(\cdot)$ is determined by a threshold, and proceed as in Theorem 4.1. Or, more generally, we can take the point of view of Theorem 10.1.4 by showing that there are ϵ-optimal controls $(u_i(\cdot), i = 1, 2, F(\cdot))$ for the limit system that are piecewise constant and finite-valued. The details of this latter approach are left to the reader.

The Bellman equation. The quasivariational inequality that is the Bellman equation is similar to (4.9), except for the different drift and cost function. Let $\mathcal{L}$ again denote the differential generator of the uncontrolled process defined by $dx(t) = bdt + dw(t)$. Then we expect that there is a threshold $\bar{d}$ such that

$$\min_{\theta} \left[\mathcal{L}V_\beta(WL) + V_{\beta,w}(WL)b_1(WL,\theta) + \sum_i f_i(WL,\theta) - \beta V_\beta(WL) \right] = 0 \tag{4.16}$$

for $WL > \bar{d}$ and

$$V_{\beta,WL}(WL) = 0 \text{ at } WL = \bar{d}.$$

The actual Bellman equation is the quasivariational inequality

$$\min \left\{ \min_{\theta} \left[\mathcal{L}V_\beta(WL) + V_{\beta,w}(WL)b_1(w,\theta) + \sum_i f_i(WL,\theta) - \beta V_\beta(WL) \right], \right.$$
$$\left. V_{\beta,w}(WL) \right\} = 0. \tag{4.17}$$

11.5 Controlled Polling with Interruptions: Introduction

In the remainder of this chapter we are concerned with the polling problem for two queues when the connections between each queue and the server are broken at random times and for random durations. The broken connections constitute the process of infrequent but long vacations of Subsections 5.6.4 and 7.3.2. The occurrence of such vacations complicates the control problem considerably, since the possibility that any queue might not be available to the server at any future time needs to be accounted for in choosing the current server allocation. For simplicity, we will suppose that there is no switchover time or cost, so that we can concentrate on the effects of the vacations. Additional discussion concerning stability, unreliable channels, and other aspects of the problem is in [3].

This problem arises in contemporary mobile communications, where the queues are in the mobile sources that generate data to be transmitted, and the server is the channel or antenna of the base station. At each time, the control action is the assignment of the channel between the sources. In fact, the channel can be simultaneously shared among the sources in a controlled way, say via CDMA. This latter possibility is covered by the formulation, since it allows any feasible allocation of the resource between the queues. Analyses of the vacation problem for some fixed control policies and a fixed system are in [234, 235, 236].

Even with linear holding costs, the classical $c\mu$-rule [243] can be far from optimal when there are vacations. As is usual with processes subject to long vacations, the weak-sense limit is a jump diffusion. But the jumps are state- and control-dependent. The weak-sense limit control problem is nonstandard, owing to the way that the control influences the jump distributions. The problem will be formulated and assumptions given in this section. The weak convergence of the intervacation sections of the workload and of the jumps due to the vacations, for a fixed set of controls, is shown in Section 6, and the correct form of the weak-sense limit control problem is described in Section 7. Section 8 concerns various extensions. The concept of relaxed Poisson measure, which is needed for the proper formulation of the weak-sense limit control problem, is dealt with in Section 9.

The Problem formulation. There are two sources and zero switchover time and cost. Each of the sources will be unavailable to be polled (be on vacation) at random times. When a source is on vacation its job creation process does not stop. When both sources are available, service can be either of the preempt–resume type, or nonpreemptive in that a job once started is completed, assuming that there is no intervening vacation. Also, the server will not idle if there is work to do on an available queue. We also allow that the channel be shared so that both queues are worked on simultaneously, with the sharing subject to control. The results will be the same for all cases. To simplify the notation, and without loss of generality, suppose that both sources are available at the initial time. Assumptions and notation analogous to those of Theorems 5.6.4 and 7.3.2 will be used.

A5.0. *For each n, $x_i^n(0), WL_i^n(0)$, $i = 1, 2$, are independent of all of the "future" driving random variables. None of the sources is on vacation at $t = 0$.*

A5.1. *For each n, i, the real time intervals between the end of the $(l-1)$st vacation, $l = 1, \ldots,$ and the start of the next one for source i are denoted by $n\tau_{i,l}^{s,n}$, $l = 1, \ldots.$ The $\tau_{i,l}^{s,n}$ are mutually independent, exponentially distributed, independent of all the other "driving" random variables and have rate $\bar{\lambda}_i^{s,n}$, where $\bar{\lambda}_i^{s,n}$ converges to $\bar{\lambda}_i^s > 0$ as $n \to \infty$. The intervals for the different sources are mutually independent.*

A5.2. *For each n, i, there are mutually independent and identically distributed random variables $\tau_{i,l}^{v,n}$, $l = 1, \ldots,$ such that the real time duration of the lth vacation interval for source i is $\sqrt{n}\tau_{i,l}^{v,n}$. Also, $\tau_{i,l}^{v,n}$ converges weakly to a random variable τ_i^v as $n \to \infty$. For each i, the $\tau_{i,l}^{v,n}$, $l = 1, \ldots,$ are independent of all other "driving" random variables. The intervals for the different sources are mutually independent. Define $\tau_{i,0}^{v,n} = 0$.*

Correlated vacations. Assumption (A5.2) can be extended to cover intervacation and vacation intervals that are correlated between the sources. The added difficulties are only algebraic and are discussed in [3]. The server controller can be based either on the numbers queued or on the work queued. We will confine ourselves to the second case, but Theorem 5.3.3 implies that the results are asymptotically (as $n \to \infty$) equivalent in that the minimum costs are the same and a good policy for one is equivalent to a good policy for the other, under heavy traffic. Thus, the server is assumed to know the entire past history of the arrivals and their work requirements.

Discussion of the control problem. Recall the definition (7.3.13):

$$\nu_{i,l}^n = \sum_{k=1}^{l} \left[\tau_{i,k}^{s,n} + \tau_{i,k-1}^{v,n}/\sqrt{n} \right],$$

the *scaled* time of the start of the lth vacation at source i. The (scaled) lth vacation interval for source i is the half-open (scaled) interval $[\nu_{i,l}^n, \nu_{i,l}^n + \tau_{i,l}^{v,n}/\sqrt{n})$.

Let $z^n(t)$ denote $1/\sqrt{n}$ times the total real time that both queues are empty *and* neither source is on vacation up to real time nt. Let $T^{v,n}(t)$ denote $1/\sqrt{n}$ times the total time up to real time nt that the server could not work due to a vacation (that is, where the contents of the available queue, if any, is zero, or where there are no available queues). Let $WL_i^n(t)$ denote the individual workload in queue i at scaled time t and define the total workload $WL^n(t) = \sum_i WL_i^n(t)$. Then (1.4) is modified as

$$\begin{aligned} WL^n(t) = WL^n(0) &+ \frac{1}{\sqrt{n}} \sum_i \sum_{l=1}^{nS_i^{a,n}(t)} \Delta_{i,l}^{d,n} \\ &- \frac{1}{\sqrt{n}} \left[\text{real time of all service by real time } nt\right], \end{aligned} \tag{5.1}$$

and the last term on the right is

$$-\left[\sqrt{n}t - z^n(t) - T^{v,n}(t)\right]. \tag{5.2}$$

We need to examine $T^{v,n}(t)$, since it will be through this term that the control affects the weak-sense limit. The independence and rate conditions (A5.1) and (A5.2) imply that as $n \to \infty$, the probability that both sources are on vacation simultaneously on any scaled time interval $[0, T]$ is $O(1/\sqrt{n})$. Thus, in the following discussion dealing with weak convergence, which is concerned with large n, it will be supposed, and without loss of generality, that at most one source can be on vacation at a time.

Consider the lth vacation of source 1. It starts at (scaled) time $\nu_{1,l}^n$, and the total workload is then $WL^n(\nu_{1,l}^n)$. Define $u^n(\cdot)$ by $u^n(t) = WL_1^n(t)$. Hence, $WL_2^n(t) = WL^n(t) - u^n(t)$. The $u^n(\cdot)$ is determined by the polling policy and will be considered to be the control. Its value will be seen to be the mechanism by which the polling policy affects the jumps. Analogously to the notation in Subsection 5.6.4 and in (1.8), let $A_{i,l}^{v,j,n}(t)$ denote $1/\sqrt{n}$

times the work arriving at queue i during the scaled time interval $[\nu^n_{j,l}, \nu^n_{j,l} + (\tau^{v,n}_{j,l} \wedge t)/\sqrt{n})$. Note that we are using the *local-fluid time scale* again to represent the process during the vacation. Then $A^{v,j,n}_{i,l} = A^{v,j,n}_{i,l}(\infty)$ is $1/\sqrt{n}$ times the work arriving at queue i during the lth vacation of source j. Modulo a residual-time error term, which is bounded by $1\sqrt{n}$ times the sum of the (unscaled) work in the first two and the last arrivals during the real time interval $(n\nu^n_{j,l}, n\nu^n_{j,l} + \sqrt{n}(\tau^{v,n}_{j,l} \wedge t)]$,

$$A^{v,j,n}_{i,l}(t) = \frac{1}{\sqrt{n}} \sum_{l=nS^{a,n}_i(\nu^n_{j,l})+2}^{nS^{a,n}_i(\nu^n_{j,l}+(\tau^{v,n}_{j,l}\wedge t)/\sqrt{n})-1} \Delta^{d,n}_{i,l}. \tag{5.3}$$

Let $\xi^{v,n}_{i,l}$ denote the change in the *total workload* during the lth vacation of source i. Until further notice, for notational simplicity in this motivational discussion, let us fix attention on the lth vacation of source 1. From Theorems 1.1 and 5.6.4, it can be supposed that work arrives continuously during the vacation, as a fluid at the mean rate. Thus, if

$$\tau^{v,n}_{1,l} < WL^n_2(\nu^n_{1,l}) + A^{v,1,n}_{2,l} = WL^n(\nu^n_{1,l}) - u^n(\nu^n_{1,l}) + A^{v,1,n}_{2,l}, \tag{5.4}$$

then the vacation ends before queue 2 is emptied, and the vacation would not have an immediate effect on the total idle time and workload and (asymptotically, as $n \to \infty$) $\xi^{v,n}_{1,l} = 0$.

On the other hand, if

$$\tau^{v,n}_{1,l} > WL^n(\nu^n_{1,l}) - u^n(\nu^n_{1,l}) + A^{v,1,n}_{2,l}, \tag{5.5}$$

then queue 2 is emptied before the vacation ends. While new arrivals to queue 2 can be served, the heavy traffic condition (A1.1) implies that the service rate is so much faster than the arrival rate of work at queue 2 that (asymptotically, as $n \to \infty$) the workload in queue 2 is zero for a nonvanishing fraction of the remaining vacation time. Thus, there is (asymptotically) forced idle time and an increase in the total workload during the vacation. This increase will depend on the value of the workload at queue 2 at the time that the vacation at queue 1 starts. In turn, that value depends on the control policy. This is the *only* way that the control policy affects the workload: via the sizes of the jumps during the vacations, which (in turn) is determined by the distribution of the total workload when the vacation starts. Obviously, one can reverse sources 1 and 2 in the above discussion.

11.6 Weak Convergence of the Intervacation Sections and Jumps

Suppose, formally, that for each l the set $\{\nu^n_{1,l}, \tau^{v,n}_{1,l}, u^n(\nu^n_{1,l}), WL^n(\nu^n_{1,l})\}$ converges weakly to a limit that we denote by $(\nu_{1,l}, \tau^v_{1,l}, u(\nu_{1,l}), WL(\nu_{1,l}-))$.

Recall that the limit process $W\!L(\cdot)$ is right continuous. We use the notation $W\!L(\nu_{1,l}-)$ for the limit, since it is the limit of the sequence of state values just at the start of the vacation. The scaled duration of the vacation goes to zero as $n \to \infty$, and if the limit process has a jump at $\nu_{1,l}$ due to the effects of the vacation, then the jump will be $W\!L(\nu_{1,l}) - W\!L(\nu_{1,l}-)$. Asymptotically, the increase in $T^{v,n}(\cdot)$ (equivalently, in the total workload) during the lth vacation of source 1 can be written as

$$\begin{aligned} \xi^v_{1,l} &= \left[(1-\rho_2)\,\tau^v_{1,l} - [W\!L(\nu_{1,l}-) - u(\nu_{1,l})]\right]^+ \\ &= \left[\rho_1\tau^v_{1,l} - [W\!L(\nu_{1,l}-) - u(\nu_{1,l})]\right]^+. \end{aligned} \tag{6.1}$$

The analogue for the lth vacation of source 2 is

$$\xi^v_{2,l} = \left[(1-\rho_1)\,\tau^v_{2,l} - u(\nu_{2,l})\right]^+ = \left[\rho_2\tau^v_{2,l} - u(\nu_{2,l})\right]^+. \tag{6.2}$$

Define the functions

$$q_1(W\!L,\gamma,\alpha) = [\rho_1\gamma - (W\!L - \alpha)]^+, \quad q_2(W\!L,\gamma,\alpha) = [\rho_2\gamma - \alpha]^+.$$

The process $u(\cdot)$, satisfying $u(t) \le W\!L(t-)$, can be considered to be the control function for the limit system (6.4), (6.5). However, there is no guarantee that any subsequence of $\{u^n(\cdot)\}$ converges weakly. In the original work [3], it was assumed that $u^n(t) = \theta^n(W\!L^n(t))$ for some function $\theta^n(\cdot)$ of $W\!L$ belonging to an equicontinuous class. While this seems quite adequate for applications, to develop the full theory and with an eye on future applications we will allow the control policy for the physical system to be an arbitrary $\mathcal{F}^n_t$-adapted process satisfying $u^n(t) \le W\!L^n(t)$ and realizable by some polling policy.

As was seen in Section 3.7, jump–diffusions can be represented in terms of a "driving" Poisson measure. When the jump is controlled, the classical concept of Poisson measure is not adequate and needs to be extended, just as the concept of control needed to be extended to that of relaxed control for nonlinear problems. The extension, which is developed in Section 9 and used in the next section, does not affect the infimum of the costs, and any solution under the extended definition can be approximated by a solution under the classical Poisson measure.

Let $\bar\nu^n_l$ and ν^n_l denote, respectively, the (scaled) time of the beginning and end of the lth intervacation interval, irrespective of the source. Thus $\bar\nu^n_1 = 0$. Define the *intervacation sections* of the workload as the functions

$$W\!L^n\left((\bar\nu^n_l + t) \wedge \nu^n_l\right), \quad t \ge 0. \tag{6.3}$$

They are constant for $t \ge \nu^n_l - \bar\nu^n_l$. Let $r^n(\cdot)$ denote the relaxed control representation of $u^n(\cdot)$.

The limit process will be constructed in sections, analogously to the procedure used in Section 10.2. First, one considers the section of the workload up to the start of the first vacation and shows that the set of first sections of the workload and the set of endpoints $\{WL^n(\nu_1^n)\}$ is tight. Then one characterizes the effect of the vacation and shows that the set $\{WL^n(\bar{\nu}_2^n)\}$ is tight. Then continue on to the second intervacation section of the workload, etc. A weakly convergent subsequence of all the parts is taken, and the limit process (6.4), (6.5) is constructed by putting the limit parts together. This procedure is adequate for dealing with the discounted cost function and avoids the problems associated with the nontightness of $WL^n(\cdot)$ in the Skorohod topology. Theorem 6.1 contains many parts, but it gives a complete characterization of the limit process.

Theorem 6.1. *Assume* (A5.0)–(A5.2), (A1.1), (A1.2′), *and suppose that* $(WL_i^n(0), i = 1,2)$ *converges weakly to* $(WL_i(0), i = 1,2)$. *Then the interjump sections satisfy* (1.5). *For each* l, *the set* $\{\Psi_l^n\}$ *is tight, where*

$$\Psi_l^n = \left\{WL^n(\nu_{i,l}^n), u^n(\nu_{i,l}^n), \tau_{i,l}^{v,n}, \tau_{i,l}^{s,n}, \xi_{i,l}^{v,n},\ i = 1,2\right\}.$$

The set of sections of the workload defined by (6.3) *and the set* $\{w^n(\cdot), z^n(\cdot), r^n(\cdot)\}$ *are also tight, and the weak-sense limits are continuous.*

Fix a weakly convergent subsequence (indexed by n*) of the various sets for all* l, *and denote the weak-sense limits by*

$$\left(WL(\nu_{i,l}-), u(\nu_{i,l}), \tau_{i,l}^{v}, \tau_{i,l}^{s}, \xi_{i,l}^{v},\ i = 1,2\right), \quad (w(\cdot), z(\cdot), r(\cdot)).$$

Then $\tau_{i,l}^s$ *is exponentially distributed with rate* $\bar{\lambda}_i^s$. *The differences* $WL_i^n(\cdot) - \bar{\Delta}_i^{d,n} x_i^n(\cdot)$ *converge weakly to the "zero" process.*

Define $WL(\cdot)$ *by piecing together the weak-sense limits of the successive intervacation sections of the workload and jumps of* $WL^n(\cdot)$. *The weak-sense limits of any weakly convergent subsequence are related by*

$$WL(t) = WL(0) + bt + w(t) + \sum_i J_i(t) + z(t), \tag{6.4}$$

where

$$J_i(t) = \sum_{l:\nu_{i,l}\le t} \xi_{i,l}^v, \quad \nu_{i,l} = \sum_{k=1}^{l} \tau_{i,k}^s. \tag{6.5}$$

Also

$$\left(WL(0), w(\cdot), \tau_{i,l}^v, \tau_{i,l}^s;\ i = 1,2, l < \infty\right) \tag{6.6}$$

are mutually independent. Let $\mathcal{F}_t$ *denote the* σ*-algebra that is generated by the limit variables up to time* t. *Then* $w(\cdot)$ *is an* $\mathcal{F}_t$*-Wiener process. The process* $z(\cdot)$ *is the reflection term at the origin. The* $\xi_{i,l}^v$ *have the representation* (6.1) *and* (6.2). *For* $i = 1,2$, *define the piecewise constant process* $p_i(\cdot)$

with jumps of values $\tau_{i,l}^v$ at times $\nu_{i,l}$, and let $N_i(\cdot)$ denote the $\mathcal{F}_t$-Poisson measure that it induces. Then for each t,

$$w(t+\cdot) - w(t), N_i(t+\cdot) - N_i(t), \ i = 1, 2, \tag{6.7}$$

is independent of

$$\left\{ w(s), r(s), N_i(s), \ s \leq t, \ i = 1, 2; \right. \\ \left. u(\nu_{i,l}) I_{\{\nu_{i,l} \leq t\}}, \xi_{i,l}^v I_{\{\nu_{i,l} \leq t\}}, \ i = 1, 2, \ l < \infty \right\}. \tag{6.8}$$

The processes $A_{i,l}^{v,j,n}(\cdot)$ converge weakly to the processes with values $\rho_i(t \wedge \tau_{j,l}^v)$.

Proof. The tightness of the set of first sections of the workload ((6.3), with $l = 1$) follows by Theorem 5.1.1, (A5.1), and (A1.2$'$). Take a weakly convergent subsequence (abusing terminology and indexing all subsequences by n). Then the set $\{WL^n(\nu_1^n)\}$ converges weakly. Suppose that the first vacation is that of queue j. Then the assertions concerning $A_{i,l}^{v,j,n}(\cdot)$ are proved in Theorems 1.1 and 5.6.4. Now continue with the set of second intervacation intervals, etc., to get (6.4) and (6.5). The independence assertions are easily seen from the assumptions. ■

11.7 The Limit Control Problem

The limit dynamical model. Let the filtration $\{\mathcal{F}_t, t \geq 0\}$ be given, and suppose that $w(\cdot)$ and $N_i(\cdot)$, $i = 1, 2$, are (respectively) an $\mathcal{F}_t$-Wiener process and an (independent) $\mathcal{F}_t$-Poisson measure as defined in Theorem 6.1. Let $WL(0)$ be $\mathcal{F}_0$-measurable. An admissible control for (6.4) and (6.5) is an $\mathcal{F}_t$-predictable process that satisfies $u(t) \leq WL(t-)$. Since the approximation theorems for the controls that are appealed to are based on controls whose range space is not state-dependent, one can always ensure this state-dependent upper bound by using $\min\{WL(t-), u(t)\}$ in lieu of $u(t)$ in the dynamical equations. In getting (6.4) and (6.5), only a convergent subsequence of the sequence of values $\{u^n(\nu_{i,l}^n)\}$ was taken for each l, since the sequence of processes $\{u^n(\cdot)\}$ was not necessarily tight. Even though the sequence $\{r^n(\cdot)\}$ was tight, the limits $u(\nu_{i,l})$ of the $u^n(\nu_{i,l}^n)$ cannot necessarily be obtained from the limits $r(\cdot)$. However, suppose that $u(\cdot)$ is a process with value $u(\nu_{i,l})$ at $\nu_{i,l}$. Then (7.1) is a representation of (6.4) and (6.5):

$$WL(t) = WL(0) + bt + w(t) + \sum_i \int_0^t \int q_i(WL(s-), \gamma, u(s)) N_i(ds\, d\gamma). \tag{7.1}$$

It turns out that (7.1) is the correct form of the general controlled limit process in that the infimal costs for the physical system (5.1) will converge to the infimal cost for (7.1), and (7.1) can be used to get good controls for (5.1). The full details involve the concepts of Section 9.

The cost function. Let $k_i(\cdot)$ be strictly increasing and continuous functions on $[0,\infty)$ with $k_i(0) = 0$, and having at most linear growth. If we wish to penalize the workloads, then use the cost rate

$$\sum_i k_i(WL_i^n(\cdot)) = k_1(u^n(\cdot)) + k_2(WL^n(\cdot) - u^n(\cdot)) \equiv k(WL^n(\cdot), u^n(\cdot)).$$

If the scaled number of jobs is to be penalized, then use the approximation

$$\sum_i k_i(x_i^n(\cdot)) \approx k_1\left(\frac{u^n(\cdot)}{\bar{\Delta}_1^{d,n}}\right) + k_2\left(\frac{WL^n(\cdot) - u^n(\cdot)}{\bar{\Delta}_2^{d,n}}\right) \equiv k(WL^n(\cdot), u^n(\cdot)).$$

We will use the discounted cost function with cost rate $k(WL^n(t), u^n(t))$:

$$W_\beta^n(WL, u^n) = E_{WL}^{u^n} \int_0^\infty e^{-\beta t} k(WL^n(s), u^n(s))ds, \quad WL^n(0) = WL. \tag{7.2}$$

Define the cost $W_\beta(WL, u)$ for the weak-sense limit system analogously, and define the infimums $V_\beta^n(\cdot)$ and $V_\beta(\cdot)$ in the usual way.

The proof of the following theorem follows from the concepts and results in Section 9, which should be read in conjunction with the remarks on the proof.

Theorem 7.1. *Let* $WL^n(0) = WL$ *and assume* (A5.0)–(A5.2), (A1.1), *and* (A1.2′). *Suppose that there are positive* K_1 *and* p *such that*

$$\sup_n \sup_{u(\cdot)} E\left|WL^n(t)\right|^2 \le K_1 t^p + K_1. \tag{7.3}$$

Then the sequence of functions whose expectations are being taken in (7.2) *is uniformly integrable, and*

$$V_\beta^n(WL(0)) \to V_\beta(WL(0)).$$

Remarks on the proof. Let $N_i^n(\cdot)$, $i = 1, 2$, be mutually independent Poisson measures with jump rates $\bar{\lambda}_i^{s,n}$, and let the jumps of $\int_0^t \int \gamma N_i^n(ds\, d\gamma)$ be the $\tau_{i,l}^{v,n}$. Modify the $N_i^n(\cdot)$ so that no jump is possible for a time gap $\tau_{i,l}^{v,n}/\sqrt{n}$ after its lth jump, $l = 1, 2 \ldots$. The modification will not affect the weak convergence. Let $\tau_i^n(t)$ denote the scaled time since the start of any vacation of queue i that is active at scaled time t. Then the effects of the vacation can be approximated by the jump process (note that $(WL^n(\cdot), u^n(\cdot))$

is continuous)

$$J^n(t) = \sum_i \int_0^{t-\tau_i^n(t)} \int q_i(WL^n(s), \gamma, u^n(s)) N_i^n(ds\, d\gamma) + \epsilon^{v,n}(t), \quad (7.4)$$

where the error term $\epsilon^{v,n}(\cdot)$ will now be defined. There are two sources of error in the integrals in (7.4). The first is that it puts the total effect of the vacation all at its end, and does not count the effects of incompleted vacations. The second is due to the fact that the effect of the vacation is approximated by its fluid limit (via the functions $q_i(\cdot)$). The term $\epsilon^{v,n}(\cdot)$ compensates for this latter error, and it converges weakly to the "zero" process. The first type of error will not affect the results. It simplifies the problem by introducing a Poisson measure, and eliminating the need to work on each intervacation segment separately. Using the approximation (7.4), write

$$WL^n(t) = WL^n(0) + b^n t + w^n(t) + J^n(t) + z^n(t) + \epsilon^{v,n}(t) + \epsilon^n(t), \quad (7.5)$$

where $\epsilon^n(\cdot)$ is a residual-time error term.

Let $r^n(\cdot)$ denote the relaxed control representation of $u^n(\cdot)$. Let $N^n_{r^n,i}(\cdot)$ be the *relaxed Poisson measure* constructed from $r^n(\cdot)$ and $N_i^n(\cdot)$ as in Section 9. The term $N^n_{r^n,i}(\cdot)$ is a counting-measure-valued process where, for Borel sets A_0 and U_0, $N^n_{r^n,i}(t, A_0, U_0) \equiv N^n_{r^n,i}([0,t] \times A_0 \times U_0)$ is the number of jumps of $N_i^n(\cdot)$ in A_0 on $[0,t]$, when $u^n(s) \in U_0$ at the jump times s. Rewrite the $J^n(\cdot)$ in (7.4) as

$$J^n(t) = \sum_i \int_0^{t-\tau_i^n(t)} \int \int_{\mathcal{U}} q_i(WL^n(s), \gamma, \alpha) N^n_{r^n,i}(ds\, d\gamma\, d\alpha) + \epsilon^{v,n}(t). \quad (7.6)$$

Then by the results in Section 9, $\{WL^n(\cdot), z^n(\cdot), r^n(\cdot), w^n(\cdot), N^n_{r^n,i}(\cdot), i = 1,2\}$ is tight, and any weak-sense limit satisfies

$$WL(t) = WL(0) + bt + w(t) + J(t) + z(t), \quad (7.7)$$

where

$$J(t) = \sum_i \int_0^t \int \int_{\mathcal{U}} q_i(WL(s-), \gamma, \alpha) N_{r,i}(ds\, d\gamma\, d\alpha), \quad (7.8)$$

where $N_{r,i}(\cdot)$, $i = 1, 2$, are relaxed Poisson measures associated with the limit relaxed control $r(\cdot)$. Let $\{\mathcal{F}_t, t \geq 0\}$ denote the filtration engendered by the weak-sense limit processes. Then $w(\cdot)$ is an $\mathcal{F}_t$-Wiener process. For each i and any Borel sets A_i and U_i, the compensated processes

$$M_{r,i}(t, A_i, U_i) = N_{r,i}(t, A_i, U_i) - \bar{\lambda}_i^s r(t, U_i) \Pi_i(A_i)$$

are $\mathcal{F}_t$-martingales and are orthogonal for disjoint $A_i \times U_i$. It follows from the weak convergence, (7.3), and Fatou's lemma, that

$$\liminf_n V_\beta^n(WL) \geq V_\beta(WL), \quad (7.10)$$

where (as in Section 9) $V_\beta(WL)$ is the infimum over the ordinary controls, which equals the infimum over the relaxed controls.

As usual, the next step is to prove

$$\limsup_n V_\beta^n(WL^n(0)) \leq V_\beta(WL(0)) \tag{7.11}$$

by the method used in (9.1.24) in Theorem 9.1.4. Given $\epsilon > 0$, there is an ϵ-optimal control for the system (9.12) and cost (9.15) of the type of the $u^\epsilon(\cdot)$ in (9.19). For processes with jumps, this is an analogue of what was used in Theorem 9.1.4. In the present context, we require that $u^n(t) \leq WL^n(t)$. But this only entails writing $\min\{u^n(t), WL^n(t)\}$ in lieu of $u^n(t)$ in the dynamics and cost rate.

There is a complication in the adaptation to the physical problem of the control rule defined by (9.19) when following the procedure of Theorem 9.1.4. The adaptation of (9.19) determines the conditional probability law of the *desired* levels of the piecewise constant control process $WL_1^n(k\delta) = u^n(k\delta)$. In the limit, any desired level can be attained instantaneously, except at the jump times. But this is not possible for the physical problem. When a vacation ends, $WL_1^n(\cdot)$ cannot be brought immediately to the new desired value. But the required (scaled) time that is required goes to zero as $n \to \infty$. Similarly, when the desired piecewise constant control level determined by (9.19) changes values, the new desired value of $WL_1^n(\cdot)$ cannot be attained immediately. But again, it can be realized after a time that goes to zero as $n \to \infty$, and this is enough for the proof of (7.11). Keep in mind that this adapted control is being used for theoretical purposes only, namely, to prove (7.11). ■

A sufficient condition for (7.3). The following inequalities are a sufficient condition for (7.3): For some $p \geq 0$,

$$\begin{aligned} \sup_n E|WL^n(0)|^2 &< \infty, \\ \sup_n E \sup_{s\leq t} |w_i^{\alpha,n}(S_i^{a,n}(s))|^2 &= O(t^p), \ \alpha = a, d, \ i = 1, 2, \end{aligned} \tag{7.12}$$

$$\sup_n \sum_j E \left| \sum_{k:\nu_{j,k}^n \leq t} \xi_{j,k}^{v,n} \right|^2 = O(t^p). \tag{7.13}$$

A sufficient condition for the second part of (7.12) under (A7.1) is given in Theorem 9.3.1. A sufficient condition for (7.13) is given in Theorem 7.2, under the following additional conditions.

A7.1. *For each n, the random variables $\left\{\Delta_{i,l}^{\alpha,n}, l < \infty\right\}$ are mutually independent and identically distributed for each $i = 1, 2$, $\alpha = a, d$, and the absolute third moments are uniformly bounded. There are positive $\bar\Delta_i^{\alpha,n}$ and $\bar\Delta_i^\alpha$ such that $E\Delta_{i,l}^{a,n} = \bar\Delta_i^{\alpha,n} \to \bar\Delta_i^\alpha$.*

A7.2. *The second moments of $\tau_{i,l}^{v,n}$ are uniformly bounded.*

Theorem 7.2. *Assume* (A5.0)–(A5.2), (A7.1), *and* (A7.2). *Then* (7.13) *holds.*

Proof. To bound (7.13), start with the expression

$$E\left|\xi_{j,k}^{v,n}\right|^2 \leq \sum_i \frac{1}{n} E \left| \sum_{l=nS_i^{a,n}(\nu_{j,k}^n)+1}^{nS_i^{a,n}(\nu_{j,k}^n+\tau_{j,k}^{v,n}/\sqrt{n})} \Delta_{i,l}^{d,n} \right|^2 . \tag{7.14}$$

Writing $\Delta_{i,l}^{d,n} = [\Delta_{i,l}^{d,n} - \bar{\Delta}_i^{d,n}] + \bar{\Delta}_i^{d,n}$ in (7.14), we obtain an upper bound as twice the sum of (the K_i are constants)

$$K_0 \sum_i E \left| w_i^{d,n}(\nu_{j,k}^n + \tau_{j,k}^{v,n}/\sqrt{n}) - w_i^{d,n}(\nu_{j,k}^n) \right|^2 \tag{7.15}$$

and

$$\sum_i \frac{[\bar{\Delta}_i^{d,n}]^2}{n} E \left[\#\text{arrivals at queue } i \text{ in real time } \left(n\nu_{j,k}^n, n\nu_{j,k}^n + \sqrt{n}\tau_{j,k}^{v,n} \right] \right]^2 . \tag{7.16}$$

By a proof like that of Theorem 9.3.1, (7.15) is bounded by

$$K_1 E \tau_{j,k}^{v,n}/\sqrt{n} = O(1/\sqrt{n}).$$

Following the idea in Theorem 9.3.1 that was used to bound $E[N^n(t)]^2$ there yields that (7.16) is bounded by

$$K_2 E \left[\tau_{j,k}^{v,n} \right]^2 + K_2 E \left[\text{a residual-time term}\right]^2 /n + K_2.$$

To complete the proof, we need to average over the number of completed vacations on real time $[0, nt]$. We do this by ignoring the vacation durations in computing the distribution of the number of vacations on any real time interval $[0, nt]$, which gives an upper bound. Then the number has a Poisson distribution for each n, with the rate (in real time) parameter being bounded by $(1/n)\sup_{n,i} \bar{\lambda}_i^{s,n}$. If there are L vacations on $[0, nt]$, then (7.13) is bounded by L^2 times the bound on the mean square value of each jump. Finally, using the dominating Poisson distribution, average over L to get (7.13) for $p = 2$. ■

The Bellman equation for the limit system. Let $\mathcal{L}$ denote the differential generator of the pure diffusion part of (7.7). The jump part of the differential operator, acting on a bounded and measurable real-valued function $f(\cdot)$, is

$$\sum_i \bar{\lambda}_i^s E \left[f(WL + \xi_i^v) - f(WL) \right], \tag{7.17}$$

where ξ_i^v is the jump due to a vacation of source i, and E denotes the expectation of the jump given the *WL* and the control just before the start of the jump. See Section 3.3. Define the function

$$H(V_\beta, WL) = \min_{u \le WL} \left\{ k(WL, u) + \sum_i \bar{\lambda}_i^s E\left[V_\beta(WL + \xi_i^v) - V_\beta(WL)\right] \right\}. \tag{7.18}$$

The formal Bellman equation is the partial-differential-integral equation

$$\mathcal{L}V_\beta(WL) - \beta V_\beta(WL) + H(V_\beta, WL) = 0, \tag{7.19}$$

with the boundary condition $V_{\beta,w}(0) = 0$.

11.8 Extensions and Comments

In special cases, the weak convergence results and the form of the limit problem suggest nearly optimal strategies for the physical problem, without much additional analysis. Two cases of current interest will be discussed.

11.8.1 Minimizing the Total Expected Workload.

Suppose that the cost rates satisfy $k_i(WL_i) = WL_i$. Then $k(WL, u) = WL$, and the control problem is the minimization of the expectation of the integral of the discounted total workload. A reasonable policy is the function $u(WL(t))$, which minimizes the mean jump

$$Q(WL, u) = \bar{\lambda}_1^s E\left[\rho_1 \tau_1^v - [WL - u]\right]^+ + \bar{\lambda}_2^s E\left[\rho_2 \tau_2^v - u\right]^+. \tag{8.1}$$

Example: Exponentially distributed vacation intervals. Suppose that τ_i^v is exponentially distributed with mean $1/v_i$. Note that for any real number y and any random variable τ that is exponentially distributed with mean $1/w$, we have

$$E(\tau - y)^+ = w \int_y^\infty e^{-wx}(x - y)dx = w \int_0^\infty e^{-w(y+z)} z dz = \frac{e^{-wy}}{w}.$$

Define, for $i = 1, 2$, $j = 1, 2$, $j \ne i$,

$$w_i = \frac{v_i}{1 - \bar{\lambda}_j^a / \bar{\lambda}_j^d}.$$

Then (8.1) can be written as

$$Q(WL, u) = \frac{\bar{\lambda}_1^s e^{-w_1(WL - u)}}{w_1} + \frac{\bar{\lambda}_2^s e^{-w_2 u}}{w_2}.$$

Thus, for each value of WL, $Q(WL, u)$ is convex with respect to u, and its minimum is obtained at the value of u for which $dQ(WL, u)/du = 0$, provided that this solution satisfies $u \in [0, WL]$. If it does not, then the minimum over $u \in [0, WL]$ is obtained on one of the boundaries. Differentiating with respect to u yields

$$\bar{\lambda}_1^s e^{-w_1(WL-u)} - \bar{\lambda}_2^s e^{-w_2 u} = 0,$$

with solution

$$u(WL) = \frac{\log(\bar{\lambda}_2^s/\bar{\lambda}_1^s)}{w_1 + w_2} + \frac{w_1}{w_1 + w_2} WL.$$

The policy $u(\cdot)$ for the limit problem is then given by

$$u^*(WL) = \left[\min\left(WL, \frac{\log(\bar{\lambda}_2^s/\bar{\lambda}_1^s)}{w_1 + w_2} + \frac{w_1}{w_1 + w_2} WL\right)\right]^+. \tag{8.2}$$

In the special case where the two sources have the same rates, it is obvious that $u(WL) = WL/2$. Numerical analysis shows that the general form of the policy defined by (8.2) holds for linear $k_i(\cdot)$ (for exponentially distributed intervals). For small values of workload, we keep one of the queues as close to zero as possible, with the choice depending on the rates and durations. Beyond some threshold, the switching point is affine in the workload. The policy (8.2) is nearly optimal, and the numerically obtained optimum is very close for Erlang distributed vacation intervals as well.

The control functions. Figures 8.1 to 8.3 illustrate various possibilities (obtained numerically) for the optimal control $u(WL)$ for different cost rates. Figure 8.1 illustrates two possibilities of the form where the cost function is linear, where the optimal $u(\cdot)$ has the form in (8.2). The duration of the vacations is exponentially distributed with rates $\bar{\lambda}_i^v$, $i = 1, 2$. For the first case in Figure 8.1, $\bar{\lambda}_2^s$ is much greater than $\bar{\lambda}_1^s$, and conversely for the second case. The control curve $c(WL) = (u(WL), WL - u(WL))$ is the optimal vector of desired queue lengths.

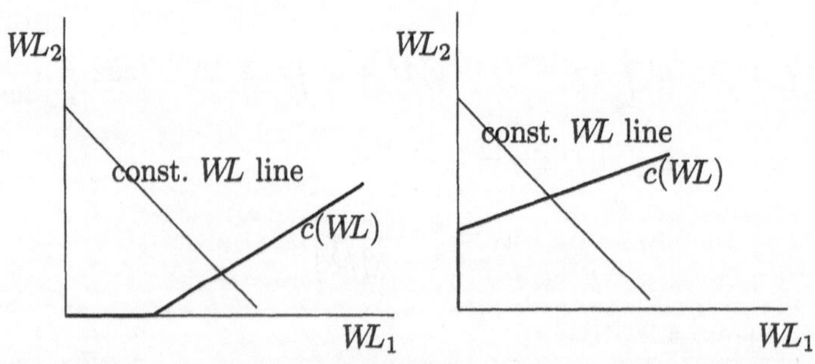

Figure 8.1. Optimal $c(WL) = (u(WL), WL - u(WL))$ with a linear cost rate function.

The motivation for the cost rate functions which are used for the cases of Figures 8.2 and 8.3 is that small sizes of queue 2 are not important, but we wish to penalize large sizes. In both cases, $\bar{\lambda}_1^s = .1$, $\bar{\lambda}_2^s = .3, \bar{\lambda}_1^v = .6$, $\bar{\lambda}_2^v = .4$.

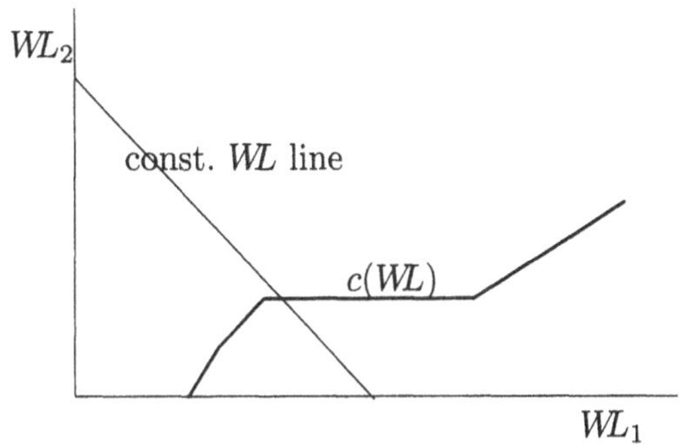

Figure 8.2. Optimal $c(WL) = (u(WL), WL - u(WL))$, cost rate function $WL_1 + [WL_2 - 2]I_{\{WL_2>2\}}$.

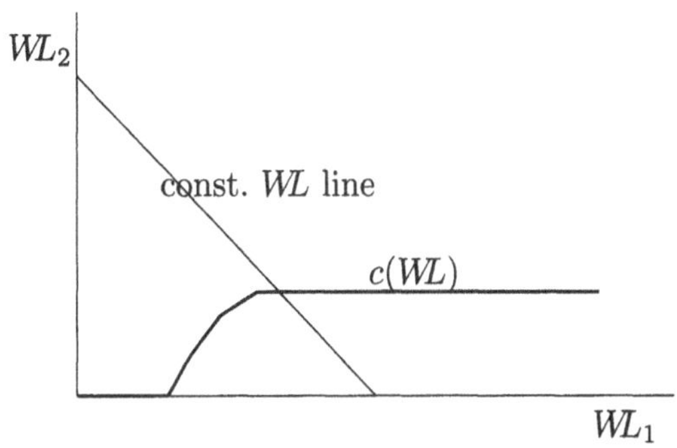

Figure 8.3. Optimal $c(WL) = (u(WL), WL - u(WL))$. Cost rate function $WL_1 + .3[WL_2 - 2]^2 I_{\{WL_2>2\}}$.

The form of the switching functions in Figures 8.2 and 8.3 are not a priori obvious. For small values of WL, $WL - u(WL) = WL_2 = 0$ even though the cost rate for WL_2 is zero there. The control function anticipates the effects of the possible vacations, the rate and duration of which are higher for queue 2.

11.8.2 No Vacations: The Asymptotic Optimality of the $c\mu$-rule

Suppose that there are no vacations, and the basic desired cost rate is $\bar{c}_1 x_1^n + \bar{c}_2 x_2^n$, where $\bar{c}_i > 0$. Write the limit form of the cost rate in terms of the workload as

$$\bar{\lambda}_1^d \bar{c}_1 u(WL) + \bar{\lambda}_2^d \bar{c}_2 [WL - u(WL)]. \tag{8.3}$$

The minimizer of (8.3) is just the $c\mu$-rule. Namely, poll source 1 if $\bar{\lambda}_1^d \bar{c}_1 > \bar{\lambda}_2^d \bar{c}_2$ and there are jobs there, and conversely if the reverse inequality holds. Under the conditions of Theorem 7.1, such a rule would be be asymptotically optimal for the physical system. In this case, the limit workload does not depend on the polling policy; only the cost rate does. This is an asymptotic form of the well-known $c\mu$-rule [243].

The $c\mu$-rule gives priority to one of the queues, and this might lead to unacceptably long waits in the nonpriority queue. This can be alleviated with a nonlinear weighting. For example, queue 1 might have a larger cost rate than queue 2 for moderate queue lengths. But to discourage the complete priority of queue 1, we might use nonlinear cost rates $\sum_i k_i(x_i)$, where the $k_i(\cdot)$ are continuous and monotonic, with at most linear growth. Then the optimum policy for the limit problem is to minimize

$$\sum_i k_i(x_i) = k_1 \left(\frac{u(WL)}{\bar{\Delta}_1^d} \right) + k_2 \left(\frac{WL - u(WL)}{\bar{\Delta}_2^d} \right). \tag{8.4}$$

Let $\bar{u}(\cdot)$ denote the minimizer, assumed continuous and nondecreasing. Then under the conditions of Theorem 7.1, the values defined by $x_1^n = \bar{u}(x_1^n \bar{\Delta}_1^{d,n} + x_2^n \bar{\Delta}_2^{d,n})/\bar{\Delta}_1^{d,n}$ would be asymptotically optimal for the physical system. See also [239].

11.9 Appendix: Relaxed Poisson Measures

When the jump term in a jump diffusion model is controlled, some new issues arise concerning the well-posedness of the problem and the proper formulation of the model to guarantee the existence and approximations of optimal controls. These issues will be addressed and resolved in this section.

11.9.1 A Difficulty with Controlled Jumps

Let $\{\mathcal{F}_t, t \geq 0\}$ be a filtration, $w(\cdot)$ a standard $\mathcal{F}_t$-Wiener process, and $N(\cdot)$ an $\mathcal{F}_t$-Poisson measure with jump rate $\lambda < \infty$ and jump distribution $\Pi(\cdot)$ with support Γ. Let the controls take values in the compact set $\mathcal{U}$. The controls $u(\cdot)$ will always be $\mathcal{F}_t$-predictable processes. The conditions (A3.5.1)–(A3.5.3) on the state space G and boundary reflection directions will be used. While we work with compact $\mathcal{U}$ for convenience, the compactness can be weakened. For example, let the control take values in the set $\mathcal{U}(x)$ when $x(t) = x$, where $\mathcal{U}(x)$ is continuous in x and compact for each x, a case of interest in Section 8. Then all of the results given below continue to hold.

A9.1. *$q(\cdot)$ is a continuous function on $G \times \Gamma \times \mathcal{U}$ and $b(\cdot)$ is continuous on $G \times \mathcal{U}$. There is $C < \infty$ such that for all α, $|q(x,\gamma,\alpha)| \leq C[1+|x|+|\gamma|]$, $|b(x,\alpha)| \leq C[1+|x|]$, and $\int_\Gamma |\gamma|^2 \Pi(d\gamma) < \infty$.*

A9.2. *$b(\cdot,\alpha)$ and $\sigma(\cdot)$ are Lipschitz continuous, uniformly in $\alpha \in \mathcal{U}$. There is $C < \infty$ such that for all α,*

$$\int_\Gamma |q(x,\gamma,\alpha) - q(y,\gamma,\alpha)|^2 \Pi(d\gamma) \leq C|x-y|^2.$$

Let $u(\cdot)$ be $\mathcal{F}_t$-predictable and $\mathcal{U}$-valued. Consider the controlled process defined by

$$\begin{aligned} dx(t) &= b(x(t),u(t))dt + \sigma(x(t))dw(t) \\ &\quad + \int_\Gamma q(x(t-),\gamma,u(t))N(dt\,d\gamma) + dz(t), \ x(t) \in G, \end{aligned} \tag{9.1}$$

where $z(t) = \sum_i d_i y_i(t)$ is the reflection term. The $y_i(\cdot)$ are nondecreasing and can increase only at t where $x(t) \in \partial G_i$ and $y_i(0) = 0$. Suppose that the jump does not take the state out of G. In Chapter 3, which did not consider the controlled jump case, the concept of relaxed control was introduced to deal with the fact that the limit of a sequence of processes $x^n(\cdot)$ with ordinary controls used was not necessarily representable as a process subject to an ordinary control. This is the so-called problem of "closure." An analogous extension is needed to handle the jump control problem. The extension and motivation will be developed in steps.

A motivating example. The following example will illustrate the underlying issue of "closure" and guide us to a reasonable solution. Let $\delta > 0$ and define a control $u^\delta(\cdot)$ for use in (9.1) as follows. It takes the two values α_i, $i = 1,2$. Divide time into intervals of length $\delta > 0$, and divide each of these into subintervals of lengths $v_1\delta, v_2\delta$, where $v_1 + v_2 = 1$. Use the control value α_1 on $[k\delta, k\delta + v_1\delta)$ and use α_2 on $[k\delta + v_1\delta, k\delta+\delta)$, $k = 0,1,\ldots$. Let $x^\delta(\cdot)$ denote the associated solution to (9.1). Let $I_i^\delta(s)$ denote the indicator function of the event that α_i is used at time s. Then the jump term in (9.1) takes the form

$$\begin{aligned} J^\delta(t) &= \int_0^t \int_\Gamma q(x^\delta(s-),\gamma,u^\delta(s))N(ds\,d\gamma) \\ &= \sum_i \int_0^t \int_\Gamma I_i^\delta(s) q(x^\delta(s-),\gamma,\alpha_i)N(ds\,d\gamma). \end{aligned} \tag{9.2}$$

Let $r^\delta(\cdot)$ denote the relaxed control representation of $u^\delta(\cdot)$. Then $r_t^\delta(\alpha_i) \equiv r_t^\delta(\{\alpha_i\}) = I_i^\delta(t)$. Let $\delta \to 0$. Then $r^\delta(\cdot)$ converges weakly to $r(\cdot)$, where $r_t(\alpha_i) = v_i$. By (A9.1) and (A9.2), for any integer l, the set (indexed by δ) of lth jumps is tight as is the set of the lth interjump sections of $x^\delta(\cdot)$.

Choose a subsequence on which all of the interjump sections and jumps converge weakly. Then (between jumps) the weak-sense limit of the chosen subsequence can be represented as the solution to the controlled reflected SDE

$$dx(t) = \int_{\mathcal{U}} b(x(t), \alpha) r_t(d\alpha) dt + \sigma(x(t)) dw(t). \tag{9.3}$$

The weak-sense limit of $J^\delta(\cdot)$ along the chosen subsequence can be expressed in the form

$$\sum_i \int_0^t \int_\Gamma q(x(s-), \gamma, \alpha_i) \bar{N}_i(ds\, d\gamma), \tag{9.4}$$

where $\bar{N}_i(\cdot)$, $i = 1, 2$, are mutually independent Poisson measures with jump distributions $\Pi(\cdot)$ and jump rates λv_i. The limit "triple" $(w(\cdot), \bar{N}_i(\cdot), i = 1, 2, r(\cdot))$ is admissible. Keep in mind that throughout the section, the $r(\cdot)$ and $u(\cdot)$ are functions of (ω, t) and are not necessarily representable as feedback controls.

It can be seen from the form (9.4) that the control value that affects the jump is the result of a *randomization,* and not an averaging such as is done on the *drift* term in (9.3). By randomization, we mean that given a jump at t, the value is chosen by selecting $q(x(t-), \gamma, \alpha_i)$ with probability $v_i / \sum_j v_j$, where γ is chosen independently with the distribution $\Pi(\cdot)$. This type of approximation and weak convergence analysis could be carried out for any number of values of the control. It can also be adapted to the case where the fractions of the intervals on which the different α_i are used are time-dependent. For example, let $r^\delta(\cdot)$ denote the relaxed control representation of an $\mathcal{F}_t$-predictable control process $u^\delta(\cdot)$ that takes only a finite number of values α_i, $i = 1, \ldots, p$. Let $x^\delta(\cdot)$ denote the associated solution and let $(x^\delta(\cdot), r^\delta(\cdot))$ converge weakly to $(x(\cdot), r(\cdot))$. Let $\{\mathcal{F}_t, t \geq 0\}$ denote the filtration that this limit process induces (perhaps augmented by a Wiener process and Poisson measure that are independent of the other processes). Then there are a standard $\mathcal{F}_t$-Wiener process $w(\cdot)$ and $\mathcal{F}_t$-adapted counting-measure-valued processes $\bar{N}_i(\cdot)$, $i \leq p$, such that (9.3) holds between jumps and the jumps are represented by (9.4), but where the former jump rate λv_i of $\bar{N}_i(\cdot)$ is replaced by the random and time varying (always $\mathcal{F}_t$-predictable) quantity $\lambda r_t(\alpha_i)$. Thus, the limit of the jump term can be represented in terms of a set of (extended) Poisson measures with jump rates depending on the derivative of the limit relaxed control. The $\bar{N}_i(\cdot)$ would not necessarily be independent, but the processes defined by

$$\begin{aligned} &\int_0^t \int_\Gamma I_i^\delta(s) q(x^\delta(s-), \gamma, \alpha_i) N(ds\, d\gamma) \\ &\qquad - \lambda \int_0^t \int_\Gamma q(x^\delta(s-), \gamma, \alpha_i) \Pi(d\gamma) r_s^\delta(\alpha_i) ds, \quad i = 1, \ldots, p, \end{aligned} \tag{9.5a}$$

are orthogonal martingales and converge weakly to the processes

$$\int_0^t \int_\Gamma q(x(s-),\gamma,\alpha_i)\bar{N}_i(ds\,d\gamma) - \lambda \int_0^t \int_\Gamma q(x(s-),\gamma,\alpha_i)\Pi(d\gamma)r_s(\alpha_i)ds, \tag{9.5b}$$

which are orthogonal $\mathcal{F}_t$-martingales.

There is an alternative representation of (9.5b) that is sometimes useful. Extend the Poisson measure as follows. Let $\Pi_0(\cdot)$ be Lebesgue measure on $[0,1]$, and let $\bar{N}(ds\,d\gamma\,d\gamma_0)$ denote the Poisson measure with jump rate λ and jump distribution $\Pi(d\gamma)\Pi_0(d\gamma_0)$. Define $\mu_0(t) = 0$ and $\mu_i(t) = \sum_{j=1}^i r_t(\alpha_j)$. Then the ith summand in (9.4) can be written in the form

$$\int_0^t \int_\Gamma \int_0^1 I_{\{\gamma_0 \in (\mu_{i-1}(t),\mu_i(t)]\}} q(x(s-),\gamma,\alpha_i)\bar{N}(ds\,d\gamma\,d\gamma_0) \tag{9.6}$$

in the sense that the distribution of the solution has not changed. The form of (9.6) emphasizes, again, that the actual realization of the jump value is determined by a randomization via the derivative of the relaxed control measure. The representations in (9.4) and (9.6) differ only in the realization of the randomization. The presence of the discontinuous indicator function in (9.6) does not affect the existence, uniqueness, or the approximation arguments, since it does not depend on the state. One could write the limit jump process as

$$\sum_i \int_0^t \int_\Gamma \int_0^1 I_{\{\gamma_0 \in (\mu_{i-1}(t),\mu_i(t)]\}} q(x(s-),\gamma,\alpha_i)\bar{N}_i(ds\,d\gamma\,d\gamma_0), \tag{9.7}$$

where the $\bar{N}_i(\cdot)$ are mutually independent and identically distributed. This representation in terms of a set of mutually independent Poisson measures works only if the control takes a finite or countable number of values.

Recapitulation. The above discussion suggests a generalization of the concept of Poisson measure that would allow the use of a continuum of control values within a well-defined framework. In preparation for this, let us summarize some of the essential points of the discussion that will be helpful for the generalization. Let $\{\mathcal{F}_t, t \geq 0\}, w(\cdot), N(\cdot)$ be as above. Let $u(\cdot)$ be an arbitrary $\mathcal{F}_t$-predictable control process with relaxed control representation $r(\cdot)$, and define the measure-valued process $N_r(ds\,d\gamma\,d\alpha)$ as follows. Let $\Gamma_0 \in \mathcal{B}(\Gamma)$ and $U_0 \in \mathcal{B}(\mathcal{U})$. Then define $N_r([0,t] \times \Gamma_0 \times U_0) \equiv N_r(t,\Gamma_0,U_0)$ to be the number of jumps of $\int_0^\cdot \int_\Gamma \gamma N(ds\,d\gamma)$ on $[0,t]$ with values in Γ_0, and where $u(s) \in U_0$ at the jump times s. The stochastic model can then be written as

$$\begin{aligned} dx(t) = &\int_{\mathcal{U}} b(x(t),\alpha)r_t(d\alpha)dt + \sigma(x(t))dw(t) \\ &+ \int_{\mathcal{U}} \int_\Gamma q(x(t-),\gamma,\alpha)N_r(dt\,d\gamma\,d\alpha) + dz(t), \quad x(t) \in G. \end{aligned} \tag{9.8}$$

The compensator of the counting-measure-valued process $N_r(\cdot)$ is the integral of

$$\lambda\Pi(d\gamma)r_t(d\alpha)dt \tag{9.9}$$

in the sense that the processes defined by

$$N_r(t, \Gamma_0, U_0) - \lambda\Pi(\Gamma_0)r(t, U_0)$$

are $\mathcal{F}_t$-martingales, and they are orthogonal for disjoint $\Gamma_0 \times U_0$. For a bounded and measurable real-valued function $\phi(\cdot)$, define

$$p(t) = \int_0^t \int_\Gamma \int_\mathcal{U} \phi(s, \gamma, \alpha) N_r(ds\, d\gamma\, d\alpha).$$

Then

$$p(t) - \int_0^t \int_\Gamma \int_\mathcal{U} \phi(s, \gamma, \alpha)\lambda\Pi(d\gamma)r_s(d\alpha)ds \tag{9.10}$$

is also an $\mathcal{F}_t$-martingale. Let $f(\cdot)$ be a bounded and continuous real-valued function. Then the compensator for $f(p(\cdot))$ is

$$A(t) = \int_0^t \int_\Gamma \int_\mathcal{U} \left[f(p(s) + \phi(s, \gamma, \alpha)) - f(p(s))\right] \lambda\Pi(d\gamma)r_s(d\alpha)ds$$

in the sense that $f(p(t)) - f(p(0)) = A(t)$ plus an $\mathcal{F}_t$-martingale.

The set of integrands can be extended. For example, the $\mathcal{F}_t$-martingale property of (9.10) continues to hold if any real-valued, bounded and $\mathcal{F}_t$-predictable process $\phi_0(\cdot)$ multiplies $\phi(\cdot)$. Recall that any left continuous and $\mathcal{F}_t$-adapted process is $\mathcal{F}_t$-predictable. Note that by its definition

$$r_t(U_0) = \lim_{\delta \to 0} \frac{r(t, U_0) - r(t - \delta, U_0)}{\delta}$$

as the limit of the sequence of predictable processes, $r_t(U_0)$ is $\mathcal{F}_t$-predictable for any $U_0 \in \mathcal{B}(\mathcal{U})$.

11.9.2 *The Relaxed Poisson Measure*

The relaxed Poisson measure. With the previous discussion, which started with the primitives $(w(\cdot), N(\cdot), r(\cdot))$, as a guide, we are now in a position to develop the needed extension of the Poisson measure, which is consistent with the motivating discussion and (9.1).

Now let us restart from the beginning. Let $\{\mathcal{F}_t, t \geq 0\}$ be a filtration and $w(\cdot)$ a standard $\mathcal{F}_t$-Wiener process and let $r(\cdot)$ be an admissible relaxed control. Let $N_r(\cdot)$ be a counting-measure-valued process with right-continuous paths and with the property that for any Borel sets Γ_0 and U_0, the processes

$$N_r(t, \Gamma_0, U_0) - \lambda\Pi(\Gamma_0)r(t, U_0) \tag{9.11}$$

are $\mathcal{F}_t$-martingales, and are orthogonal for disjoint $\Gamma_0 \times U_0$. (Note that we are not assuming that $N_r(\cdot)$ is derived from a Poisson measure as was done in the previous subsection.) This martingale property and the fact that $r_t(\cdot)$ is $\mathcal{F}_t$-predictable constitute the definition of *admissibility*. Such an $N_r(\cdot)$ will be called a *relaxed Poisson measure* or a relaxed $\mathcal{F}_t$-Poisson measure. Then $(w(\cdot), r(\cdot), N_r(\cdot))$ is said to be an *admissible triple* or, equivalently, $r(\cdot)$ is said to be admissible. The martingale property, the form of the compensator and the fact that $N_r(\cdot)$ is a counting-measure-valued process with right-continuous paths specify the distribution of $N_r(\cdot)$ uniquely. The appropriate weak topology is to be used on the space of measures, whatever the type.

Write the stochastic differential equation with controlled jumps in terms of the relaxed Poisson measure as

$$x(t) = x(0) + \int_0^t \int_{\mathcal{U}} b(x(s), \alpha) r_s(d\alpha) ds + \int_0^t \sigma(x(s)) dw(s) + J(t) + z(t), \tag{9.12}$$

where $x(t) \in G$ and

$$J(t) = \int_0^t \int_\Gamma \int_{\mathcal{U}} q(x(s-), \gamma, \alpha) N_r(ds\, d\gamma\, d\alpha). \tag{9.13}$$

Under (A9.1), (A9.2), and (A3.5.1)–(A3.5.3), for each admissible triple $(w(\cdot), r(\cdot), N_r(\cdot))$, Theorem 3.5.2 can be extended, via the Picard iteration method, to yield that there is a strong-sense-unique solution to (9.12) and (9.13).

Recall the motivating problem where the jumps were represented by either (9.2) or the first term of (9.5a), and define the counting-measure-valued process $N_{r^\delta}(ds\, d\gamma d\alpha)$ as above (9.8) (but with $r^\delta(\cdot)$ replacing the $r(\cdot)$ used there), and let $r(\cdot)$ be the weak-sense limit of $r^\delta(\cdot)$. Then $N_{r^\delta}(\cdot)$ converges weakly to a relaxed Poisson measure $N_r(\cdot)$ associated with the limit relaxed control $r(\cdot)$. Also, the process with values

$$\int_0^t \int_\Gamma \int_{\mathcal{U}} q(x^\delta(s-), \gamma, \alpha) N_{r^\delta}(ds\, d\gamma\, d\alpha)$$

converges weakly to the process with values

$$\int_0^t \int_\Gamma \int_{\mathcal{U}} q(x(s-), \gamma, \alpha) N_r(ds\, d\gamma\, d\alpha).$$

In general, if $r_t(\cdot)$ is concentrated on a finite number of points $\{\alpha_i, i \le p\}$, then the jump processes can be represented in terms of mutually independent and identically distributed Poisson measures as in (9.7) or with a single Poisson measure as in (9.6). The form of the representation does not change the distribution of the solution. Suppose that $r(\cdot)$ is the relaxed control representation of an ordinary (predictable, of course) control that is piecewise constant. Then there is an $\mathcal{F}_t$-Poisson measure $N(\cdot)$ such that (9.12) and (9.13) reduce to (9.1).

Implications of the martingale property. By the martingale property of (9.11), for bounded and measurable $\phi(\cdot)$,

$$\int_0^t \int_\Gamma \int_{\mathcal{U}} \phi(s,\gamma,\alpha) N_r(ds\, d\gamma\, d\alpha) - \lambda \int_0^t \int_\Gamma \int_{\mathcal{U}} \phi(s,\gamma,\alpha)\Pi(d\gamma) r_s(d\alpha) ds$$

is an $\mathcal{F}_t$-martingale. This implies that the conditional (on $\mathcal{F}_t$) probability of a jump in any interval $[t, t+\delta)$ is $\lambda\delta + o(\delta)$, and the probability of more than one jump is $o(\delta)$. If $\psi(\cdot)$ is an $\mathcal{F}_t$-adapted process with paths in $D(\mathbb{R}^r; 0, \infty)$, then the process defined by

$$\begin{aligned} &\int_0^t \int_\Gamma \int_{\mathcal{U}} q(\psi(s-),\gamma,\alpha) N_r(ds d\gamma d\alpha) \\ &\quad -\lambda \int_0^t \int_\Gamma \int_{\mathcal{U}} q(\psi(s-),\gamma,\alpha)\Pi(d\gamma) r_s(d\alpha) ds \end{aligned} \tag{9.14}$$

is an $\mathcal{F}_t$-martingale. If the lth jump occurs at τ, then the conditional (on $\mathcal{F}_{\tau-}$ and that the jump occurs at τ) distribution of the jump is the distribution of $q(\psi(\tau-),\gamma,\alpha)$, where γ is distributed as $\Pi(d\gamma)$ and α as $r_\tau(d\alpha)$, independently. Thus, once again, we see that the relaxed control plays the role of a randomization.

Next, let $\{\mathcal{F}_t^\epsilon, t \geq 0\}, w^\epsilon(\cdot), r^\epsilon(\cdot), N_{r^\epsilon}(\cdot)$, be a sequence of filtrations, standard $\mathcal{F}_t^\epsilon$-Wiener processes, admissible controls, and relaxed $\mathcal{F}_t^\epsilon$-Poisson measures. We have the following limit theorem.

Theorem 9.1. *Assume* (A9.1), (A9.2), *and* (A3.5.1)–(A3.5.3). *Suppose that the set of reflection directions at any edge or corner are linearly independent. Then the set*

$$\{x^\epsilon(\cdot), y^\epsilon(\cdot), w^\epsilon(\cdot), r^\epsilon(\cdot), N_{r^\epsilon}(\cdot)\}$$

is tight. The limit of any weakly convergent subsequence satisfies (9.12) *and* (9.13). *Let* $\{\mathcal{F}_t, t \geq 0\}$ *denote the filtration induced by the limit processes. Then* $w(\cdot)$ *is a standard* $\mathcal{F}_t$*-Wiener process,* $r(\cdot)$ *is admissible, and* $N_r(\cdot)$ *is a relaxed* $\mathcal{F}_t$*-Poisson measure with compensator process defined by* (9.9).

Comments on the proof. Only a few comments will be made. Since $\{EN_{r^\epsilon}(\cdot)\}$ is tight, by Theorem 2.7.1 the set $\{N_r^\epsilon(\cdot)\}$ is also tight. The set $\{r^\epsilon(\cdot), w^\epsilon(\cdot)\}$ is always tight. The tightness of $\{y^\epsilon(\cdot)\}$ follows from a modification of Theorem 3.6.1 that allows for jumps, and the fact that $z^\epsilon(\cdot)$ determines $y^\epsilon(\cdot)$ continuously and uniquely by the linear independence of the set of reflection directions at any edge or corner. Suppose that (abusing terminology) ϵ indexes a weakly convergent subsequence, with weak-sense limit denoted by $(x(\cdot), y(\cdot), w(\cdot), r(\cdot), N_r(\cdot))$ and let $\{\mathcal{F}_t, t \geq 0\}$ denote the filtration engendered by the limit process. The nonanticipativity, the Wiener and martingale properties, and the admissibility of $r(\cdot)$ follow by standard weak convergence arguments. Note, in particular, that $N_r(\cdot)$ is a relaxed $\mathcal{F}_t$-Poisson measure associated with $r(\cdot)$. By the properties of $q(\cdot)$ in (A9.1) and (A9.2), the sequence of processes defined by

$$J^\epsilon(t) = \int_0^t \int_\Gamma \int_{\mathcal{U}} q(x^\epsilon(s-), \gamma, \alpha) N_{r^\epsilon}(ds\, d\gamma\, d\alpha)$$

converges weakly to the $J(\cdot)$ in (9.13). Now piece the interjump limits and jump limits together to get (9.12) and (9.13).

Existence of an optimal control. Let $\beta > 0$, $c_i \geq 0$, and let $k(\cdot)$ be continuous and real-valued and satisfy $|k(x, \alpha)| \leq C[1 + |x|]$ for some $C < \infty$. We will work with the discounted cost function

$$W_\beta(x, u) = E_x^u \int_0^\infty \int_{\mathcal{U}} e^{-\beta t} \left[k(x(t), \alpha) r_t(d\alpha) dt + c' dy(t)\right]. \tag{9.15}$$

Define $V_\beta(x) = \inf_r W(x, r)$, where the infimum is over the relaxed admissible controls and the system is (9.12) and (9.13).

The weak convergence and the fact that the $r^\epsilon(\cdot)$ can be chosen to be ϵ-optimal controls in Theorem 9.1 imply the following theorem. Without additional restrictions, it is possible that $x(t)$ will grow fast enough as $t \to \infty$ to outweigh the discounting $e^{-\beta t}$. This is the reason for the boundedness and growth rate conditions in the theorem.

Theorem 9.2. *Assume* (A9.1), (A9.2), *and* (A3.5.1)–(A3.5.3). *Suppose that the set of reflection directions at any edge or corner are linearly independent. Let* $b(\cdot)$ *and* $\sigma(\cdot)$ *be bounded and suppose that there is a constant* C *such that* $|q(x, \gamma, \alpha)| \leq C(1 + |\gamma|)$. *Then there is an optimal control of the relaxed problem.*

Comment on the proof. Let $(x^\epsilon(\cdot), y^\epsilon(\cdot), r^\epsilon(\cdot), w^\epsilon(\cdot), N_{r^\epsilon}(\cdot))$ denote a minimizing sequence and, abusing terminology, let ϵ index a weakly convergent subsequence with limit $(x(\cdot), y(\cdot), r(\cdot), w(\cdot), N_r(\cdot))$. Then Theorem 3.5.3 can be extended to yield

$$\sup_\epsilon E\left|y^\epsilon(t)\right|^2 = O(t^2). \tag{9.16}$$

Then the bound on the growth rate of $k(\cdot)$ and the weak convergence yield

$$W_\beta(x, r^\epsilon) \to W_\beta(x, r) = \inf_r W_\beta(x, r) \equiv V_\beta(x), \tag{9.17}$$

which is the conclusion of the theorem.

Approximating the optimal relaxed control and the relaxed Poisson measure. The weak-sense uniqueness of the solution to (9.12), (9.13) implies that the jump terms and control can be approximated. Let $\rho > 0$ and divide $\mathcal{U}$ into a finite number of disjoint connected subsets U_i^ρ, $i \leq p_\rho$, with diameters less than ρ and let α_i^ρ be a point in U_i^ρ. Given an admissible relaxed control $r(\cdot)$, define $r^\rho(\cdot)$ by $r^\rho(t, \alpha_i^\rho) = r(t, U_i^\rho)$, $i \leq p_\rho$. Thus, $r(\cdot)$ is approximated by an admissible relaxed control with values in a finite set. The measure $N_{r^\rho}(\cdot)$ is now constructed from $N_r(\cdot)$ in the obvious way.

The following theorems, whose proofs are left to the reader, are used to extend the approximation results in [167, Subsection 9.1.2 and Section 10.3] and in Section 3.8 to the problem with controlled jumps.

Theorem 9.3. *Assume the conditions of Theorem 9.2. For $\rho > 0$, let*

$$(x^\rho(\cdot), y^\rho(\cdot), r^\rho(\cdot), w^\rho(\cdot), N_{r^\rho}(\cdot))$$

solving (9.12) and (9.13) be given, where $r^\rho(\cdot)$ is constructed as above the theorem. Then the set converges weakly to the solution of (9.12), (9.13), and $W_\beta(x, r^\rho) \to W_\beta(x, r)$.

Theorem 9.4. *Assume the conditions of Theorem 9.2, that $\mathcal{U}$ has only finitely many points $\{\alpha_i,\ i \leq p\}$, and let $\Delta > 0$. Let $(w(\cdot), N(\cdot), r(\cdot))$ be admissible with respect to a filtration $\{\mathcal{F}_t, t \geq 0\}$. Define the piecewise constant control $u^\Delta(\cdot)$ as follows. For $l = 1, \ldots$ and $i \leq p$, define $\tau_i^\Delta(l) = r(l\Delta, \alpha_i) - r(l\Delta - \Delta, \alpha_i)$ and divide each interval $[l\Delta, l\Delta + \Delta)$ into subintervals of lengths $\tau_1^\Delta(l), \ldots, \tau_p^\Delta(l)$. Then use the control values α_i, $i \leq p$, on the subintervals successively. Let $r^\Delta(\cdot)$ denote the relaxed control representation of $u^\Delta(\cdot)$. Let $N_{r^\Delta}(\cdot)$ denote the associated relaxed Poisson measure that is obtained from $N(\cdot)$ and $r^\Delta(\cdot)$, and let $(x^\Delta(\cdot), y^\Delta(\cdot))$ denote the solution to (9.12) and (9.13). Then $(x^\Delta(\cdot), y^\Delta(\cdot), w(\cdot), r^\Delta(\cdot), N_{r^\Delta}(\cdot))$ converges weakly to $(x(\cdot), y(\cdot), w(\cdot), r(\cdot), N_r(\cdot))$, solving (9.12) and (9.13).*

Infimums over ordinary controls. Theorems 9.3 and 9.4 imply that the infimum of the costs over the ordinary admissible controls equals the infimum over the relaxed controls. (In fact, each ordinary control in the minimizing sequence need have only finitely many values.) Thus, the extension of the model via the introduction of the relaxed Poisson measure does not affect the infimum of the cost function.

Representation in terms of a standard Poisson measure. Let $u(\cdot)$ be admissible, piecewise constant, and take only finitely many values $\{\alpha_i$, $i \leq p\}$, as in Theorem 9.4. Then the jump term can be represented in terms of an $\mathcal{F}_t$-Poisson measure. Let $I_i(t)$ denote the indicator function of the event that $u(t) = \alpha_i$, which we can take to be a predictable process for each i. Then the jump term can be represented as

$$\sum_i \int_0^t \int_\Gamma I_i(s) q(x(s-), \gamma, \alpha_i) N(ds\, d\gamma), \tag{9.18}$$

for a standard Poisson measure with jump rate λ and jump distribution $\Pi(\cdot)$.

An analogue of Theorem 3.8.1. It follows from Theorem 9.4 that for each $\epsilon > 0$ there are a $\delta > 0$, a finite set of points $\mathcal{U}^\epsilon = \{\alpha_i,\ i \le p\}$, and an ϵ-optimal control $u^\epsilon(\cdot)$ for (9.12) and (9.15) that is constant on the intervals $[k\delta, k\delta + \delta)$ and is $\mathcal{U}^\epsilon$-valued. Use the representation (9.18) for the jumps and let $x^\epsilon(\cdot), w^\epsilon(\cdot)$ and $N^\epsilon(\cdot)$ denote the solution, driving Wiener process and Poisson measure, respectively. Then the control $u^\epsilon(\cdot)$ can be constructed from its conditional probability law as in (3.8.3). That is,

$$\begin{aligned} P\left\{u^\epsilon(k\delta) = \alpha \middle| x(0), u^\epsilon(i\delta), i < k,\ w^\epsilon(s), N^\epsilon(s), s \le k\delta\right\} \\ = q_k\left(\alpha; x(0), u^\epsilon(i\delta),\ i < k,\ \ w^\epsilon(p\theta), N^\epsilon(p\theta, \Gamma_j^q),\ j \le q,\ p\theta < k\delta\right), \end{aligned} \tag{9.19}$$

where $q_k(\cdot), \theta$ and Γ_j^q are as in Theorem 3.8.1.

12

Assignment and Scheduling: Many Classes and Processors

Chapter Outline

This chapter concerns the problem of scheduling service in multiclass and multiprocessor systems, and discusses key ideas in recent work in this area. The possibility of effective treatment of multiclass problems enlarges the scope of applications considerably. There will be no switchover time or cost or processor interruption. Sections 5.3 and 6.4 dealt with some specific multiclass systems, but where the priorities were fixed beforehand. Those examples well illustrated the considerable advantages of the workload formulation. The advantages of the workload formulation were also seen in Chapter 11, which dealt with scheduling problems for simpler systems, but where many of the basic issues (averaging theorems, switchover time, vacations) were different. A significant advantage of the workload formulation in Sections 5.3 and 6.4 was that the workload did not (at least asymptotically) depend on the priorities. It was well behaved and did not exhibit the rapid movement of the individual queues. In this sense the workload was an "invariant" of the system. The advantages of the workload formulation for multiclass systems will also be apparent for the classes of problems in this chapter.

Throughout this chapter, if a processor is working on a job from queue i, and a job with a higher priority for that processor arrives, then the processor can either finish the work or else stop it, and continue later. In the latter case, the time required for completion of the work is assumed to be just the residual time. Some terminology will be redefined. If a term is not redefined, than the definition is as in Chapters 5–7. The assignment is always done at the time that service is to begin.

In the examples of Section 6.4 or of Chapter 11 the natural definition of workload was essentially obvious. This will not usually be the case here, and a substantial part of the problem concerns the proper definition of workload so that it is a useful "system invariant" and can be used to determine the asymptotically optimal scheduling policies. The way that it is defined and the consequent advantages are among the most important recent developments in heavy traffic analysis for multiclass systems. For simplicity of exposition of the basic ideas, the development in this chapter starts with the simpler formulations. Section 1 is concerned with the model where there are many input job classes, each queued separately, and a single processor, that must allocate its time to the various queues. In this case, the definition of workload is just that of Sections 5.3 and 6.4, namely the total work in the system. In Chapters 8–11, the cost criteria of interest were expectations of pathwise costs. In this chapter, to illustrate a common alternative point of view, a pathwise optimization approach is taken. Thus we seek to find a scheduling or control policy that (asymptotically) minimizes in a pathwise sense an integral of a cost rate (that depends on the queue sizes) over an arbitrary time interval. The resulting scheduling rule is simple: Suppose that the current value of the workload is $WL^n(t) = WL$, and that one could instantaneously shift the work among the individual queues in any way at all provided only that the workload is not changed. Then there are functions $g_i(WL)$, $i \leq K$, that are the scaled sizes of the queues that would minimize the cost rate at workload level WL. An asymptotically optimal policy is to serve next the queue where $x_i^n(t) - g_i(WL^n(t))$ is most positive. The control policy is an asymptotic and nonlinear version of the $c\mu$-rule. With appropriate conditions on the driving processes, such pathwise results are easily converted to results on the convergence of, say, mean discounted cost criteria, as in Chapters 9–11.

If there is more than one processor, then some of the job classes might be servable by more than one of them, and the service time of a job will depend on the processor to which it is assigned. We seek a definition of workload that does not (asymptotically) depend on the control policy so that it is an (asymptotic) system invariant, but the natural definition is no longer a priori obvious. The issues concerning the multiclass and multiprocessor problem are introduced in Section 2 via a specific example of a two-class and two-processor system. A straightforward analysis allows us to pick out the unique linear combination of the individual queue sizes that is an (asymptotic) system invariant and is an appropriate definition of workload, and can be used to get an asymptotically optimal scheduling policy. Section 3 extends the above analysis to a large family of single-stage multiclass and multiprocessor problems. The same notion of workload as an (asymptotic) system invariant leads to a linear program that yields a natural definition, and an asymptotically optimal policy.

A large part of the difficulty in determining asymptotically optimal policies in the multiclass and multiprocessor case concerns avoiding idle time on all processors when the workload is not near zero. Properly dealing with this issue would require a diversion into the fine structure of "reserve" inventories, as done in [10, 93, 94, 255]. We allow nonlinear cost functions, and restrict consideration to those cases for which all of the optimal queue sizes are nonzero when the workload is nonzero. This greatly simplifies the problem of finding asymptotically optimal policies, and allows us to focus on the intuitively important ideas and not get diverted into excessive detail. Keep in mind that the suggested policies are only a guide and serve mainly to illustrate some general principles. Individual cases will certainly profit from refinement. The ideas are extended to a class of feedforward networks in Section 4. The entire area is under active development at this time. The notation in each section is selected to simplify the discussion in that section, and differs slightly from section to section.

12.1 Many Input Classes and a Single Processor

Consider a system with H input streams (or job classes) A_i^n, $i \le H$, with each queued separately. The control problem is the scheduling of the work at a single processor, and the assignment is made at the moment that processing is to begin. The processor works if there is work to do. Let $x_i^n(t)$ denote $1/\sqrt{n}$ times the number at queue i at real time nt, and let the interarrival and service intervals be $\Delta_{i,l}^{a,n}$ and $\Delta_{i,l}^{d,n}$, $l = 1, \ldots$, respectively, with centering constants $\bar{\Delta}_i^{a,n}$ and $\bar{\Delta}_i^{d,n}$, converging to $\bar{\Delta}_i^a = 1/\bar{\lambda}_i^a$ and $\bar{\Delta}_i^d = 1/\bar{\lambda}_i^d$, respectively, as in Section 6.1. Define $\rho_i = \bar{\lambda}_i^a/\bar{\lambda}_i^d$. Let $WL_i^n(t)$ denote the workload in queue i at real time nt. Unless otherwise noted, the terminology is that of Section 6.1.

Assumptions.

A1.1. (A6.1.0) *holds and* (A6.1.1) *holds for the* $w_i^{\alpha,n}(\cdot)$, $\alpha = a, d$, $i \le H$. *The set of initial conditions* $\{x^n(0)\}$ *is tight. The heavy traffic condition is that there is a real* b *such that*

$$\sqrt{n}\left[\sum_i \frac{\bar{\Delta}_i^{d,n}}{\bar{\Delta}_i^{a,n}} - 1\right] = b^n \to b. \tag{1.1}$$

The state equation is

$$x_i^n(t) = x_i^n(0) + A_i^n(t) - D_i^n(t), \tag{1.2}$$

where $D_i^n(t)$ has the usual decomposition

$$D_i^n(t) = w_i^{d,n}(S_i^{d,n}(t)) + \frac{1}{\sqrt{n}\bar{\Delta}_i^{d,n}} \sum_{l=1}^{nS_i^{d,n}(t)} \Delta_{i,l}^{d,n}. \tag{1.3}$$

The right-hand term in (1.3) can be represented as

$$\frac{\sqrt{n}}{\bar{\Delta}_i^{d,n}} T_i^{d,n}(S_i^{d,n}(t)). \tag{1.4}$$

The workload $WL_i^n(t)$ in queue i can be defined to be either the actual work in queue i or its asymptotic equivalent $\bar{\Delta}_i^d x_i^n(t)$. For specificity, in this section the *latter value* will be used. By Theorem 5.3.2, the sequence of workload processes $\{WL^n(\cdot)\}$ defined by $WL^n(t) = \sum_i WL_i^n(t)$ is tight, $S_i^{a,n}(\cdot)$ converges weakly to the process, with values $\bar{\lambda}_i^a t$, and $S_i^{d,n}(\cdot) - S_i^{a,n}(\cdot)$ converges weakly to the "zero" process. The weak-sense limit equation for the workload is

$$WL(t) = WL(0) + bt + w(t) + z(t), \tag{1.5}$$

where

$$w(t) = \sum_i \bar{\Delta}_i^d \left[w_i^a(\bar{\lambda}_i^a t) - w_i^d(\bar{\lambda}_i^a t) \right],$$

and the reflection term is the limit of the scaled idle time and can increase only at t where $WL(t) = 0$. The weak-sense limit is not affected by the scheduling policy.

A cost function. Let the cost rate $k(\cdot)$ be a nonnegative and continuous function. Define $\bar{k}(\cdot)$ by

$$\bar{k}(WL) = \min_{x \geq 0} \left\{ k(x) : \sum_i \bar{\Delta}_i^d x_i = WL,\ x_i \geq 0 \right\}. \tag{1.6}$$

Let $g(WL) = \{g_i(WL), i \leq H\}$ denote the minimizing value of x and assume the following:

A1.2. *$g(\cdot)$ is unique and continuous.*

Pathwise asymptotically optimal control. The point of view to optimality or almost optimality that will be taken differs slightly from that of Chapters 9–11. To follow the lines of argument of much current work on the scheduling problem [10, 93, 94, 96, 100, 255], a pathwise rather than an expected value approach will be used. The weak-sense limit of $WL^n(\cdot)$ does not depend on the scheduling policy. In this sense, $WL^n(\cdot)$ is an "invariant" or "asymptotic invariant" of the system. Suppose that $x^n(t)$ is arbitrarily close to $g(WL^n(t))$ on some time interval, as $n \to \infty$. Then the pathwise cost rate $k(x^n(t))$ is asymptotically *minimum* on that interval. In this sense, a control that approximates the minimizing values $g(WL^n(t))$ is asymptotically *pathwise optimal.* It differs from what was done in Chapters 9-11, where the cost function to be minimized was an *expectation* of a functional over the infinite interval. The latter problem is more difficult, since one must consider issues of uniform integrability as well as of weak convergence. The integrability difficulties are only technical, and with appropriate and reasonable conditions, of the types used in those chapters, similar results can be obtained for the problems of this chapter.

When an assignment is to be made, the solution is simply to assign from the queue i for which $x_i^n(t) - g_i(WL^n(t))$ is most positive. This policy will attain the desired state value $g(WL^n(\cdot))$ arbitrarily closely (as $n \to \infty$) after an arbitrarily small interval of time, and then track $g(WL^n(\cdot))$ arbitrarily well, all with an arbitrarily high probability on any bounded time interval. It will next be shown that one can essentially attain the minimizing point in a scaled time of order $O(1/\sqrt{n})$. The argument uses a local-fluid approximation analogous to what was done in Subsection 5.6.4 and Sections 7.3 and 11.1, and serves as an introduction to the development for the more complicated problems in subsequent sections. The argument is similar to that used in Section 11.1 and uses the facts that the workload changes "smoothly" and that $\sum_i \bar{\Delta}_i^d [x_i^n(t) - g_i(WL^n(t)] = 0$.

An approximating control problem in the local-fluid time scale. Consider the following problem. Start at time $t = 0$. We wish to drive all of the individual differences $\delta x_i^n(t) = g_i(WL^n(t)) - x_i^n(t)$, $i \le H$, to zero in the smallest possible time (asymptotically). The analysis of the vacation and polling problems in Subsection 5.6.4 and Theorem 11.1.1 suggests that this will take $O(1/\sqrt{n})$ units of (scaled) time. Thus, an analysis of the paths would be similar to what was done in those theorems, and we will work on the so-called local-fluid time scale that was used there. This scaling is defined by the "barred" process $\bar{x}^n(t) = x^n(t/\sqrt{n})$.

In this local-fluid time scaling, consider an interval $[0, \tau]$, with a total time τ_i being used for processing queue i, where $\sum_i \tau_i \le \tau$, and where the "distribution" of processing time among the queues can be done "continuously." Then asymptotically,

$$\bar{x}_i^n(\tau) = x_i^n(0) + \left[\bar{\lambda}_i^a \tau - \bar{\lambda}_i^d \tau_i\right].$$

Recall that by the definition of $g(\cdot)$, $\sum_i \bar{\Delta}_i^d \delta x_i^n(t) = 0$. With these facts in mind, proceed as follows. Let δx_i, $i \le H$, satisfy $\sum_i \bar{\Delta}_i^d \delta x_i = 0$. Then get the minimum τ and associated τ_i that solve

$$\bar{\lambda}_i^a \tau - \bar{\lambda}_i^d \tau_i = \delta x_i, \; \tau_i \ge 0, \; i \le H, \quad \sum_i \tau_i \le \tau. \tag{1.7}$$

This can be formalized as the linear program

$$\text{minimize } \tau, \quad \text{subject to (1.7).} \tag{1.8}$$

Alternatively, the constraints can be written as

$$\rho_i \tau - \tau_i = \bar{\Delta}_i^d \delta x_i,\ \tau_i \geq 0,\ i \leq H, \quad \sum_i \tau_i \leq \tau. \tag{1.9}$$

Since

$$\sum_i \bar{\Delta}_i^d \delta x_i = 0 \text{ and } \sum_i \rho_i = 1, \tag{1.10}$$

the constraint $\sum_i \tau_i \leq \tau$ is actually an equality and is redundant. It will be dropped. The solution is a "fluid" way of rearranging the work among the competing queues to ensure a desired (asymptotic) change in state of value δx_i, $i \leq H$, without changing the workload.

There is a feasible solution to (1.8), as can be seen from the dual program:

$$\text{maximize } \sum_i (\bar{\Delta}_i^d \delta x_i) \pi_i, \tag{1.11}$$

subject to the constraints (shown for the case $H = 3$)

$$\begin{bmatrix} \rho_1 & \rho_2 & \rho_3 \\ -1 & 0 & 0 \\ 0 & -1 & 0 \\ 0 & 0 & -1 \end{bmatrix} \begin{bmatrix} \pi_1 \\ \pi_2 \\ \pi_3 \end{bmatrix} \leq \begin{bmatrix} 1 \\ 0 \\ 0 \\ 0 \end{bmatrix}, \tag{1.12}$$

where the π_i are unsigned. In fact, (1.12) implies that $\pi_i \geq 0$ and $\sum_i \rho_i \pi_i \leq 1$.

Suppose that the largest positive $\bar{\Delta}_i^d \delta x_i / \rho_i$ is attained at $i = 1$. Then the maximum in (1.11) is attained at $\pi_1 = 1/\rho_1$, $\pi_i = 0$, $i > 1$, and the optimal value for the dual program is $\bar{\Delta}_1^d \delta x_1 / \rho_1$. This is also the minimal value of τ for the primal program. Then dividing (1.9) by ρ_i, we get

$$\frac{\bar{\Delta}_1^d \delta x_1}{\rho_1} - \frac{\bar{\Delta}_i^d \delta x_i}{\rho_i} = \frac{\tau_i}{\rho_i}. \tag{1.13}$$

Clearly, $\tau_1 = 0$, as expected. We just let queue 1 grow to its desired value without any service. Since $\bar{\Delta}_1^d \delta x_1 / \rho_1$ is the largest positive ratio, $\tau_i \geq 0$. If there are only two queues, then (1.10) and (1.13) yield

$$\tau_2 = \frac{\rho_2 \bar{\Delta}_1^d \delta x_1}{\rho_1} - \bar{\Delta}_2^d \delta x_2 = \frac{\rho_2 \bar{\Delta}_1^d \delta x_1}{\rho_1} + \bar{\Delta}_1^d \delta x_1 = \frac{\bar{\Delta}_1^d \delta x_1}{\rho_1} = \tau,$$

as expected.

The above policy for asymptotically achieving a change δx can be realized in many ways. One possibility is to divide time into intervals of maximal length T_0 (in the local-fluid time scale) and reschedule at the beginning of each interval. On each interval, next serve the nonempty queue i with the most positive discrepancy between the scheduled time τ_i and the actual time already allocated to it, stop at $\min\{\tau, T_0\}$, and then repeat, but do not serve an empty queue. A simpler alternative, which is asymptotically equivalent, is to next serve the queue with the most positive value of $x_i^n(t) - g_i(WL^n(t))$. The latter strategy will guarantee that $x_i^n(\cdot) - g_i(WL^n(\cdot))$, $i \leq H$, converges to the "zero" process on $[\epsilon_n, \infty)$, where ϵ_n is any sequence satisfying $\epsilon_n \to 0$ and $\sqrt{n}\epsilon_n \to \infty$.

Suppose that $k_i(x_i) = c_i x_i$, $c_i > 0$. Then the nearly optimal solution is to serve the nonempty queue with the largest value of $c_i \bar{\lambda}_i^d$. This is a form of the well-known $c\mu$-rule [243]. The solution for the nonlinear problem is a generalization of this rule, and was first pointed out in [239].

12.2 A One-Stage, Two-Class and Two-Processor Problem

The problem in the previous section was simple in that there was only a single processor, so that the definition of workload was obvious, and it was easy to guarantee that there would not be any idling if there was work to be done. When there is more than one processor, the appropriate definition of workload is not obvious. We would like to define the workload such that it is a "system invariant" in the sense that for any scheduling policy for which there is no idle time for any processor if there is any work to do in the system, the workload does not otherwise depend on the policy. Such a definition will allow us to define a minimal cost rate in terms of the workload, analogously to (1.6), and an asymptotically optimal scheduling policy can be based on attempting to attain this minimal cost rate at each time point, without regard for the future. Thus, one does not need to solve an optimization problem. This is the great simplification of the workload concept. In this section the ideas will be introduced for a special, yet informative, case. A more general system is dealt with in the next section.

Consider a system with two input classes A_i^n, $i = 1, 2$, each queued separately until served, and two processors P_i, $i = 1, 2$. Stream A_2^n can be served only by P_2, but A_1^n can be served by either processor. See Figure 2.1. This is the same system that was used in [93, 94] to illustrate the concept of "resource pooling," where some of the main issues and ideas were developed for the problem of this section as well as for the more general single-stage problem of Section 3. The workload will be defined as a "system invariant," which will be used to get an asymptotically pathwise optimal policy. The way that this "invariant" is obtained is a little indirect in comparison with the obvious definition of Section 1. Also, the attainment of the pathwise optimal policy is more subtle here owing to the connection of the two processors via their association with input stream 1.

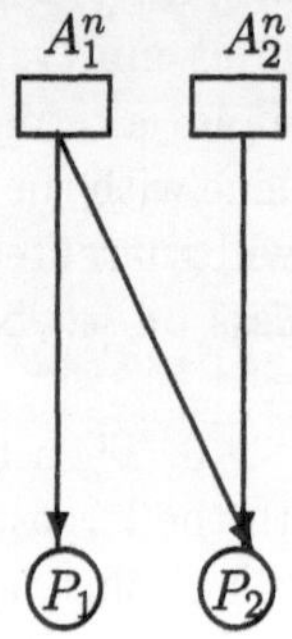

Figure 2.1. A two-processor example.

The interarrival and service intervals for A_2^n are denoted by $\Delta_{2,l}^{a,n}$, $\Delta_{22,l}^{d,n}$, $l = 1, \ldots,$ and $\bar{\Delta}_{22}^{d,n}$ denotes the centering constant for the $\Delta_{22,l}^{d,n}$. The interarrival intervals for A_1^n are $\Delta_{1,l}^{a,n}$, $l = 1, \ldots,$ but the service intervals depend on the processor to which the jobs are assigned. The service time for the lth processing of a member of A_1^n at P_j is denoted by $\Delta_{1j,l}^{d,n}$, with centering constants $\bar{\Delta}_{1j}^{d,n}$. Define the processes

$$w_{ij}^{d,n}(t) = \frac{1}{\sqrt{n}} \sum_{l=1}^{nt} \left[1 - \frac{\Delta_{ij,l}^{d,n}}{\bar{\Delta}_{ij}^{d,n}} \right].$$

The $w_i^{a,n}(\cdot)$ are defined as in Theorem 6.1.1. Define $\bar{\lambda}_i^{a,n} = 1/\bar{\Delta}_i^{a,n}$ and $\bar{\lambda}_{ij}^{d,n} = 1/\bar{\Delta}_{ij}^{d,n}$.

Assumptions.

A2.1. (A6.1.0) *holds,* $\{x^n(0)\}$ *is tight, and* (A6.1.1) *holds for the* $w_i^{a,n}(\cdot)$, $w_{ij}^{d,n}(\cdot)$, $\bar{\Delta}_i^{a,n}$, *and* $\bar{\Delta}_{ij}^{d,n}$. *Write* $\bar{\Delta}_i^a = 1/\bar{\lambda}_i^a$ *and* $\bar{\Delta}_{ij}^d = 1/\bar{\lambda}_{ij}^d$. *The heavy traffic condition is that there is a real* b *such that*

$$\sqrt{n} \left[\frac{\bar{\lambda}_1^{a,n} - \bar{\lambda}_{11}^{d,n}}{\bar{\lambda}_{12}^{d,n}} + \frac{\bar{\lambda}_2^{a,n}}{\bar{\lambda}_{22}^{d,n}} - 1 \right] = b^n \to b. \tag{2.1}$$

Also, $\sqrt{n}[\bar{\Delta}_i^{a,n} - \bar{\Delta}_i^a]$ *and* $\sqrt{n}[\bar{\Delta}_{ij}^{d,n} - \bar{\Delta}_{ij}^d]$ *are bounded.*

Remark on (2.1). To ensure that both processors play an important role in serving A_1^n, let $\bar{\lambda}_1^a - \bar{\lambda}_{11}^d > 0$. Then loosely speaking, the first term in the brackets in (2.1) is the scaled rate of arrival of work in A_1^n that must be done at P_2, and the second term is the (scaled) rate of arrival of work in A_2^n.

The system equations. Let $D_{ij}^n(t)$ denote $1/\sqrt{n}$ times the number of completed services from input stream A_i^n at P_j by real time nt, and write $S_{ij}^{d,n}(t) = D_{ij}^n(t)/\sqrt{n}$. Then the queue equations are

$$x_1^n(t) = x_1^n(0) + A_1^n(t) - D_{11}^n(t) - D_{12}^n(t),$$
$$x_2^n(t) = x_2^n(0) + A_2^n(t) - D_{22}^n(t).$$

The $A_i^n(\cdot)$ are represented in the usual way in terms of a "pre"-Wiener process and a drift term:

$$A_i^n(t) = \frac{1}{\sqrt{n}} \sum_{l=1}^{nS_i^{a,n}(t)} 1 = \frac{1}{\sqrt{n}} \sum_{l=1}^{nS_i^{a,n}(t)} \left[1 - \frac{\Delta_{i,l}^{a,n}}{\bar{\Delta}_i^{a,n}}\right] + \frac{1}{\sqrt{n}\bar{\Delta}_i^{a,n}} \sum_{l=1}^{nS_i^{a,n}(t)} \Delta_{i,l}^{a,n}.$$

As in Theorem 5.1.1, $S_i^{a,n}(\cdot)$ converges weakly to the process with values $\bar{\lambda}_i^a t$.

Representation of the service processes. Let $x_i^n(t)$ denote $1/\sqrt{n}$ times the number of A_i^n-jobs that are in the system at real time nt. The development is suggested by a scheme that was used in [173], in which the scheduling policy has two stages: First, there is a prior assignment of time on P_2 to the two queues, and then a reallocation of this time according to the evolving need. The resulting form suggests an appropriate definition of workload, which is the one pioneered in [93, 94]. The motivation behind the development and the definition of workload is that the workload should be a linear combination of the $x_i^n(t)$ that is a "system invariant" for reasonable assignment policies in that it will have (asymptotically) no dependence on the policy. The reallocation process is assumed to be admissible in that it depends only on the available data, which consists of the initial condition and the past history of arrivals, assignments, and service times. Let $p_{ji} > 0$ (with $\sum_j p_{ji} = 1$) be the fraction of time on P_i that is a priori assigned to A_j^n. The a priori allocation is just the fraction of time that is required to handle the mean load. Thus, $p_{12} = [\bar{\lambda}_1^a - \bar{\lambda}_{11}^d]/\bar{\lambda}_{12}^d$, $p_{22} = \bar{\lambda}_2^a/\bar{\lambda}_{22}^d$. Let $G_{ij}^n(t)$ denote $1/\sqrt{n}$ times the time *initially* allocated by P_2 to input stream i but reallocated to input stream $j \neq i$ on the scaled time interval $[0, t]$. Not all of the initially allocated or later reallocated time might be used. Let $T_{ij}^{d,n}(t)$ denote $1/\sqrt{n}$ times the *total* time that was allocated to input stream A_i^n on P_j but was not used due to an empty queue and could not be used elsewhere, by scaled time t. Thus, $T_{11}^n(\cdot)$ can increase only at t where $x_1^n(t) \leq 1/\sqrt{n}$, and $T_{12}^n(\cdot)$ and $T_{22}^n(\cdot)$ can increase only at t where both $x_2^n(t) = 0$ and $x_1^n(t) \leq 1\sqrt{n}$.

Write

$$D_{1i}^n(t) = \frac{1}{\sqrt{n}} \sum_{l=1}^{nS_{1i}^{d,n}(t)} \left[1 - \frac{\Delta_{1i,l}^{d,n}}{\bar{\Delta}_{1i}^{d,n}}\right] + \frac{1}{\sqrt{n}\bar{\Delta}_{1i}^{d,n}} \sum_{l=1}^{nS_{1i}^{d,n}(t)} \Delta_{1i,l}^{d,n}.$$

For $i = 1$, the right-hand term is (modulo a residual-time error term)

$$\frac{1}{\bar{\Delta}_{11}^{d,n}} \left[\sqrt{n}t - T_{11}^{d,n}(t)\right].$$

For $i = 2$, it can be written as (modulo a residual-time error term)

$$\frac{1}{\bar{\Delta}_{12}^{d,n}} \left[\sqrt{n}p_{12}t + G_{21}^n(t) - G_{12}^n(t) - T_{12}^{d,n}(t)\right].$$

Expand $D_{22}^n(t)$ as

$$D_{22}^n(t) =$$

$$\frac{1}{\sqrt{n}} \sum_{l=1}^{nS_{22}^{d,n}(t)} 1 = \frac{1}{\sqrt{n}} \sum_{l=1}^{nS_{22}^{d,n}(t)} \left[1 - \frac{\Delta_{22,l}^{d,n}}{\bar{\Delta}_{22}^{d,n}}\right] + \frac{1}{\sqrt{n}\bar{\Delta}_{22}^{d,n}} \sum_{l=1}^{nS_{22}^{d,n}(t)} \Delta_{22,l}^{d,n}.$$

The right-hand term is (modulo a residual-time error term)

$$\frac{1}{\bar{\Delta}_{22}^{d,n}} \left[\sqrt{n}p_{22}t + G_{12}^n(t) - G_{21}^n(t) - T_{22}^{d,n}(t)\right].$$

Using these representations, modulo residual-time error terms, the state equations can be written as

$$\begin{aligned} x_1^n(t) = x_1^n(0) &+ w_1^{a,n}(S_1^{a,n}(t)) - w_{11}^{d,n}(S_{11}^{d,n}(t)) - w_{12}^{d,n}(S_{12}^{d,n}(t)) \\ &+ \sqrt{n}t\left[\bar{\lambda}_1^{a,n} - \bar{\lambda}_{11}^{d,n} - p_{12}\bar{\lambda}_{12}^{d,n}\right] + \left[\bar{\lambda}_{11}^{d,n}T_{11}^{d,n}(t) + \bar{\lambda}_{12}^{d,n}T_{12}^{d,n}(t)\right] \\ &- \bar{\lambda}_{12}^{d,n}\left[G_{21}^n(t) - G_{12}^n(t)\right], \end{aligned} \tag{2.2a}$$

$$\begin{aligned} x_2^n(t) = x_2^n(0) &+ w_2^{a,n}(S_2^{a,n}(t)) - w_{22}^{d,n}(S_{22}^{d,n}(t)) + \sqrt{n}t\left[\bar{\lambda}_2^{a,n} - p_{22}\bar{\lambda}_{22}^{d,n}\right] \\ &+ \bar{\lambda}_{22}^{d,n}T_{22}^{d,n}(t) + \bar{\lambda}_{22}^{d,n}\left[G_{21}^n(t) - G_{12}^n(t)\right]. \end{aligned} \tag{2.2b}$$

Define

$$\begin{aligned} w^n(t) = \bar{\Delta}_{12}^{d,n} &\left[w_1^{a,n}(S_1^{a,n}(t)) - w_{11}^{d,n}(S_{11}^{d,n}(t)) - w_{12}^{d,n}(S_{12}^{d,n}(t))\right] \\ &+ \bar{\Delta}_{22}^{d,n}\left[w_2^{a,n}(S_2^{a,n}(t)) - w_{22}^{d,n}(S_{22}^{d,n}(t))\right]. \end{aligned} \tag{2.3}$$

The workload equation. Recall that the aim is to find a (positive) linear combination of the $x_i^n(\cdot)$, $i = 1, 2$, that does not (asymptotically, and for reasonable control policies) depend on either the initial allocations or the later reallocations of time and that will have a continuous weak-sense limit. The forms (2.2a) and (2.2b) suggest the definition

$$WL^n(t) = \Delta_{12}^{d,n} x_1^n(t) + \Delta_{22}^{d,n} x_2^n(t), \tag{2.4}$$

which is the one used in [94], and it satisfies (modulo a residual-time error term)

$$WL^n(t) = WL^n(0) + w^n(t) + b^n t + \left[T_{12}^{d,n}(t) + \bar{\Delta}_{12}^{d,n} \bar{\lambda}_{11}^{d,n} T_{11}^{d,n}(t)\right] + T_{22}^{d,n}(t). \tag{2.5}$$

The idle-time terms can increase (asymptotically) only when the state path is arbitrarily close to the boundary where $x_1 = 0$.

Remark. The workload $WL^n(t)$ defined by (2.4) is (asymptotically) the work for P_2 if it is to handle all of the jobs in the system at scaled time t. The definition is reasonable, since only P_2 needs to be scheduled.

The cost function. The desired cost rate is $k(x)$, a nonnegative and continuous function. Define $\bar{k}(\cdot)$ by

$$\bar{k}(WL) = \min_{x \geq 0} \left\{ k(x) : \bar{\Delta}_{12}^{d} x_1 + \bar{\Delta}_{22}^{d} x_2 = WL,\ x_i \geq 0 \right\}. \tag{2.6}$$

Let $g(WL) = \{g_i(WL),\ i = 1, 2\}$ denote the minimizing value of x.

Suppose for the moment that, asymptotically, the $T_{ij}^{d,n}(\cdot)$ in (2.5) can increase only at t where $WL^n(t)$ is arbitrarily close to zero. Then the resulting $WL^n(\cdot)$ can be considered to be an "invariant" of the system, as it was for the problem of Section 1, since no other terms in (2.5) are affected by the policy in the limit. Then under this temporary assumption on the $T_{ij}^{d,n}(\cdot)$, the solution to (2.5) yields (asymptotically) a lower bound for the linear combination in (2.4) over *all* control policies, whatever their idling times, and $\bar{k}(WL^n(\cdot))$ is asymptotically the smallest possible cost rate. Thus, the aim is to schedule so that the idle times can increase only when $WL^n(t)$ is arbitrarily close to zero *and* the asymptotically minimizing path $g(WL^n(\cdot))$ is followed as closely as possible. There is one more assumption:

A2.2. *The function $g(\cdot)$ is unique and continuous. Also,*

$$g_i(WL) > 0,\ i = 1, 2, \text{ when } WL > 0. \tag{2.7}$$

Remark on (A2.2). The references [10, 94, 255] use linear $k(\cdot)$ for which (2.7) would not hold, since then $g_i(WL) = 0$ for some i even if $WL > 0$. These references devoted a substantial amount of effort to the problem of avoiding idle time when some component $g_i(WL)$ of the minimizer has zero value, required the introduction of reserve or buffer inventories of the (unscaled) order of $\log n$, and utilized the theory of large deviations to justify the $\log n$ level. As such it was more sophisticated than the approach taken here, and dealt with significant issues, even if (as is usually the case) one does not know the distributions of the interarrival and service intervals well enough to have confidence in the large-deviations estimates that are used to get the "reserve" inventories. The condition (2.7), while more restrictive than one would like, simplifies the development, since it avoids the problem of possible idle time when the ideal $g(WL^n(t))$ is on the boundary of the state space but $WL^n(t) > 0$. It is used for convenience, to show that one can proceed in a reasonable and systematic way to fruitfully exploit the possibilities allowed by the workload concept. Usually, the appropriate form of the cost function is somewhat arbitrary. There are rarely clearly defined "cost rates," and one selects the cost function to balance various criteria of interest, such as relative priorities and queue sizes, actual inventory holding costs, and matters of "fairness." In this sense, one might wish to avoid giving some queues absolute priority, irrespective of the sizes of the others, as would happen when $k(\cdot)$ is linear.

Theorem 2.1. *Assume the conditions of this section and let $WL^n(0)$ converge weakly to $WL(0)$. Then for any assignment policy such that the $T_{ij}^{d,n}(\cdot)$ can increase only at t where $WL^n(t)$ is arbitrarily close to zero, $WL^n(\cdot)$ converges weakly to $WL(\cdot)$, defined by*

$$WL(t) = WL(0) + bt + w(t) + z(t), \tag{2.8}$$

where the reflection term $z(\cdot)$ is continuous and can increase only when $WL(t) = 0$, and

$$\begin{aligned} w(t) = \bar{\Delta}_{12}^d \left[w_1^a(\bar{\lambda}_1^a t) - w_{11}^d(\bar{\lambda}_{11}^d t) - w_{12}^d((\bar{\lambda}_1^a - \bar{\lambda}_{11}^d)t) \right] \\ + \bar{\Delta}_{22}^d \left[w_2^a(\bar{\lambda}_2^a t) - w_{22}^d(\bar{\lambda}_2^a t) \right]. \end{aligned} \tag{2.9}$$

Consider the policy where if possible P_2 serves the queue with the largest value of $x_i^n(t) - g_i(WL^n(t))$ when it selects the next job. It is asymptotically optimal, and the $T_{ij}^{d,n}(\cdot)$ can increase only at t where $WL^n(t)$ is arbitrarily close to zero. Let $0 < \epsilon_n \to 0$ such that $\sqrt{n}\epsilon_n \to \infty$. Then for any $\gamma > 0$ and $T < \infty$,

$$\lim_n P\left\{ \sup_{\epsilon_n \le t \le T} |x^n(t) - g(WL^n(t))| \le \gamma \right\} = 1 \tag{2.10}$$

and

$$\lim_n P\left\{ \sup_{\epsilon_n \le t \le T} \left|k(x^n(t)) - \bar{k}(WL^n(t))\right| \le \gamma \right\} = 1. \tag{2.11}$$

Comments on the proof of the first part of the theorem. The $w_{ij}^{d,n}(S_{ij}^{d,n}(\cdot))$ are always asymptotically continuous, no matter what the policy. Consider a policy such that the idle times can increase only when the workload is arbitrarily close to zero. Then (as in Theorem 5.1.1) the idle-time terms $T_{ij}^{d,n}(\cdot)$ are asymptotically continuous, as is $WL^n(\cdot)$. Consequently, for any $T < \infty$, $\sup_{t\le T}|x^n(t)|$ is bounded in probability, which implies that $S_{22}^{d,n}(\cdot)$ converges weakly to the process with values $\bar{\lambda}_2^a t$, and $S_{11}^{d,n}(\cdot) + S_{12}^{d,n}(\cdot)$ converges weakly to the process with values $\bar{\lambda}_1^a t$. Hence, the $S_{1i}^{d,n}(\cdot)$, $i = 1, 2$, are asymptotically continuous. Define

$$\mathcal{T}_{ij}^{d,n}(t) = \frac{1}{n}\sum_{l=1}^{nt} \Delta_{ij,l}^{d,n}.$$

Then

$$\mathcal{T}_{11}^{d,n}(S_{11}^{d,n}(t)) = t - \frac{T_{11}^{d,n}(t)}{\sqrt{n}},$$

$$\mathcal{T}_{12}^{d,n}(S_{12}^{d,n}(t)) + \mathcal{T}_{22}^{d,n}(S_{22}^{d,n}(t)) = t - \frac{T_{12}^{d,n}(t) + T_{22}^{d,n}(t)}{\sqrt{n}}.$$

The sequence $\mathcal{T}_{ij}^{d,n}(\cdot)$ converges weakly to the process with values $t/\bar{\lambda}_{ij}^d$. These equations, the convergence of $T_{ij}^{d,n}(\cdot)/\sqrt{n}$ to the "zero" process, and the heavy traffic condition (2.1) imply that $S_{12}^{d,n}(\cdot)$ (respectively, $S_{11}^{d,n}(\cdot)$) converges weakly to the process with values $(\bar{\lambda}_1^a - \bar{\lambda}_{11}^d)t$ (respectively, $\bar{\lambda}_{11}^d t$).

The second part of the theorem: The nearly optimal assignment. An important concern is that (possibly due to an "aggressive" policy where P_2 takes too many jobs from queue 1) P_1 might have some idle time when $WL^n(t)$ is not arbitrarily close (asymptotically) to zero. (In such a case, $WL^n(\cdot)$ would not converge weakly to the process $WL(\cdot)$ defined by (2.8).) If it is possible to avoid this situation, then any asymptotically optimal policy must avoid it. Processor P_1 will always work on queue 1 if there is more work to be done there. The simplest acceptable policy for P_2, which both avoids the idle time problem and tracks $g(WL^n(\cdot))$, is for P_2 to serve the queue with the most positive value of $x_i^n(t) - g_i(WL^n(t))$ when it selects the next job. Define $\delta x^n(t) = g(WL^n(t)) - x^n(t)$. Since by the definition of $g(\cdot)$, $\sum_i \bar{\Delta}_{i2}^d \delta x_i^n(t) = 0$, either $\delta x^n(t) = 0$ or one component is negative and the other positive.

As for the problem in Section 1, an approximation in the local-fluid time scale is useful to exhibit the local behavior around $t = 0$, where $x^n(0)$ would not equal $g(WL^n(0))$ in general. Recall that by (2.7), $g_i(WL) > 0$, $i = 1, 2$, if $WL > 0$. Set $x = x^n(0)$ and let δx be such that

$$\bar{\Delta}_{12}^d \delta x_1 + \bar{\Delta}_{22}^d \delta x_2 = 0. \tag{2.12}$$

Then find $p_{i2}(\delta x)$, $i = 1, 2$, and the largest $\alpha > 0$ such that $p_{i2}(\delta x) \geq 0$, $p_{12}(\delta x) + p_{22}(\delta x) = 1$ and

$$\begin{aligned} \alpha \delta x_1 &= \bar{\lambda}_1^a - \bar{\lambda}_{11}^d - \bar{\lambda}_{12}^d p_{12}(\delta x), \\ \alpha \delta x_2 &= \bar{\lambda}_2^a - \bar{\lambda}_{22}^d p_{22}(\delta x). \end{aligned} \tag{2.13}$$

By the heavy traffic condition and (2.12), there is a unique solution. Suppose, for example, that $\delta x_2 > 0$. Then $p_{22}(\delta x) = 0$ and $\alpha = \bar{\lambda}_2^a / \delta x_2$. Note that $p_{22}(0) = \bar{\lambda}_2^a / \bar{\lambda}_{22}^d$. In the local-fluid time scale, the solution takes us (asymptotically) in a straight-line path from x to $x + \delta x$ in $1/\alpha$ units of time. For large n, this path will not touch the boundary, unless $x = 0$. The policy of having P_2 take its next job from the queue with the largest value of $x_i^n(t) - g_i(WL^n(t))$ is asymptotically equivalent to following this path for a scaled time $1/[\sqrt{n}\alpha]$ and then repeating the procedure. Indeed, asymptotically, after scaled time $1/[\sqrt{n}\alpha]$, $x^n(t)$ is arbitrarily close to $g(WL^n(t))$.

The suggested policy would not work if $g_i(WL) = 0$ for some i and $WL > 0$ without some modification to avoid possible idle time when $WL^n(t)$ is not asymptotically arbitrarily close to zero.

12.3 Many Input Classes and Servers: One Stage

Now return to the the model of Section 1 where there are H queues, but let there be $K > 1$ processors. Because of the numerous possible connections between the various queues and processors, the efficient notational scheme of [96, 100] will be used. This is the current notation of choice for such problems. The queues are to be called *job classes*. Not all job classes can be handled by all processors. The possible service of a job class by a processor is to be called an *activity* or service activity. Thus, each activity is the *unique* connection between a particular job class and a particular processor. The number of physically feasible activities is denoted by J, and is at most HK. Let A_{kj} denote the indicator of the fact that processor k does activity j, and let C_{ij} be the indicator of the fact that activity j involves job class i. Define the matrices $A = \{A_{kj}; k \leq K, j \leq J\}$ and $C = \{C_{ij}\ i \leq H, j \leq J\}$. Each job class can be handled by at least one processor and each processor can handle at least one job class. By the definition of C_{ij}, $\sum_i C_{ij} = 1$. The cost rate functions $k_i(\cdot)$ are as in Section 1, and we continue to take the pathwise asymptotically optimal approach of the previous sections.

The power of the development in Sections 1 and 2 was due to the availability of the "asymptotically invariant" process $WL^n(\cdot)$, which allowed us to obtain an asymptotic lower bound to the cost rate. The fact that for reasonable control policies the chosen form of $WL^n(\cdot)$ converged weakly to a continuous diffusion with a reflection at the origin was essential, as was the fact that the weak-sense limit did not depend on the policy. In the case of Section 1, the correct definition of workload was obvious, since there was only one processor. The proper definition of workload was not hard to find for the case of Section 2, since there was only one linear combination of the queue lengths which did not contain either the original allocation or the reallocation of time to the various queues by P_2, and it could be obtained by inspection. The general approach of Section 2 will work for the problem of this section, but a useful definition of workload (as a linear combination of the queue lengths $x_i^n(\cdot)$) is not a priori obvious. The approach of [95, 96, 100, 255] provides an answer via an analysis in the local-fluid time scale. This analysis will provide the definition of a workload process with the desired properties, yield the associated heavy traffic condition and lead us to an asymptotically optimal assignment policy.

Notation. Owing to the use of "activity" to denote the connection of a queue to a processor, some of the notation for the *service intervals* will differ from that in the previous sections or in Section 6.4, where we wished to keep some "pictorial" similarity between the system and the notation. Redefine $D_j^n(t)$ to be $1/\sqrt{n}$ times the number of jobs serviced by *activity* j by real time nt. The symbol $\Delta_{j,l}^{d,n}$ (with centering constant $\bar{\Delta}_j^{d,n} = 1/\bar{\lambda}_j^{d,n}$) denotes the time required for the service of the lth job that is serviced by activity j. Using this definition, redefine

$$w_j^{d,n}(t) = \frac{1}{\sqrt{n}} \sum_{l=1}^{nt} \left[1 - \frac{\Delta_{j,l}^{d,n}}{\bar{\Delta}_j^{d,n}} \right].$$

The term $\mathcal{T}_j^{d,n}(t)$ will denote $1/n$ times the total service time required for the first nt completed jobs by activity j, and $S_j^{d,n}(t)$ will denote $1/n$ times the number of jobs that activity j has completed by nt. The following assumption will be used.

A3.1. (A6.1.0) *holds*, $\{x^n(0)\}$ *is tight, and* (A6.1.1) *holds for the* $w_i^{a,n}(\cdot)$, $w_j^{d,n}(\cdot)$, $\bar{\Delta}_i^{a,n}$, *and* $\bar{\Delta}_j^{d,n}$.

The system equations. The equations for the scaled queues are, for $i \le H$,

$$x_i^n(t) = A_i^n(t) - \sum_j C_{ij} D_j^n(t). \tag{3.1}$$

The right-hand side can be expanded as

$$\begin{aligned} &w_i^{a,n}(S_i^{a,n}(t)) + \frac{\sqrt{n}}{\bar{\Delta}_i^{a,n}} \mathcal{T}_i^{a,n}(S_i^{a,n}(t)) \\ &\quad - \sum_j C_{ij} w_j^{d,n}(S_j^{d,n}(t)) - \sum_j C_{ij} \frac{\sqrt{n}}{\bar{\Delta}_j^{d,n}} \mathcal{T}_j^{d,n}(S_j^{d,n}(t)). \end{aligned} \tag{3.2}$$

As in the previous sections, we make the allocation of time to the activities in two steps: First, there is an a priori allocation of time to activity j and then a reallocation when desired. With this in mind, let $n\beta_j t$ denote the time initially allocated to activity j by real time nt. Let $G_j^n(t)$ denote $1/\sqrt{n}$ times the time reallocated to activity j by real time nt. The $G_j^n(t)$ can take any sign. Let $T_j^n(t)$ denote $1/\sqrt{n}$ times the total time allocated to activity j by real time nt that is unused owing to there being no work in activity j and no way of reassigning it so that it can be used. It will be assumed below in (A3.2) that there is a unique $\{\beta_j\}$ that can handle the average load, analogously to the situation in Section 2.

Now modulo a residual-time error term,

$$\sqrt{n}T_j^{d,n}(S_j^{d,n}(t)) = \sqrt{n}\beta_j t + G_j^n(t) - T_j^{d,n}(t), \tag{3.3}$$

subject to the constraints

$$\sum_j A_{kj}G_j^n(t) = 0, \ \sum_j A_{kj}\beta_j = 1, \ k = 1, \ldots, K. \tag{3.4}$$

Note that

$$\sum_j A_{kj}T_j^{d,n}(t) \tag{3.5}$$

can increase only when it is not possible to reallocate time so that processor k has work to do. It will be seen that under appropriate conditions, there is an asymptotically optimal policy for which (3.5) can increase only when $WL^n(t)$ is arbitrarily close to zero, for large n and all k.

An approximation in the local-fluid time scale. In order to determine the appropriate a priori allocation of time and asymptotically optimal policy, it is useful to examine the processes in the local-fluid time scale. Define the "barred" processes $\bar{A}_i^n(t) = A_i^n(t/\sqrt{n})$, $\bar{D}_j^n(t) = D_j^n(t/\sqrt{n})$, and $\bar{x}_i^n(t) = x_i^n(t/\sqrt{n})$. Let $\bar{S}_j^{d,n}(t)$ denote $1/\sqrt{n}$ times the number of completed services by activity j by real time $\sqrt{n}t$, and define

$$\bar{T}_j^{d,n}(t) = \frac{1}{\sqrt{n}} \sum_{l=1}^{\sqrt{n}t} \Delta_{j,l}^{d,n}.$$

Then in the local-fluid time scale, the state equation is

$$\bar{x}_i^n(t) = \bar{A}_i^n(t) - \sum_j C_{ij}\bar{D}_j^n(t), \ i \le H,$$

and

$$\bar{D}_j^n(t) = w_j^{d,n}(\bar{S}_j^{d,n}(t)/\sqrt{n}) + \frac{1}{\bar{\Delta}_j^{d,n}} \bar{T}_j^{d,n}(\bar{S}_j^{d,n}(t)).$$

The nominal or a priori time assignment. Now return to the question of the a priori assignment of time. Suppose for the moment that there are

$\beta_j \geq 0$, $j \leq J$, such that the asymptotic fraction of (actually used) time that the associated processor assigns to activity j is β_j. Then

$$\bar{T}_j^{d,n}(\bar{S}_j^{d,n}(t)) \rightarrow \beta_j t.$$

Consequently, since $\bar{T}_j^{d,n}(\cdot)$ converges weakly to the process with values $\bar{\Delta}_j^d t$, $\bar{S}_j^{d,n}(\cdot)$ converges weakly to the process with values $\bar{\lambda}_j^d \beta_j t$, and this yields the limit system in the local -fluid time scale:

$$\dot{\bar{x}}_i(t) = \bar{\lambda}_i^a - \sum_j C_{ij} \bar{\lambda}_j^d \beta_j + \text{idle-time effects} \; , \; i \leq H, \tag{3.6a}$$

$$\sum_j A_{kj} \beta_j \leq 1, \quad k \leq K. \tag{3.7}$$

Define $R_{ij} = C_{ij} \bar{\lambda}_j^d$ and the matrix $R = \{R_{ij}; i, j\}$. In more compact notation, (3.6a) can be written as

$$\dot{\bar{x}}(t) = \bar{\lambda}^a - R\beta + \text{idle-time effects}, \tag{3.6b}$$

where $\bar{\lambda}^a = (\bar{\lambda}_i^a,\, i \leq H)$. The limit equation (3.6) and constraint (3.7) will be used to get the weights in the linear combinations that will yield the appropriate definition of workload, that is, the sought-for invariant.

The heavy traffic assumptions. Henceforth, we will use the following assumption.

A3.2. *There is a unique $\beta = (\beta_j,\, j \leq J)$, called $\bar{\beta}$, under which $\dot{\bar{x}} = 0$ or, equivalently, for which*

$$\bar{\lambda}^a = R\beta \tag{3.8}$$

and

$$\sum_j A_{kj} \beta_j = 1, \text{ for all } k. \tag{3.9}$$

If the allocation rates β_j, $j \leq J$, satisfy (3.8) and (3.9), then following [96, 100, 255], it is said that the fluid system is *balanced*. The condition (A3.2) implies that there is only one possible assignment of time, except for a fraction that goes to zero as $n \to \infty$, under which the local-fluid limit is constant. Equivalently, in the limit there is a unique fractional assignment of time under which the input rate equals the service rate for each queue (hence, the queues are balanced). The uniqueness condition would not hold, for example, if some pair of servers were completely interchangeable in that they could handle the same activities in the same way. (In that particular case, the two identical servers can be pooled and the same analysis used.) If (3.8) and (3.9) hold, then [96, 100, 255] say that the system is in *heavy traffic*. Let $\bar{e}$ be the vector all of whose components are unity. Then (3.9) can be written as $A\beta = \bar{e}$. The fact that $A\bar{\beta} = \bar{e}$ implies that the asymptotic fraction of time that the processors are idle under the balanced policy is zero, and that there is no idle time in the limit (all in the local-fluid time scale).

Defining heavy traffic via a linear program. It turns out that the system being in heavy traffic in the above sense is equivalent to the following [255, Lemma 3.3]: *There is a unique solution to the linear program*

$$\min\left\{\tau : R\beta = \bar{\lambda}^a,\ A\beta \le \tau\bar{e},\ \beta_j \ge 0\right\}, \tag{3.10}$$

and it is such that $\tau = 1$ *and* $A\beta = \bar{e}$. *In fact,* $\beta = \bar{\beta}$. Two more assumptions will be made.

A3.3. *The heavy traffic condition: The sequence*

$$\sqrt{n}\left[\frac{1}{\bar{\Delta}_i^{a,n}} - \sum_j \frac{C_{ij}\beta_j}{\bar{\Delta}_j^{d,n}}\right] = b_i^n \tag{3.11}$$

is bounded for each $i \le H$, *and there is a real* b *such that*

$$\sum_i \bar{v}_i^n b_i^n = b^n \to b, \tag{3.12}$$

where $\bar{v}^n = \{\bar{v}_i^n\}$ *is to be defined above* (3.15).

The basic definitions given above allow the possibility that the optimal solution is such that the queues and processors can each be divided into, say, $L > 1$ disjoint subgroups such that the balanced solution involves assigning only the processors in the lth subgroup to the queues in the lth subgroup. In such a case, connections between the queues in one group and a processor in another might be possible or not. To avoid dealing with a variety of special cases, let us make the final assumption that the division into such subgroups is not possible. This is phrased as follows.

A3.4. *In any basic solution of the linear program* (3.10), *there are exactly* $H + K - 1$ *components of* β *which are positive.*

Discussion of the linear program (3.10). The linear program (3.10) resembles that of a transportation or multicommodity flow problem, with link gains. Identify the queues as sources, and the processors as destinations. Each destination has the same storage capacity τ. Source i has $\bar{\lambda}_i^a$ units of supply. The activities are the feasible links, and an assignment of capacity β_j to link j actually means a transfer of $\beta_j\bar{\lambda}_j^d$ units. Then find the minimal storage capacity. Under our conditions there is a unique solution, the optimal $\tau = 1$, and each destination uses all of its capacity. Concepts from multicommodity flow problems are applied to the assignment problem in [126, 178], which were, in fact, the origin of many of the linear programming ideas used here, and contain many examples that provide a great deal of insight into the scheduling problem. Under (A3.2), the minimization in (3.10) is redundant, since there is only one feasible solution. Nevertheless, it will be seen that the theory of linear programming provides valuable insights as well as leading to a natural definition of workload.

There are $H + K$ constraints, at most $J = HK$ activities, and the constraints are linearly independent. By the theory of linear programming [105], any basic solution has $H + K$ basic activities. Since the variable τ is always in the basis, any basic solution contains at most $H + K - 1$ components of $\{\beta_j\}$. In fact, under (A3.4), these $H + K - 1$ components are positive. These basic activities can be viewed as the links connecting the queues to the processors to which the basic solution gives a nonzero flow. The set of such links is a tree. It is connected, and there are no loops. Furthermore, if the directions of the links are ignored, then there is a path connecting any two of the $H + K$ queues and processors.

The left-hand example in Figure 3.1 illustrates a model of the type dealt with in Section 2, where the system is balanced under the heavy traffic condition. The values of $\bar{\beta}_j$ for the links to P_2 are both 0.5. The right-hand case illustrates a three-queue and three-processor example, where the solution is balanced and the links not exhibited serve too slowly to be of use. The values of $\bar{\beta}_j$ for the links connecting Q_3 to P_1 and P_2 are both 0.5.

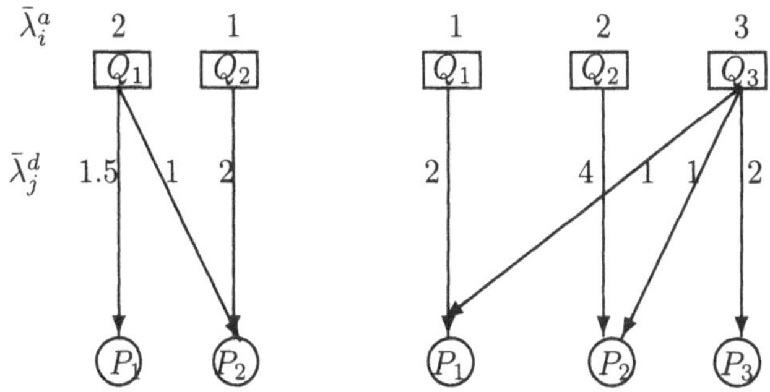

Figure 3.1. Two examples of networks and their basic activities.

The dual of the linear program. Now return to the state equation (3.1). The contribution of the β_j and $G_j^n(\cdot)$ to the state equation is embedded in the form of (3.2) and (3.3). Recall that we seek a (nonnegative) linear combination $v'x^n(\cdot)$ of the $x_i^n(\cdot)$ that does not contain the $\{\beta_j\}$ or $\{G_i^n(\cdot)\}$ for any $\{\beta_j\}$ or $\{G_i^n(\cdot)\}$ for which (3.4) holds. Such a linear combination would be a possible candidate for the workload process. More precisely, we would like to find a vector v such that for the limit problem,

$$v'R\psi = 0 \text{ for any } \psi \text{ that satisfies } A\psi = 0. \tag{3.13}$$

In expressions $R\psi, A\psi$, it is always assumed that $\psi_i = 0$ if i is a nonbasic coordinate. Equivalently, we would like to find the vectors (with nonnegative components) in the orthogonal complement $\mathcal{N}^\perp$ of the set

$$\mathcal{N} = \{u : u = R\psi \text{ for } \psi \text{ such that } A\psi = 0\}.$$

There is an elegant characterization of the orthogonal complement, due to [95, 96], in terms of the dual of the linear program (3.10), namely,

$$\max\left\{v'\bar{\lambda}^a : v'R \le z'A,\ z'\bar{e} \le 1,\ z_i \ge 0,\ v_i \text{ unsigned}\right\}. \tag{3.14}$$

The dual variables v and z correspond to the constraints $R\beta = \bar{\lambda}^a$ and $A\beta \le \tau\bar{e}$, respectively. The following result, specialized to our case, is proved in [95] (see also [255]).

Theorem 3.1. *Assume* (A3.2) *and* (A3.4). *Let* $(\bar{v}^i, \bar{z}^i)$, $i = 1, \dots m$, *be the (extreme point) optimal solutions to the dual program* (3.14). *Then* $\mathcal{N}^\perp$ *is the linear space spanned by* $\{\bar{v}^i,\ i \le m\}$. *In the case of the problem of this section, the solution is unique, is denoted by* $(\bar{v}, \bar{z})$, *and* $\bar{v}$ *has positive components. Thus, the orthogonal complement of* $\mathcal{N}$ *is one-dimensional.*

Remarks on the proof. It will be proved only that $\bar{v} \in \mathcal{N}^\perp$. Let $(\bar{v}, \bar{z})$ be a dual optimal solution. Then

$$\begin{aligned} \bar{v}'R - \bar{z}'A &\le 0, \\ \bar{z}'\bar{e} &= 1. \end{aligned}$$

Since the optimal τ equals unity, by dual optimality $\bar{v}'\bar{\lambda}^a = 1$. Write $\beta = \bar{\beta} + \psi$, where $\bar{\beta}_j = \psi_j = 0$ for the nonbasic components j. Recall that $\bar{\beta}_j > 0$ for the basic components and let $\beta_j \ge 0$. Then $[\bar{v}'R - \bar{z}'A]\,\beta \le 0$, or

$$\bar{v}'R(\bar{\beta} + \psi) - \bar{z}'A(\bar{\beta} + \psi) \le 0,$$

which, together with (3.10) and (A3.2), implies that

$$\bar{v}'\bar{\lambda}^a + \bar{v}'R\psi - \bar{z}'\bar{e} - \bar{z}'A\psi \le 0.$$

Since $\bar{v}'\bar{\lambda}^a = 1$ and $\bar{z}'\bar{e} = 1$, we have $\bar{v}'R\psi - \bar{z}'A\psi \le 0$. Thus, $\bar{v}'R\psi \le 0$ for all ψ such that $A\psi = 0$. Since for small enough ψ, $-\psi$ can replace ψ, we have $\bar{v}'R\psi = 0$. Thus, $\bar{v} \in \mathcal{N}^\perp$. The details of the converse, that any element of $\mathcal{N}^\perp$ is in the span of the dual optimal solution $\bar{v}$, is left to the reader.

The component $\bar{v}_i$ of the dual variable has the interpretation of the marginal rate of increase in τ when $\bar{\lambda}_i^a$ is increased.

The definition of workload. Define R^n and $\mathcal{N}^n$ as R and $\mathcal{N}$ were defined, but using the quantities $\bar{\lambda}_j^{d,n}$. Then, for large n, there are vectors $\bar{v}^n \to \bar{v}$ such that the orthogonal complement of $\mathcal{N}^n$ is the linear span of $\bar{v}^n$.

Now we are prepared to define the workload as

$$WL^n(t) = [\bar{v}^n]' x^n(t). \tag{3.15}$$

Finally, using (3.1)–(3.3) and the fact that $[\bar{v}^n] \in [\mathcal{N}^n]^\perp$ yields that (modulo a residual-time error term),

$$WL^n(t) = WL^n(0) + b^n t + \sum_i \bar{v}_i^n w_i^n(t) + \sum_{i,j} \frac{\bar{v}_i^n C_{ij}}{\bar{\Delta}_j^{d,n}} T_j^{d,n}(t), \tag{3.16}$$

where

$$w_i^n(t) = w_i^{a,n}(S_i^{a,n}(t)) - \sum_j C_{ij} w_j^{d,n}(S_j^{d,n}(t)).$$

The cost function and control problem. Let $k(x)$ denote the cost rate at x, where $k(\cdot)$ is as in the previous sections. Redefine $\bar{k}(\cdot)$ as

$$\bar{k}(WL) = \min_x \left\{ k(x) : \sum_i \bar{v}_i x_i = WL \right\}, \tag{3.17}$$

and let $g(WL)$ denote the minimizer.

An asymptotically optimal control. Assume (A2.2). It will be seen that there is a scheduling policy that is asymptotically optimal. It ensures that the idle time cannot increase (asymptotically) unless $WL^n(t)$ is arbitrarily close to zero and it tracks the minimizer $g(WL^n(\cdot))$. The policy to be described can certainly be improved in many ways. But it does capture the essential features of any acceptable method. The precise variation which would be used will depend on the details of the particular application. The aim is simply to show that there is some asymptotically optimal strategy.

Time is divided into scheduling intervals, with the time allocation policy being constant in each one. For each scheduling interval there are β_j, $j \leq J$, and an associated "ideal" allocation of time which is such that activity j "continuously" gets a fraction $A_{kj}\beta_j$ of processor k's time. The actual policy is that on each scheduling interval, each processor selects its next job from the class for which the difference between its ideal allocation and the actual allocation up to that time in the scheduling interval is most positive, with ties broken in any way at all. If the queue to be served is empty, then choose the queue for which the difference is next largest, etc. The β_j and the durations of the scheduling intervals are chosen to track the minimizer $g(WL^n(\cdot))$.

Under the policy to be described in the next few paragraphs, $|g(WL^n(t)) - x^n(t)|$ can be "large" only for $t \approx 0$. Loosely speaking, if $|g(WL^n(t)) - x^n(t)|$

is not very large at the beginning of a scheduling interval, then at the end of the interval, $x_i^n(\cdot)$ will take its minimizing value modulo minor errors. The first error is just a residual-time error term, the second error is the change in $w_i^n(\cdot)$ over the scheduling interval, and the third is due to whatever idle time there is. Asymptotically, there will be no increase in the idle-time terms $T_j^{d,n}(\cdot)$ during such a scheduling interval unless $WL^n(t)$ converges to zero at some time in the interval. Additionally, the first two types of errors become unimportant. If $|g(WL^n(0))-x^n(0)|$ is large, then several scheduling intervals might be needed to bring $x^n(t) \approx g(WL^n(t))$, after which time it stays arbitrarily close. Thus, the minimizer is tracked. A key point is that (A2.2) implies that (asymptotically) the desired paths keep the state away from the boundary on any interval where $WL^n(t) \geq \epsilon_1$ for any $\epsilon_1 > 0$.

We will next show how the allocations and durations of the scheduling intervals are determined. Choose the valuer $T_0 > 0$, that will be the maximum duration of a scheduling interval in the local-fluid time scale. For notational simplicity, shift time so that the scheduling interval of concern starts at time zero, and compute $\delta x = g(WL^n(0)) - x^n(0)$, the desired "instantaneous" change in the state. The strategy is to be determined by an analysis in the local-fluid time scale. Essentially, the solution will yield an allocation of time for which (as $n \to \infty$ and in the local-fluid time scale) $x^n(\cdot)$ moves arbitrarily close to the straight line connecting $x^n(0)$ and $g(WL^n(0))$. The analysis also establishes that the desired change can be realized arbitrarily well (asymptotically) in a finite time interval *in the local-fluid time scale.*

By the definition of $g(\cdot)$, $\bar{v}'\delta x = 0$. Consider the limit (in the local-fluid time scale) model. Thus we wish to find an interval τ and a vector $\beta = \{\beta_j\}$ (with $\beta_j = 0$ for the nonbasic coordinates j) such that

$$\begin{aligned} \delta x &= \bar{\lambda}^a \tau - R\beta\tau, \\ A\beta &= \bar{e}. \end{aligned} \tag{3.18}$$

Define $\delta\hat{x} = \delta x/\tau$, temporarily set $\tau = 1$ and replace (3.18) by

$$\begin{aligned} \delta\hat{x} &= \bar{\lambda}^a - R\beta, \\ A\beta &= \bar{e}. \end{aligned} \tag{3.19}$$

Write $\beta = \bar{\beta} - \psi$, where $\psi_i = 0$ for the nonbasic coordinates i, and solve

$$\begin{aligned} \delta\hat{x} &= R\psi, \\ A\psi &= 0. \end{aligned} \tag{3.20}$$

Since $\bar{v}'\delta\hat{x} = 0$, we have $\delta\hat{x} \in \mathcal{N}$. Thus, there is a solution ψ to (3.20).

Now find the largest $\alpha = \bar{\alpha} > 0$ such that $\bar{\beta} - \alpha\psi \geq 0$ (that is, all components are nonnegative). Then get τ such that

$$\delta x = \left[\bar{\lambda}^a - R(\bar{\beta} - \bar{\alpha}\psi)\right]\tau. \tag{3.21}$$

The above analysis yields a set of time allocation rates $\beta_j = \bar{\beta}_j - \bar{\alpha}\psi_j$, $j \leq J$, and a time τ. In the limit of the local-fluid scale limit, the desired transfer would take place in time τ, under the given allocation rates. Now use these allocation rates until time (in the local-fluid time scale) $\min\{\tau, T_0\}$ has passed. But do not serve an empty queue. Then repeat. One can use $\bar{v}^n, R^n$ in lieu of $\bar{v}, R$. If time cannot be divided and allocated "continuously," then let each P_j take its next job from the queue for which its current discrepancy in the scheduling interval between the ideal allocation and actual allocation of time is most positive, while ignoring empty queues. The above rule and (A2.2) ensure that, asymptotically, the $T_j^{d,n}(\cdot)$ can increase only at t where $WL^n(t)$ is arbitrarily close to zero. The formal details are close to what was done in Theorem 11.1.1. Some additional comments appear after the following theorem, which summarizes the discussion.

Theorem 3.2. *Assume* (A2.2) *and* (A3.1)–(A3.4). *Then under the policy described above, Theorem* 2.1 *holds with*

$$w(t) = \sum_i \bar{v}_i w_i(t) = \sum_i \bar{v}_i \left[w_i^a(\bar{\lambda}_i^a t) - \sum_j C_{ij} w_j^d(\bar{\lambda}_j^d \beta_j t) \right].$$

Comments on the proof. Restrict attention to the scaled time interval $[0, T]$, $T < \infty$. Suppose that for some $\epsilon_1 > 1$, $WL^n(t) \geq \epsilon_1$ for $t \leq T$. Let σ_ν^n denote the scaled time of the start of the νth scheduling interval and recall that $\sqrt{n}[\sigma_{\nu+1}^n - \sigma_\nu^n] \leq T_0$. In the local-fluid time scale in the νth scheduling interval, the evolution equation for the state can be written as

$$\begin{aligned} \bar{x}_i^{\nu,n}(t) = x_i^n(\sigma_\nu^n) &+ \frac{1}{\sqrt{n}} \sum_{l=nS_i^{a,n}(\sigma_\nu^n)+1}^{nS_i^{a,n}((\sigma_\nu^n + t/\sqrt{n})\wedge\sigma_{\nu+1}^n)} 1 \\ &- \sum_j C_{ij} \frac{1}{\sqrt{n}} \sum_{l=nS_j^{d,n}(\sigma_\nu^n)+1}^{nS_j^{d,n}((\sigma_\nu^n + t/\sqrt{n})\wedge\sigma_{\nu+1}^n)} 1. \end{aligned} \tag{3.22}$$

Define

$$\begin{aligned} \delta\bar{w}_i^{\nu,n}(t) &= w_i^{a,n}(S_i^{a,n}(\sigma_\nu^n + t/\sqrt{n})) - w_i^{a,n}(S_i^{a,n}(\sigma_\nu^n)) \\ &- \sum_j C_{ij} \left[w_j^{d,n}(S_j^{d,n}(\sigma_\nu^n + t/\sqrt{n})) - w_j^{d,n}(S_j^{d,n}(\sigma_\nu^n)) \right]. \end{aligned} \tag{3.23}$$

The coefficient of C_{ij} in (3.22) can be expanded into the coefficient of C_{ij} in (3.23) plus the term

$$\frac{1}{\sqrt{n}\bar{\Delta}_j^{d,n}} \sum_{l=nS_j^{d,n}(\sigma_\nu^n)+1}^{nS_j^{d,n}((\sigma_\nu^n + t/\sqrt{n})\wedge\sigma_{\nu+1}^n)} \Delta_{j,l}^{d,n}.$$

Let $\beta^\nu = \{\beta_j^\nu, j \leq J\}$ denote the selected value of β in the νth scheduling interval and let τ^ν be the associated value of τ in (3.21). By the scheduling rule, the last expression is (modulo a residual-time error term and neglecting idle time)

$$\beta_j^\nu [t \wedge (\sigma_{\nu+1}^n - \sigma_\nu^n)\sqrt{n}]/\bar{\Delta}_j^{d,n}. \tag{3.24}$$

Thus, for $t \leq \sqrt{n}[\sigma_{\nu+1}^n - \sigma_\nu^n] \leq T_0$, and ignoring possible idle time and modulo a residual-time error term,

$$\bar{x}_i^{\nu,n}(t) = x_i^n(\sigma_\nu^n) + \left[\frac{1}{\bar{\Delta}_i^{a,n}} - \sum_j C_{ij}\frac{\beta_j^\nu}{\bar{\Delta}_j^{d,n}}\right] t + \delta\bar{w}_i^{\nu,n}(t),$$

which we can approximate by

$$\bar{x}_i^{\nu,n}(t) = x_i^n(\sigma_\nu^n) + \left[\bar{\lambda}_i^a - \sum_j R_{ij}\beta_j^\nu\right] t + \delta\bar{w}_i^{\nu,n}(t). \tag{3.25}$$

For simplicity, suppose that $T_0 \geq \tau$ on the scaled time intervals of concern. The case where this does not hold is treated similarly. Then (3.25) implies that, neglecting idle time, $x^n(\sigma_{\nu+1}^n) = g(WL^n(\sigma_\nu^n))$ plus an error that goes to zero uniformly in the scheduling interval, with an arbitrarily high probability for large n. Similarly, in the local-fluid time scale, the paths connecting the beginning and the desired end points are arbitrarily close to the straight-line path that connects them, again uniformly in the scheduling interval, with an arbitrarily high probability for large n. These considerations, together with the fact that there is $\epsilon_2 > 0$ such that $g(WL^n(t)) \geq \epsilon_2$ for $t \leq T$, implies that there is no idle time in the limit for $t \leq T$. It follows that tracking is asymptotically arbitrarily close with an arbitrarily high probability.

Linear costs: Let $k(x) = \sum_i c_i x_i, c_i \geq 0$. Then all $g_i(WL)$ equal zero, except for the component or components for which $c_i/\bar{v}_i$ is smallest. Then the policy discussed above will not work, since the optimal point will be on the boundary of the state space even if $WL^n(t) > 0$. One then needs a policy that (asymptotically) avoids idle time when $WL^n(t) > 0$, and this is discussed in detail in [10, 94, 255]. The controls are similar when none of the $x_i^n(t)$ are close to zero, but with modifications otherwise.

12.4 Feedforward Networks

The general scheme and notation that were developed in the last section make sense for feedforward networks, provided that analogous assumptions are made. Let q_{ij}^n denote the (conditioned on the data to the time of routing) probability that a job of class i becomes a job of class j on completing service. To facilitate the development of the extensions and main ideas, we concentrate on a special class of systems and cost functions. It is assumed that there are K processing stations with each having only one processor. Each processor has a unique set of job classes assigned to it, and each job class can be served at only one station. Let $k(i)$ denote the unique processor that is responsible for the class i jobs. For this model, the job class i is identified uniquely with the activity i. Thus, $C_{ij} = 1$ if $i = j$, and it is zero otherwise. Also, $H = J$. Following the time allocation scheme that was used in the previous section, let β_i denote the a priori fraction of its time that processor $k(i)$ allocates to class i. Let $G_i(t)$ denote $1/\sqrt{n}$ times the time that is reallocated to job class i by scaled time t (it can be positive or negative) and let $T_i^n(t)$ denote $1/\sqrt{n}$ times the allocated time (by scaled time t) that is unused and not reallocated. The time might be unused either because the queue is empty and there is no way of reallocating, or because we have deliberately *chosen* not to use it. The multistage network problem differs from the single-stage network problem of the last section in that here the actions of the earlier stages affect the content of the queues in the latter stages. If it is much more expensive to store jobs in some downstream queue than in an upstream queue that eventually feeds it, it might be better not to serve the upstream queue when the downstream queue has sufficient work in it.

Assumptions.

A4.1. *$\{x^n(0)\}$ is tight. The set of all service and interarrival times are mutually independent and independent of the initial condition. Let $\Delta_{i,l}^{\alpha,n}$, $\alpha = a, d$, denote the exogenous interarrival (if applicable) and service times, respectively, for the lth member of job class i. The centering variables $\bar{\Delta}_i^{\alpha,n} = 1/\bar{\lambda}_i^{d,n}$ converge to $\bar{\Delta}_i^{\alpha} = 1/\bar{\lambda}_i^{\alpha}$, $i \leq H$, $\alpha = a, d$, and*

$$\left\{ \left| \frac{\Delta_{i,l}^{\alpha,n}}{\bar{\Delta}_i^{\alpha,n}} \right|^2 ; \ \alpha = a, d; n, l; \ i \leq J \right\}$$

is uniformly integrable. There are $\sigma_{\alpha,i}^2$, $\alpha = a, d$, $i \leq J$, such that

$$\lim_n E\left[1 - \Delta_{i,l}^{\alpha,n}/\bar{\Delta}_i^{\alpha,n}\right]^2 = \sigma_{\alpha,i}^2,$$

where the convergence is uniform in l.

The system equations. The definition (3.15) of the workload is an asymptotic "invariant" of the system in Section 3. The procedure that led to it will now be repeated for the current problem and will lead to an appropriate definition of workload and policy. The first step is the representation of the state equation in a form that suggests the linear combinations of the state components that should be used for the workload definition. The state equation is

$$x_i^n(t) = x_i^n(0) + A_i^n(t) - D_i^n(t) + \sum_j D_{ji}^n(t), \tag{4.1}$$

where $D_{ji}^n(t)$ denotes $1/\sqrt{n}$ times the number of completed class j jobs that have become class i jobs by scaled time t. Let $I_{ij,l}^n$ denote the indicator function of the event that the lth completed class i job joins the queue for class j. Then q_{ij}^n is the conditional probability (given the past data) that $I_{ij,l}^n = 1$. Let $S_i^{d,n}(\cdot)$ denote $1/n$ times the number of class i jobs that have been served by real time nt. Recall the definition (6.1.11):

$$w_{ij}^{r,n}(t) = \frac{1}{\sqrt{n}} \sum_{l=1}^{nS_i^{d,n}(t)} \left[I_{ij,l}^n - q_{ij}^n \right].$$

As usual, expand

$$D_i^n(t) = \frac{1}{\sqrt{n}} \sum_{l=1}^{nS_i^{d,n}(t)} \left[1 - \frac{\Delta_{i,l}^{d,n}}{\bar{\Delta}_i^{d,n}} \right] + \frac{1}{\sqrt{n}} \sum_{l=1}^{nS_i^{d,n}(t)} \frac{\Delta_{i,l}^{d,n}}{\bar{\Delta}_i^{d,n}}. \tag{4.2}$$

The right-hand term is, modulo a residual-time error term,

$$\frac{1}{\bar{\Delta}_i^{d,n}} \left[\sqrt{n}\beta_i t + G_i^n(t) - T_i^n(t) \right].$$

Then modulo a residual-time error term, (4.2) can be written as

$$w_i^{d,n}(S_i^{d,n}(t)) + \sum_v \frac{C_{iv}}{\bar{\Delta}_v^{d,n}} \left[\sqrt{n}\beta_v t + G_v^n(t) - T_v^n(t) \right].$$

Expand

$$D_{ji}^n(t) = \frac{1}{\sqrt{n}} \sum_{l=1}^{nS_j^{d,n}(t)} I_{ji,l}^n = w_{ji}^{r,n}(S_j^{d,n}(t)) + q_{ji}^n D_j^n(t).$$

Then

$$\begin{aligned} D_{ji}^n(t) &= w_{ji}^{r,n}(S_j^{d,n}(t)) + q_{ji}^n w_j^{d,n}(S_j^{d,n}(t)) \\ &\quad + \frac{q_{ji}^n}{\bar{\Delta}_j^{d,n}} \left[\sqrt{n}\beta_j t + G_j^n(t) - T_j^n(t) \right]. \end{aligned}$$

With the definition $R^n_{ij} = C_{ij}/\bar{\Delta}^{d,n}_i = C_{ij}\bar{\lambda}^{d,n}_i$, we can write

$$\begin{aligned}\sum_j D^n_{ji}(t) &= \sum_j w^{r,n}_{ji}(S^{d,n}_j(t)) + \sum_j q^n_{ji} w^{d,n}_j(S^{d,n}_j(t)) \\ &\quad + \sum_{j,v} q^n_{ji} R^n_{jv}\left[\sqrt{n}\beta_v t + G^n_v(t) - T^n_v(t)\right].\end{aligned}$$

Define

$$\begin{aligned}w^n_i(t) &= w^{a,n}_i(S^{a,n}_i(t)) - w^{d,n}_i(S^{d,n}_i(t)) \\ &\quad + \sum_j w^{r,n}_{ji}(S^{d,n}_j(t)) + \sum_j q^n_{ji} w^{d,n}_j(S^{d,n}_j(t)).\end{aligned}$$

Then for $i \le H$ and modulo asymptotically negligible terms, the state equation can be written as

$$\begin{aligned}x^n_i(t) &= x^n_i(0) + w^n_i(t) + \sqrt{n}\bar{\lambda}^{a,n}_i t - \sum_v R^n_{iv}\left[\sqrt{n}\beta_v t + G^n_v(t) - T^n_v(t)\right] \\ &\quad + \sum_{j,v} q^n_{ji} R^n_{jv}\left[\sqrt{n}\beta_v t + G^n_v(t) - T^n_v(t)\right].\end{aligned} \tag{4.3a}$$

Define the matrices $R^n = \{R^n_{ij};\ i,j \le H\}$ and $Q_n = \{q^n_{ij};\ i,j \le H\}$. Set $w^n(\cdot) = \{w^n_i(\cdot),\ i \le H\}$, and define the vectors $\bar{\lambda}^{a,n}$, β, $T^n(\cdot)$, and $G^n(\cdot)$ analogously. Then a compact form of (4.3a) is

$$x^n(t) = x^n(0) + w^n(t) + \sqrt{n}\bar{\lambda}^{a,n}t - [I - Q'_n]\, R^n\left(\sqrt{n}\beta t + G^n(t) - T^n(t)\right). \tag{4.3b}$$

The local-fluid time scale. Dropping the idle-time and G-terms for the moment, the limit state equation in the local-fluid time scale is

$$\begin{aligned}\dot{x}(t) &= \bar{\lambda}^a - [I - Q']\, R\beta, \\ A\beta &\le \bar{e}.\end{aligned} \tag{4.4}$$

As in Section 3, if there is a β such that $\bar{\lambda}^a = [I - Q']\, R\beta$, then we say that the system is *balanced*. The next assumption ensures that each job class gets a nonzero proportion of its processor's time. If it holds, then we say that the system is in *heavy traffic*.

A4.2. *There is a unique vector β, called $\bar{\beta}$, that balances the system and for which all $\bar{\beta}_i > 0$ and $A\bar{\beta} = \bar{e}$.*

The heavy traffic condition. Define

$$\Lambda^n_i = \sqrt{n}\left[\bar{\lambda}^{a,n}_i - \bar{\lambda}^{d,n}_i\bar{\beta}_i + \sum_j q^n_{ji}\bar{\lambda}^{d,n}_j\bar{\beta}_j\right]$$

and $\Lambda^n = \{\Lambda_i^n, i \le H\}$. For the $\bar{v}_Q^{i,n}$ defined above (4.7), define $\left[\bar{v}_Q^{i,n}\right]' \Lambda^n = b_i^n$.

A4.3. *For each $i \le H$, the sequence $\{\Lambda_i^n\}$ is bounded. There are real b_i such that $\lim_n b_i^n = b_i$.*

The second part of (A4.3) is needed only if we want an equation for the limit $WL(\cdot)$. Define $b = \{b_i\}$.

The workload equations and linear program. Following the idea in Sections 2 and 3, but for the system (4.3), to get the asymptotic system invariants or workload processes, we start by seeking all vectors v_Q such that

$$v_Q' [I - Q'] R\psi = 0 \text{ for } \psi \text{ such that } A\psi = 0. \tag{4.5}$$

Define $\bar{\lambda}_Q^a = [I - Q']^{-1}\bar{\lambda}^a = \{\bar{\lambda}_{Q,i}^a, i \le H\}$. The quantity $\bar{\lambda}_{Q,i}^a$ can be considered to be the effective (asymptotic) scaled exogenous arrival rate of jobs that eventually become class i jobs, when the network structure is taken into account. Similarly, the quantity $\bar{\lambda}_{Q,i}^a/\bar{\lambda}_i^d$ can be considered to be the effective asymptotic exogenous arrival rate of class i work. By comparing (4.4) with (3.6), we see that the linear program (3.10) is replaced by

$$\min\left\{\tau : R\beta = \bar{\lambda}_Q^a, \, A\beta \le \tau\bar{e}, \, \beta_i \ge 0\right\}. \tag{4.6}$$

There are $H + K$ constraints in (4.6). But $K - 1$ of the constraints are linearly dependent on the others, hence redundant. Since τ is always a basic variable, any basic solution to the linear program must contain H of the components of β. Since β is H-dimensional, all components of β are in the basis for our model, as expected. By (A4.2), the system is balanced and there is a unique solution $\beta = \bar{\beta}$, $\tau = 1$, $\bar{\beta}_i > 0$, to the linear program. With $\bar{\lambda}_Q^a$ replacing $\bar{\lambda}^a$, Theorem 3.1 still holds, except that there are K linearly independent solutions to the dual program, a consequence of the fact that $K - 1$ of the constraint equations are redundant.

Let $(\bar{v}^k, \bar{z}^k)$, $k \le K$, denote the (extreme point) optimal solutions to the dual program, and define $\bar{v}_Q^i = [I - Q]^{-1}\bar{v}^i = \{\bar{v}_{Q,j}^i, j \le H\}$. Define $\mathcal{N}$ as above (3.14) and $\mathcal{N}^n$ as above (3.15). Then for large n, there are vectors $\bar{v}^{i,n} \to \bar{v}^i$, $i \le K$, such that $[\mathcal{N}^n]^\perp$ is the span of $\{\bar{v}^{i,n}, i \le K\}$. Define $\bar{v}_Q^{i,n} = [I - Q_n]^{-1}\bar{v}^{i,n} = \{\bar{v}_{Q,j}^{i,n}, j \le H\}$. Analogously to the case in Section 3, the (vector-valued) workload is defined by

$$WL_i^n(t) = [\bar{v}_Q^{i,n}]' x^n(t), \quad i \le K. \tag{4.7}$$

For state vector x, define $WL_i = [\bar{v}_Q^i]' x$. Using $\bar{\beta}$ in (4.3) together with the fact that $AG^n(t) = 0$ yields, for $i \le K$,

$$WL_i^n(t) = WL_i^n(0) + \left[\bar{v}_Q^{i,n}\right]' w^n(t) + b_i^n t + \left[\bar{v}_Q^{i,n}\right]' [I - Q_n'] R^n T^n(t). \tag{4.8}$$

The last term on the right is $[\bar{v}^{i,n}]' R^n T^n(t)$. Since R is diagonal with entries $\bar{\lambda}_i^d$, $i \leq H$, and $\bar{\beta} = R^{-1}\bar{\lambda}_Q^a$, the equality $A\bar{\beta} = \bar{e}$ implies that

$$\sum_i A_{ki} \frac{\bar{\lambda}_{Q,i}^a}{\bar{\lambda}_i^d} = 1, \quad k \leq K. \tag{4.9}$$

The characterization of the weights $\bar{v}^i$. The equation (4.7) is just what one would get (asymptotically) with the direct method of Section 6.4; namely, $WL_i^n(t)$ is (asymptotically) the queued work or potential work anywhere in the system that is eventually to be done by processor i. This can be seen from the following argument. The matrix R is diagonal with diagonal entries $\bar{\lambda}_i^d$, $i \leq H$. Let processor k be responsible for m_k job classes. The matrix A has K rows, and by reordering if necessary we can suppose that the first m_1 entries of row 1 are unity, the second m_2 entries of row 2 are unity, etc., with the other entries being zero. The dual constraints are $v'R \leq z'A$, $z'\bar{e} \leq 1$, and $v'\bar{\lambda}_Q^a$ is to be maximized. The constraints $v'R \leq z'A$ can be rewritten as

$$\begin{aligned} &v_i \bar{\lambda}_i^d \leq z_1, \ i \leq m_1, \\ &v_i \bar{\lambda}_i^d \leq z_2, \ m_1 < i \leq m_1 + m_2, \end{aligned}$$

etc. The kth extreme point optimal solution of the dual problem is obtained by setting $z_k = 1$. Suppose that $z_1 = 1$, $z_i = 0$, $i > 1$. Then for $i \leq m_1$, the optimizing values are $\bar{v}_i^1 = 1/\bar{\lambda}_i^d$, with $\bar{v}_i^1 = 0$ for $i > m_1$. The associated (the maximum) dual cost is

$$\sum_{i \leq m_1} \bar{\lambda}_{Q,i}^a / \bar{\lambda}_i^d = \sum_i A_{1i} \bar{\lambda}_{Q,i}^a / \bar{\lambda}_i^d,$$

which equals unity by (4.9). The result is analogous if $z_i = 1$, for any $i > 1$. Equivalently, asymptotically, the scaled rate of exogenous arrival of work for each processor (accounting for the network structure) is unity, the maximum rate of working. Furthermore, with these values for the $\bar{v}_i^k$, $i \leq H$, $x'[I-Q]^{-1}\bar{v}^k = x'\bar{v}_Q^k$ is just the total amount of work that processor k will have to do on all of the jobs currently in the system, when the current state is x.

The workload cost function. Let $k(\cdot)$ be a continuous nonnegative cost rate function. The possibility that deliberate idling is part of an optimal or asymptotically optimal strategy must be taken into account. Let $F_j^n(\cdot)$ denote the scaled time that was allocated to job class j when its queue as not empty, but which was neither used nor reallocated due to deliberate idling. It is a (singular) control process. As in Section 3, the workloads $WL_i^n(\cdot)$ are used as "system invariants," but are now subject to the controls

$F^n(\cdot)$. Owing to the possibility of deliberate idling, the definition (3.17) of the minimizing cost must be modified. Let $F = \{F_j, j \leq H\}$ be a vector with nonnegative components. Define

$$\bar{k}(WL) = \min_{x,F\geq 0} \left\{ k(x) : \left[\bar{v}_Q^i\right]' x = WL_i + \left[\bar{v}_Q^i\right]' [I - Q'] RF,\ i \leq K \right\}. \tag{4.10}$$

In (4.10), if the minimizing F is not zero, then some component(s) of WL will increase. One can view (4.10) as two-step process. First, get the new WL, by choosing F, and then rearrange the state x optimally, given this new WL. Alternatively, one can idle some queues and reallocate time for others simultaneously.

The following assumption is more restrictive than one would like. But it is a natural analogue of (A2.2). It facilitates getting an approximating control, since it ensures that the paths $x_j^n(\cdot)$ will not (asymptotically) become zero on any finite time interval if the components of $WL^n(\cdot)$ that depend on it are strictly positive on that interval.

A4.4. *Let $g(WL) = \{g_i(WL), i \leq H\}$ denote the minimizing x in (4.10), and let $h(WL) = \{h_i(WL), i \leq H\}$ denote the minimizing F. They are unique and continuous functions of WL. If $WL_i > 0$, then $g_j(WL) > 0$ for all components j that are involved in the definition of WL_i.*

An asymptotically optimal policy. The procedure is similar to that of Section 3. It is intended to be suggestive, and will certainly be modified to suit any particular application. Write $F = h(WL)$ and define $\delta x = g(WL) - x$. Formally, applying F, working in in the local-fluid time scale, and ignoring the boundary reflection, if any, yields the new workload and state vectors $\bar{WL}$ and $\bar{x}$, respectively, where

$$\begin{aligned} \bar{WL}_i &= WL_i + \left[\bar{v}_Q^i\right]' [I - Q'] RF, \\ \bar{x} &= x + [I - Q']RF, \end{aligned}$$

where $g(WL) = g(\bar{WL})$. It is possible that some component of $\bar{x}$ will be negative. Now (4.10) can be replaced by

$$\bar{k}(WL) = \min_{x\geq 0} \left\{ k(x) : \left[\bar{v}_Q^i\right]' x = \bar{WL}_i,\ i \leq K \right\}.$$

Define $\delta\bar{x} = g(\bar{WL}) - \bar{x}$. In lieu of (3.18) write

$$\delta\bar{x} = \bar{\lambda}^a \tau - [I - Q'] R \left[\bar{\beta} - \psi\right] \tau, \tag{4.11}$$

$$A\psi = 0. \tag{4.12}$$

Define $\delta\hat{x} = \delta\bar{x}/\tau$, temporarily set $\tau = 1$, and use the fact that $\bar{\lambda}^a = [I - Q']R\bar{\beta}$ to replace (4.11) by

$$\delta\hat{x} = [I - Q']R\psi.$$

There is a solution ψ satisfying (4.12). Next, find the largest $\alpha = \bar{\alpha} > 0$ such that $\bar{\beta} - \alpha[\psi + F] \geq 0$. Now choose τ such that δx is given by

$$\delta x = \left(\bar{\lambda}^a - [I - Q'] R \left[\bar{\beta} - \bar{\alpha}(\psi + F)\right]\right) \tau, \tag{4.13}$$

the analogue of (3.21). The length of the current scheduling interval in the local-fluid time scale is $\min\{\tau, T_0\}$, as in Section 3.

The details of proof of the asymptotic optimality use the ideas of Theorem 11.1.1 and of the previous section, appropriately modified to deal with the fact that there is more than one stage, and are left to the reader. Loosely speaking, the asymptotic path is the straight line connecting x and $g(WL)$, as seen from (4.13). Hence for any j, the x_j component will not hit zero if any of the WL_i that depend on it are positive.

Example. The criss-cross system. The following example is a concrete illustration of the definition of workload. Consider the so-called criss-cross system that was dealt with in [173, 190].

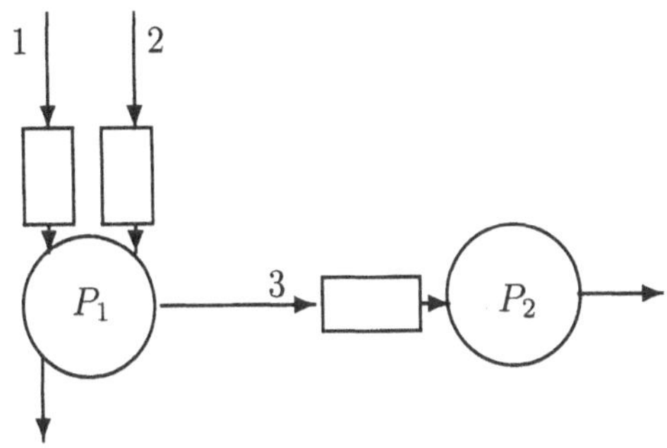

Figure 4.1. The criss-cross system.

For this example, job classes 1 and 2 are queued at processor 1. On completion of service, class 1 jobs leave the system, and class 2 jobs become class 3 jobs and go to processor 2. Thus,

$$Q = \begin{bmatrix} 0 & 0 & 0 \\ 0 & 0 & 1 \\ 0 & 0 & 0 \end{bmatrix},$$

$$[I - Q']^{-1} = I + Q' = \begin{bmatrix} 1 & 0 & 0 \\ 0 & 1 & 0 \\ 0 & 1 & 1 \end{bmatrix},$$

and, since $\bar{\lambda}_3^a = 0$,

$$[I - Q']^{-1}\bar{\lambda}^a = \bar{\lambda}_Q^a = (\bar{\lambda}_1^a, \bar{\lambda}_2^a, \bar{\lambda}_2^a).$$

The heavy traffic condition implies that $\bar{\lambda}_1^a/\bar{\lambda}_1^d + \bar{\lambda}_2^a/\bar{\lambda}_2^d = 1$ and $\bar{\lambda}_2^a/\bar{\lambda}_3^d = 1$. The constraints for the linear program are

$$\begin{aligned} \bar{\lambda}_1^d \beta_1 &= \bar{\lambda}_1^a, \\ \bar{\lambda}_2^d \beta_2 &= \bar{\lambda}_2^a, \\ \bar{\lambda}_3^d \beta_3 &= \bar{\lambda}_2^a, \\ \beta_1 + \beta_2 &\le \tau, \\ \beta_3 &\le \tau. \end{aligned}$$

The unique solution is $\bar{\beta}_1 = \bar{\lambda}_1^a/\bar{\lambda}_1^d$, $\bar{\beta}_2 = \bar{\lambda}_2^a/\bar{\lambda}_2^d$, $\bar{\beta}_3 = 1$.

The dual problem is

$$\max \left\{ \bar{\lambda}_1^a v_1 + \bar{\lambda}_2^a v_2 + \bar{\lambda}_2^a v_3 \right\}$$

subject to $v'R \le z'A$, $z'\bar{e} \le 1$. Since

$$R = \begin{bmatrix} \bar{\lambda}_1^d & 0 & 0 \\ 0 & \bar{\lambda}_2^d & 0 \\ 0 & 0 & \bar{\lambda}_3^d \end{bmatrix}, \quad A = \begin{bmatrix} 1 & 1 & 0 \\ 0 & 0 & 1 \end{bmatrix},$$

the dual constraints are

$$\bar{\lambda}_1^d v_1 \le z_1, \quad \bar{\lambda}_2^d v_2 \le z_1, \quad \bar{\lambda}_3^d v_3 \le z_2; \quad z_1 + z_2 \le 1.$$

The two dual optimal solutions correspond to $z_1 = 1$ and and $z_2 = 1$, yielding $\bar{v}^1 = (1/\bar{\lambda}_1^d, 1/\bar{\lambda}_2^d, 0)$ and $\bar{v}^2 = (0, 0, 1/\bar{\lambda}_3^d)$. The associated (optimal) cost is unity for each. Now the multipliers that are used to define the workload are $\bar{v}_Q^{i,n} = [I - Q]^{-1}\bar{v}^{i,n}$, $i = 1, 2$: that is,

$$\bar{v}_Q^{1,n} = (1/\bar{\lambda}_1^{d,n}, 1/\bar{\lambda}_2^{d,n}, 0), \quad \bar{v}_Q^{2,n} = (0, 1/\bar{\lambda}_3^{d,n}, 1/\bar{\lambda}_3^{d,n}),$$

and

$$WL_1^n(t) = \bar{\Delta}_1^{d,n} x_1^n(t) + \bar{\Delta}_2^{d,n} x_2^n(t), \quad WL_2^n(t) = \bar{\Delta}_3^{d,n} [x_2^n(t) + x_3^n(t)].$$

This yields (asymptotically) the actual limit workload as it would be defined in Section 6.4.

This example illustrates the general principle. The $WL_i^n(t)$ is always (asymptotically) the total work in the system for processor i. A detailed analysis of the optimal control problem for a linear cost rate $\sum_i c_i x_i$, $c_i > 0$, is in [173, 190]. In that problem, for each value of workload at least one of the optimizing values of x_i is zero. The approximately optimal policies are similar to what we obtain here, except for the extra care that is needed to ensure that there is no undesired idle time when a state value is near zero. If c_3 is sufficiently greater than c_2, then the (approximately) optimal policy would not serve queue 2 if there is some "minimal" amount of work still in queue 3. Thus, if $x_1^n(t) = 0$, P_1 might idle even if $WL_1^n(t) > 0$.

References

[1] E. Altman, D. Kofman, and U. Yechiali. Discrete time queues with delayed information. *Queueing Systems*, 19:361–376, 1995.

[2] E. Altman and H.J. Kushner. Admission control for combined guaranteed performance and best effort communications systems under heavy traffic. *SIAM J. Control and Optimiz.*, 37:1780–1807, 1999.

[3] E. Altman and H.J. Kushner. Control of polling in presence of vacations in heavy traffic with applications to satellite and mobile radio systems. To appear, SIAM J. on Control and Optimiz., 2001.

[4] E. Altman and A. Shwartz. Optimal priority assignment: a time sharing approach. *IEEE Trans. on Automatic Control*, 34:1098–1102, 1989.

[5] D. Anick, D. Mitra, and M.M. Sondhi. Stochastic theory of a data handling system with multiple sources. *Bell Systems Technical J.*, 61:1971–1894, 1982.

[6] S. Asmussen. *Applied Probability and Queues.* Wiley, New York, 1987.

[7] F. Baccelli and S. Foss. Stability of Jackson-type queueing networks. *Queueing Systems*, 17:5–72, 1994.

[8] F. Baccelli and P. Bremaud. *Elements of Queueing Theory.* Springer-Verlag, Berlin and New York, 1994.

[9] F. Baskett, K.M. Chandy, R.R. Muntz, and F.G. Palacios. Open, closed and mixed classes of queues with different classes of customers. *J. of the ACM*, 22:248–260, 1975.

[10] S.L. Bell and R.J. Williams. Dynamic scheduling of a system with two parallel servers in heavy traffic with complete resource pooling; asymptotic optimality of a continuous review threshold policy. Preprint, Math. Dept., University of California, La Jolla, 1999.

[11] V.E. Benes. Existence of finite invariant measures for Markov processes. *Proc. Amer. Math. Soc.*, 18:1058–1061, 1967.

[12] V.E. Benes. Finite regular invariant measures for Feller processes. *J. Appl. Probab.*, 5:203–209, 1968.

[13] A. Bensoussan. *Perturbation Methods in Optimal Control.* Wiley, New York, 1989.

[14] A. Bensoussan and J-L. Lions. *Contrôle Impulsionel et Inéquations Quasi-variationelles.* Dunod, Paris, 1981.

[15] P. Billingsley. *Convergence of Probability Measures.* Wiley, New York, 1968.

[16] J.M. Bismut. *Théorie Probabiliste du Contrôle des Diffusions.* Mem. Amer. Math. Soc. no. 167, Providence, 1976.

[17] G. Blankenship and G.C. Papanicolaou. Stability and control of systems with wide band noise disturbances. *SIAM J. Appl. Math.*, 34:437–476, 1978.

[18] G. Bolch, S. Greiner, H. de Meer, and K.S. Trivedi. *Queueing Networks and Markov Chains.* Wiley-Interscience, New York, 1998.

[19] V.S. Borkar. *Optimal Control of Diffusion Processes.* Longman Scientific and Technical, Harlow, Essex, UK, 1989.

[20] A.A. Borovkov. Some limit theorems in the theory of mass service I. *Theory of Probability and its Applications*, 9:550–565, 1964.

[21] A.A. Borovkov. Some limit theorems in the theory of mass service II. *Theory of Probability and its Applications*, 10:375–400, 1965.

[22] A.A. Borovkov. On limit laws for service processs in multi channel systems. *Siberian Math. Journal*, 8:746–763, 1967.

[23] A.A. Borovkov. On the convergence to diffusion processes. *Theory of Probability and its Applications*, 12:405–431, 1967.

[24] A.A. Borovkov. *Stochastic Processes in the Theory of Mass Service.* Nauk, Moscow, 1972.

[25] A.A. Borovkov. *Asymptotic Methods in Queueing Theory.* Wiley, New York, 1984.

[26] A.A. Borovkov. Limit theorems for queueing networks. *Theory of Probability and its Applications*, 31:413–427, 1986.

[27] O.J. Boxma and H. Tagaki. Editors: Special issue on polling systems. *Queueing Systems*, 11, 1992.

[28] M. Bramson. Instability of FIFO queueing networks. *Ann. Appl. Probab.*, 4:414–431, 1994.

[29] M. Bramson. Convergence to equilibria for fluid models of head-of-the-line processor sharing queueing networks. *Queueing Networks*, 23:1–26, 1996.

[30] M. Bramson. State space collapse with application to heavy traffic limits for multiclass queueing networks. *Queueing Systems*, 30:89–148, 1998.

[31] M. Bramson and J.G. Dai. Heavy traffic limits for some queueing networks. Preprint, 1999.

[32] L. Breiman. *Probability Theory*, volume 7 of *Classics in Applied Mathematics*, A reprint of the 1968 Addison-Wesley edition. SIAM, Philadelphia, 1992.

[33] P. Brémaud. *Point Processes and Queues.* Springer-Verlag, New York and Berlin, 1981.

[34] M. Butto, E. Cavallero, and A. Tonietti. Effectiveness of the "leaky bucket" policing mechanism in ATM networks. *IEEE J. in Selected Areas of Communications*, 9:335–342, 1991.

[35] S. Chandrasekhar. Stochastic problems in physics and astronomy. In N. Wax, editor, *Noise and Stochastic Processes.* Dover, New York, 1994.

[36] H. Chen, O. Kella, and G. Weiss. Fluid approximations for a processor sharing queue. *Queueing Systems*, 27:99–125, 1997.

[37] H. Chen and A. Mandlebaum. Discrete flow networks, diffusion approximations and bottlenecks. *Ann. Probab.*, 19:1463–1519, 1991.

[38] H. Chen and W. Whitt. Diffusion approximations for open queueing networks with service interruptions. *Queueing Systems*, 13:335–359, 1993.

[39] H. Chen and H. Zhang. Stability of multiclass queueing networks under priority service disciplines. *Operations Research*, 48:26–37, 1998.

[40] P.B. Chevalier and L.M. Wein. Scheduling networks of queues: heavy traffic analysis of a closed multistation network. *Operations Research*, 41:743–758, 1993.

[41] P.L. Chow, J.-L. Menaldi, and M. Robin. Additive control of stochastic linear systems with finite horizons. *SIAM J. Control Optim.*, 23:858–899, 1985.

[42] E.G. Coffman, Jr., A.A. Puhalskii, and M.I. Reiman. Polling systems with zero switchover time: a heavy traffic averaging princple. *Ann. Appl. Probab.*, 5:781–719, 1995.

[43] E.G. Coffman, Jr., A.A. Puhalskii, and M.I. Reiman. Polling systems with zero switchover time: a Bessel process limit. *Math of Oper. Res.*, 23:257–304, 1998.

[44] C. Costantini. *The Skorokhod oblique reflection problem and a diffusion approximation for a class of transport processes.* Ph.D. thesis, University of Wisconsin, Madison, 1987.

[45] R.W. Cottle. Completely-$\mathcal{Q}$ matrices. *Math. Program.*, 19:347–351, 1980.

[46] D.R. Cox and W.L. Smith. *Queues.* Methuen, London, 1961.

[47] J.G. Dai. *Steady state analysis of reflected Brownian motions: characterization, numerical methods and queueing applications.* Ph.D thesis, Operations Research Dept., Stanford University, 1990.

[48] J.G. Dai and J.M. Harrison. Steady state analysis of RBM in a rectangle: numerical methods and a queueing application. *Ann. Appl. Probab.*, 1:16–35, 1991.

[49] J.G. Dai and J.M. Harrison. Reflected Brownian motion in an orthant: numerical methods for steady state analysis. *Ann. Appl. Probab.*, 2:65–86, 1992.

[50] J.G. Dai and J.M. Harrison. The QNET method for two moment analysis of closed manufacturing systems. *Ann. Appl. Probab.*, 3:968–1012, 1993.

[51] J.G. Dai and Y. Wang. Nonexistence of Brownian models for certain multiclass queueing networks. *Queueing Systems*, 13:41–46, 1993.

[52] J.G. Dai. A fluid–limit model criterion for instability of multiclass queueing networks. *Ann. of Appl. Prob.*, 6:751–757, 1996.

[53] J.G. Dai and Viên Nguyen. On the convergence of multiclass queueing networks in heavy traffic. *Ann. Appl. Probab.*, 4:26–42, 1994.

[54] J.G. Dai and G. Weiss. Stability and instability of fluid models for reentrant lines. *Math. of Oper. Res.*, 21:115–135, 1995.

[55] J.G. Dai, D.H. Yeh, and C. Zhou. The QNET method for reentrant queueing networks with priority disciplines. *Operations Research*, 45:610–623, 1997.

[56] W.B. Davenport and W.L. Root. *Random Signals and Noise.* McGraw-Hill, New York, 1958.

[57] M.H.A. Davis and P. Varaiya. Dynamic programming conditions for partially observable stochastic systems. *SIAM J. on Control*, 11:226–261, 1973.

[58] J.L. Doob. Asymptotic properties of Markov transition probabilities. *Trans. Amer. Math. Soc.*, 63:393–421, 1948.

[59] J.L. Doob. *Stochastic Processes.* Wiley, New York, 1953.

[60] D. Down and S.P. Meyn. Piecewise linear test functions for stability and instability of queueing networks. *Queueing Systems*, 27:205–226, 1997.

[61] N. Dunford and J.T. Schwartz. *Linear Operators, Part 1: General Theory.* Wiley-Interscience, New York, 1966.

[62] P. Dupuis and H. Ishii. On oblique derivative problems for fully nonlinear second-order elliptic partial differential equations on nonsmooth domains. *Nonlinear Anal.*, 15:1123–1138, 1990.

[63] P. Dupuis and H. Ishii. On Lipschitz continuity of the solution mapping to the Skorokhod problem, with applications. *Stochastics and Stochastics Rep.*, 35:31–62, 1991.

[64] P. Dupuis and H. Ishii. SDE's with oblique reflection on nonsmooth domains. *Ann. Probability*, 21:554–580, 1993.

[65] P. Dupuis and K. Ramanan. A Skorokhod problem formulation and large deviation analysis of a processor sharing model. *Queueing Systems*, 28:109–124, 1998.

[66] P. Dupuis and K. Ramanan. Convex duality and the Skorokhod problem; Part i. *Probability Theory and Related Fields*, 115:153–195, 1999.

[67] P. Dupuis and K. Ramanan. Convex duality and the Skorokhod problem; Part ii. *Probability Theory and Related Fields*, 115:197–236, 1999.

[68] E.B. Dynkin. *Markov Processes.* Springer-Verlag, Berlin and New York, 1965.

[69] R. Elliott. *Stochastic Calculus and Applications.* Springer-Verlag, Berlin and New York, 1982.

[70] A.I. Elwalid and D. Mitra. Fluid models for the analysis and design of statistical multiplexing with loss priorities on multiple classes of bursty traffic. In *Proc. IEEE INFOCOM'92*, pages 415–425, New York, 1992. IEEE Press.

[71] S.N. Ethier and T.G. Kurtz. *Markov Processes: Characterization and Convergence.* Wiley, New York, 1986.

[72] W. Feller. Diffusion processes in genetics. In *Proc. Second Berk. Symp. on Probab.and Statist.*, pages 227–246, Berkeley, 1951. Univ. of California Press.

[73] K.W. Fendick and M.A. Rodrigues. A heavy traffic comparison of shared and segregated buffer schemes for queues with head-of-the-line processor sharing discipline. *Queueing Systems*, 9:163–190, 1991.

[74] K.W. Fendick and M.A. Rodrigues. Asymptotic analysis of adaptive rate control for diverse sources with delayed feedback. *IEEE Trans. on Information Theory*, 40:2008–2025, 1994.

[75] L. Flatto. Two parallel queues created by arrivals with two demands, II. *SIAM J. Appl. Math.*, 45:861–878, 1985.

[76] W.F. Fleming. Generalized solutions in optimal stochastic control. In P.T. Liu, E. Roxin and R. Sternberg, editors, *Differential Games and Control Theory: III*, pages 147–165. Marcel Dekker, 1977.

[77] W.H. Fleming and H.M. Soner. Asymptotic expansions for Markov processes with Lèvy generators. *Appl. Math. Optimization*, 19:203–223, 1989.

[78] W.H. Fleming and H.M. Soner. *Controlled Markov Processes and Viscosity Solutions.* Springer-Verlag, New York, 1992.

[79] G.J. Foschini. Equilibria for diffusion models for pairs of communicating computers–symmetric case. *IEEE Trans. Inf. Theory*, 28:273–284, 1982.

[80] G.J. Foschini and J. Salz. A basic dynamic routing problem and diffusion approximation. *IEEE Transactions on Communications*, 26:320–327, 1978.

[81] E. Gelenbe and G. Pujolle. *Introduction to Queueing Networks; Second Edition.* Wiley, New York, 1997.

[82] I.I. Gihman and A.V. Skorohod. *Introduction to the Theory of Random Processes.* Saunders, Philadelphia, 1965.

[83] I.I. Gihman and A.V. Skorohod. *Stochastic Differential Equations.* Springer-Verlag, Berlin and New York, 1972.

[84] P.W. Glynn and W. Whitt. A new view of the heavy traffic limit theorem for infinite server queues. *Adv. in Applied Prob.*, 23:188–209, 1991.

[85] W.J. Gordon and G.F. Newell. Closed queueing systems with exponential servers. *Oper. Res.*, 15:254–265, 1967.

[86] H. Gross and C. Harris. *Fundamentals of Queueing Theory, Second Edition.* Wiley, New York, 1985.

[87] J.M. Harrison and L.M. Wein. Scheduling networks of queues: heavy traffic analysis of a simple open network. *Queueing Systems*, 5:265–280, 1989.

[88] J.M. Harrison. The heavy traffic approximation for single server queues in series. *J. Appl. Prob.*, 10:613–629, 1973.

[89] J.M. Harrison. A limit theorem for priority queues in heavy traffic. *J. Appl. Prob.*, 10:907–912, 1973.

[90] J.M. Harrison. The diffusion approximation for tandem queues in heavy traffic. *Adv. in Appl. Prob.*, 10:886–905, 1978.

[91] J.M. Harrison. *Brownian Motion and Stochastic Flow Systems.* Wiley, New York, 1985.

[92] J.M. Harrison. Brownian models of queueing networks with heterogeneous customer populations. In *Proc. of the IMA*, volume 10, pages 147–186, Berlin and New York, 1988. Springer-Verlag. Edited by P.-L. Lions and W. Fleming.

[93] J.M. Harrison. The BIGSTEP approach to flow management in stochastic processing networks. In F.P. Kelly, S. Zachary, and I. Ziedins, editors, *Stochastic Networks: Theory and Applications.* Oxford University Press, Oxford, 1996.

[94] J.M. Harrison. Heavy traffic analysis of a system with parallel servers: asymptotic optimality of discrete review policies. *Ann. Appl. Probab.*, 8:822–848, 1998.

[95] J.M. Harrison. Brownian models of open processing networks: canonical representation of workload. To appear in Ann. Appl. Probab., 2000.

[96] J.M. Harrison and M.J. Lopez. Heavy traffic resource pooling in parallel server systems. To appear: Queueing Systems, 1999.

[97] J.M. Harrison and Viên Nguyen. The QNET method for two moment analysis of open queueing networks. *Queueing Systems*, 6:1–32, 1990.

[98] J.M. Harrison and M.I. Reiman. Reflected Brownian motion on an orthant. *Ann. Probab.*, 9:302–308, 1981.

[99] J.M. Harrison and M. Taksar. Instantaneous control of a Brownian motion. *Math. of Oper. Res.*, 8:439–453, 1983.

[100] J.M. Harrison and J. Van Mieghem. Dynamic control of Brownian networks: state space collapse and equivalent workload formulations. *Ann. Appl. Prob.*, 7:747–771, 1997.

[101] J.M. Harrison and L.M. Wein. Scheduling networks of queues: heavy traffic analysis of a two-station closed network. *Oper. Res.*, 38:1052–1064, 1990.

[102] J.M. Harrison and R.J. Williams. Brownian models of open queueing networks with homogeneous customer populations. *Stochastics and Stochastics Rep.*, 22:77–115, 1987.

[103] J.M. Harrison, R.J. Williams, and H. Chen. Brownian models of closed queueing networks with homogeneous customer populations. *Stochastics and Stochastics Rep.*, 29:37–74, 1990.

[104] J.M. Harrison, R.J. Williams, and H. Chen. Brownian models of closed queueing networks with homogeneous customer populations. *Stochastics and Stochastics Rep.*, 29:37–74, 1990.

[105] F.S. Hillier and G.J. Lieberman. *Introduction to Operations Research; Sixth Edition.* McGraw Hill, New York, 1995.

[106] D.L. Iglehart. Limit diffusion approximations for the many server queue and the repairman problem. *J. Appl. Prob.*, 2:429–441, 1965.

[107] D.L. Iglehart. Limit theorems for queues with traffic intensity one. *Ann. Math. Statist.*, 36:1437–1449, 1965.

[108] D.L. Iglehart. Weak convergence in queueing theory. *Advances in Applied Probability*, 5:570–594, 1973.

[109] D.L. Iglehart and H.M. Taylor. Weak convergence of a sequence of quickest detection problems. *Ann. Math. Statist.*, 39:2149–2153, 1968.

[110] D.L. Iglehart and W. Whitt. Multiple channel queues in heavy traffic I. *Adv. Appl. Prob.*, 2:150–177, 1970.

[111] D.L. Iglehart and W. Whitt. Multiple channel queues in heavy traffic II: sequences, networks and batches. *Adv. Appl. Prob.*, 2:355–369, 1970.

[112] D.L. Iglehart and W. Whitt. Multiple channel queues in heavy traffic III: random server selection. *Adv. Appl. Prob.*, 2:370–375, 1970.

[113] N. Ikeda and S. Watanabe. *Stochastic Differential Equations and Diffusion Processes*, First Ed. North-Holland, Amsterdam, 1981.

[114] N. Ikeda and S. Watanabe. *Stochastic Differential Equations and Diffusion Processes*, Second Ed. North-Holland, Amsterdam, 1989.

[115] J. Jackson. Networks of waiting lines. *Operations Research*, 4:518–524, 1957.

[116] J. Jackson. Jobshop-like queueing systems. *Management Science*, 10:131–142, 1963.

[117] J. Jacod. *Calcul Stochastique et Problèmes de Martingales.* Springer-Verlag, New York, 1979.

[118] J. Jacod and A.N. Shiryaev. *Limit Theorems for Stochastic Processes.* Springer-Verlag, Berlin and New York, 1987.

[119] Dai J.G and S. Meyn. Stability and convergence of moments for multiclass queueing networks via fluid limit models. *IEEE Trans on Aut. Control*, 40:1889–1904, 1995.

[120] D.F. Johnson. *Diffusion approximations for the optimal filtering of a jump process and for queueing networks.* Ph.D thesis, University of Wisconsin, 1983.

[121] I. Karatzas. A class of singular stochastic control problems. *Adv. in Appl. Probab.*, 15:225–254, 1983.

[122] I. Karatzas and S.E. Shreve. *Brownian Motion and Stochastic Calculus.* Springer-Verlag, New York, 1988.

[123] O. Kella and W. Whitt. Diffusion approximations for queues with server vacations. *Adv. in Appl. Prob.*, 22:706–729, 1990.

[124] F.P. Kelly. Routing and capacity allocation in networks with trunk reservation. *Math. Oper. Res.*, 15:771–793, 1990.

[125] F.P. Kelly. Loss networks. *Ann. Appl. Prob.*, 1:319–377, 1991.

[126] F.P. Kelly and C.N. Laws. Dynamic routing in open queueing networks: Brownian models, cut constraints and resource pooling. *Queueing Systems*, 13:47–86, 1993.

[127] D.G. Kendall. Some problems in the theory of dams. *J. Royal Statist. Soc. Ser B*, 19:207–212, 1957.

[128] J.F.C. Kingman. The single serve queue in heavy traffic. *Proc. Camb. Phil. Soc.*, 57:902–904, 1961.

[129] J.F.C. Kingman. On queues in heavy traffic. *J. Royal Statist. Soc., Ser B*, 25:383–392, 1962.

[130] J.F.C. Kingman. Approximations for queues in heavy traffic. In R. Cruon, editor, *Queueing Theory: Recent Developments and Applications.* Elsevier, New York, 1965.

[131] J.F.C. Kingman. The heavy traffic approximation in the theory of queues. In W. Smith and W. Wilkinson, editors, *Proceedings of the Symposium on Congestion Theory*, pages 137–159. North Carolina Press, Chapel Hill, 1965.

[132] L. Kleinrock. *Queueing Systems: Vol. I: Theory.* Wiley, New York, 1975.

[133] L. Kleinrock. *Queueing Systems: Vol. II: Computer Applications.* Wiley, New York, 1976.

[134] C. Knessl. On the diffusion approximation to a fork and join queue. *SIAM J. Appl. Math.*, 51:160–171, 1991.

[135] C. Knessl. On the diffusion approximation to two parallel queues with processor sharing. *IEEE Trans. on Automatic Control*, 36:1356–1367, 1991.

[136] C. Knessl and J.A. Morrison. Heavy traffic analysis of a data handling system with multiple sources. *SIAM J. Appl. Math.*, 51:187–213, 1991.

[137] C. Knessl and C. Tier. Asymptotic expansions for large closed queueing networks. *J. ACM*, 37:144–174, 1990.

[138] Y. Kogan and E.V. Krichagina. Closed exponential queueing networks with blocking in heavy traffic. In H.G. Perros and T. Altiok, editors, *Queueing Networks With Blocking*. Elsevier, North-Holland, Amsterdam, 1989.

[139] Y. Kogan and R. Liptser. Limit non-stationary behavior of large closed queueing networks with bottlenecks. *Queueing Systems*, 14:33–55, 1993.

[140] Y. Kogan, R.S.Liptser, and A.V. Smorodinskii. Gaussian diffusion approximation of closed Markov models of computer networks. *Problems Inform. Transmission*, 22:38–51, 1986.

[141] Y. Kogan and A. Yakovlev. Asymptotic analysis for closed multichain queueing networks with bottlenecks. *Queueing Systems*, 23:235–258, 1996.

[142] P. Kokotovic, H. Khalil, and J. O'Reilly. *Singular Perturbation Methods in Control: Analysis and Design*. Academic Press, New York, 1986.

[143] K.R. Krishnan and T.J. Ott. An improved scheme for state dependent routing of telephone traffic. Technical report, Bellcore, 1986. TM-ARH-008-410.

[144] P.R. Kumar and S.P. Meyn. Stability of queueing networks and scheduling policies. *IEEE Trans. on Automatic Control*, 40:251–260, 1995.

[145] P.R. Kumar and T.I. Seidman. Dynamic instabilities and stabilization methods in distributed real time scheduling policies. *IEEE Trans. on Automatic Control*, 35:289–298, 1990.

[146] H. Kunita and S. Watanabe. On square integrable martingales. *Nagoya Math. J.*, 30:209–245, 1967.

[147] T.G. Kurtz. Semigroups of conditional shifts and approximation of Markov processes. *Ann. Probab.*, 4:618–642, 1975.

[148] T.G. Kurtz. *Approximation of Population Processes*, volume 36 of *CBMS-NSF Regional Conf. Series in Appl. Math.* SIAM, Philadelphia, 1981.

[149] H. J. Kushner. Control and optimal control of assemble to order manufacturing systems under heavy traffic. *Stochastics and Stochastics Rep.*, 66:233–272, 1999.

[150] H. J. Kushner. Stability of single class queueing networks. 2000.

[151] H.J. Kushner. The Cauchy problem for a class of degenerate parabolic equations and asymptotic properties of the related diffusion process. *J. Math. Anal. Appl.*, 6:209–231, 1969.

[152] H.J. Kushner. Stability and existence of diffusions with discontinuous or rapidly growing terms. *J. Math. Anal. Appl.*, 11:156–168, 1972.

[153] H.J. Kushner. Existence results for optimal stochastic controls. *J. Optimiz. Theory and Applic.*, 15:347–359, 1975.

[154] H.J. Kushner. Probabilistic methods for finite difference approximation to degenerate elliptic and parabolic equations with Neumann and Dirichlet boundary conditions. *J. Math. Anal. Appl.*, 53:644–668, 1976.

[155] H.J. Kushner. *Probability Methods for Approximations in Stochastic Control and for Elliptic Equations.* Academic Press, New York, 1977.

[156] H.J. Kushner. Optimality conditions for the average cost per unit time problem with a diffusion model. *SIAM J. Control Optim.*, 16:330–346, 1978.

[157] H.J. Kushner. *Approximation and Weak Convergence Methods for Random Processes with Applications to Stochastic Systems Theory.* MIT Press, Cambridge, Mass., 1984.

[158] H.J. Kushner. Direct averaging and perturbed test function methods for weak convergence. *Lecture Notes in Control and Information Sciences*, 81:412–426, 1985.

[159] H.J. Kushner. Numerical methods for stochastic control problems in continuous time. *SIAM J. Control Optim.*, 28:999–1048, 1990.

[160] H.J. Kushner. *Weak Convergence Methods and Singularly Perturbed Stochastic Control and Filtering Problems*, volume 3 of *Systems and Control.* Birkhäuser, Boston, 1990.

[161] H.J. Kushner. Approximations of large trunk line systems under heavy traffic. *Adv. in Appl. Probab.*, 26:1063–1094, 1994.

[162] H.J. Kushner. Analysis of controlled multiplexing systems via numerical stochastic control techniques. *IEEE J. on Selected Areas in Communications*, 13:1207–1218, 1995.

[163] H.J. Kushner. Control of trunk line systems in heavy traffic. *SIAM J. Control Optim.*, 33:765–803, 1995.

[164] H.J. Kushner. Existence of optimal controls for variance control. In W.M. McEneaney, G. Yin, and Q. Zhang, editors, *Stochastic Analysis, Control, Optimization and Applications: A Volume in Honor of W.H. Fleming*. Birkhäuser, Boston, 1998.

[165] H.J. Kushner. Heavy traffic analysis of controlled multiplexing systems. *Queueing Systems*, 28:79–107, 1998.

[166] H.J. Kushner and Y.N. Chen. Optimal control of assignment of jobs to processors under heavy traffic. *Stochastics and Stochastics Rep.*, 68:177–228, 2000.

[167] H.J. Kushner and P. Dupuis. *Numerical Methods for Stochastic Control Problems in Continuous Time*. Springer-Verlag, Berlin and New York, 1992. Second edition, 2001.

[168] H.J. Kushner and D. Jarvis. Codes for optimal stochastic control: documentation and users guide. Technical report, Brown University, Lefschetz Center for Dynamical Systems, Division of Applied Math., 1994.

[169] H.J. Kushner, D. Jarvis, and J. Yang. Controlled and optimally controlled multiplexing systems: A numerical exploration. *Queueing Systems*, 20:255–291, 1995.

[170] H.J. Kushner and L.F. Martins. Numerical methods for stochastic singular control problems. *SIAM J. Control Optim.*, 29:1443–1475, 1991.

[171] H.J. Kushner and L.F. Martins. Heavy traffic analysis of a data transmission system with many independent sources. *SIAM J. Appl. Math.*, 53:1095–1122, 1993.

[172] H.J. Kushner and L.F. Martins. Numerical methods for controlled and uncontrolled multiplexing and queueing systems. *Queueing Systems*, 16:241–285, 1994.

[173] H.J. Kushner and L.F. Martins. Heavy traffic analysis of a controlled multi class queueing network via weak convergence theory. *SIAM J. on Contral and Optimiz.*, 34:1781–1797, 1996.

[174] H.J. Kushner and K.M. Ramachandran. Optimal and approximately optimal control policies for queues in heavy traffic. *SIAM J. Control Optim.*, 27:1293–1318, 1989.

[175] H.J. Kushner and J. Yang. Numerical methods for controlled routing in large trunk line systems via stochastic control theory. *ORSA J. Computing*, 6:300–316, 1994.

[176] H.J. Kushner and J. Yang. An effective numerical method for controlling routing in large trunk line networks. *Math. Computation Simulation*, 38:225–239, 1995.

[177] H.J. Kushner and G. Yin. *Stochastic Approximation Algorithms and Applications.* Springer-Verlag, Berlin and New York, 1997.

[178] C.N. Laws. Resource pooling in queueing networks with dynamic routing. *Adv. in Appl. Probab.*, 24:699–726, 1992.

[179] J.P. Lehoczky and S.E. Shreve. Absolutely continuous and singular stochastic control. *Stochastics and Stochastics Rep.*, 17:91–110, 1986.

[180] A.J. Lemoine. Networks of queues: A survey of weak convergence results. *Management Sci.*, 24:1175–1193, 1978.

[181] H. Levy and M. Sidi. Polling models: applications, modelling and optimization. *IEEE Trans. on Communications*, 38:1750–1760, 1991.

[182] W. Lin and P.R. Kumar. Optimal control of a queueing system with two heterogeneous servers. *IEEE Trans. on Automatic Control*, AC-29:696–703, 1984.

[183] P.L. Lions and A.S. Sznitman. Stochastic differential equations with reflecting boundary conditions. *Comm. Pure and Appl. Math.*, 37:511–553, 1984.

[184] R. Liptser and A.N. Shiryaev. *Statistics of Random Processes.* Springer-Verlag, Berlin and New York, 1977.

[185] A. Mandlebaum, W.A. Massey, M.I. Reiman, and B. Rider. Time varying multiserver queues with abandonment and retrials. Preprint: Bell Labs, Lucent Technologies, 1999.

[186] A. Mandlebaum and G. Pats. State-dependent stochastic networks. part i: Approximations and applications with continuous diffusion limits. *Annals of Applied Probability*, 8:569–646, 1998.

[187] D. Markowitz. *A unified approach to single machine scheduling: Heavy traffic analysis of dynamic cyclic policies.* Ph.D thesis, M.I.T: Sloan School of Management, Cambridge, Mass., 1996.

[188] D.M. Markowitz, M.I. Reiman, and L.M. Wein. The stochastic economic lot size scheduling problem: heavy traffic analysis of dynamic cyclic policies. Report, MIT, School of Management, 1996.

[189] L.F. Martins and H.J. Kushner. Routing and singular control for queueing networks in heavy traffic. *SIAM J. Control Optim.*, 28:1209–1233, 1990.

[190] L.F. Martins, S.E. Shreve, and H.M. Soner. Heavy traffic convergence of a controlled, multi-class queueing system. *SIAM J. on Control and Optimiz.*, 34:2133–2171, 1996.

[191] J. McKenna, D. Mitra, and K.G. Ramakrishnan. A class of closed queueing networks; integral representations asymptotic expansions and generalizations. *Bell System Technical J.*, 60:599–641, 1981.

[192] E.J. McShane and R.B. Warfield. On Fillopov's implicit function theorem. *Proc. Amer. Math. Soc.*, 18:41–47, 1967.

[193] J.-L. Menaldi and M. Taksar. Optimal correction problem for a multidimensional stochastic system. *Automatica*, 25:223–232, 1989.

[194] P. Meyer. *Probability and Potentials.* Blaisdell, Waltham, Massachusetts, 1966.

[195] P. Meyer. Intégrals stochastiques. In *Lecture Notes in Mathematics, 39*, pages 72–141. Springer-Verlag, Berlin and New York, 1967.

[196] S.P. Meyn and R.I. Tweedie. *Markov Chains and Stochastic Stability.* Springer-Verlag, Berlin and New York, 1994.

[197] D. Mitra. Stochastic theory of a fluid model of producers and consumers coupled by buffer. *Adv. Appl. Probab.*, 20:646–676, 1988.

[198] D. Mitra and R.J. Gibbens. State dependent routing on symmetric loss networks with trunk reservations: Asymptotics, optimal design. *Ann. Oper. Res.*, 35:3–30, 1992.

[199] D. Mitra, R.J. Gibbens, and B.D. Huang. Analysis and optimal design of aggregated-least-busy-alternative routing on symmetric loss networks with trunk reservation. In A. Jensen and V.B. Jensen, editors, *Teletraffic and Datatraffic in a Period of Change.* North-Holland–Elsevier, 1991.

[200] J.A. Morrison. Diffusion approximations for head-of-the-line processor sharing for two parallel queues. *SIAM J. Appl. Math.*, 53:471–490, 1993.

[201] J. Neveu. *Mathematical Foundations of the Calculus of Probability.* Holden-Day, San Francisco, 1965.

[202] G.F. Newell. Approximate stochastic behavior of n-server service stations. In *Lecture Notes in Economics and Mathematical Systems, Vol 87.* Springer-Verlag, Berlin and New York, 1973.

[203] Viên Nguyen. Processing networks with parallel and sequential tasks: heavy traffic analysis and Brownian limits. *Ann. Appl. Prob.*, 3:28–55, 1993.

[204] B. Øksendal. *Stochastic differential equations.* Springer-Verlag, Berlin and New York, 1995.

[205] A.K. Parekh and R.G. Gallagher. A generalized processor sharing approach to flow control in integrated services networks: the single node case. *IEEE/ACM Trans. on Networking*, 1:344–357, 1993.

[206] W.P. Peterson. Diffusion approximations for networks of queues with multiple customer types. *Math. Oper. Res.*, 9:90–118, 1951.

[207] B. Pittel. Closed exponential networks of queues with saturation: the Jackson type stationary distribution and its asymptotic analysis. *Math. of Oper. Res.*, 4:357–378, 1979.

[208] M.I. Reiman. Some diffusion approximations with state space collapse. In F. Baccelli and G. Fayolle, editors, *Proc. Int. Seminar on Modelling and Performance Evaluation Methodology*, pages 209–240. Springer-Verlag, 1983.

[209] M.I. Reiman. A multiclass feedback queue in heavy traffic. *Advances in Appl. Prob.*, 20:179–207, 1998.

[210] M.I. Reiman, R. Rubio, and L.M. Wein. Heavy traffic analysis of the dynamic routing stochastic inventory-routing problem. 1996.

[211] M.I. Reiman and L.M. Wein. Dynamic scheduling of a two class queue with setups. *Operations Research*, 46:532–547, 1998.

[212] M.I. Reiman and R.J. Williams. A boundary property of semimartingale reflecting Brownian motions. *Prob. Theory Rel. Fields*, 77:87–97, 1988.

[213] M.R. Reiman. Open queueing networks in heavy traffic. *Math. Oper. Res.*, 9:441–458, 1984.

[214] M.R. Reiman. Asymptotically optimal trunk reservation for large trunk groups. In *Proceedings of the 28th Conference on Decision and Control*, New York, 1989. IEEE.

[215] M.R. Reiman. Optimal trunk reservations for a critically loaded line. In A. Jensen and V.B. Jensen, editors, *Teletraffic and Datatraffic in a Period of Change.* North–Holland-Elsevier, 1991.

[216] R. Rishel. Necessary and sufficient dynamic programming conditions for continuous time stochastic optimal control. *SIAM J. Control*, 8:559–571, 1970.

[217] A.N. Rybko and A.L. Stolyar. On the ergodicity of stochastic processes describing open queueing networks. *Problems Inform. Transmission*, 28:199–220, 1992.

[218] Y. Saisho. Stochastic differential equations for multi-dimensional domain with reflecting boundary. *Probab. Theory Related Fields*, 74:455–477, 1987.

[219] Z. Schuss. *Theory and Applications of Stochastic Differential Equations*. Wiley, New York, 1988.

[220] R. Serfozo. *Introduction to Queueing Networks*. Springer-Verlag, Berlin and New York, 1999.

[221] M. Shashiashvili. A lemma on variational distance between maximal functions with application to the Skorokhod problem in a nonnegative orthant with state dependent refelctions. *Stochastics and Stochastics Rep.*, 48:161–194, 1994.

[222] K. Sigman. The stability of open queueing networks. *Stochastic Processes, Appl.*, 35:11–25, 1990.

[223] A.V. Skorohod. Limit theorems for stochastic processes. *Theory Probab. Appl.*, pages 262–290, 1956.

[224] A.V. Skorokhod. Stochastic equations for diffusions in a bounded region. *Theory Probab. Appl.*, 6:264–274, 1961.

[225] H.M. Soner and S.E. Shreve. Regularity of the value function for a two-dimensional singular stochastic control problem. *SIAM J. Control Optim.*, 27:876–907, 1989.

[226] R. Srikant. Diffusion approximations for models of congestion control in high speed networks. In *Proceedings of the 38-th Conference on Decision and Control*, New York, 1998. IEEE Press.

[227] D.W. Stroock and S.R.S. Varadhan. Diffusion processes with boundary conditions. *Comm. Pure Appl. Math.*, 24:147–225, 1971.

[228] D.W. Stroock and S.R.S. Varadhan. *Multidimensional Diffusion Processes*. Springer-Verlag, New York, 1979.

[229] H. Takagi. *Stochastic Analysis of Computer and Communications Systems*. North-Holland, Amsterdam, 1990.

[230] H. Takagi. Queueing analysis of polling models, an update. In J.H. Dshalalow, editor, *Frontiers in Queueing, Models, Methods and Problems*. CRC press, Boca Raton, Fla, 1994.

[231] M. Taksar. Average optimal singular control and a related stopping problem. *Math. Oper. Res.*, 10:63–81, 1985.

[232] X. Tan and C. Knessl. A fork-join queueing model: diffusion approximation, integral representations and asymptotics. *Queueing Systems*, 22:287–322, 1996.

[233] H. Tanaka. Stochastic differential equations with reflecting boundary conditions in convex regions. *Hiroshima Math. J.*, 9:163–177, 1979.

[234] L. Tassiulas and A. Ephremides. Stability properties of constrained queueing systems and scheduling policies for maximum throughput in multihop radio networks. *IEEE Trans, on Automatic Control*, 37:1936–1948, 1992.

[235] L. Tassiulas and A. Ephremides. Dynamic server allocation to parallel queues with randomly varying connectivity. *IEEE Trans. on Automatic Control*, 39:466–478, 1993.

[236] L. Tassiulas and S. Papavassiliou. Optimal anticipative scheduling with asynchronous transmission opportunities. *IEEE Trans. on Automatic Control*, 40:2052–2062, 1995.

[237] L.M. Taylor and R.J. Williams. Existence and uniqueness of semimartingale reflecting Brownian motions in an orthant. *Probability Theory and Related Fields*, 96:283–317, 1993.

[238] R.D. van der Mei and H. Levy. Polling systems in heavy traffic: exhaustiveness of service policies. *Queueing Systems*, 27:227–250, 1997.

[239] J.A. van Mieghem. Dynamic scheduling with convex delay costs: The generalized $c\mu$ rule. *Ann. Appl. Prob.*, 5:809–833, 1995.

[240] A. Varma and D. Stiliadis. Hardware implementation of fair queueing algorithms for asynchronous transfer mode networks. *IEEE Communications Magazine*, 35:54–68, 1997.

[241] S Varma. *Heavy and light traffic approximations for queues with synchronization constraints.* Ph.D thesis, University of Maryland, Systems Research Center, 1990.

[242] A.F. Veinott, Jr. Discrete dynamic programming with sensitive discount optimality criteria. *Ann. Math. Statist.*, 40:1635–1660, 1969.

[243] J. Walrand. *An Introduction to Queuing Networks.* Prentice–Hall, Englewood Cliffs, 1988.

[244] J. Warga. Relaxed variational problems. *J. Math. Anal. Appl.*, 4:111–128, 1962.

[245] L.M. Wein. Optimal control of a two station Brownian network. *Math. Oper. Res.*, 15:215–242, 1990.

[246] L.M. Wein. Scheduling networks of queues: heavy traffic analysis of a two-station network with controllable inputs. *Oper. Res.*, 38:1065–1078, 1990.

[247] L.M. Wein. Dynamic scheduling of a multiclass make-to-stock queue. *Operations Research*, 40:724–735, 1992.

[248] L.M. Wein. Scheduling networks of queues: heavy traffic analysis of a multistation network with controllable inputs. *Oper. Res.*, 40:S312–S334, 1992.

[249] W. Whitt. Weak convergence theorems for priority queues: Preemptive-resume discipline. *J. Appl. Prob.*, 8:74–94, 1971.

[250] W. Whitt. Mathematical methods in queueing theory: A survey. In *Lecture Notes in Economics and Mathematical Systems,* Vol 98, pages 307–350. Springer-Verlag, Berlin and New York, 1974.

[251] W. Whitt. Departure from a queue with many busy servers. *Math. Operations Research*, 9:534–544, 1984.

[252] W. Whitt. Heavy traffic approximations with blocking. *AT&T Bell Laboratories Technical Journal*, 63:689–708, 1984.

[253] W. Whitt. Open and closed models for networks of queues. *Bell System Technical Journal*, 63:1911–1979, 1984.

[254] R.J. Williams. Diffusion approximations for open multiclass queueing networks: sufficient conditions involving state space collapse. *Queueing Systems*, 30:27–88, 1998.

[255] R.J. Williams. On dynamic scheduling of a parallel server system with complete resource pooling. Fields Institute Workshop on Analysis and Simulation of Communication Networks., 1998.

[256] K. Yamada. Diffusion approximations for storage processes with general release rules. *Math. of Oper. Res.*, pages 459–470, 1984.

[257] L.C. Young. Generalized curves and the existence of an attained absolute minimum in the calculus of variations. *Compt. Rend. Soc. Sci. et Lettres Varsovie CI III*, 30:212–234, 1937.

Symbol Index

Index

Applications of Mathematics

(continued from page ii)

33 Embrechts/Klüppelberg/Mikosch, **Modelling Extremal Events** (1997)
34 Duflo, **Random Iterative Models** (1997)
35 Kushner/Yin, **Stochastic Approximation Algorithms and Applications** (1997)
36 Musiela/Rutkowski, **Martingale Methods in Financial Modeling: Theory and Application** (1997)
37 Yin/Zhang, **Continuous-Time Markov Chains and Applications** (1998)
38 Dembo/Zeitouni, **Large Deviations Techniques and Applications**, Second Ed. (1998)
39 Karatzas/Shreve, **Methods of Mathematical Finance** (1998)
40 Fayolle/Iasnogorodski/Malyshev, **Random Walks in the Quarter Plane** (1999)
41 Aven/Jensen, **Stochastic Models in Reliability** (1999)
42 Hernández-Lerma/Lasserre, **Further Topics on Discrete-Time Markov Control Processes** (1999)
43 Yong/Zhou, **Stochastic Controls: Hamiltonian Systems and HJB Equations** (1999)
44 Serfozo, **Introduction to Stochastic Networks** (1999)
45 Steele, **Stochastic Calculus and Financial Applications** (2000)
46 Chen/Yao, **Fundamentals of Queueing Networks: Performance, Asymptotics, and Optimization** (2001)
47 Kushner, **Heavy Traffic Analysis of Controlled Queueing and Communication Networks** (2001)

Applications of Mathematics

(continued from page ii)

33 Embrechts/Klüppelberg/Mikosch, Modelling Extremal Events (1997)
34 Duflo, Random Iterative Models (1997)
35 Kushner/Yin, Stochastic Approximation Algorithms and Applications (1997)
36 Musiela/Rutkowski, Martingale Methods in Financial Modelling: Theory and Application (1997)
37 Yin/Zhang, Continuous-Time Markov Chains and Applications (1998)
38 Dembo/Zeitouni, Large Deviations Techniques and Applications, Second Ed. (1998)
39 Karatzas/Shreve, Methods of Mathematical Finance (1998)
40 Fayolle/Iasnogorodski/Malyshev, Random Walks in the Quarter Plane (1999)
41 Aven/Jensen, Stochastic Models in Reliability (1999)
42 Hernández-Lerma/Lasserre, Further Topics on Discrete-Time Markov Control Processes (1999)
43 Yong/Zhou, Stochastic Controls. Hamiltonian Systems and HJB Equations (1999)
44 Serfozo, Introduction to Stochastic Networks (1999)
45 Steele, Stochastic Calculus and Financial Applications (2000)
46 Chen/Yao, Fundamentals of Queueing Networks: Performance, Asymptotics, and Optimization (2001)
47 Kushner, Heavy Traffic Analysis of Controlled Queueing and Communication Networks (2001)